JUDSON

INTERNATIONAL UNION OF CRYSTALLOGRAPHY
MONOGRAPHS ON CRYSTALLOGRAPHY

INTERNATIONAL UNION OF CRYSTALLOGRAPHY
BOOK SERIES

This volume forms part of a series of books sponsored by the International Union of Crystallography (IUCr) and published by Oxford University Press. There are three IUCr series: IUCr Monographs on Crystallography, which are in-depth expositions of specialized topics in crystallography; and IUCr Texts on Crystallography, which are more general works intended to make crystallographic insights available to a wider audience than the community of crystallographers themselves; and IUCr Crystallographic Symposia, which are essentially the edited proceedings of workshops or similar meetings supported by the IUCr.

IUCr Monographs on Crystallography
1 *Accurate molecular structures: Their determination and importance*
   A. Domenicano and I. Hargittai, *editors*

IUCr Texts on Crystallography
1 *The solid state: From superconductors to superalloys*
   A. Guinier and R. Jullien, *translated by* W. J. Duffin
2 *Fundamentals of crystallography*
   C. Giacovazzo, *editor*

IUCr Crystallographic Symposia
1 *Patterson and Pattersons: Fifty years of the Patterson function*
   J. P. Glusker, B. K. Patterson, and M. Rossi, *editors*
2 *Molecular structure: Chemical reactivity and biological activity*
   J. J. Stezowski, J. Huang, and M. Shao, *editors*
3 *Crystallographic computing 4: Techniques and new technologies*
   N. W. Isaacs and M. R. Taylor, *editors*
4 *Organic crystal chemistry*
   J. Garbarczyk and D. W. Jones, *editors*
5 *Crystallographic computing 5: From chemistry to biology*
   D. Moras, A. D. Podjarny, and J. C. Thierry, *editors*

Accurate Molecular Structures

Their Determination and Importance

Edited by

ALDO DOMENICANO

University of L'Aquila and CNR
Institute of Structural Chemistry

and

ISTVÁN HARGITTAI

Hungarian Academy of Sciences
and Technical University of Budapest

INTERNATIONAL UNION OF CRYSTALLOGRAPHY
OXFORD UNIVERSITY PRESS
1992

Oxford University Press, Walton Street, Oxford OX2 6DP
Oxford New York Toronto
Delhi Bombay Calcutta Madras Karachi
Petaling Jaya Singapore Hong Kong Tokyo
Nairobi Dar es Salaam Cape Town
Melbourne Auckland
and associated companies in
Berlin Ibadan

Oxford is a trade mark of Oxford University Press

Published in the United States
by Oxford University Press, New York

A catalogue record for this book is
available from the British Library

Library of Congress Cataloging in Publication Data
Accurate molecular structures : their determination and importance /
edited by Aldo Domenicano and István Hargittai.
p. cm. — (International Union of Crystallography monographs
on crystallography; 1)
Includes bibliographical references and index.
1. Molecular structure. 2. Crystallography. I. Domenicano,
Aldo. II. Hargittai, István. III. Series.
QD461.A197   1992   541.2'2 — dc20   90–22037
ISBN 0–19–855556–3

Typeset by Colset Private Limited, Singapore
Printed in Great Britain by
Biddles Ltd, Guildford & King's Lynn

Preface

This book summarizes the potentials and limitations of the various techniques that are used to determine accurately the structure of molecules, the areas of their concerted applications, and the significance of accurate structural information in current chemical research.

Today's physical and computational techniques are capable of determining bond lengths and bond angles in the best cases with formal precisions better than a few thousandths of an ångström and a few tenths of a degree, respectively. Such precisions, however, may be misleading if they are not accompanied by a rigorous error analysis in terms of accuracy and a careful examination of the physical meaning of the parameters. Information on molecular structure may suffer not only from experimental error, but also bears the consequences of the different nature of the physical phenomena involved in its determination; of the different averaging process over the various motions of the species under investigation; and of the structural differences in the states of aggregation in which various investigations may be carried out. For example, in order to have a meaningful comparison of the geometry of a molecule determined in the gaseous phase by microwave spectroscopy and in the crystal by X-ray diffraction, all the above points have to be considered. If we are aiming to determine the influence of intermolecular interactions in the crystal on the geometry of the molecule, the consequences of all the other differences must be accounted for before anything about gas/crystal structural differences can be reliably concluded. Once this is carried out, however, then such conclusions may prove highly valuable.

The uncertainty of an accurately determined geometrical parameter is in many cases as important as the value of the parameter itself. Unfortunately scientific papers have quite different approaches to the problem of determining and reporting uncertainties in structural results. Details about the procedure used are not always given. Even the experienced reader may find it difficult to ascertain whether the reported uncertainty is an estimate of total error or merely a least-squares standard deviation, i.e. a measure of internal consistency of the data used. With the aim of improving this situation, we have asked our contributors to report total errors as error limits, e.g. 1.399 ± 0.002 Å, and least-squares standard deviations in parentheses as units of the last digit, e.g. $1.3994(4)$ Å.

Experience shows that the information content of accurate molecular structures is often underestimated in chemical research. The purpose of this book is to demonstrate the significance and applicability of accurate structural information, following a rigorous discussion of the demands and caveats in their determination. Some guidelines are also established for the accuracy requirements in answering broadly varying questions that may occur in structural problems.

The creation of this book was preceded by more than a decade of co-operation in the field of accurate molecular structure determination on the part of the editors. The reliability and importance of small structural differences have been a focal point of our interest. The need to have all the information together that we have gathered in this volume has been increasingly felt over the years. In the late seventies and early eighties we had a couple of minisymposia in Budapest on the effects of substitution on molecular geometry and, more generally, on small structural differences. By the time of the 12th International Congress of Crystallography (Ottawa, 1981) the thought of a larger meeting along these lines was formulated. Our duo expanded to a trio by including Peter Murray-Rust. The matter was brought to Tom Blundell's attention who then provided a framework for it within the International School of Crystallography of the 'E. Majorana' Centre for Scientific Culture. A course entitled 'Static and Dynamic Implications of Precise Structural Information' took place in the spring of 1985 in Erice, Italy. The grand organizer of the School, Lodovico Riva di Sanseverino, made an extraordinary effort and ensured that the event became memorable in the best sense. The present book is an outgrowth of that course; it retains much of its original flavor and motivation, but it is by no means a proceedings volume, nor is it otherwise associated with the course. While most of the original lecturers have contributed a chapter, other scientists have also been invited to do so according to the needs of the subject matter of the book.

The organization of the volume was meant to embrace the whole field, and this gave a natural sequence for the chapters. It starts with a general historical introduction followed by a discussion of the potential energy surface. Then come three chapters on gas-phase molecular structure determination by microwave and infrared spectroscopies and electron diffraction. There are four chapters on solid-state research by X-ray diffraction, including two on the problem of extracting information on the dynamical properties of molecules from the diffraction experiment. Then a chapter follows on the experimental determination of the electron-density distribution in crystals. One chapter deals with neutron crystallography and one with liquid-crystal NMR spectroscopy. Three chapters are devoted to non-experimental techniques, namely, molecular orbital calculations, molecular mechanics, and the use of data banks for solving structural problems. The last six chapters deal with various applications. In addition to general chapters on organic and inorganic chemistry, there are case studies of reaction pathways, substituent effects, effects of crystal environment, and structural variations in metal cluster compounds.

It is our sad duty to report that the senior author of the introductory chapter, Massimo Simonetta, passed away before the preparation of the book began. We are grateful to Angelo Gavezzotti for creating this chapter on the basis of Professor Simonetta's Erice lecture, preserving the purpose and flavor of the original presentation. At the proof stage of the book we learned of the death of another author, Fred Hirshfeld. Professor Hirshfeld not only prepared his chapter and corrected its proofs but assisted us in other ways as well. Both Fred Hirshfeld and Massimo Simonetta were strong advocates of and important contributors to the ideas and concepts around which this book was built.

Most of the manuscripts have been refereed internally, i.e. by authors of other chapters, but in some cases also externally. We express our gratitude to the outstanding internal and external referees for their conscientious and invaluable assistance. We list here only the names of the external referees: R. Destro, E. Longoni, Z. B. Maksić, P. L. Mandolini, P. Müller, G. Portalone, D. W. H. Rankin, and A. L. Segre.

We express our thanks to Clara Marciante for preparing a number of figures out of sketches provided by authors, and to Anna Rita Campanelli, Gustavo Portalone and Fabio Ramondo for help with proof-reading.

The book was initiated and prepared having our students in mind, and it is only natural to dedicate it to them.

Spring 1991   Aldo Domenicano and István Hargittai
L'Aquila and Budapest

Contents

Contributors

Albano, Vincenzo G. Department of Chemistry 'G. Ciamician', University of Bologna, Via Selmi 2, I-40126 Bologna, Italy

Allen, Frank H. Crystallographic Data Centre, University Chemical Laboratory, Lensfield Road, Cambridge CB2 1EW, England

Allinger, Norman L. Department of Chemistry, The University of Georgia, Athens, Georgia 30602, USA

Bernstein, Joel Department of Chemistry, Ben-Gurion University of the Negev, Beer Sheva, 84105 Israel

Boggs, James E. Department of Chemistry, The University of Texas, Austin, Texas 78712, USA

Boudon, Stéphane URA n.422 du CNRS, Institut de Chimie, 4, Rue Blaise Pascal, F-67000 Strasbourg, France

Braga, Dario Department of Chemistry 'G. Ciamician', University of Bologna, Via Selmi 2, I-40126 Bologna, Italy

Burdett, Jeremy K. Chemistry Department and James Franck Institute, The University of Chicago, Chicago, Illinois 60637, USA

Bürgi, Hans-Beat Laboratory of Chemical and Mineralogical Crystallography, University of Bern, Freiestrasse 3, CH-3012 Bern, Switzerland

Diehl, Peter Physics Department, The University of Basel, Klingelbergstrasse 82, CH-4056 Basel, Switzerland

Domenicano, Aldo Department of Chemistry, Chemical Engineering and Materials, University of L'Aquila, Via Assergi 4, I-67100 L'Aquila, Italy (and CNR Institute of Structural Chemistry, I-00016 Monterotondo Stazione, Italy)

Dubler-Steudle, Katharina C. Laboratory of Chemical and Mineralogical Crystallography, University of Bern, Freiestrasse 3, CH-3012 Bern, Switzerland

Ferretti, Valeria Laboratory of Chemical and Mineralogical Crystallography, University of Bern, Freiestrasse 3, CH-3012 Bern, Switzerland (Present address: Department of Chemistry, University of Ferrara, I-44100 Ferrara, Italy)

Gavezzotti, Angelo Department of Physical Chemistry and Electrochemistry and CNR Centre, University of Milano, Via Golgi 19, I-20133 Milano, Italy

Glusker, Jenny P. The Institute for Cancer Research, The Fox Chase Cancer Center, 7701 Burholme Avenue, Philadelphia, Pennsylvania 19111, USA

Gramaccioli, Carlo M. Department of Earth Sciences, University of Milano, Via Botticelli 23, I-20133 Milano, Italy

Graner, Georges Laboratoire d'Infrarouge (associated to the CNRS), University of Paris-Sud, Bâtiment 350, F-91405 Orsay Cédex, France

Hargittai, István Structural Chemistry Research Group of the Hungarian Academy of Sciences, Eötvös University, Pf. 117, H-1431 Budapest, Hungary (and Technical University of Budapest, Szt. Gellért tér 4, H-1521 Budapest, Hungary)

Hirshfeld, Fred L. Department of Structural Chemistry, Weizmann Institute of Science, P.O. Box 26, Rehovot, 76100 Israel

Jeffrey, George A. Department of Crystallography, University of Pittsburgh, Pittsburgh, Pennsylvania 15260, USA

Kuchitsu, Kozo Department of Chemistry, Faculty of Science, The University of Tokyo, Bunkyo-ku, Tokyo 113, Japan (Present address: Department of Chemistry, Nagaoka University of Technology, Nagaoka 940-21, Japan)

Seiler, Paul Organic Chemistry Laboratory, Swiss Federal Institute of Technology, ETH-Zentrum, CH-8092 Zürich, Switzerland

Simonetta, Massimo Department of Physical Chemistry and Electrochemistry and CNR Centre, University of Milano, Via Golgi 19, I-20133 Milano, Italy

Trueblood, Kenneth N. Department of Chemistry and Biochemistry, University of California, Los Angeles, California 90024, USA

Van Eijck, Bouke P. Department of Crystal and Structural Chemistry, University of Utrecht, Padualaan 8, 3584 CH Utrecht, The Netherlands

Wipff, Georges URA n.422 du CNRS, Institut de Chimie, 4, Rue Blaise Pascal, F-67000 Strasbourg, France

1

Structural chemistry

Massimo Simonetta† and Angelo Gavezzotti

1.1 Outlook and purpose

Chemistry has a very important part in the scientific and technological wonders of our times. The composition of planetary atmospheres and interstellar clouds has been probed, superconductivity has been obtained at easily accessible temperatures, fibres a few millimeters thick can lift tons of weight, millions of tiny charges can be manipulated in microscopic chips to store and process information. The codes of life and evolution have been cracked, and the genes of micro-organisms can be cut and spliced to induce the production of useful substances. Much of this has been made possible by our detailed knowledge of the intimate structure of matter at a molecular level, which has in turn produced the well-established theories of modern chemistry.

At the same time, comprehensive and relatively simple theories have been constructed to account for the quantum behaviour of atoms and molecules, and phenomena like tunneling or Pauli forces are now explained. The non-classical, elusive properties of electron and nuclear spin are widely exploited for chemical characterization.

While qualitative and quantitative analyses have been revolutionized by the advent of mass spectrometry and chromatography, most major breakthroughs in structural chemistry have been fostered by the ability to understand the interaction of electromagnetic radiation with matter. Absorption and emission spectroscopies, as well as diffraction methods, have been and

---

† Deceased January 6, 1986.

still are the experimental cornerstones of structural science. In the beginning, bulk properties were recognized first, and then rationalized after structure analysis. Now the challenge proceeds the other way around, as well-established structural concepts are being used to predict bulk behaviour, from thermodynamic, optical, magnetic, and other physical properties, to chemical reactivity in gaseous and condensed media. Quantum chemistry is the natural complement of experiment in this undertaking.

We now know to a high degree of accuracy the vibrational patterns of hundreds of small molecules, the interatomic and intermolecular spacings in crystals and polymeric materials, the pitch of the DNA helices, and the conformational requirements for mutual recognition in neural receptors. The purpose of this chapter is to provide an introduction to the most accurate techniques of structural investigation of which the twentieth-century chemist may avail himself. For obvious reasons, this chapter cannot be an exhaustive one. It should, however, provide the reader with the right perspective to appreciate the large amount of information which is contained in the subsequent, more specialized chapters.

1.2   A little history: chemistry in two dimensions

Chemists have always been interested in the structure of molecules. Of course, the word *structure* has changed its meaning as more and more sophisticated methods of looking at molecules have been developed. Elemental analysis and molecular weight measurements were among the first probes of the molecular world, and helped to establish that benzene was made of six carbon and six hydrogen atoms—no small achievement in the early times, when chemists were satisfied with the determination of the atomic composition of a molecule. They had to work a long way to ascertain, by using the laws of gases and density measurements, that for example the reaction of formation of hydrochloric acid from hydrogen and chlorine was not, say, $H + Cl = HCl$ or $H_4 + Cl_4 = 2H_2Cl_2$, but $H_2 + Cl_2 = 2HCl$.

Later, chemists established the existence of chemical bonds holding together atoms in a molecule and wanted to know which atoms were bonded to each other and which were not. Structural formulae changed from an enumeration of the atoms in a molecule to sketches of their mutual connectivity; so, in methane, CH_4, each hydrogen is bound to carbon, but there are no hydrogen–hydrogen bonds. In times when spectrometers and diffractometers were not even to be dreamed of, the only way to test the mutual arrangement of atoms and bonds was to painstakingly make chemical derivatives, and count how many different ones were there.

To give just one example of the beautiful chemistry implied in such manipulations, we take up the classical proof, put forward through the combined efforts of a number of nineteenth-century chemists (among them

Ladenburg, Wroblewsky, Hübner, and Körner), of the equivalence of the six carbon atoms of benzene, after Kekulé's proposal. Aniline, **1**, can be transformed into benzenediazonium chloride **2** and then into bromobenzene **3**:

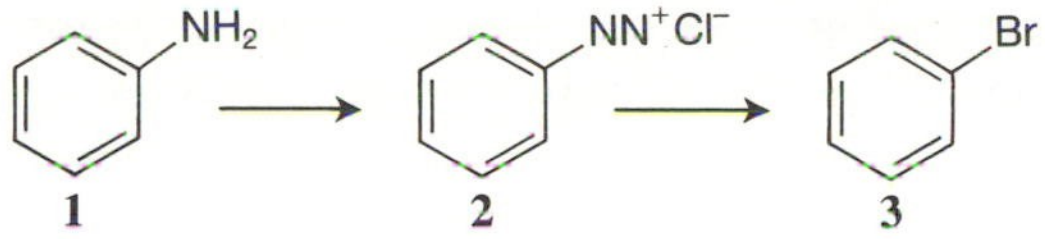

Scheme 1.1

Aniline can also be brominated to *p*-bromoaniline **4**:

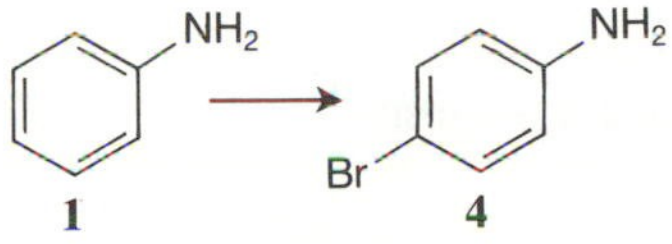

Scheme 1.2

Further treatment (Scheme 1.3) leads to the bromobenzene **3′**, in which the bromine atom is attached to a carbon which is formally different from that in **3**. However, product **3′** proved to be identical to **3**. Bromoaniline **4**, as obtained in Scheme 1.2, can be nitrated and then transformed back to aniline **1′** (Scheme 1.4):

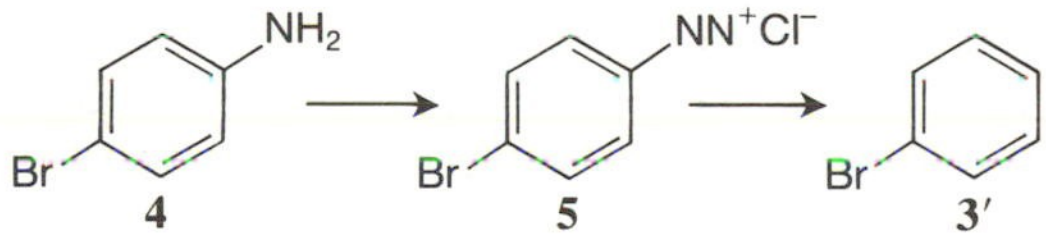

Scheme 1.3

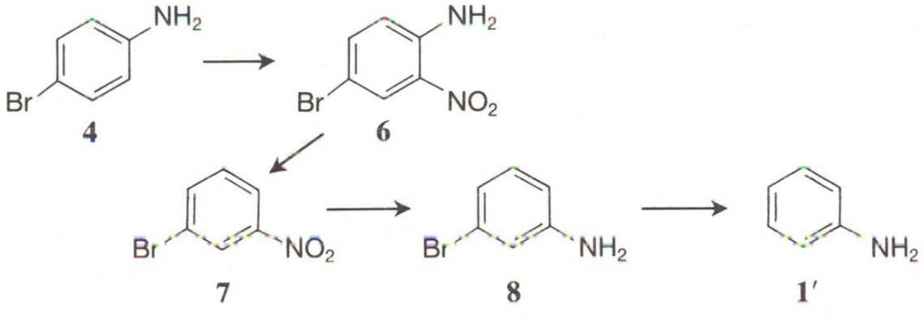

Scheme 1.4

Aniline **1'** was used as the starting reactant in Scheme 1.1 to give a bromo-benzene with bromine in a position still different from that of previous treatments, but the physical and chemical properties of the product were again the same. By a similar chain of reactions, the equivalence of the three remaining positions could be proved.

Moreover, other ingenious reaction chains, leading to formally different isomers, e.g. **9** and **9'**, unequivocally proved that such compounds were identical.

9   **9'**

1.3   Chemistry in three dimensions

Due to the existence of only one CH_3X, CH_2X_2, or CHX_3 derivative of methane, it can be argued that the four $C-H$ bonds must be equivalent, and the four hydrogen atoms must consequently be located in equivalent positions in space. This simple argument reflects a fact of enormous importance: the molecule of methane cannot be planar, since the only geometrical arrangement which is compatible with chemical evidence has the four hydrogens at the corners of a regular tetrahedron with the carbon atom at its centre.

The next issue in structural chemistry then became the study of the position of atoms in three-dimensional space, better known as stereochemistry. This led to the discovery of enantiomerism: when the hydrogen atoms in methane are replaced by four different substituents, two different molecules are obtained, one being the mirror image of the other. They can be distinguished by their different behaviour when interacting with polarized light, and in many cases they can also be separated.

From the qualitative information about the location of the atoms of a molecule in space it is natural to try to proceed to quantitative information, that is, to ask for the complete geometrical characterization of the molecule, by measuring bond distances, bond angles and torsion angles. In this step from qualitative pictures to quantitative measurement, the ingenious arguments based on chemical behaviour become insufficient: for the determination of precise geometrical parameters, physical methods or theoretical procedures become necessary. Also, having determined the molecular structure, one is confronted with the fact that for some molecules two or more conformations may have similar or even equal stability, and the energy barriers between them can be small enough so that, even at room temperature, the molecule fluctuates among all the possible conformations. The determination of the thermodynamic and kinetic parameters of this dynamic behaviour then becomes of interest.

Traditional chemistry can still provide a powerful tool of investigation, since reactions from the different conformations can lead to different products, but again the use of physical or theoretical methods is much more straightforward in the study of such conformational equilibria. Finally, since no molecule can be totally rigid, even for molecules that exist in only one energy minimum, the atoms oscillate around the equilibrium positions, and one becomes interested in the thermal behaviour of these vibrations.

1.4 Spectroscopy and diffraction

Spectroscopy is the art of measuring and interpreting the response of molecular systems to the stimuli of electromagnetic radiation, and of deriving structural data from these responses. Infrared (IR), microwave (MW) and ultraviolet–visible (UV–VIS) radiation can probe molecular vibrations, rotations, and electronic excitations, respectively. An IR spectroscopist has to go through the sometimes very awkward stage of band assignment, before proceeding to set up a mathematical model of molecular vibration, usually in the harmonic approximation. The model depends on molecular geometry and on the potential energy surface of the molecule. In MW spectroscopy, molecular moments of inertia are the crucial information. By matching the observed spectra to the model, and using critically the interpretative power of chemical intuition, one can quite often retrieve a complete and detailed picture of molecular structure. UV–VIS spectroscopy is of course very indirect if the reciprocal positions of the nuclei are at stake, but provides information on such electronic properties as, for instance, π-electron delocalization and hyperconjugation.

The spectroscopies of spin (nuclear magnetic resonance (NMR), electron spin resonance (ESR), and derived techniques, in the liquid, liquid-crystal, or crystalline state), can quite often yield relevant information on molecular structure and electronic effects, but not without hazard when the molecules (and hence the spectra) are very complex. Here, again, the stumbling block is band assignment, a procedure not always completely immune from prejudice.

Diffraction methods rely on the match between a measured and a computed scattering function, which depends on molecular geometry and vibrational behaviour. The scattering function is continuous in gas-phase electron diffraction (ED), and discontinuous in X-ray diffraction from crystals. For gas-phase ED, only molecules with enough thermal stability and vapour pressure to be obtained in sufficient concentration in the vapour phase are eligible. Also, the amount of information that can be retrieved from the scattering function is inherently small, and this poses an upper limit to the molecular complexity if a detailed picture is sought. It must be said, however, that it is only in the range of relatively small molecules where the fundamentals of structural chemistry can most easily be appreciated and studied. Some of the most accurate structure determinations on the small molecules which are the building blocks of organic chemistry have been carried out by taking advantage of the synergy of I R, M W, and E D techniques.

Electron spectroscopy for chemical analysis (ESCA) and ultraviolet photoelectron spectroscopy (UPS) probe the energies of core and valence molecular electrons, respectively. They can be coupled with quantum chemistry calculations that provide the energies of molecular orbitals or, even better, of molecular electronic states. The connection with quantitative molecular structure information is, however, rather indirect. These techniques suffer from the additional handicap that ultra-high vacuum chambers are required, and the involved apparatus is rather expensive and critical.

By far the most popular and reliable method of structure analysis for larger molecules (up to biopolymers, proteins and nucleic acids) is X-ray diffraction. Over 70 000 such analyses have been carried out to date, and the results are stored in the Cambridge Structural Database for consultation by computer. The analysis is carried out on crystalline solids or, preferably, on single crystals. Fully automated X-ray diffractometers are available, although still rather expensive. Difficulties come from the fact that single crystals may not be available for all substances, that sometimes static or dynamic disorder is present in the crystal, and that the geometry one obtains is determined not only by intramolecular forces, but also by the inter-molecular interactions. The latter are much weaker than the former, but in some cases they can appreciably modify the geometry with respect to the 'isolated' molecule (an abstract entity, most closely approximated by the gas-phase molecule). Besides the solid-state geometry, information on thermal motion, usually expressed by means of thermal ellipsoids, can be obtained by single-crystal X-ray analysis. More refined analyses of experimental data yield information about the electron-density distribution for molecules in crystals. Neutron diffraction turns out to be complementary to X-ray diffraction in these studies, since X rays are scattered by electrons, and neutrons by nuclei.

A critical assessment of the relative merits of these techniques of structure analysis, and the choice of the most appropriate one for each chemical problem, depend on a number of concomitant factors. The required degree of accuracy should be established first. When dealing with fundamental issues, such as details of electronic structure, chemical bonding or vibrational behaviour, a bond length determined to the thousandth of an ångström may be required; structure–reactivity relationships are usually sensitive to the hundredth of an ångström; while a low-resolution X-ray analysis, perhaps without even caring for bond lengths, may do for quick recognition of molecular connectivity or conformation. Also, quite often chemical research is geared so that the issue is approached from the other end — many structural chemists specialize in one of these techniques and then look for problems that can be solved by it. In any case, the reader will find numerous examples and detailed discussions of these topics in the following chapters.

1.5 Quantum chemistry and other theoretical methods

Besides experimental methods, theoretical approaches turn out to be very useful for molecular structure determination. Sometimes, pure symmetry considerations from group theory alone allow the determination of the number of possible conformations and of their mutual relationships. Empirical calculations, known as molecular mechanics or force-field calculations, can be used to obtain the equilibrium geometries and energies of isolated molecules. These approaches rely on a generalization of a few potential and structural parameters, mostly of spectroscopic origin, embodied in what in modern terminology would be called an 'expert system'. Similar methods can be applied to the study of the structure of molecular crystals, and the influence of the crystal field on molecular conformation.

In principle, any observable property of an atomic or molecular system can be obtained from the Schrödinger equation, as the expectation value of the corresponding quantum mechanical operator. The Hartree–Fock–Roothaan equations are the usual link between the intractable first-principles equations and the derivation of electronic energies for any nuclear arrangement. As a result, the minimum-energy conformation of a molecule can also be obtained by means of quantum chemical calculations, when the geometry of a molecule is optimized through an *ab initio* MO calculation. This kind of approach is especially useful for unstable species, which are labile or cannot be obtained in sufficient quantity to allow an experimental investigation. Of course, quantum chemistry is the only viable method for the study of non-existing species, since it can as well predict the structure of molecules that have not (yet) been synthesized.

The development of applied quantum chemistry is closely tied to the development of computer availability. Upper limits are encountered when the number of electrons, together with the degree of sophistication (e.g. the size of the basis set) exceed the computer speed, or the memory storage requirements for the necessary integrals. We do not venture a prediction on the actual placement of these limits, since such predictions are usually obsolete by the time they are printed.

1.6   A few examples

The rest of this chapter will be mostly devoted to the illustration—through a few examples—of the concepts and ideas which have been previously discussed. Their research interests have certainly biased the authors towards problems that require the use of, or can be solved by, accurate X-ray single-crystal structure determination or quantum chemical calculations. The reader must be aware that equally interesting and chemically relevant examples could have been drawn from spectroscopy or gas-phase diffraction, as is made clear in the rest of this book.

The first case we wish to consider is that of the structure of the cation $C_{12}H_{11}^{+}$, some salts of which were synthesized by Vogel and coworkers (Grimme *et al.* 1965, 1966). From the path followed in the preparation, and from chemical and physical properties including the UV spectrum, it was concluded that the cation could be described as bicyclo[5.4.1]dodecapentaenylium. However, Masamune and coworkers (Kemp-Jones *et al.* 1973) concluded from ^{13}C NMR spectroscopic data that the compound should be described as a benzohomotropylium cation. Since the difference between the two proposed structures (see Fig. 1.1 (a, b)) lies in the presence or absence of the $C(1)-C(6)$ bond, the most practical way to solve the dilemma was the direct measurement of the $C(1)-C(6)$ distance. Samples of three different salts, namely the tetrafluoroborate, the hexafluoroantimonate and

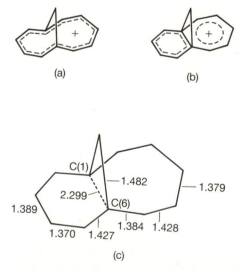

(a)

(b)

(c)

Fig. 1.1 $C_{12}H_{11}^{+}$ cation: (a) Dodecapentaenylium, or [11]annulenium, structure; (b) Benzohomotropylium structure; (c) A scheme of the cation geometry, as found by X-ray crystallography (Destro *et al.* 1976). Bond lengths (Å) are averaged assuming mirror symmetry. Estimated standard deviations are 0.004 to 0.008 Å.

the hexafluorophosphate, were examined by single-crystal X-ray diffractometry (Destro *et al.* 1976; Destro and Simonetta 1979). All of them produced disordered crystals so that the accurate determination of bond
distances and angles required more effort than is usual in X-ray work, but
eventually the geometry shown in Fig. 1.1(c) resulted unequivocally. The
C(1)—C(6) distance is a typical non-bonded distance, and the compound is
correctly described as a dodecapentaenylium, or [11]annulenium, derivative.

The same problem is found in another bridged species, **10**. This compound crystallizes in a triclinic form with two independent molecules in the
asymmetric unit. Again the C(1)—C(6) distance is the crucial structural
feature, since the compound is an annulene derivative, 11,11-dimethyl-
1,6-methano[10]annulene (**10a**), in the absence of this bond, while, if a
bond between these two atoms were established, the compound would be
11,11-dimethyltricyclo[4.4.1.0^{1,6}]undeca-2,4,7,9-tetraene (**10b**).

the crystal structure of **10** was solved at room temperature (Bianchi *et al.*
1973) and the surprising result was a C(1)—C(6) distance of 1.836(7) and
1.780(7) Å (libration-corrected values) in the two crystallographically independent molecules. These values are too large for a carbon–carbon single
bond, but much too short for a non-bonding situation. One might think that
the crystal is made of two kinds of molecules, one with a regular C(1)—C(6)
bond about 1.5–1.6 Å long, and the other with no such bond and a
C(1) $\cdots$ C(6) distance of about 2.2–2.3 Å, and that what one sees in the
diffraction experiment is an average, weighted by different intermolecular
interactions at the two sites. In this case, however, thermal vibration ellipsoids for C(1) and C(6) should show some anomalies, in particular a large
elongation along the C(1)—C(6) axis, but this was not the case. Still, one may
not trust experimentally determined libration parameters, since the anisotropic displacement parameters are obtained as a kind of by-product at the
end of a least-squares refinement procedure, where all the errors due to the
many approximations included in the treatment of the experimental data tend
to accumulate. Therefore, the atomic displacement parameters were obtained
by a completely independent procedure, using a lattice-dynamical calculation. In these calculations, the molecules are generally assumed to be rigid,
and the crystal potential is calculated by a summation over atom–atom pairwise interactions between all the atoms of the central molecule and the atoms

of the surrounding ones. Buckingham-type formulae (see Chapter 19, eqn (19.5)) represent the atom–atom potential. The equilibrium position and orientation of the molecule in the crystal field is determined first; then the equations of motion are written for the molecular librations, and the solution goes through the diagonalization of a dynamical matrix, from which the rigid-body libration tensors can be obtained. Both these tensors and the derived individual anisotropic displacement parameters were found to compare favourably with the experimental ones (Filippini *et al.* 1974). This result further legitimates the conclusion that the anomalous values found for the C(1)—C(6) distance are real, and that they are not average values.

The question now arises, whether or not these distances indicate the presence of a bonding situation between C(1) and C(6). To answer this question, one has to resort to quantum chemical calculations. An *ab initio* MO treatment (HF/STO-3G level) was used to obtain the wavefunction, using experimental geometries. To analyse it, the electron density and its gradient were calculated. The orthogonal trajectories (Collard and Hall 1977) for such gradients were then obtained and the critical points of this vector field (Bader *et al.* 1981) were investigated (Gatti *et al.* 1985). The results for the two independent molecules found in the crystal are shown in Fig. 1.2. When the C(1)—C(6) distance is 1.77 Å, there is a trajectory connecting the two atoms, with a bond-path critical point on it and a ring critical point within the C(1)—C(6)—C(11) triangle. When the C(1)—C(6) distance is 1.83 Å, on the other hand, the two critical points disappear, indicating the absence of

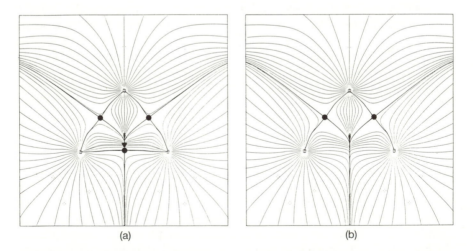

(a)   (b)

Fig. 1.2 Display of the gradient paths in the C(1)—C(11)—C(6) plane for the two independent molecules in the crystal of **10**: (a) Shorter C(1)—C(6) distance; (b) Longer C(1)—C(6) distance. The bond paths are indicated by heavy lines. The bond critical points are marked with •; ▼ is a ring critical point. After Gatti *et al.* (1985).

bonding. The crystal could then be considered as a mixture of two different compounds, **10a** and **10b**. Since these critical points are present in all calculations for similar compounds with shorter $C(1)-C(6)$ distances, and are absent for other derivatives with larger distances, it appears that the structural situation in **10** is very close to a so-called catastrophe point, separating two regions of stability pertaining to the two structures.

The theoretical approach to structure determination is the only resort for very short-lived species not amenable to experimental studies and for species whose existence has not yet been proved. As an example of the former case, we may mention CH_5^+, whose existence has been demonstrated by Olah *et al.* (1969). The cation was synthesized by reaction of dissolved methane with 'magic acids', such as $FSO_3H.SbF_5$, but nothing could be said about its three-dimensional structure. Quantum mechanical calculations (see, for example, Gamba *et al.* 1969; Raghavachari *et al.* 1981) predicted a stable species with C_s symmetry, in which three hydrogen atoms are bound to carbon by normal bonds, while the other two form a two-electron-three-centre bond with the carbon atom (see Fig. 1.3(a)). It was also argued that two such bonds could be formed by a carbon atom, leading to the CH_6^{++} dication. *Ab initio* MO calculations (Lammertsma *et al.* 1982, 1983) showed the existence of a true minimum in the energy hypersurface for CH_6^{++}, with a geometry very close to expectation (see Fig. 1.3(b)). Moreover, since the barriers to dissociation to the thermodynamically favoured $CH_5^+ + H^+$ and $CH_3^+ + H_3^+$ systems are quite high, it should be possible to observe the dication in the gas phase.

Another example is $C_2H_6^{++}$, for which three true minima in energy have been found, with a diborane-like structure (D_{2h} symmetry, Fig.

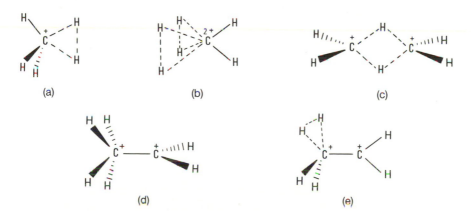

(a)   (b)   (c)

(d)   (e)

Fig. 1.3 Structures of protonated alkanes as found by quantum mechanical calculations: (a) CH_5^+ ion, C_s symmetry; (b) CH_6^{++} ion, C_{2v} symmetry; (c) $C_2H_6^{++}$ ion, D_{2h} symmetry; (d) $C_2H_6^{++}$ ion, C_{2v} symmetry; (e) $C_2H_6^{++}$ ion, C_s symmetry.

1.3(c)), a carbonium–carbenium structure (C_{2v} symmetry, Fig. 1.3(d)), and a hydrogenated-$C_2H_4^{++}$ structure (C_s symmetry, Fig. 1.3(e)) (Lammertsma *et al*. 1982; Olah and Simonetta 1982; Schleyer *et al*. 1982). The last minimum is calculated to be the most stable, with the carbonium centre involved in a two-electron-three-centre bond (Lammertsma *et al*. 1982). Preliminary evidence for the actual existence of this cation has been obtained by mass spectrometry (Stahl and Maquin 1983).

1.7 Bulk *vs.* molecular properties

Condensed media are now being considered from the molecular structure vantage point. Their properties are examined at a molecular level, and theories and models are being developed—thanks also to the advent of the new generations of electronic computers—to calculate bulk properties starting from molecular structure. In principle, the best description of liquids and crystals is based on a full molecular dynamics treatment, whereby an assembly of molecules is relaxed to equilibrium under the action of a given potential and at a given temperature and pressure. Although complex, the equations of motion for such systems can be written and integrated. Often simpler arguments, based on molecular shape parameters, suffice for a preliminary analysis of the packing of organic crystals (Gavezzotti 1985).

It may also be mentioned that with the advent of reliable ultra-high vacuum apparatus the chemisorption of organic molecules on metal surfaces in the limit of zero pressure can be reliably accomplished, and spectroscopic and diffraction experiments can be carried out on such systems to measure structural parameters. Theoretical chemistry is now also being used to predict these structures. The vast province of chemical catalysis is thus being colonized these days by structural chemists; there, too, the structural approach is very quickly taking over and making bulk techniques appear more and more obsolete.

1.8 Perspective

The concepts of structural chemistry have thus evolved in a century, to the point that chemistry has now a very sound theory of molecular structure whose successes are already evident, and with far-reaching potential applications. The above examples have hopefully provided the reader with a hint of how sophisticated structure analysis can be at the present time, and more detail is to be found in subsequent chapters. Whatever the experimental or theoretical approach one may adopt, we do wish to highlight the need, in modern chemistry, for accurate structure determination. As said previously, the level of accuracy needed depends on the nature of the chemical problem one is confronted with, but we may go as far as to state that no piece of chemical reasoning will survive, and withstand future criticism, unless it is based on a structural foundation at a molecular level.

References

Bader, R. F. W., Nguyen-Dang, T. T., and Tal, Y. (1981). *Rep. Prog. Phys.*, **44**, 893–948.

Bianchi, R., Morosi, G., Mugnoli, A., and Simonetta, M. (1973). *Acta Crystallogr.*, **B29**, 1196–208.

Collard, K. and Hall, G. G. (1977). *Int. J. Quantum Chem.*, **12**, 623–37.

Destro, R. and Simonetta, M. (1979). *Acta Crystallogr.*, **B35**, 1846–52.

Destro, R., Pilati, T., and Simonetta, M. (1976). *J. Am. Chem. Soc.*, **98**, 1999–2000.

Filippini, G., Gramaccioli, C. M., Simonetta, M., and Suffritti, G. B. (1974). *Chem. Phys. Lett.*, **26**, 301–4.

Gamba, A., Morosi, G., and Simonetta, M. (1969). *Chem. Phys. Lett.*, **3**, 20–1.

Gatti, C., Barzaghi, M., and Simonetta, M. (1985). *J. Am. Chem. Soc.*, **107**, 878–87.

Gavezzotti, A. (1985). *J. Am. Chem. Soc.*, **107**, 962–7.

Grimme, W., Hoffmann, H., and Vogel, E. (1965). *Angew. Chem. Int. Ed. Engl.*, **4**, 354–5.

Grimme, W., Kaufhold, M., Dettmeier, U., and Vogel, E. (1966). *Angew. Chem. Int. Ed. Engl.*, **5**, 604–5.

Kemp-Jones, A. V., Jones, A. J., Sakai, M., Beeman, C. P., and Masamune, S. (1973). *Can. J. Chem.*, **51**, 767–71.

Lammertsma, K., Olah, G. A., Barzaghi, M., and Simonetta, M. (1982). *J. Am. Chem. Soc.*, **104**, 6851–2.

Lammertsma, K., Barzaghi, M., Olah, G. A., Pople, J. A., Schleyer, P. v. R., and Simonetta, M. (1983). *J. Am. Chem. Soc.*, **105**, 5258–63.

Olah, G. A. and Simonetta, M. (1982). *J. Am. Chem. Soc.*, **104**, 330–1.

Olah, G. A., Klopman, G., and Schlosberg, R. H. (1969). *J. Am. Chem. Soc.*, **91**, 3261–8.

Raghavachari, K., Whiteside, R. A., Pople, J. A., and Schleyer, P. v. R. (1981). *J. Am. Chem. Soc.*, **103**, 5649–57.

Schleyer, P. v. R., Kos, A. J., Pople, J. A., and Balaban, A. T. (1982). *J. Am. Chem. Soc.*, **104**, 3771–3.

Stahl, D. and Maquin, F. (1983). *Chimia*, **37**, 87.

2

The potential energy surface and the meaning of internuclear distances

Kozo Kuchitsu

2.1 Introduction: potential energy function

The theoretical framework for the concept of molecular structure is based on the Born–Oppenheimer approximation (Born and Oppenheimer 1927; Ballhausen and Hansen 1972). Except in the case of strong vibronic interaction, the electronic motion is solved for a set of fixed nuclear positions; the electronic wavefunction contains the molecule-fixed nuclear coordinates as parameters, instead of variables, because the influence of the momentum of the nuclear motion operating on the electronic wavefunction is ignored. The electronic energy in a certain electronic state, represented as a function of the nuclear coordinates, corresponds to the potential energy function for the nuclear motion (i.e. the intramolecular vibration).

In this Born–Oppenheimer scheme, the potential function for the nuclear motion is assumed to be independent of the nuclear masses and, therefore, the potential function should be independent of isotopic substitution. Detailed spectroscopic studies of diatomic potential functions have confirmed that the isotopic differences are indeed small (Huber and Herzberg 1979). For example, the following isotopic differences have been reported on the energy minimum for an excited electronic state B relative to the ground electronic state X:

$$T_e(H_2) - T_e(D_2) = 2.8 \text{ cm}^{-1} \qquad (B^1\Sigma_u^+), \qquad (2.1)$$

$$T_e(BH) - T_e(BD) = -12.2 \text{ cm}^{-1} \qquad (B^1\Sigma^+), \qquad (2.2)$$

where T_e denotes, following the general convention in spectroscopy, the electronic energy of the potential minimum for state B and $1 \text{ cm}^{-1} = 1.1963 \times 10^{-2} \text{ kJ mol}^{-1}$. The equilibrium internuclear distance for HCl is shown to be very slightly mass-dependent:

$$r_e = 1.27460 \left[1 + 0.000077(\mu/u)^{-1}\right] \text{Å}, \qquad (2.3)$$

where $\mu = m_H m_{Cl}/(m_H + m_{Cl})$ represents the reduced mass, expressed in atomic mass units u. The slight breakdown of the Born–Oppenheimer approximation has been discussed theoretically (e.g. Bunker 1968, 1970, 1972; Coxon 1986).

2.2 Diatomic molecules

2.2.1 Determination of the potential function

A reliable method for determining the potential function, known as the *Rydberg–Klein–Rees* (RKR) *method* (see, for example, Zare 1964; Barrow *et al.* 1974; Wright 1988), has been used widely. This semiclassical method uses the observed vibrational energy levels $G(v)$ and the rotational constants $B(v)$ (see Chapter 3) for $v = 0, 1, \ldots v_{max}$, and determines the $r_{min}(E)$ and $r_{max}(E)$ up to $G(v_{max})$, where r_{min} and r_{max} denote the classical turning points, on which the horizontal line at energy E crosses the potential energy curve. The analysis is programmed in such a way that

1. $G(v)$ and $B(v)$ are expanded in power series of $(v + \frac{1}{2})$.

2. $f(v)$ and $g(v)$ defined by

$$f(v) = (\hbar/4\pi c\mu)^{\frac{1}{2}} \int_{-\frac{1}{2}}^{v} [G(v) - G(u)]^{-\frac{1}{2}} \, du, \qquad (2.4)$$

$$g(v) = (4\pi c\mu/\hbar)^{\frac{1}{2}} \int_{-\frac{1}{2}}^{v} B(u)[G(v) - G(u)]^{-\frac{1}{2}} \, du, \qquad (2.5)$$

where c denotes the velocity of light in vacuum, are calculated numerically, the convergence at the upper limit of the integral being taken into account (Fleming and Rao 1972).

3. The classical turning points are then determined by

$$r_{min} = [f(v)^2 + f(v)/g(v)]^{\frac{1}{2}} - f(v), \qquad (2.6)$$

$$r_{max} = [f(v)^2 + f(v)/g(v)]^{\frac{1}{2}} + f(v) . \qquad (2.7)$$

4. The turning points calculated for a number of the v values are connected smoothly, and the potential energy curve is obtained.

5. This RKR curve may be extrapolated by use of appropriate model functions such as

$$V^{in}(r) = c_1 r^{-12} + c_2 \quad \text{and} \quad V^{out}(r) = c_3 r^n + D_e, \qquad (2.8)$$

where superscripts in and out represent $r < r_e$ and $r > r_e$, respectively, and D_e is the bond dissociation energy, as discussed in the next section.

2.2.2 Approximate diatomic potentials

Even when spectroscopic data, $G(v)$ and $B(v)$, are insufficient, one can still estimate the potential function by use of a model function. One of the simplest and the best known is the Morse function (Morse 1929),

$$V(r) = D_e[1 - \exp(-ax)]^2 , \qquad (2.9)$$

where D_e is the bond dissociation energy measured from the potential minimum, x is $r - r_e$ (r_e being the equilibrium internuclear distance), and a parameter a represents the deviation from a harmonic oscillator, i.e. 'anhar-monicity' in the bond-stretching vibration. The potential energy curve for the ground electronic state of I_2 and the corresponding Morse function (broken curve) are compared in Fig. 2.1 as an example (Verma 1960; LeRoy 1970).

It is sometimes convenient to use an expansion formula derivable from eqn (2.9) near the equilibrium position (Kuchitsu 1967a),

$$V(x) = \frac{1}{2}fx^2 - \frac{1}{2}afx^3 + \ldots, \qquad (2.10)$$

where

$$f = 2a^2D_e \qquad (2.11)$$

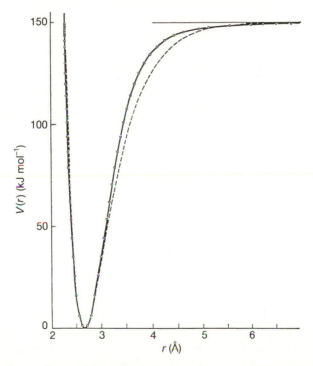

Fig. 2.1 Comparison of the potential energy curves for the ground electronic state of I_2 molecule (after Verma 1960). Solid curve: derived from the RKR analysis. Broken curve: the Morse potential fitted near the bottom of the curve.

is a quadratic force constant. For many diatomic molecules, the order of magnitude of the a parameter, sometimes denoted as a_3 (3 representing the cubic term), is 2 Å^{-1}. By a first-order perturbation the vibrational average value of the internuclear distance is shown to be (Bartell 1955)

$$\langle r \rangle = r_{\mathrm{e}} + \frac{3}{2} a \langle x^2 \rangle + \ldots \qquad (2.12)$$

This equation is used frequently for estimating an average internuclear distance of a chemical bond in a polyatomic molecule.

Besides the Morse function, various other model functions have been proposed for suitable analytical expressions of the potential energy functions for diatomic molecules (Steele *et al.* 1962; Murrell *et al.* 1984; Wright 1988).

2.2.3   Present status of diatomic potential functions

Recent experimental techniques, particularly laser spectroscopy, have enabled accurate determination of the potential functions for many electronic states, including weakly-bound or repulsive states, of many stable or short-lived diatomic molecules or ions (Huber and Herzberg 1979; Hirota 1985). In many excited electronic states the potential energy curves have complicated crossings, and various problems of molecular dynamics, such as dissociation, predissociation, bound–free transition, are being studied (see, for example, Levine and Bernstein 1987).

On the other hand, accurate *ab initio* MO calculations have been made on the electronic energies with fixed nuclear positions, and potential energy functions have been obtained for many electronic states of many diatomic species with accuracy comparable with those derived from experimental studies (see, for example, Dykstra 1988).

2.3   Polyatomic molecules

The Born–Oppenheimer approximation is generally also applicable to polyatomic molecules. Therefore, the electronic energy at a given set of nuclear positions corresponds to the potential energy which controls the nuclear displacements (i.e. intramolecular vibration). The nuclear framework which corresponds to the minimum potential energy is called the equilibrium (r_{e}) structure. The potential energy surface of an N-atomic molecule has a dimension of $3N-6$ (or $3N-5$ for a linear molecule); unlike a diatomic molecule, it is a formidable task to determine a complete potential energy hypersurface accurately from the equilibrium to the dissociation limit for even a symmetric triatomic molecule, XY_2. For a semirigid molecule, however, it is sufficient for most structural studies to describe the potential energy surface near the equilibrium structure (Mills 1972). For this purpose the potential energy function is expanded in terms of a set of nuclear displacement coordinates around the equilibrium structure. Thus the 'structure' of a free molecule usually represents the following structural parameters:

(1) total binding energy (stability);

(2) equilibrium nuclear positions (geometry);

(3) quadratic and higher-order potential derivatives around the equilibrium structure (force constants).

2.3.1 Internal coordinates

The displacement coordinates which are the most appealing to our chemical intuition are the 'internal coordinates', defined as the displacements in the internuclear distances r, bond angles φ, etc. from their equilibrium values. A set of these internal coordinates, Δr, $\Delta \varphi$, etc., are in general denoted as Я (Plíva 1974) (also denoted as R (Hoy *et al.* 1972)). The potential energy measured from the potential minimum can be expanded as

$$V = \frac{1}{2}\sum_i \sum_j f_{ij} Я_i Я_j + \frac{1}{6}\sum_i \sum_j \sum_k f_{ijk} Я_i Я_j Я_k + \ldots, \qquad (2.13)$$

where the force constants, f_{ij}, f_{ijk}, etc. are defined as

$$f_{ij} = (\partial^2 V / \partial Я_i \partial Я_j)_e \qquad (2.14)$$

$$f_{ijk} = (\partial^3 V / \partial Я_i \partial Я_j \partial Я_k)_e \qquad (2.15)$$

etc.

The first-order force constants are all zero, because the expansion is made at the equilibrium. The physical meaning of the diagonal second-order force constants, f_{ii}, is easily understood, because they represent the valence forces. The off-diagonal terms, f_{ij} ($i \neq j$), the third-order force constants, f_{ijk}, and constants of still higher orders represent complicated interactions. These internal coordinates are defined by the nuclear positions only and are independent of the nuclear masses. Therefore, these force constants remain unchanged, in the scheme of the Born–Oppenheimer approximation, even when one or more of the nuclei composing the molecule are replaced by their isotopes.

2.3.2 Normal coordinates

The kinetic energy of the intramolecular motion, here denoted as T, is given in terms of the molecule-fixed Cartesian coordinates (principal axes of inertia) by

$$T = \frac{1}{2}\sum_n m_n(\delta\dot{x}_n^2 + \delta\dot{y}_n^2 + \delta\dot{z}_n^2) \, , \tag{2.16}$$

where the Cartesian displacement coordinates of nucleus n with mass m_n are denoted as $\delta\alpha_n$ ($\alpha = x, y, z$). The potential energy may also be written, if desired, as a function of these Cartesian displacement coordinates. The normal coordinates, Q_r, are defined as a linear combination of the Cartesian displacement coordinates (Wilson $et\,al.$ 1955; Meal and Polo 1956) (see Section 2.3.4)

$$\delta\alpha_n = \sum_r (L_x)_{nr}Q_r = \sum_r (m_n^{-\frac{1}{2}}\,l_{\alpha n}^r)Q_r, \tag{2.17}$$

where the $\mathbf{L}_x$ matrix is often used in vibrational spectroscopy to visualize the normal modes of vibration in the molecule-fixed coordinate system. A set of the $l_{\alpha n}^r$ coefficients is known as the $\mathbf{l}$ matrix (Nielsen 1951). By use of the normal coordinates, it is possible to express the kinetic and potential energy terms as

$$T = \frac{1}{2}\sum_r \dot{Q}_r^2, \tag{2.18}$$

$$V = \frac{1}{2}\sum_r \lambda_r Q_r^2 + \sum\sum\sum_{r \leq s \leq t} K_{rst}Q_rQ_sQ_t + \ldots$$

$$= \frac{1}{2}\sum_r \lambda_r Q_r^2 + \frac{1}{6}\sum_r\sum_s\sum_t \Phi_{rst}Q_rQ_sQ_t + \ldots, \tag{2.19}$$

the expressions for K_{rst} and Φ_{rst} being related to restricted and unrestricted sums, respectively (Mills 1972).

For a semirigid molecule, the anharmonic terms of V (i.e. all but the first terms in eqn (2.19)) are usually so small that the equation of motion can be separated into that of a one-dimensional harmonic oscillator for each Q_r and the anharmonic terms can be treated as perturbations. Thus the normal-coordinate analysis rests on the choice of the $rectilinear$ coordinates in order to simplify the kinetic energy terms.

It is convenient to take a set of the dimensionless normal coordinates defined by

$$q_r = (4\pi^2 c\omega_r/h)^{\frac{1}{2}} Q_r, \tag{2.20}$$

where ω_r is the wavenumber of the r-th normal mode, and

$$\omega_r = \lambda_r^{\frac{1}{2}}/2\pi c. \tag{2.21}$$

It is then shown that (Hoy *et al.* 1972)

$$V/hc = \frac{1}{2}\sum_r \omega_r q_r^2 + \sum_{r \leq s \leq t}\sum\sum k_{rst} q_r q_s q_t + \dots$$

$$= \frac{1}{2}\sum_r \omega_r q_r^2 + \frac{1}{6}\sum_r\sum_s\sum_t \phi_{rst} q_r q_s q_t + \dots \tag{2.22}$$

One should note that the use of the normal coordinates is not based on the assumption that the vibrational amplitudes are infinitesimal; the normal coordinates **Q** and **q** are simply derived from a linear transformation of the Cartesian displacement coordinates, irrespective of whether the molecule is semirigid or non-rigid. However, the convergence of the potential function in eqns (2.19) and (2.22) is more rapid if the molecule is semirigid and the amplitudes of the nuclear displacements are small. One should also note that **Q** and **q** are defined under the 'Eckart conditions' (Wilson *et al.* 1955; Meal and Polo 1956) by which the molecule-fixed coordinate axes are defined (i.e. elimination of the overall translation and rotation of the molecule in space). Therefore, **Q** and **q** both depend on the nuclear masses and, unlike the internal coordinates, the anharmonic constants, k_{rst} and ϕ_{rst} in eqn (2.22), are isotope-dependent. In particular, the k_{rst} and ϕ_{rst} constants for a hydride and those for the corresponding deuteride are usually very different (see, for example, H_2O and D_2O in Table 2.5).

2.3.3 Curvilinear vs. rectilinear internal coordinates

The internal coordinates **ℜ** are non-linear functions of the normal coordinates **Q**. In general, **ℜ** can be expanded in terms of the normal coordinates as

$$\Re_i = \sum_r L_i^r Q_r + \frac{1}{2}\sum_r\sum_s L_i^{rs} Q_r Q_s + \frac{1}{6}\sum_r\sum_s\sum_t L_i^{rst} Q_r Q_s Q_t + \dots, \tag{2.23}$$

where L_i^r, L_i^{rs}, and L_i^{rst} are called the L-tensor elements. The method for calculating these elements is described in detail by Hoy *et al.* (1972). By substitution of eqn (2.23) into eqn (2.13), the coefficients of eqn (2.13) can be related to those in eqn (2.19) by

$$\lambda_r = \sum_i \sum_j f_{ij} L_i^r L_j^r, \qquad (2.24)$$

$$\Phi_{rst} = \sum_i \sum_j \sum_k f_{ijk} L_i^r L_j^s L_k^t$$

$$+ \sum_i \sum_j f_{ij} (L_i^{rs} L_j^t + L_i^{rt} L_j^s + L_i^{st} L_j^r). \qquad (2.25)$$

Thus the cubic potential constants, k_{rst} and ϕ_{rst}, contain both the second- and third-order force constants, f_{ij} and f_{ijk}. Equations (2.24) and (2.25) are often used in an analysis of anharmonic constants, i.e. transformation from f_{ijk} to k_{rst} or ϕ_{rst}, and vice versa. For example, one can derive k_{rst} from spectroscopy, from which f_{ijk} can be estimated in this scheme, or one can estimate f_{ij} and f_{ijk} by an *ab initio* MO calculation (e.g. Ermler *et al.* 1985; Fogarasi and Pulay 1985, 1986; Schaefer and Yamaguchi 1986; Handy *et al.* 1987; Botschwina 1988; Clabo *et al.* 1988; Hargiss and Ermler 1988; see Chapter 13, Section 13.5), from which k_{rst} or ϕ_{rst} can be derived. See Section 2.3.5 and Morino (1969), Hoy *et al.* (1972) for further details.

For a semirigid molecule, the internal coordinates $\mathcal{R}$ introduced in Section 2.3.1 are often approximated by the first term of its expansion in terms of the normal coordinates Q, eqn (2.23); these 'linearized' coordinates are denoted as R:

$$R_i \equiv \sum_r L_i^r Q_r. \qquad (2.26)$$

Particularly when the vibration is assumed to be purely harmonic, this approximation is sufficient, because the second- and higher-order terms of eqn (2.23) do not contribute to λ_r (see eqns (2.13) and (2.19)). Therefore, $\mathcal{R}$ and R are often confused in textbooks and papers. However, the *curvilinear* internal coordinates $\mathcal{R}$ should never be confused with the *rectilinear* internal coordinates R whenever anharmonicity is taken into account. A simple example is a linear XY_2 molecule (Fig. 2.2). The stretching and bending displacements in the $\mathcal{R}$ scheme, Δr_n ($n = 1$ or 2) and $\Delta \varphi$, respectively, can be written in terms of the Cartesian (rectilinear) displacements in the R scheme, Δz_n and $\Delta \rho$, as

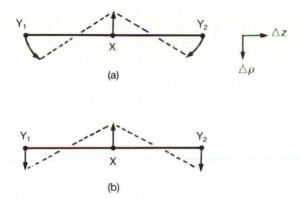

(a)

(b)

Fig. 2.2 $\mathbf{Я}$ and $\mathbf{R}$ coordinates representing the bending displacements of a linear XY_2 molecule. (a) $\Delta r_1 = \Delta r_2 = 0$ in the $\mathbf{Я}$ scheme, and (b) $\Delta z_1 = \Delta z_2 = 0$ in the $\mathbf{R}$ scheme.

$$\Delta r_n = \Delta z_n + \Delta\rho^2/2r_e - \Delta z_n \Delta\rho^2/2r_e^2 + \ldots, \qquad (2.27)$$

$$r_e\Delta\varphi = 2\Delta\rho - \Delta\rho(\Delta z_1 + \Delta z_2)/r_e$$
$$+ \Delta\rho(\Delta z_1^2 + \Delta z_2^2)/r_e^2 - 2\Delta\rho^3/3r_e^2 + \ldots, \qquad (2.28)$$

where r_e denotes the equilibrium internuclear distance $X-Y$. As shown in Fig. 2.2, a perpendicular (rectilinear) displacement ($\Delta z_n = 0, \Delta\rho \neq 0$) stretches the $X-Y$ bond, while a pure bending displacement ($\Delta r_n = 0$, $\Delta\varphi \neq 0$) makes the rectilinear stretching coordinate Δz_n non-zero (negative). Thus, the second- and higher-order terms in eqns (2.27) and (2.28) represent the non-linearity of Δr_n and $\Delta\varphi$. The $\mathbf{R}$ coordinates are isotope-dependent, whereas the $\mathbf{Я}$ coordinates are defined to be isotope-independent, because

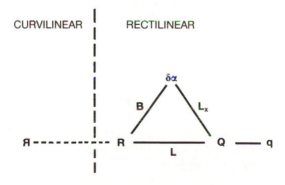

Fig. 2.3 Relationship among the coordinates discussed in the present chapter.

Table 2.1 Curvilinear and rectilinear internal symmetry coordinates derived from the $\mathfrak{R}$ and $\mathbf{R}$ coordinates, respectively, for XY_2 molecules[a]

Curvilinear coordinates	Rectilinear coordinates	Symmetry
Linear XY_2 molecules ($D_{\infty h}$ symmetry)		
$S_1^{\mathfrak{R}} = (\Delta r_1 + \Delta r_2)/\sqrt{2}$	$S_1 = (\Delta z_1 + \Delta z_2)/\sqrt{2}$	Σ_g^+
$S_2^{\mathfrak{R}} = r_e \Delta\varphi$	$S_2 = \Delta\rho$	Π_u
$S_3^{\mathfrak{R}} = (\Delta r_1 - \Delta r_2)/\sqrt{2}$	$S_3 = (\Delta z_1 - \Delta z_2)/\sqrt{2}$	Σ_u^+
Bent XY_2 molecules (C_{2v} symmetry)		
$S_1^{\mathfrak{R}} = (\Delta r_1 + \Delta r_2)/\sqrt{2}$	$S_1 = (\Delta z_1 + \Delta z_2)/\sqrt{2}$	A_1
$S_2^{\mathfrak{R}} = r_e \Delta\varphi$	$S_2 = \Delta x_1 + \Delta x_2$	A_1
$S_3^{\mathfrak{R}} = (\Delta r_1 - \Delta r_2)/\sqrt{2}$	$S_3 = (\Delta z_1 - \Delta z_2)/\sqrt{2}$	B_2

[a] See Figs. 2.2 and 2.5 for the internal coordinates for linear and bent XY_2 molecules, respectively.

the second- and higher-order terms in eqn (2.23) are in general isotope-dependent. A diagram showing the relationship among the coordinates and their transformation is given in Fig. 2.3. For a molecule with elements of symmetry, one can take the 'internal symmetry coordinates', either curvilinear or rectilinear, by a suitable linear combination of the $\mathfrak{R}$ or $\mathbf{R}$ coordinates (Wilson *et al.* 1955). The rectilinear internal symmetry coordinates derived from $\mathbf{R}$ are often denoted as $\mathbf{S}$. An example of the internal symmetry coordinates for linear and bent XY_2 molecules is shown in Table 2.1.

2.3.4 *Quadratic force constants: analysis of normal modes*

The quadratic part of the potential energy function, eqns (2.13) and (2.19), can be written in a matrix form by substitution of eqn (2.26) into eqn (2.13) (Wilson *et al.* 1955) (see also eqn (2.24)):

$$V_2 = \frac{1}{2} \mathbf{R}^T \mathbf{F} \mathbf{R} = \frac{1}{2} \mathbf{Q}^T \mathbf{\Lambda} \mathbf{Q}, \tag{2.29}$$

$$\mathbf{\Lambda} = \mathbf{L}^T \mathbf{F} \mathbf{L}, \tag{2.30}$$

where the elements of Wilson's $\mathbf{F}$ matrix are f_{ij} given in eqn (2.14), those of the $\mathbf{L}$ matrix are L_i^r defined in eqn (2.26), and the elements of a diagonal matrix $\mathbf{\Lambda}$ are, as shown in eqn (2.21),

$$\lambda_r = 4\pi^2 c^2 \omega_r^2. \tag{2.31}$$

On the other hand, the kinetic energy can be written in terms of Wilson's **G** matrix as

$$T = \frac{1}{2}(\delta\dot{\alpha})^T \mathbf{M}(\delta\dot{\alpha}) = \frac{1}{2}\dot{\mathbf{R}}^T \mathbf{G}^{-1}\dot{\mathbf{R}} = \frac{1}{2}\dot{\mathbf{Q}}^T\dot{\mathbf{Q}}. \qquad (2.32)$$

The elements of the **G** matrix can be derived from the geometry of the molecule under study by use of Wilson's **s** vectors (Wilson *et al.* 1955). The following useful equations exist among the coordinates and their transformation matrices:

$$\mathbf{G} = \mathbf{BM}^{-1}\mathbf{B}^T = \mathbf{LL}^T, \qquad (2.33)$$

$$\mathbf{R} = \mathbf{B}\delta\alpha = \mathbf{LQ} = \mathbf{BM}^{-\frac{1}{2}}\mathbf{IQ}, \qquad (2.34)$$

$$\mathbf{I} = \mathbf{M}^{-\frac{1}{2}}\mathbf{B}^T(\mathbf{L}^T)^{-1}, \qquad (2.35)$$

where **M** is a diagonal matrix composed of the nuclear masses. The **B** matrix is composed of the coefficients of expansion of the rectilinear internal

Table 2.2 Experimental sources of information on the harmonic and anharmonic force constants

Technique[a]	Parameters[b]
Harmonic force constants	
SP	ω_s, ω_s^i (normal wavenumbers)
	D_J, D_{JK}, D_K, etc. (centrifugal distortion constants)
	$\zeta_{ss'}$ (Coriolis coupling constants)
	Δ (inertial defect for a planar molecule)
	q (l-type doubling constant for a linear molecule)
ED	l (mean amplitudes)
	δr (shrinkage)
Anharmonic force constants	
SP	$X_{ss'}$, $g_{tt'}$ (vibrational anharmonicity)
	α_s, $\gamma_{ss'}$ (vibration–rotation constants)
	q_t, r_t (l-type doubling constants)
	H_{JJJ}, H_{JKK}, etc. (sextic centrifugal distortion constants)
ED	δr_z (isotope effect)
	κ (asymmetry parameter)

[a] SP: high-resolution spectroscopy; ED: gas-phase electron diffraction.
[b] s, s': numbering of a normal mode; i: isotopic species; J, K: rotational quantum numbers; t, t': numbering of a degenerate normal mode.

coordinates **R** in terms of the Cartesian displacement coordinates; the elements of this matrix are the components of the s vectors mentioned above. It is shown by use of eqns (2.33) and (2.35) that

$$\mathbf{l}^T\mathbf{l} = \mathbf{E}, \tag{2.36}$$

where **E** is the unit matrix.

Equations (2.32) and (2.33) lead to

$$\mathbf{GFL} = \mathbf{L\Lambda}, \text{ or} \tag{2.37}$$

$$|\mathbf{GF} - \mathbf{E\Lambda}| = 0. \tag{2.38}$$

If the quadratic force field **F** is given, the wavenumbers of the normal modes, ω_r, can be calculated by this secular equation. The wavenumbers derived from vibrational spectroscopy are the most important source of information on the quadratic force constants f_{ij}. As listed in Table 2.2, other spectroscopic constants are also related to the f_{ij} constants, and they are also used as valuable additional experimental sources for determination of the quadratic force constants (Duncan 1975).

2.3.5 Cubic constants

The number of independent anharmonic constants is determined by the number of atoms in the molecule in question and its symmetry (Henry and Amat 1960). Typical examples are given in Table 2.3. It is seen that the number of independent cubic constants increases rapidly with the number of atoms and with the loss of symmetry.

Important experimental sources of information on the cubic constants are

Table 2.3 Total numbers of independent force constants

Type of molecule	Number of force constants	
	Second-order	Third-order
Diatomic	1	1
Linear XY_2	3	3
Linear XYZ	4	6
Bent XY_2	4	6
Bent XYZ	6	10
Planar XY_3	5	9
Pyramidal XY_3	6	14
Planar X_2YZ	10	22

listed in Table 2.2. The α constants representing the dependence of the rotational constants on the vibrational quantum numbers (see Chapter 3, eqn (3.18)) are the most important source. The cubic constants for linear and bent XY_2 molecules can be determined uniquely from the α constants alone. Except for these types, there are very few polyatomic molecules for which the cubic constants have been determined (see Appendix).

The cubic potential constants, k_{rst} or ϕ_{rst}, in the normal-coordinate system [eqn (2.22)] can be converted to the third-order force constants, f_{ijk}, in the non-linear internal coordinates (eqn (2.15)) by non-linear transformation (eqn (2.25)) (Kuchitsu and Morino 1965a, b, 1966; Morino et al. 1968; Morino 1969; Plíva 1974; Hoy et al. 1972; Suzuki 1975). As shown in the relationship between the k_{rst} and f_{ijk} constants (Table 2.4), one can

Table 2.4 Relationships among the spectroscopic constants α, the cubic potential constants k, and the third-order force constants f and F (after Morino et al. 1968)

(a) Linear XY_2 molecules

α	k	F	f	Term	n
α_1	111	111	rrr	Δr_1^3	2
α_3	133	133	rrr'	$\Delta r_1^2 \Delta r_2$	2
α_2	122	122	$r\varphi\varphi$	$\Delta r_1 \Delta\varphi^2$	2

(b) Bent XY_2 molecules

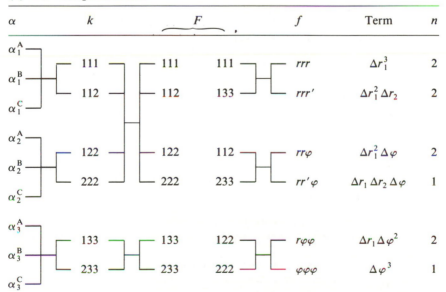

α	k	F	f	Term	n
α_1^A, α_1^B, α_1^C	111, 112	111, 112	111 / 133 → rrr / rrr'	Δr_1^3 / $\Delta r_1^2 \Delta r_2$	2 / 2
α_2^A, α_2^B, α_2^C	122, 222	122, 222	112 / 233 → $rr\varphi$ / $rr'\varphi$	$\Delta r_1^2 \Delta\varphi$ / $\Delta r_1 \Delta r_2 \Delta\varphi$	2 / 1
α_3^A, α_3^B, α_3^C	133, 233	133, 233	122 / 222 → $r\varphi\varphi$ / $\varphi\varphi\varphi$	$\Delta r_1 \Delta\varphi^2$ / $\Delta\varphi^3$	2 / 1

(c) Pyramidal XY_3 molecules

α	k	F		f	Term	n
α_1^B	111	111	111	(rrr)	Δr_1^3	3
α_1^C	112	112	(133)	(rrr')	$\Delta r_1^2 \Delta r_2$	6
α_2^B	122	122	(333)	$(rr'r'')$	$\Delta r_1 \Delta r_2 \Delta r_3$	1
α_2^C	222	222	112	$(rr\varphi)$	$\Delta r_1^2 \Delta \varphi_1$	3
α_3^B	133	(133)	(233)	$(rr\varphi')$	$\Delta r_1^2 \Delta \varphi_2$	6
α_3^C	233	(233)	(134)	$(rr'\varphi)$	$\Delta r_1 \Delta r_2 \Delta \varphi_1$	6
α_4^B	144	(144)	(334)	$(rr'\varphi'')$	$\Delta r_1 \Delta r_2 \Delta \varphi_3$	3
α_4^C	244	(244)	122	$(r\varphi\varphi)$	$\Delta r_1 \Delta \varphi_1^2$	3
	(134)	(134)	(144)	$(r'\varphi\varphi)$	$\Delta r_1 \Delta \varphi_2^2$	6
	(234)	(234)	(234)	$(r\varphi\varphi')$	$\Delta r_1 \Delta \varphi_1 \Delta \varphi_2$	6
q_3	(333)	(333)	(344)	$(r\varphi'\varphi'')$	$\Delta r_1 \Delta \varphi_2 \Delta \varphi_3$	3
	(334)	(334)	222	$(\varphi\varphi\varphi)$	$\Delta \varphi_1^3$	3
q_4	(344)	(344)	(244)	$(\varphi\varphi\varphi')$	$\Delta \varphi_1^2 \Delta \varphi_2$	6
	(444)	(444)	(444)	$(\varphi\varphi'\varphi'')$	$\Delta \varphi_1 \Delta \varphi_2 \Delta \varphi_3$	1

The functional (all linear) dependences between α and k, k and F, and F and f are indicated by horizontal lines; see eqns (2.13), (2.22), Tables 2.5 and 2.6 for the definitions of the k and f constants. The F constants are defined on the basis of the internal symmetry coordinates; see Table 2.1 for a typical example of the internal symmetry coordinates of a bent XY_2 molecule. The α constants represent the dependences of the rotational constants on the vibrational quantum numbers (Nielsen 1951). The entry 'Term' represents a typical term in the potential function, with the internal coordinates indicated, involving the f constant listed on the same row, and n represents the total number of such terms. For example, the following relationships exist for a bent XY_2 molecule (Kuchitsu and Morino 1966):

1. Each of the α_3^A, α_3^B, α_3^C constants is composed of a harmonic part, which is independent of the anharmonic constants, and a linear combination of k_{133} and k_{233}.
2. Each of the k_{133} and k_{233} constants is represented as a linear combination of F_{133} and F_{233}, with an additional contribution dependent on the second-order force constants.

3. Each of the F_{111} and F_{133} constants is represented as a linear combination of f_{rrr} and $f_{rrr'}$, by use of the relationship between Δr and S coordinates shown in Table 2.1.
4. There are two equivalent terms including $f_{rrr} : f_{rrr} \Delta r_1^3$ and $f_{rrr} \Delta r_2^3$, as shown in the explanatory legend of Table 2.5.

For linear and bent XY_2 molecules shown in (a) and (b), three and six independent α constants, respectively, are sufficient to determine the three and six independent f constants uniquely. For pyramidal XY_3 molecules shown in (c), however, none of the fourteen independent f constants can be determined by use of all the experimental α and q constants; the constants which cannot be determined uniquely from experiment are enclosed in parentheses.

Table 2.5 The f and k constants for H_2O and D_2O molecules[a]

Second-, third-, and fourth-order force constants				Cubic and quartic potential constants		
$ijkl$	f		Unit	$rstu$	k (cm^{-1})	
	H_2O	D_2O			H_2O	D_2O
rr	8.454	8.454	10^2 J m^{-2}	111	−319.4(17)	−193.3(11)
rr'	−0.101	−0.101	10^2 J m^{-2}	112	39.6(12)	7.6(8)
$\varphi\varphi$	0.697	0.697	10^{-18} J	122	255.4(157)	191.1(124)
$r\varphi$	0.219	0.219	10^{-8} J m^{-1}	133	−921.7(8)	−632.1(90)
rrr	−59.9(2)	−63.8(14)	10^{12} J m^{-3}	222	−61.9(13)	−33.4(11)
rrr'	−0.66(17)	1.0(5)	10^{12} J m^{-3}	233	147.3(2)	93.7(11)
$rr\varphi$	0.32(3)	0.44(6)	10^2 J m^{-2}			
$rr'\varphi$	−0.66(1)	−0.46(2)	10^2 J m^{-2}	1111	38.5(7)	18.8(4)
$r\varphi\varphi$	0.3(2)	0.6(3)	10^{-8} J m^{-1}	1122	−140.4(80)	−96.8(68)
$\varphi\varphi\varphi$	−0.78(3)	−0.72(5)	10^{-18} J	1133	209.2(13)	135.7(33)
$rrrr$	403(5)	485(20)	10^{22} J m^{-4}	2222	11.2(21)	9.9(18)
$rrrr'$	5(2)	−19(7)	10^{22} J m^{-4}	2233	−122.2(38)	−74.3(29)
$rrr'r'$	6(3)	0(9)	10^{22} J m^{-4}	3333	35.0(1)	26.0(12)
$rr\varphi\varphi$	−7(2)	−11(3)	10^2 J m^{-2}			
$rr'\varphi\varphi$	−1(2)	−3(3)	10^2 J m^{-2}			
$\varphi\varphi\varphi\varphi$	0(2)	0(2)	10^{-18} J			

For a bent XY_2 molecule, eqn (2.13) can be written as (see Fig. 2.5): $V = \frac{1}{2} f_{rr}$ $(\Delta r_1^2 + \Delta r_2^2) + f_{rr'} \Delta r_1 \Delta r_2 + \frac{1}{2} f_{\varphi\varphi} \Delta\varphi^2 + f_{r\varphi} (\Delta r_1 + \Delta r_2) \Delta\varphi + \frac{1}{6} f_{rrr} (\Delta r_1^3 + \Delta r_2^3)$ $+ \frac{1}{2} f_{rrr'} (\Delta r_1 + \Delta r_2) \Delta r_1 \Delta r_2 + \frac{1}{2} f_{rr\varphi} (\Delta r_1^2 + \Delta r_2^2) \Delta\varphi + f_{rr'\varphi} \Delta r_1 \Delta r_2 \Delta\varphi + \frac{1}{2} f_{r\varphi\varphi}$ $(\Delta r_1 + \Delta r_2) \Delta\varphi^2 + \frac{1}{6} f_{\varphi\varphi\varphi} \Delta\varphi^3 +$ fourth-order terms, and eqn (2.22) as: $V/hc =$ $\frac{1}{2}(\omega_1 q_1^2 + \omega_2 q_2^2 + \omega_3 q_3^2) + k_{111} q_1^3 + k_{112} q_1^2 q_2 + k_{122} q_1 q_2^2 + k_{133} q_1 q_3^2 + k_{222} q_2^3 +$ $k_{233} q_2 q_3^2 +$ quartic terms, and the corresponding equation using the ψ constants. The relationships between the k and ϕ constants are: $\phi_{111} = 6 k_{111}$; $\phi_{122} = 2 k_{122}$; $\phi_{1111} = 24 k_{1111}$; $\phi_{1122} = 4 k_{1122}$; etc.

[a] Based on data listed in Kuchitsu and Morino (1965b). See Hoy et al. (1972) for a refined set of the f and ϕ constants for H_2O.

Table 2.6 The f, k, and ϕ constants for the $^{12}C^{16}O_2$ molecule[a]

Second-, third-, and fourth-order force constants			Cubic and quartic potential constants		
$ijkl$	f	Unit	$rstu$	$k\,(cm^{-1})$	$\phi\,(cm^{-1})$
rr	16.030	10^2 $J\,m^{-2}$	111	−45.07	−270.4
rr'	1.268	10^2 $J\,m^{-2}$	122	73.81	147.6
$\varphi\varphi$	0.783	10^{-18} J	133	−252.91	−505.8
rrr	−114.7	10^{12} $J\,m^{-3}$			
rrr'	−3.02	10^{12} $J\,m^{-3}$	1111	1.25	30.0
$r\varphi\varphi$	−1.26	10^{-8} $J\,m^{-1}$	1122	−9.11	−36.4
$rrrr$	621.4	10^{22} $J\,m^{-4}$	1133	20.37	81.5
$rrrr'$	−14.6	10^{22} $J\,m^{-4}$	2222	2.16	51.8
$rrr'r'$	−23.9	10^{22} $J\,m^{-4}$	2233	−27.81	−111.2
$rr\varphi\varphi$	4.8	10^2 $J\,m^{-2}$	3333	6.63	159.2
$rr'\varphi\varphi$	6.4	10^2 $J\,m^{-2}$			
$\varphi\varphi\varphi\varphi$	0.6	10^{-18} J			

For a linear XY_2 molecule, eqn (2.13) can be written as (see Fig. 2.2): $V = \frac{1}{2} f_{rr} (\Delta r_1^2 + \Delta r_2^2) + f_{rr'}\Delta r_1 \Delta r_2 + \frac{1}{2} f_{\varphi\varphi}\Delta\varphi^2 + \frac{1}{6} f_{rrr}(\Delta r_1^3 + \Delta r_2^3) + \frac{1}{2} f_{rrr'}(\Delta r_1 + \Delta r_2)\Delta r_1\Delta r_2 + \frac{1}{2} f_{r\varphi\varphi}(\Delta r_1 + \Delta r_2)\Delta\varphi^2 +$ fourth-order terms, and eqn (2.22) as: $V/hc = \frac{1}{2}(\omega_1 q_1^2 + \omega_2 q_2^2 + \omega_3 q_3^2) + k_{111} q_1^3 + k_{122} q_1 q_2^2 + k_{133} q_1 q_3^2 +$ quartic terms, and the corresponding equation using the ϕ constants.

[a] Based on data listed in Kuchitsu and Morino (1965a).

account for the origin of the cubic constants by expansion of k_{rst} or ϕ_{rst} in terms of f_{ij} and f_{ijk}. Two examples, H_2O/D_2O and CO_2, are listed in Tables 2.5 and 2.6, respectively, where the quartic constants, k_{rstu} or ϕ_{rstu}, as well as the cubic constants are listed. For example, the negative sign of the $f_{r\varphi\varphi}$ constant for CO_2 means that the C=O bond is stretched slightly when it is bent, as shown in Fig. 2.4.

The origin of the k_{rst} or ϕ_{rst} constants derived from spectroscopic experiments has been studied systematically, and simple model functions for estimating these cubic constants have been proposed (Kuchitsu and Morino 1965a, b, 1966; Morino et al. 1968; Morino 1969). For example, a model which assumes that each bond-stretching displacement is an independent diatomic oscillator and each angle-bending displacement follows a quadratic potential seems to be a good starting point for this purpose. Further details have been discussed by Morino et al. (1968), Plíva (1974), and Suzuki (1975).

2.3.6 Average values of normal coordinates

The linear average values of the normal coordinates for a vibrational state v and for thermal equilibrium at temperature T are calculated up to the

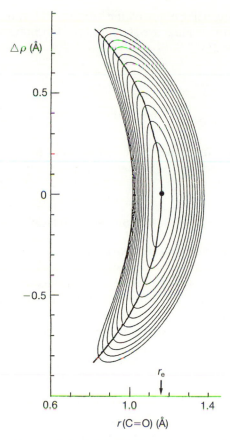

Fig. 2.4 Contour map of the potential energy surface for CO_2 (after Pariseau *et al.* 1965 and Suzuki 1975). The axes represent the Cartesian displacements of one O atom from its equilibrium position. The other O atom and the C atom are constrained at their equilibrium positions. Contours are drawn at intervals of 2×10^{-20} J, which correspond to 1.5 quanta of the bending mode. Note that the valley of the potential curve is slightly opened up with respect to the arc centered at the C atom (i.e. the locus of the O atom in a hypothetical bending vibration with the C=O bond length fixed to its equilibrium value).

first-order perturbation of the harmonic oscillator to be (Toyama *et al.* 1964):

$$\langle q_r \rangle_v = -(1/\omega_r) \left[3k_{rrr} \left(v_r + \frac{1}{2} \right) + \sum_{s \neq r} g_s k_{rss} \left(v_s + \frac{1}{2} \right) \right], \qquad (2.39)$$

$$\langle q_r \rangle_T = -(1/2\omega_r) [3k_{rrr} \coth(hc\omega_r/2kT) \\ + \sum_{s \neq r} g_s k_{rss} \coth(hc\omega_s/2kT)], \qquad (2.40)$$

where g_s denotes the degeneracy of the s-th normal mode. The corresponding quadratic average values are shown in the harmonic approximation to
be (Wilson *et al.* 1955)

$$\langle q_r^2 \rangle_v = v_r + \frac{1}{2}, \tag{2.41}$$

$$\langle q_r^2 \rangle_T = \frac{1}{2} \coth(hc\omega_r / 2kT). \tag{2.42}$$

These equations are fundamental for calculation of the average values of
the Cartesian displacement coordinates by use of eqn (2.26), as described
below.

2.4 Internuclear distance: definitions and analysis

The geometrical parameters of free polyatomic molecules can be determined
with high precision by rotational spectroscopy and by electron diffraction
(see Chapters 3–5). In order to take full advantage of such precision, one
should carefully consider the influence of molecular vibration on the distance parameters derived from different experimental methods, because a
difference in the definition is likely to be a significant source of systematic
error in the structure analysis (Kuchitsu and Cyvin 1972; Kuchitsu 1981;
Callomon *et al.* 1987). Important definitions of the internuclear distance are
listed below, and examples of experimental values are shown in Table 2.7.

2.4.1 Definitions of the internuclear distance

2.4.1.1 Equilibrium nuclear positions (r_e distance)   The internuclear distance corresponding to that between the nuclear positions at the potential
minimum (i.e. the hypothetical vibrationless structure) is denoted as r_e.
The optimized geometry derived from an *ab initio* MO calculation (Dykstra
1988) corresponds to this structure.

2.4.1.2 Distance average (r_g distance)   The internuclear distance averaged over thermal vibration is denoted as r_g (originally defined by Bartell
(1955) as $r_g(0)$). Let an instantaneous displacement of an internuclear distance r be denoted as Δr, and its projection onto the z axis be taken along
the equilibrium direction of the nuclei in question (see the local Cartesian
coordinates shown in Fig. 2.5); it can then be shown that

$$\begin{aligned}
r_g &= r_e + \langle \Delta r \rangle_T \\
&= \langle [(r_e + \Delta z)^2 + \Delta x^2 + \Delta y^2]^{\frac{1}{2}} \rangle_T \\
&= r_e + \langle \Delta z \rangle_T + (\langle \Delta x^2 \rangle_T + \langle \Delta y^2 \rangle_T)/2r_e + \dots
\end{aligned} \tag{2.43}$$

Table 2.7   Internuclear distances determined by experiment

Molecule	Distance type	Value (Å)	References
CH_4	r_g (C—H)	1.107(1)	Bartell *et al.* (1961)
	r_z (^{12}C—H)	1.09912	Nakata and Kuchitsu (1986)
	r_e (C—H)	1.0862(12)	Bartell and Kuchitsu (1978)
		1.0858 ± 0.001	Gray and Robiette (1979)
		1.0870 ± 0.0007	Hirota (1979)
CD_4	r_g (C—D)	1.103(1)	Bartell *et al.* (1961)
	r_z (^{12}C—D)	1.09552	Nakata and Kuchitsu (1986)
	r_e (C—D)	1.0875(13)	Bartell and Kuchitsu (1978)
BF_3	r_g (B—F)	1.3133(10)	Kuchitsu and Konaka (1966)
	r_α (B—F)	1.3109(10)	Kuchitsu and Konaka (1966)
	r_z (^{11}B—F)	1.31103(5)	Yamamoto *et al.* (1986)
	r_z (^{10}B—F)	1.31110(6)	Yamamoto *et al.* (1986)
	r_e (B—F)	1.3070(1)	Yamamoto *et al.* (1986)
Cl_2CO	r_g (C=O)	1.184 ± 0.003	Nakata *et al.* (1980*a*)
	r_z (C=O)	1.1789(12)	Yamamoto *et al.* (1985*a*)
	r_e (C=O)	1.1766(22)	Yamamoto *et al.* (1985*a*)
	r_s (C=O)	1.1841 ± 0.0004	Nakata *et al.* (1980*b*)
	r_g (C—Cl)	1.744 ± 0.001	Nakata *et al.* (1980*a*)
	r_z (C—Cl)	1.7423(6)	Yamamoto *et al.* (1985*a*)
	r_e (C—Cl)	1.7365(12)	Yamamoto *et al.* (1985*a*)
	r_s (C—Cl)	1.73656 ± 0.00005	Nakata *et al.* (1980*b*)
	$y_m{}^a$	2.1502 ± 0.0007	Nakata *et al.* (1980*b*)
	$y_e{}^a$	2.1502 ± 0.0033	Nakata *et al.* (1980*b*)
	Angles (degrees)		
	$\angle_z$ Cl—C—Cl	111.85(5)	Yamamoto *et al.* (1985*a*)
	$\angle_e$ Cl—C—Cl	111.91(12)	Yamamoto *et al.* (1985*a*)
	$\angle_s$ Cl—C—Cl	112.21 ± 0.01	Nakata *et al.* (1980*b*)

^{a} Defined as $y = r(C=O) + r(C—Cl) \cos(\frac{1}{2} \angle Cl—C—Cl)$.

The quadratic average values over thermal equilibrium, denoted by the subscript T, can be calculated using the quadratic force constants f_{ij} by eqns (2.17), (2.20), (2.35), and (2.42). The linear average value can also be calculated by use of eqns (2.26) and (2.40), but in this case the cubic constants must be known at least approximately (Kuchitsu 1967*b*). In order to estimate the r_g value for a bond distance, a simple calculation assuming that the bond is a diatomic oscillator (eqn (2.12)) is often used.

The r_a distance, which appears in the fundamental equation of gas-phase electron diffraction (see Chapter 5, eqn (5.21)) as an argument of the sine function, can be related to r_g by

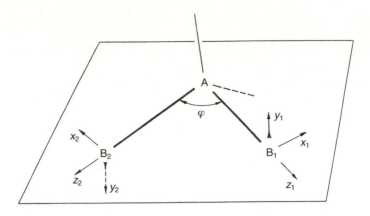

Fig. 2.5 An example of the local Cartesian coordinates.

$$r_g = r_a + \langle \Delta r^2 \rangle_T / r_e + \dots \tag{2.44}$$

Therefore, the r_g distance is often reported in papers on gas electron diffraction.

2.4.1.3 Position average (r_z and r_α distances) The distance between the nuclear positions averaged over the zero-point vibration (i.e. the average nuclear positions at the ground vibrational state) is called the r_z distance (Oka 1960; Morino *et al.* 1962). The local Cartesian coordinate axes being taken as above, it is shown that

$$r_z = \left[(r_e + \langle \Delta z \rangle_0)^2 + \langle \Delta x \rangle_0^2 + \langle \Delta y \rangle_0^2 \right]^{\frac{1}{2}}$$
$$= r_e + \langle \Delta z \rangle_0 + (\langle \Delta x \rangle_0^2 + \langle \Delta y \rangle_0^2)/2r_e + \dots, \tag{2.45}$$

where the subscript 0 denotes the vibrational average over the zero-point vibration. Since $\langle \Delta x \rangle_0$ and $\langle \Delta y \rangle_0$ are usually much smaller than 0.01 Å, eqn (2.45) can be approximated by

$$r_z = r_e + \langle \Delta z \rangle_0. \tag{2.46}$$

The rotational constants derived from spectroscopy can be converted, after small corrections for harmonic vibration (Oka 1960; Morino *et al.* 1962; Kuchitsu and Cyvin 1972; Kuchitsu 1981), to those corresponding to the inverse of the moments of inertia for the zero-point average nuclear positions, so that the r_z structure can be determined by a combination of these rotational constants (often by a further combination with those for isotopic species and vibrationally excited states, see Nemes (1984)).

The distance between the nuclear positions averaged over the thermal vibration at a given temperature T (i.e. at thermal equilibrium) is called the r_α distance. By a calculation similar to eqn (2.46), it is shown that

$$r_\alpha = \sum_v w_v(T) \langle r \rangle_v = r_e + \langle \Delta z \rangle_T, \tag{2.47}$$

where $w_v(T)$ is the Boltzmann factor.

The r_g distance derived from electron diffraction can be converted to r_α by using eqns (2.43) and (2.47) as

$$r_\alpha = r_g - (\langle \Delta x^2 \rangle_T + \langle \Delta y^2 \rangle_T)/2r_e. \tag{2.48}$$

If necessary, a small correction for centrifugal distortion may also be made (Iwasaki and Hedberg 1962; Toyama *et al.* 1964).

The r_α distance can further be converted to r_z by approximate extrapolation to zero kelvin,

$$r_\alpha^0 = r_z = r_g - (\langle \Delta x^2 \rangle_0 + \langle \Delta y^2 \rangle_0)/2r_e - (\langle \Delta r \rangle_T - \langle \Delta r \rangle_0). \tag{2.49}$$

The last term requires approximate knowledge of the anharmonic potential function. For a bond distance, a simple approximation that the bond in question is like a diatomic oscillator (eqn (2.12)) is often sufficient for this purpose (Kuchitsu and Cyvin 1972; Kuchitsu 1981). A more sophisticated treatment is required for a non-bonded distance (Nakata *et al.* 1983; Yamamoto *et al.* 1985*b*; Kuchitsu *et al.* 1988).

The r_g and r_z (or r_α) structures have merits and demerits. A bond length can be better represented by r_g, because it is a real vibrational average, as shown in eqn (2.43), but r_z and r_α are projected averages (eqns (2.46) and (2.47)). On the other hand, a bond angle or a non-bonded distance can be better represented by r_z or r_α, because they refer to the average nuclear positions. The non-bonded distance expressed in r_g is slightly inconsistent with the bond distances, because these distances do not satisfy the cosine rule; this is known as the Bastiansen–Morino shrinkage effect (Cyvin 1968) (see Table 2.8). For example, twice the $r_g(C{=}O)$ in CO_2 is 0.0054 Å longer than the $r_g(O \cdots O)$, although the molecule is linear. A small correction for this shrinkage effect is necessary for a consistent analysis of electron diffraction data; this correction is equivalent to the conversion from r_g to r_α (eqn (2.48)).

Table 2.8 Examples of shrinkage effects, calculated from harmonic force fields

---

Linear shrinkage effect:
CO_2: $2r_g(C{=}O) - r_g(O{\cdots}O) = 0.0054(1)$ Å (Yamamoto *et al.* 1985*b*)

Non-linear shrinkage effect:
BF_3: $\sqrt{3}r_g(B{-}F) - r_g(F{\cdots}F) = 0.0028 \pm 0.0011$ Å (Kuchitsu and Konaka 1966)

2.4.1.4 Other definitions   The r_0 structure, which appeared frequently in the past in papers on experimental spectroscopy, is based on a simple analysis disregarding the effect of zero-point vibration. Careful studies (e.g. Nemes 1984) have shown that it is hard to eliminate or estimate significant systematic error in the r_0 structure caused by the vibrational effect. Therefore, recent papers on microwave spectroscopy often report the substitution (r_s) structure derived from a combination of many isotopic rotational constants (see Chapter 3, Section 3.4).

If sufficient isotopic rotational constants are available, then the r_s structure can be refined further by determining the r_m structure formulated by Watson (1973) and by Smith and Watson (1978), and still further, the r_c structure, which is expected to be very close to the r_e structure (Nakata *et al.* 1980*b*, 1981, 1984; Nakata and Kuchitsu 1986).

The structural parameters for Cl_2CO are given in Table 2.7, those for OCl_2 and SO_2 in Table 2.9, as typical examples.

Table 2.9 Comparison of the modified-mass-dependence (r_c) and equilibrium (r_e) structures for OCl_2 and SO_2 (Nakata *et al.* 1981)[a]

Molecule	Parameter	r_c structure	r_e structure
OCl_2	$r(O{-}Cl)$	1.69587 ± 0.00007	1.69591 ± 0.00013
	$\angle Cl{-}O{-}Cl$	110.886 ± 0.006	110.876 ± 0.015
SO_2	$r(S{=}O)$	1.43071 ± 0.00004	1.43077 ± 0.00004
	$\angle O{=}S{=}O$	119.332 ± 0.006	119.329 ± 0.006

[a] Distances are given in Å, angles in degrees.

2.4.2 Spectroscopy-diffraction-theory joint analysis for determination of potential constants

As shown above, the primary sources of information on the r_e structure and the f_{ij} and f_{ijk} constants are data obtained by high-resolution spectroscopy. However, these data are seldom sufficient for unambiguous determination of the potential constants except for a very limited number of simple molecules (see Appendix). In all other cases, electron diffraction intensity provides valuable additional information, because it contains accurate experimental data related directly to the probability distribution functions for all the internuclear distances (bond and non-bonded) in the molecule (Kuchitsu 1972, 1984; see also Chapter 5).

One of the great advantages of the use of electron diffraction data is that the systematic error in the r_z structure caused by its isotopic dependence is diminished. The r_z structure depends on isotopic substitution only very slightly but critically; the differences in the r_z distances are of the order of 10^{-3} Å when H/D substitution is made, and 10^{-4} Å or less in other cases. However, a small uncertainty in this isotopic difference is amplified because of strong parameter correlation when the r_z structure is derived from an analysis of isotopic rotational constants; this causes a systematic error of the order of 10^{-2} Å. In most cases, the isotopic dependence of r_z cannot be estimated with sufficient accuracy by spectroscopy or by theory (Kuchitsu and Oyanagi 1977). This difficulty can be remedied by introduction of electron diffraction data into the analysis, because much of the correlation can then be removed by a joint analysis of all the available rotational constants, electron diffraction intensities, and sometimes accurate theoretical potential constants. Then the r_z structure and its isotopic dependence can be determined accurately. The latter information can be used for estimation of the anharmonic potential constants and, furthermore, the r_e distances can be estimated by use of the r_z distances and the anharmonic constants.

Such a joint analysis has been made for many molecules, such as Cl_2CS (Nakata *et al.* 1982), Cl_2CO (Nakata *et al.* 1980*a*, *b*; Yamamoto *et al.* 1985*a*), $CCl_2{=}CH_2$ (Nakata and Kuchitsu 1982), and cyclopropane (Yamamoto *et al.* 1985*b*), where the r_e structure and the important cubic constants have been determined by use of the rotational constants for the ground vibrational states of several isotopic species and those for several excited vibrational states, plus electron diffraction intensities. It is often desirable to reinforce this analysis by the potential constants derived from an *ab initio* MO calculation (see, for example, Schäfer 1983; van Nuffel *et al.* 1984; Schäfer *et al.* 1988), because the above-mentioned experimental data may still be insufficient for complete determination of the structure; the advancement of theoretical techniques has now enabled reliable prediction of the force constants, cubic as well as quadratic, for simple polyatomic molecules (relevant references are given in Section 2.3.3).

2.5 Large-amplitude cases

The foregoing discussion is mainly concerned with semirigid molecules. However, many molecules of great chemical importance have at least one large-amplitude intramolecular motion, such as internal rotation, inversion, ring puckering, and pseudorotation, while other molecules have weak chemical bonds, which result in large-amplitude vibrations. For example, carbon suboxide, $O=C=C=C=O$, is a typical quasilinear molecule, which has a very anharmonic, low-frequency vibrational mode, ν_7, related to the $C=C=C$ bending; an average amplitude of this bending displacement is estimated to be more than $20°$ (Ohshima *et al.* 1988). Complexes bound by hydrogen bonds or van der Waals' forces are also very non-rigid.

An analysis of such a large-amplitude motion needs caution, because (*i*) the expansion of the potential function in terms of the rectilinear coordinates (eqn (2.19)) converges only very slowly, (*ii*) the effective mass which appears in the kinetic energy term of the equation of motion depends significantly on the amplitude of the displacement so that it can no longer be regarded as constant, and (*iii*) complicated couplings among the large-amplitude motion in question with other large-amplitude or small-amplitude motions need to be taken into account; therefore, a multi-dimensional large-amplitude analysis is often necessary for a complete analysis of the potential function of such a 'floppy' molecule. There remains much to be done in the future, in spite of a number of excellent studies in this field.

2.6   Concluding remarks

The molecular force field (mainly quadratic) has long been one of the most important and well-established areas of research in structural chemistry, particularly in vibrational spectroscopy (see, for example, Shimanouchi 1970; Duncan 1975; Matsuura and Tasumi 1983). This area is now being extended to studies of the potential energy surfaces including anharmonicity, keeping pace with the recent advances in laser spectroscopy and quantum chemistry. The problems discussed in the present chapter, i.e. the structure of a bound electronic state of a molecule in the gas phase, are related closely to studies of the potential energy surfaces encountered in studies of reaction dynamics (e.g. Bruna and Peyerimhoff 1987), where remarkable development is being made in recent years through the collaboration of experimentalists and theoreticians (e.g. Hirst 1985; Levine and Bernstein 1987).

2.7   Acknowledgements

The author is grateful to Drs P. R. Bunker and G. Graner for their valuable comments and to Drs S. Yamamoto and H. Ito for their assistance in the preparation of the manuscript.

References

Ballhausen, C. J. and Hansen, A. E. (1972). Electronic spectra. In *Annual review of physical chemistry*, Vol. 23 (ed. H. Eyring), pp. 15–38. Annual Reviews Inc., Palo Alto.

Barrow, R. F., Clark, T. C., Coxon, J. A., and Yee, K. K. (1974). *J. Mol. Spectrosc.*, **51**, 428–49.

Bartell, L. S. (1955). *J. Chem. Phys.*, **23**, 1219–22.

Bartell, L. S. and Kuchitsu, K. (1978). *J. Chem. Phys.*, **68**, 1213–5.

Bartell, L. S., Kuchitsu, K., and deNeui, R. J. (1961). *J. Chem. Phys.*, **35**, 1211–8.

Born, M. and Oppenheimer, R. (1927). *Ann. Physik*, **84**, 457–84.

Botschwina, P. (1988). *J. Chem. Soc. Faraday Trans. II*, **84**, 1263–76.

Bruna, P. J. and Peyerimhoff, S. D. (1987). Excited state potentials. In *Ab initio methods in quantum chemistry*, Part I, Advances in Chemical Physics, Vol. 67 (ed. K. P. Lawley), pp. 1–97. Wiley-Interscience, Chichester.

Bunker, P. R. (1968). *J. Mol. Spectrosc.*, **28**, 422–43.

Bunker, P. R. (1970). *J. Mol. Spectrosc.*, **35**, 306–13.

Bunker, P. R. (1972). *J. Mol. Spectrosc.*, **42**, 478–94.

Callomon, J. H., Hirota, E., Iijima, T., Kuchitsu, K., and Lafferty, W. J. (1987). *Structure data of free polyatomic molecules*, Landolt-Börnstein, Numerical data and functional relationships in science and technology (New series), Group II, Vol. 15. Springer, Berlin.

Clabo Jr., D. A., Allen, W. D., Remington, R. B., Yamaguchi, Y., and Schaefer III, H. F. (1988). *Chem. Phys.*, **123**, 187–239.

Coxon, J. A. (1986). *J. Mol. Spectrosc.*, **117**, 361–87.

Cyvin, S. J. (1968). *Molecular vibrations and mean square amplitudes*, Chapters 14 and 15. Universitetsforlaget, Oslo, and Elsevier, Amsterdam.

Duncan, J. L. (1975). *Mol. Spectrosc.*, **3**, 104–62.

Dykstra, C. E. (1988). *Ab initio calculation of the structures and properties of molecules*. Elsevier, Amsterdam.

Ermler, W. C., Rosenberg, B. J., and Shavitt, I. (1985). Ab initio SCF and CI studies on the ground state of the water molecule. III. Vibrational analysis of potential energy and property surfaces. In *Comparison of ab initio quantum chemistry with experiment for small molecules* (ed. R. J. Bartlett), pp. 171–215. Reidel, Dordrecht.

Fleming, H. E. and Rao, K. N. (1972). *J. Mol. Spectrosc.*, **44**, 189–93.

Fogarasi, G. and Pulay, P. (1985). *Ab initio* calculation of force fields and vibrational spectra. In *Vibrational spectra and structure*, Vol. 14 (ed. J. R. Durig), pp. 125–219. Elsevier, Amsterdam.

Fogarasi, G. and Pulay, P. (1986). *J. Mol. Struct.*, **141**, 145–52.

Gray, D. L. and Robiette, A. G. (1979). *Mol. Phys.*, **37**, 1901–20.

Handy, N. C., Gaw, J. F., and Simandiras, E. D. (1987). *J. Chem. Soc. Faraday Trans. II*, **83**, 1577–93.

Hargiss, L. O. and Ermler, W. C. (1988). *J. Phys. Chem.*, **92**, 300–6.

Henry, L. and Amat, G. (1960). *J. Mol. Spectrosc.*, **5**, 319–25.

Hirota, E. (1979). *J. Mol. Spectrosc.*, **77**, 213–21.

Hirota, E. (1985). *High-resolution spectroscopy of transient molecules*. Springer, Berlin.

Hirst, D. M. (1985). *Potential energy surfaces: Molecular structure and reaction dynamics*. Taylor and Francis, London.

Hoy, A. R., Mills, I. M., and Strey, G. (1972). *Mol. Phys.*, **24**, 1265–90.

Huber, K. P. and Herzberg, G. (1979). *Molecular spectra and molecular structure*, Vol. 4, *Constants of diatomic molecules*. Van Nostrand-Reinhold, New York.

Iwasaki, M. and Hedberg, K. (1962). *J. Chem. Phys.*, **36**, 2961–3.

Kuchitsu, K. (1967*a*). *Bull. Chem. Soc. Jpn.*, **40**, 498–504.

Kuchitsu, K. (1967*b*). *Bull. Chem. Soc. Jpn.*, **40**, 505–10.

Kuchitsu, K. (1972). Gas electron diffraction. In *Molecular structure and properties*, MTP International Review of Science, Physical Chemistry Ser. 1, Vol. 2 (ed. G. Allen), pp. 203–40. Butterworths, London, and University Park Press, Baltimore.

Kuchitsu, K. (1981). Geometrical parameters of free molecules: Their definitions and determination by gas electron diffraction. In *Diffraction studies on noncrystalline substances* (ed. I. Hargittai and W. J. Orville-Thomas), pp. 63–116. Elsevier, Amsterdam, and Akadémiai Kiadó, Budapest.

Kuchitsu, K. (1984). Gas electron diffraction: Theory and application to structure determination. In *Methods and applications in crystallographic computing* (ed. S. R. Hall and T. Ashida), pp. 441–9. Clarendon Press, Oxford.

Kuchitsu, K. and Cyvin, S. J. (1972). Representation and experimental determination of the geometry of free molecules. In *Molecular structures and vibrations* (ed. S. J. Cyvin), pp. 183–211. Elsevier, Amsterdam.

Kuchitsu, K. and Konaka, S. (1966). *J. Chem. Phys.*, **45**, 4342–7.

Kuchitsu, K. and Morino, Y. (1965*a*). *Bull. Chem. Soc. Jpn.*, **38**, 805–13.

Kuchitsu, K. and Morino, Y. (1965*b*). *Bull. Chem. Soc. Jpn.*, **38**, 814–24.

Kuchitsu, K. and Morino, Y. (1966). *Spectrochim. Acta*, **22**, 33–46.

Kuchitsu, K. and Oyanagi, K. (1977). *Faraday Discuss. Chem. Soc.*, **62**, 20–8.

Kuchitsu, K., Nakata, M., and Yamamoto, S. (1988). Joint use of electron diffraction and high-resolution spectroscopic data for accurate determination of molecular structure. In *Stereochemical applications of gas-phase electron diffraction* (ed. I. Hargittai and M. Hargittai), Part A, pp. 227–63. VCH, New York.

LeRoy, R. J. (1970). *J. Chem. Phys.*, **52**, 2683–9.

Levine, R. D. and Bernstein, R. B. (1987). *Molecular reaction dynamics and chemical reactivity*. Oxford University Press.

Matsuura, H. and Tasumi, M. (1983). Force fields for large molecules. In *Vibrational spectra and structure*, Vol. 12 (ed. J. R. Durig), pp. 69–143. Elsevier, Amsterdam.

Meal, J. H. and Polo, S. R. (1956). *J. Chem. Phys.*, **24**, 1119–25.

Mills, I. M. (1972). Infrared spectra. Vibration–rotation structure in asymmetric- and symmetric-top molecules. In *Molecular spectroscopy: Modern research* (ed. K. N. Rao and C. W. Mathews), pp. 115–40. Academic Press, New York.

Morino, Y. (1969). *Pure Appl. Chem.*, **18**, 323–38.

Morino, Y., Kuchitsu, K., and Oka, T. (1962). *J. Chem. Phys.*, **36**, 1108–9.

Morino, Y., Kuchitsu, K., and Yamamoto, S. (1968). *Spectrochim. Acta*, **24A**, 335–52.

Morse, P. M. (1929). *Phys. Rev.*, **34**, 57–64.

Murrell, J. N., Carter, S., Farantos, S. C., Huxley, P., and Varandas, A. J. C. (1984). *Molecular potential energy functions*. Wiley, Chichester.

Nakata, M. and Kuchitsu, K. (1982). *J. Mol. Struct.*, **95**, 205–14.

Nakata, M. and Kuchitsu, K. (1986). *J. Chem. Soc. Jpn.*, 1446–50.

Nakata, M., Kohata, K., Fukuyama,T., and Kuchitsu, K. (1980*a*). *J. Mol. Spectrosc.*, **83**, 105–17.

Nakata, M., Fukuyama, T., Kuchitsu, K., Takeo, H., and Matsumura, C. (1980*b*). *J. Mol. Spectrosc.*, **83**, 118–29.

Nakata, M., Sugie, M., Takeo, H., Matsumura, C., Fukuyama, T., and Kuchitsu, K. (1981). *J. Mol. Spectrosc.*, **86**, 241–9.

Nakata, M., Fukuyama, T., and Kuchitsu, K. (1982). *J. Mol. Struct.*, **81**, 121–9.

Nakata, M., Yamamoto, S., Fukuyama, T., and Kuchitsu, K. (1983). *J. Mol. Struct.*, **100**, 143–59.

Nakata, M., Kuchitsu, K., and Mills, I. M. (1984). *J. Phys. Chem.*, **88**, 344–8.

Nemes, L. (1984). Vibrational effects in spectroscopic geometries. In *Vibrational spectra and structure*, Vol. 13 (ed. J. R. Durig), pp. 161–221. Elsevier, Amsterdam.

Nielsen, H. H. (1951). *Rev. Modern Phys.*, **23**, 90–136.

Ohshima, Y., Yamamoto, S., and Kuchitsu, K. (1988). *Acta Chem. Scand.*, **A42**, 307–17.

Oka, T. (1960). *J. Phys. Soc. Jpn.*, **15**, 2274–9.

Pariseau, M. A., Suzuki, I., and Overend, J. (1965). *J. Chem. Phys.*, **42**, 2335–44.

Plíva, J. (1974). Anharmonic force fields. In *Critical evaluation of chemical and physical structural information* (ed. D. R. Lide Jr. and M. A. Paul), pp. 289–311. National Academy of Sciences, Washington.

Schaefer III, H. F. and Yamaguchi, Y. (1986). *J. Mol. Struct. (Theochem)*, **135**, 369–90.

Schäfer, L. (1983). *J. Mol. Struct.*, **100**, 51–73.

Schäfer, L., Ewbank, J. D., Siam, K., Chiu, N.-S., and Sellers, H. L. (1988). Molecular orbital constrained electron diffraction (MOCED) studies: The concerted use of electron diffraction and quantum chemical calculations. In *Stereochemical application of gas-phase electron diffraction* (ed. I. Hargittai and M. Hargittai), Part A, pp. 301–19. VCH, New York.

Shimanouchi, T. (1970). Molecular force field. In *Physical chemistry, an advanced treatise*, Vol. 4 (ed. H. Eyring, D. Henderson, and W. Jost), pp. 233–306. Academic Press, New York.

Smith, J. G. and Watson, J. K. G. (1978). *J. Mol. Spectrosc.*, **69**, 47–52.

Steele, D., Lippincott, E. R., and Vanderslice, J. T. (1962). *Rev. Modern Phys.*, **34**, 239–51.

Suzuki, I. (1975). Anharmonic potential functions in polyatomic molecules as derived from their vibrational and rotational spectra. In *Applied spectroscopy reviews*, Vol. 9 (ed. E. G. Brame Jr.), pp. 249–301. Dekker, New York.

Toyama, M., Oka, T., and Morino, Y. (1964). *J. Mol. Spectrosc.*, **13**, 193–213.

Van Nuffel, P., van den Enden, L., van Alsenoy, C., and Geise, H. J. (1984). *J. Mol. Struct.*, **116**, 99–118.

Verma, R. D. (1960). *J. Chem. Phys.*, **32**, 738–49.

Watson, J. K. G. (1973). *J. Mol. Spectrosc.*, **48**, 479–502.

Wilson, E. B., Decius, J. C., and Cross, P. C. (1955). *Molecular vibrations*. McGraw-Hill, New York.

Wright, J. S. (1988). *J. Chem. Soc. Faraday Trans. II*, **84**, 219–26.

Yamamoto, S., Nakata, M., and Kuchitsu, K. (1985*a*). *J. Mol. Spectrosc.*, **112**, 173–82.

Yamamoto, S., Nakata, M., Fukuyama, T., and Kuchitsu, K. (1985*b*). *J. Phys. Chem.*, **89**, 3298–302.

Yamamoto, S., Kuwabara, R., Takami, M., and Kuchitsu, K. (1986). *J. Mol. Spectrosc.*, **115**, 333–52.

Zare, R. N. (1964). *J. Chem. Phys.*, **40**, 1934–44.

Appendix

A list of polyatomic molecules for which equilibrium structures and anharmonic potential constants have been determined experimentally up to 1988

Compiled by Satoshi Yamamoto, Department of Astrophysics, Faculty of Science, Nagoya University, Chikusa-ku, Nagoya 464–01, Japan.

Molecule	Structure, potential constants	References
Bent XY$_2$ *molecules*		
H$_2$O	r_e, cubic, quartic	Hoy, A. R., Mills, I. M., and Strey, G. (1972). *Mol. Phys.*, **24**, 1265–90.
		Hoy, A. R. and Bunker, P. R. (1979). *J. Mol. Spectrosc.*, **74**, 1–8.
H$_2$S	r_e, cubic, quartic*	Kuchitsu, K. and Morino, Y. (1965). *Bull. Chem. Soc. Jpn.*, **38**, 814–24.
		Cook, R. L., De Lucia, F. C., and Helminger, P. (1975). *J. Mol. Struct.*, **28**, 237–46.
H$_2$Se	r_e, cubic	Kuchitsu, K. and Morino, Y. (1965). *Bull. Chem. Soc. Jpn.*, **38**, 814–24.
O$_3$	r_e, cubic, quartic	Tanaka, T. and Morino, Y. (1970). *J. Mol. Spectrosc.*, **33**, 538–51.
		Carney, G. D., Curtiss, L. A., and Langhoff, S. R. (1976). *J. Mol. Spectrosc.*, **61**, 371–81.
SO$_2$	r_e, cubic, quartic	Saito, S. (1969). *J. Mol. Spectrosc.*, **30**, 1–16.
SeO$_2$	r_e, cubic	Takeo, H., Hirota, E., and Morino, Y. (1970). *J. Mol. Spectrosc.*, **34**, 370–82.
		Takeo, H., Hirota, E., and Morino, Y. (1972). *J. Mol. Spectrosc.*, **41**, 420–2.
OF$_2$	r_e, cubic	Morino, Y. and Saito, S. (1966). *J. Mol. Spectrosc.*, **19**, 435–53.

*Partly determined.

SF_2	r_e, cubic	Endo, Y., Saito, S., Hirota, E., and Chikaraishi, T. (1979). *J. Mol. Spectrosc.*, **77**, 222–34.
SiF_2	r_e, cubic	Shoji, H., Tanaka, T., and Hirota, E. (1973). *J. Mol. Spectrosc.*, **47**, 268–74.
GeF_2	r_e, cubic	Takeo, H. and Curl Jr., R. F. (1972). *J. Mol. Spectrosc.*, **43**, 21–30.
OCl_2	r_e, cubic	Sugie, M., Takeo, H., and Matsumura, C. (1980). Presentation at the *Symposium on Structural Chemistry*, Tokyo, Japan. Abstracts, Part I, p. 117 (cited in Nakata, M., Yamamoto, S., Fukuyama, T., and Kuchitsu, K. (1983). *J. Mol. Struct.*, **100**, 143–59).
NO_2	r_e, cubic	Hardwick, J. L. and Brand, J. C. D. (1976). *Can. J. Phys.*, **54**, 80–91.
		Morino, Y., Tanimoto, M., Saito, S., Hirota, E., Awata, R., and Tanaka, T. (1983). *J. Mol. Spectrosc.*, **98**, 331–48.
		Morino, Y. and Tanimoto, M. (1984). *Can. J. Phys.*, **62**, 1315–22.
		Morino, Y. and Tanimoto, M. (1986). *J. Mol. Spectrosc.*, **115**, 442–5.
ClO_2	r_e, cubic	Jones, H. and Brown, J. M. (1981). *J. Mol. Spectrosc.*, **90**, 222–48.
		Miyazaki, K., Tanoura, M., Tanaka, K., and Tanaka, T. (1986). *J. Mol. Spectrosc.*, **116**, 435–49.
CH_2	r_e, cubic*	Bunker, P. R. and Jensen, P. (1983). *J. Chem. Phys.*, **79**, 1224–8.

Linear XY_2 *molecules*

CO_2	r_e, cubic–sextic	Chedin, A. (1979). *J. Mol. Spectrosc.*, **76**, 430–91.
		Chedin, A. and Teffo, J.-L. (1984). *J. Mol. Spectrosc.*, **107**, 333–42.
		Yamamoto, S., Nakata, M., Fukuyama, T., and Kuchitsu, K. (1985). *J. Phys. Chem.*, **89**, 3298–302.
		Graner, G., Rossetti, C., and Bailly, D. (1986). *Mol. Phys.*, **58**, 627–36.
CS_2	r_e, cubic, quartic	Giguere, J., Wang, V. K., Overend, J., and Cabana, A. (1973). *Spectrochim. Acta*, **29A**, 1197–206.
		Lacy, M. and Whiffen, D. H. (1981). *Mol. Phys.*, **43**, 1205–17.

Bent XYZ *molecules*

FNO	r_e, cubic	Degli Esposti, C., Cazzoli, G., and Favero, P. G. (1985). *J. Mol. Spectrosc.*, **109**, 229–38.
ClNO	r_e, cubic, quartic	Mirri, A. M., Cervellati, R., and Cazzoli, G. (1978). *J. Mol. Spectrosc.*, **71**, 386–98.

Molecule	Structure, potential constants	References
		Cazzoli, G., Degli Esposti, C., Palmieri, P., and Simeone, S. (1983). *J. Mol. Spectrosc.*, **97**, 165–85.
BrNO	cubic, quartic	Mirri, A. M., Cervellati, R., and Cazzoli, G. (1978). *J. Mol. Spectrosc.*, **71**, 386–98.
HO$_2$	r_e, cubic	Yamada, C., Endo, Y., and Hirota, E. (1983). *J. Chem. Phys.*, **78**, 4379–84.
		Hirota, E. (1986). *J. Mol. Struct.*, **146**, 237–52.
HNO	r_e, cubic	Hirota, E. (1986). *J. Chem. Soc. Jpn.*, 1438–45.
HCO	r_e, cubic*	Hirota, E. (1986). *J. Mol. Struct.*, **146**, 237–52.
HCF	r_e, cubic*	Suzuki, T. and Hirota, E. (1988). *J. Chem. Phys.*, **88**, 6778–84.
HOF	r_e, cubic*	Halonen, L. and Ha, T.-K. (1988). *J. Chem. Phys.*, **89**, 4885–8.
		Thiel, W., Scuseria, G., Schaefer III, H. F., and Allen, W. D. (1988). *J. Chem. Phys.*, **89**, 4965–75.
HOCl	r_e, cubic*	Deeley, C. M. (1987). *J. Mol. Spectrosc.*, **122**, 481–9.
		Halonen, L. and Ha, T.-K. (1988). *J. Chem. Phys.*, **88**, 3775–9.
FSN	cubic	Degli Esposti, C., Cazzoli, G., and Favero, P. G. (1988). *J. Mol. Struct.*, **190**, 327–42.

Linear X Y Z *molecules*

Molecule	Structure, potential constants	References
HCN	r_e, cubic, quartic	Strey, G. and Mills, I. M. (1973). *Mol. Phys.*, **26**, 129–38.
		Murrell, J. N., Carter, S., and Halonen, L. O. (1982). *J. Mol. Spectrosc.*, **93**, 307–16.
		Quapp, W. (1987). *J. Mol. Spectrosc.*, **125**, 122–7.
FCN	r_e, cubic	Whiffen, D. H. (1978). *Spectrochim. Acta*, **34A**, 1165–71.
HCP	r_e, cubic, quartic	Murrell, J. N., Carter, S., and Halonen, L. O. (1982). *J. Mol. Spectrosc.*, **93**, 307–16.
FCP	r_e, cubic	Whiffen, D. H. (1985). *J. Mol. Spectrosc.*, **111**, 62–5.
NNO	r_e, cubic–sextic	Griggs Jr., J. L., Rao, K. N., Jones, L. H., and Potter, R. M. (1968). *J. Mol. Spectrosc.*, **25**, 34–61.
		Lacy, M. and Whiffen, D. H. (1982). *Mol. Phys.*, **45**, 241–52.
		Kobayashi, M. and Suzuki, I. (1987). *J. Mol. Spectrosc.*, **125**, 24–42.

OCS	r_e, cubic, quartic*– sextic*	Morino, Y. and Nakagawa, T. (1968). *J. Mol. Spectrosc.*, **26**, 496–523. Whiffen, D. H. (1980). *Mol. Phys.*, **39**, 391–405.
HBS	r_e, cubic, quartic	Turner, P. and Mills, I. M. (1982). *Mol. Phys.*, **46**, 161–70.

Linear XYYX molecules

C_2H_2	r_e, cubic, quartic	Baldacci, A., Ghersetti, S., Hurlock, S. C., and Rao, K. N. (1976). *J. Mol. Spectrosc.*, **59**, 116–25. Strey, G. and Mills, I. M. (1976). *J. Mol. Spectrosc.*, **59**, 103–15. Kostyk, E. and Welsh, H. L. (1980). *Can. J. Phys.*, **58**, 912–20.

Pyramidal XY₃ molecules

NH_3	r_e, cubic, quartic*	Morino, Y., Kuchitsu, K., and Yamamoto, S. (1968). *Spectrochim. Acta*, **24A**, 335–52.
NF_3	r_e, cubic*	Otake, M., Matsumura, C., and Morino, Y. (1968). *J. Mol. Spectrosc.*, **28**, 316–24.
PF_3	r_e, cubic*	Hirota, E. and Morino, Y. (1970). *J. Mol. Spectrosc.*, **33**, 460–73. Hirota, E. (1971). *J. Mol. Spectrosc.*, **37**, 20–32.

Planar XY₃ molecules

CH_3	r_e, cubic*	Hirota, E. and Yamada, C. (1982). *J. Mol. Spectrosc.*, **96**, 175–82.
BF_3	r_e, cubic	Yamamoto, S., Kuwabara, R., Takami, M., and Kuchitsu, K. (1986). *J. Mol. Spectrosc.*, **115**, 333–52.

Tetrahedral XY₄ molecules

CH_4	r_e, cubic	Gray, D. L. and Robiette, A. G. (1979). *Mol. Phys.*, **37**, 1901–20. Hirota, E. (1979). *J. Mol. Spectrosc.*, **77**, 213–21.

Planar X₂YZ molecules

H_2CO	r_e, cubic, quartic*	Tanaka, Y. and Machida, K. (1977). *J. Mol. Spectrosc.*, **64**, 429–37.
Cl_2CO	r_e, cubic	Yamamoto, S., Nakata, M., and Kuchitsu, K. (1985). *J. Mol. Spectrosc.*, **112**, 173–82.

XY₃Z molecules

CH_3F	r_e, cubic	Kondo, S. (1984). *J. Chem. Phys.*, **81**, 5945–51. Egawa, T., Yamamoto, S., Nakata, M., and Kuchitsu, K. (1987). *J. Mol. Struct.*, **156**, 213–28.
CH_3Cl	cubic	Kondo, S., Koga, Y., Nakanaga, T., and Saëki, S. (1983). *J. Mol. Spectrosc.*, **100**, 332–42.

Molecule	Structure, potential constants	References
CH_3Br	r_e, cubic	Graner, G. (1981). *J. Mol. Spectrosc.*, **90**, 394–438. Kondo, S., Koga, Y., Nakanaga, T., and Saëki, S. (1983). *J. Mol. Spectrosc.*, **100**, 332–42.
$SiHF_3$	r_e, cubic*	Hoy, A. R., Bertram, M., and Mills, I. M. (1973). *J. Mol. Spectrosc.*, **46**, 429–47.
XY_2Z_2 *molecules*		
CH_2F_2	r_e, cubic*	Hirota, E. (1978). *J. Mol. Spectrosc.*, **71**, 145–59.
XY_2ZW *molecules*		
CHF_2Cl	cubic, quartic*	Brown, A., McKean, D. C., and Duncan, J. L. (1988). *Spectrochim. Acta*, **44A**, 553–65.

3

Reliability of structure determinations by microwave spectroscopy

Bouke P. van Eijck

3.1   Introduction

The microwave region was the first part of the electromagnetic spectrum where essentially monochromatic radiation sources were used for molecular spectroscopy. This field was opened in the years around 1950 after the development of radar techniques in the second world war. Thus it became possible to measure the frequencies of rotational transitions in gaseous substances with great precision. A typical wavelength would be 1 cm, corresponding to a frequency of 30 000 MHz. At a typical gas pressure of a few Pa the lines are quite narrow and a transition frequency can easily be measured within 0.1 MHz (Fig. 3.1). Nowadays the frequency range which is available has been considerably extended. Millimeter wave spectra can be obtained relatively easily, and laser techniques have made many infrared measurements nearly as accurate as microwave ones. Still, as far as our interest in structure determinations is concerned, the interpretation of the results uses the same methods. We shall see that, unfortunately, much of the accuracy is lost when one tries to extract structural information from the observed transition frequencies.

In principle, microwave spectroscopy could be a wonderful analytical technique, since every polar substance has a quite unique spectrum which becomes appreciably different even upon a structural change which is as slight as an isotopic substitution. This potential analytical power has not been realized due to the lack of direct and obvious correlation between spectrum and structure. Extensive frequency tabulations would be necessary to allow identification of an unknown gaseous substance from its microwave 'fingerprint'.

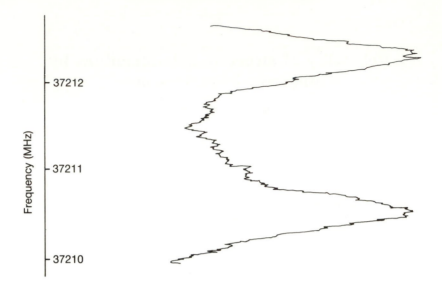

Fig. 3.1 A rotational transition of $^{35}Cl-CH_2-CO-^{35}Cl$ at a gas pressure of about 5 Pa. A typical feature, not discussed in the text, is the doublet splitting caused by effects of the ^{35}Cl nuclear spins. The average frequency (37211.33 ± 0.05 MHz) is assigned to the transition $25_{2,23} \leftarrow 25_{1,24}$.

For structural analysis too a molecule must usually possess a permanent dipole moment to produce an observable spectrum. Then it is necessary to *assign* the spectrum, i.e. to determine the rotational quantum numbers which define the transitions. The number of atoms in the molecule should not be too large; a structure containing more than about 10 non-hydrogen atoms would be very difficult to determine. Furthermore, it is not a quick technique: the time required for recording the spectrum and obtaining the first assignment might vary from a week to several months. After that one may decide to spend more time on the preparation of isotopic derivatives which are necessary for a complete structure determination as will be explained below. This discussion will concentrate on the determination of molecular geometries, but often dipole moments, centrifugal distortion parameters, nuclear quadrupole coupling constants and barriers to internal rotation can also be found.

General textbooks on rotational spectroscopy have been written by Townes and Schawlow (1955), Wollrab (1967), Kroto (1974), and by Gordy and Cook (1984). Structural data on particular molecules have been compiled by Harmony *et al.* (1979), Demaison *et al.* (1982), Brown *et al.* (1983), and by Callomon *et al.* (1987). The problems of accurate structure determination are exposed in detail by several authors in the book edited by Lide and Paul (1974), and also by Lide (1975), and by Nemes (1984).

3.2 Rigid molecules

For the ideal, and non-existing, rigid molecule a perfect method for structure determination can be developed. The principle can be illustrated by the well-known case of a diatomic molecule with atomic masses m_1 and m_2, where the energy levels are given by:

$$E/h = B\,J(J+1),\qquad(3.1)$$

and the rotational frequencies obey the selection rule $\Delta J = 1$:

$$\nu_{J+1\leftarrow J} = 2\,B\,(J+1).\qquad(3.2)$$

Here $J = 0, 1, 2, 3, \ldots$ and B is the *rotational constant*:

$$B = \frac{h}{8\pi^2 I},\qquad(3.3)$$

where I is the *inertial moment*:

$$I = \frac{m_1 m_2}{m_1 + m_2}\,r^2.\qquad(3.4)$$

Obviously the interatomic distance r follows directly from the observed frequencies. These are equidistantly spaced with intervals $2B$.

For a non-linear molecule there are three inertial moments:

$$I_x = \sum_j m_j(y_j^2 + z_j^2)\qquad(3.5)$$

and likewise for I_y and I_z. Here the summation is over all the atoms in the molecule. The coordinates x_j, y_j, z_j must be measured in the *principal axes system* of the molecule, which is defined by the relations:

$$\sum_j m_j x_j = \sum_j m_j y_j = \sum_j m_j z_j = 0,\qquad(3.6)$$

$$\sum_j m_j x_j y_j = \sum_j m_j y_j z_j = \sum_j m_j z_j x_j = 0.\qquad(3.7)$$

It is customary to replace x, y, z by a, b, c in such a way that:

$$I_a \leq I_b \leq I_c . \tag{3.8}$$

To these inertial moments correspond rotational constants A, B, C as in eqn (3.3). For example, using the present values of the fundamental constants:

$$A I_a = 505\ 379\ \mathrm{MHz\,u\,\AA^2} . \tag{3.9}$$

The three rotational constants are the only parameters that determine the energy levels of our hypothetical rigid molecule. To find the transition intensities the dipole moment components in the principal axes system are also required, but a rough guess is often all that is needed. Predicting a spectrum from a molecular model is thus a straightforward (albeit numerically slightly complex) procedure. The energy levels are specified by the rotational quantum number J and two sub-indices K_a and K_b. The spectrum is generally enormously more complex than the series of equidistant lines characteristic of a linear molecule.

In practice the reverse procedure should be followed. The first task is to extract the rotational constants from an observed spectrum, the second one to extract structural information from the rotational constants. Both stages present problems which are not trivial. As a real molecule is not a simple rigid rotor, the assignment may be complicated by effects of centrifugal distortion (briefly discussed below), nuclear quadrupole coupling (doublet splitting in Fig. 3.1) or large-amplitude motions like internal rotation. However, for our discussion it is sufficient to realize that the spectrum contains a lot of redundant information and in the end only three parameters (A, B, C) emerge. In the course of the spectrum identification this is an advantage, since it is extremely unlikely that one will come up with an incorrect assignment. But in the structure analysis it will become a disadvantage since all but the most simple molecules have more than three geometry parameters to be determined.

In this stage it is often possible to determine the point-group symmetry from the transition intensities, the inertial moments, and sometimes the Stark effect. If one is interested only in the molecular conformation, reasonable values for bond lengths and bond angles must be assumed and the one or two unknown dihedral angles can be adjusted to obtain the best agreement with the observed inertial moments. An example is given in Fig. 3.2, where the dihedral angle in cis-2-pentene was estimated to be 125 ± 5° (van Eijck 1981). Occasionally two or even more conformations can be studied separately. For instance, Fig. 3.3 shows two forms of formic acid. From intensity ratios the cis conformer was found to be 16.3 ± 0.4 kJ mol⁻¹ higher in energy than the trans conformer (Hocking 1976). Details of the structure will be discussed in Section 3.5.

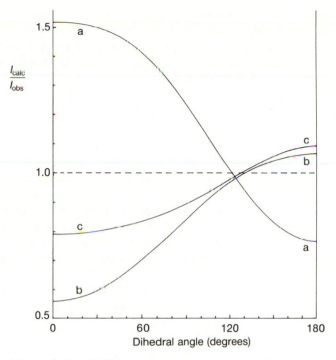

Fig. 3.2 Variation of the inertial moment ratios I_{calc}/I_{obs} with the dihedral angle $C(2)-C(3)-C(4)-C(5)$ in *cis*-2-pentene, $CH_3-CH=CH-CH_2-CH_3$. After van Eijck (1981).

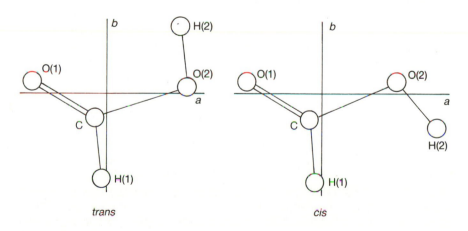

Fig. 3.3 The *trans* and *cis* conformations of formic acid, drawn in their respective principal axes systems.

The only way to obtain more detailed information is by *isotopic substitution*. Let us assume for the hypothetical rigid molecule discussed here that the geometry is independent of the nuclear masses; the situation in real molecules will be discussed in the following sections. Then the isotopically substituted molecule will provide additional observed inertial moments, whereas the number of geometric parameters remains the same. To see more clearly what happens it is convenient to introduce the *planar moments of inertia*:

$$P_x = \sum_j m_j x_j^2 = \tfrac{1}{2}(I_y + I_z - I_x) \tag{3.10}$$

and likewise P_y and P_z. By eqns (3.9) and (3.10) these planar moments can be directly calculated from the observed rotational constants A, B, C; accordingly, they can also be denoted by P_a, P_b, P_c. Although the name might suggest differently, the planar moments are defined for any molecule, whether planar or not.

Now suppose that one atom (k) is substituted by an isotope with a corresponding mass change Δm_k. Temporarily neglecting the small shifts and rotations of the axes system which are required to satisfy eqns (3.6) and (3.7) also for the substituted molecule, we have:

$$P_x' = P_x + x_k^2 \Delta m_k \quad \text{(etc.)}, \tag{3.11}$$

where primed quantities refer to the substituted molecule and unprimed ones to the parent. Since both P and P' are experimentally accessible quantities, we can find the coordinates of the substituted atom:

$$x_k^2 = \frac{P_x' - P_x}{\Delta m_k} = \frac{\Delta P_x}{\Delta m_k} \quad \text{(etc.)}. \tag{3.12}$$

It can be shown that the exact solution of the problem is:

$$x_k^2 = \frac{\Delta P_x}{\Delta m_k}\left(1 + \frac{\Delta m_k}{M}\right)\left(1 + \frac{\Delta P_y}{P_y - P_x}\right)\left(1 + \frac{\Delta P_z}{P_z - P_x}\right), \tag{3.13}$$

where the first correction factor accounts for the shift of the axes system (M being the total molecular mass of the parent molecule) and the last two factors account for its rotation. Equation (3.13) was introduced by Kraitchman (1953). Note that it is sufficient to know the inertial moments of parent and monosubstituted molecules to obtain the coordinates of the substituted atom, defined in the principal axes system of the parent molecule. The *differences* ΔP dominate the result, the values P themselves only contributing slightly except when two planar moments happen to be nearly equal (Kuczkowski *et al.* 1976; Mazur and Kuczkowski 1977). If necessary, corresponding equations for multiple substitutions can be developed or found in the literature (Wilson and Smith 1981).

To determine the complete molecular structure the following procedure should be followed: each atom *in turn* should be replaced by an isotope and the corresponding spectrum should be analyzed. Application of eqn (3.13) will then give the coordinates of all atoms in the molecule. Actually only the squares are determined, but the relative sign choice seldom presents serious ambiguities and when it does, multiple substitution can be used to solve the dilemma.

3.3 Vibrational effects

In real molecules the atoms are always subject to vibrations, so the concept of a molecular structure needs some careful definition. The most satisfactory one is probably the *equilibrium structure*, which corresponds to a minimum in the potential energy surface. Within the framework of the Born–Oppenheimer approximation this structure is invariant for isotopic substitution, and the equations in Section 3.2 should be relevant. Further discussion is given in Chapters 2 and 4.

The importance of the molecular vibrations can be immediately seen if one uses the observed inertial moments to obtain the atomic coordinates by means of eqn (3.13). A check on a complete set of such coordinates is provided by eqns (3.5)–(3.7). It is found that eqns (3.6) and (3.7) are usually well satisfied, but the inertial moments calculated from eqn (3.5) are usually 0.3–0.5 per cent smaller than the observed values. The difference, which may seem small, is quite outside experimental error. The limitations of the rigid-rotor assumption become very apparent for *planar molecules*: here all c coordinates are zero, and so we should have

$$P_c = \sum_j m_j c_j^2 = 0. \tag{3.14}$$

This is a quick test for planarity, and isotopic substitution should confirm it for each c coordinate. However, in practice one finds a non-zero value for P_c (often even negative!) and the c coordinates can easily be 0.1 Å or apparently imaginary. An example is discussed in Section 3.5. The observed quantity $-2P_c$ is called the *inertial defect* and it is predominantly vibrational in origin. Obviously an understanding of such effects is essential for the assessment of the reliability of structure determinations by this method, so a short discussion will be given.

The most common vibrational effects are already apparent in the case of the diatomic molecule, where eqn (3.1) should be replaced by:

$$E/h = B_v\, J(J + 1) - D_v\, J^2(J + 1)^2 + \ldots \tag{3.15}$$

The last term represents the effect of *centrifugal distortion*: one may imagine the molecule to stretch a little upon rotation, so the effective rotational constant will decrease with J. The consequence is a small shift of the spectral lines. For the non-linear molecule the phenomenon is more complex, five parameters being generally needed to describe it. We shall only note that these parameters are related to the harmonic force field and that their effects occasionally complicate the spectrum identification considerably.

More serious is the effect of the vibration on the rotational constants. For the diatomic molecule we have:

$$B_v = B_e - \alpha(v + 1/2) + \ldots, \tag{3.16}$$

where v is the vibrational quantum number ($v = 0, 1, 2, 3, \ldots$), B_v is the effective rotational constant in the vibrational state v and B_e is the equilibrium rotational constant which is related to the *equilibrium distance* r_e as in eqns (3.3) and (3.4). The vibration–rotation interaction constant α contains contributions from the harmonic force field as well as from the cubic terms in the potential energy expansion:

$$\alpha = \alpha_{\text{harm}} + \alpha_{\text{cubic}}. \tag{3.17}$$

Usually the latter term is somewhat larger than the former.

So the spectrum consists of a ground-state spectrum corresponding to B_0, accompanied by weaker satellite spectra whose intensities decrease with v as a geometric series due to the Boltzmann factor for the vibrational levels (Fig. 3.4). If at least one such satellite is observed, α can be measured and the correction from B_0 to B_e can be easily performed. But otherwise we have to make do with B_0 from which we calculate the *effective bond length* r_0 which may differ up to about 0.01 Å from r_e, and also from one isotopic species to the other.

A non-linear molecule with N atoms has $3N - 6$ modes of normal vibration, so a set of quantum numbers v_s ($s = 1, 2, \ldots 3N - 6$) is needed to define the vibrational state. Now eqn (3.16) is generalized to:

$$A_v = A_e - \sum_{s=1}^{3N-6} \alpha_s^A(v_s + 1/2) + \ldots \tag{3.18}$$

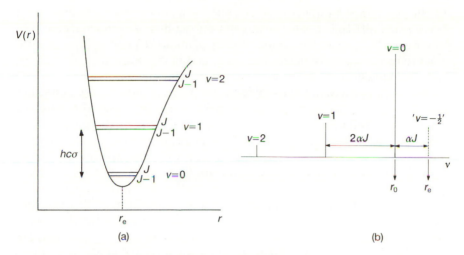

Fig. 3.4 (a) Schematic drawing of the potential energy versus interatomic distance in a diatomic molecule. Only two rotational levels are shown for each vibrational state. (b) Contribution of the $J \leftarrow J - 1$ transitions to the spectrum. Notice the extrapolation to the hypothetical equilibrium state and the intensities of the rotational lines which decrease following eqn (3.19).

and likewise for B_v and C_v. Thus there are many satellite spectra, one for each vibrational mode $\{v_1 v_2 \ldots v_{3N-6}\}$. The intensities of these spectra are proportional to the Boltzmann factor

$$\exp\left(-hc\sum_s v_s \sigma_s / kT\right) \tag{3.19}$$

which decreases quickly with increasing infrared wavenumbers σ_s, and only a few spectra will be observable.

Some vibrational modes may be degenerate, and their contributions in eqn (3.18) should be considered together. This refinement shall be worked out in Chapter 4 (eqn (4.1), where the next higher term is also given explicitly). For highly accurate work as discussed in Chapter 4 such terms are important, but for this introduction to the problems of structure determination we shall neglect them.

3.4 Structure determination

To determine an equilibrium structure *all* α constants must be known for all relevant isotopic species. However, the observation of all corresponding satellite spectra is hardly ever practicable except for a few small molecules. It is already difficult enough to obtain the ground-state rotational constants of the isotopic species. (An isotopic purity of about 20 per cent is often sufficient to produce an assignable spectrum, and in exceptional cases ^{13}C in its natural abundance of 1 per cent has been used. But more often difficult chemistry is involved in the isotopic substitutions.) Thus in normal practice one uses eqn (3.13) with planar moments deduced from the observed ground-state rotational constants A_0, B_0, C_0, neglecting the α corrections which should have been applied to allow the proper use of these equations. The resulting values for x, y, z (or a, b, c) are called *substitution coordinates*, and they are combined to give *substitution bond lengths* (r_s) or even complete *substitution structures*.

Costain (1958, 1966) has shown by multiple substitutions that this procedure gives significantly more consistent results than the straightforward use of eqn (3.5), which would give a so-called r_0 *structure*. This success can be explained if one assumes that the vibrational contributions cancel to a large extent in the planar moment differences. Table 3.1 is intended to give a rough idea of the orders of magnitude involved.

Table 3.1   Orders of magnitude of planar moments, P_g, and their isotopic differences, ΔP_g^a

	P_g (u Å^2)	ΔP_g (u Å^2)
Value	0 . . . 200	(0 . . . 3) Δm
Vibrational contribution	0.3	0.003 Δm
Experimental accuracy		0.001

^{a} The subscript g is one of x, y, z (or a, b, c).

Note that a small P_x signifies that most of the molecular mass is found in the y, z plane, whereas ΔP_x refers mainly to the x coordinate of the substituted atom. The residual vibrational contribution to the planar moment differences has been assumed, rather arbitrarily, to be 1 per cent per unit mass change. The experimental error in P is always very much smaller than the vibrational contribution, but the two effects may become comparable in the difference ΔP. Complications due to centrifugal distortion, nuclear quadrupole splitting, large-amplitude internal motions or the occurrence of 'heavy' atoms like Br could very well increase the experimental errors to 0.005 u Å^2 or more.

From the above estimate and eqn (3.12) we immediately find the order of magnitude of the uncertainty $\sigma(x)$ in a substitution coordinate:

$$\sigma(x) = \frac{\sigma(\Delta P_x)}{2x\,\Delta m} = \frac{0.0015}{x}, \tag{3.20}$$

where x is measured in Å. Equation (3.20) is known as *Costain's rule*. (Costain (1966) originally proposed 0.0012 rather than the now more generally adopted value of 0.0015.) From a survey of observed ΔP values for substitution in a mirror plane or on a twofold axis, where ΔP should be zero by symmetry, van Eijck (1982) has concluded that eqn (3.20) is adequate or even too pessimistic except for hydrogen atoms where the uncertainties must be multiplied by a factor of at least two.

It follows immediately from eqn (3.20) that *small coordinates are unreliable*: for $x < 0.15$ Å the expected error is larger than 0.01 Å. Fortunately we can also use eqns (3.6) and (3.7), which provide up to six additional relations to be used for problematic atoms where a small coordinate occurs or no isotopic substitution is available. Only as a last resort, e.g. for atoms like F, P, I which have no stable isotopes, should one use eqn (3.5) since these relations are known to be unreliable below the 1 per cent level. Except for hydrides with one 'heavy' atom eqn (3.5) should never be used to locate hydrogen atoms since their contribution to the inertial moments may be comparable to the vibrational effect. The problems caused by small coordinates are illustrated in Section 3.5 for *trans* formic acid.

Many investigators prefer the use of a least-squares method to the use of Kraitchman's equations. Observables to be fitted are P (or I) and ΔP (or ΔI) to obtain r_0 and r_s parameters, respectively. Obviously, the latter ones should be used where available, and a complete set of ΔP values should give results which are entirely equivalent to a substitution structure. The advantages of the least-squares formalism are that multiple substitutions can be treated easily and objectively, that eqn (3.5) can be added with a properly adjusted weight factor, and that bond lengths and bond angles can be used directly as parameters. If necessary, assumptions about molecular geometry can be introduced as additional 'observations' with clearly defined 'error' margins. As with all computerized procedures, the danger is that possible problems with light atoms, small coordinates or unsubstituted atoms are no longer obvious, and could lead to unstable results if the program is used blindly. In all cases a fair assessment of error limits to derived molecular parameters is a difficult task. The number of observations will never exceed the number of parameters by a factor which is large enough to make the least-squares error limits statistically significant, and direct error propagation from the observations is recommended instead. Our least-squares routine uses ΔI values with experimental errors augmented by vibrational contributions of $0.005\,u\text{Å}^2$ ($0.010\,u\text{Å}^2$ for H/D substitutions), while I values are omitted or, if unavoidable, are multiplied by 0.995 and given an 'error' limit of 0.5 per cent. This procedure is just as arbitrary as the use of eqn (3.20), but it is easier to implement and we feel that both methods give a reasonable order of magnitude for the possible deviations from the equilibrium structure.

Hydrogen atoms form an especially difficult subject. The relative mass change upon deuteration is so large that the cancelling of vibrational effects, on which the substitution method is based, can no longer be taken for granted. As stated, the uncertainties calculated from eqn (3.20) must be multiplied by a factor of at least 2. Even then, cases are known where hydrogen atoms were found at quite unreasonable geometric positions without any hint of a special vibrational effect that could be held responsible (van Eijck *et al.* 1983). Yet these are very probably vibrational artefacts since occasionally quite large imaginary coordinates occur which can never be explained in terms of abnormal structures (van Zoeren and van Eijck 1983). Interestingly, Penn and Olsen (1976) have found that a previously determined $O-H$ bond length (Penn and Curl 1971) of 1.14 Å in 2-aminoethanol, $NH_2-CH_2-CH_2-OH$, should be corrected to 1.00 ± 0.02 Å! The original determination of the hydrogen position by means of deuterium substitution was contaminated by an accompanying change in the N and O coordinates. Since these atoms are much heavier a small change is magnified in the hydrogen position. More generally it is found that hydrogen atoms involved in intramolecular hydrogen bonding tend to behave erratically.

To complicate matters further, hydrogen atoms are also frequently involved in large-amplitude motions like internal rotation of a methyl group or inversion of an amino group (Lister *et al.* 1978). Often a good description of the potential energy barriers for such internal motions can be obtained from special line-splitting effects in the microwave spectra, and then a correction for the influence of the large-amplitude motion on the effective rotational constants can be applied. Unfortunately this is not always done properly.

An approximation to the α constants needed to correct the effective inertial moments to their equilibrium values (eqn (3.18)) can sometimes be obtained from other techniques. Infrared spectroscopy nowadays can provide the required rotational constants of excited vibrational states with sufficient accuracy (see Chapter 4). Further, the increasing power of *ab initio* MO methods is beginning to allow the calculation of reliable harmonic and cubic force fields for small molecules, from which the α constants can be calculated (see Chapter 13, Section 13.5). In favorable cases the combination of all available data from different techniques will provide a reasonable approximation to the equilibrium structure.

Interesting results can still be obtained if only the harmonic force field is known (from infrared studies or *ab initio* MO calculations), so the correction from the harmonic part of the α constants (eqn (3.17)) can be applied. It can be shown that the resulting rotational constants (A_z, B_z, C_z) correspond to the *average structure* (distances r_z) which can also be obtained from electron diffraction if the proper corrections are also performed in that technique. (For more details see Chapter 2, Section 2.4.1.3, and Chapter 5,

Section 5.8; see also Kuchitsu and Oyanagi (1977).) In this way the results from microwave spectroscopy and electron diffraction become comparable, and can be combined to give a molecular structure that is sometimes superior to what can be obtained from either technique separately. The r_z structure is geometrically consistent (in contrast to the set of *average distances* which forms the r_g structure) and should exhibit no inertial defect since the average structure of a planar molecule is planar. However, it is not invariant for isotopic substitution. So to derive this structure the effect of isotopic substitution on the r_z distances must be estimated. This is usually done by using the equation

$$r_z = r_e + \frac{3}{2} a \langle \Delta z^2 \rangle - \frac{\langle \Delta x^2 + \Delta y^2 \rangle}{2r}, \qquad (3.21)$$

where $\langle \Delta z^2 \rangle$ is the mean-square amplitude of vibration in the bond direction and $\langle \Delta x^2 + \Delta y^2 \rangle$ the corresponding perpendicular amplitude (Kuchitsu *et al.* 1969). This equation is based on the Morse function and a is the corresponding anharmonicity parameter which is about 2 Å^{-1} for most bonds. Of course, this correction is only approximate (all effects on bond angles are neglected) and probably only justified for isotopic bond length *differences* between parent and substituted molecules:

$$\delta r_z = \frac{3}{2} a \delta \langle \Delta z^2 \rangle - \frac{\delta \langle \Delta x^2 + \Delta y^2 \rangle}{2r}. \qquad (3.22)$$

Table 3.2 Orders of magnitude of uncertainties in various structure types

Uncertainty (degrees)		0.2	0.5	1.0	2.0	5.0	
(Å)		0.002	0.005	0.010	0.020	0.050	
Classification		A	B	C	D	E	X
r_e	M	———					
	H	————					
r_s	M	———————————					
	H	——————————————— · · · · · · · · · · · · · ·					
r_0	M	———————————					
	H	———————————					

'Uncertainty' is defined as deviation from the true r_e structure; M denotes an atomic weight 10 or higher. A small coordinate tends to place the uncertainty in the right hand part of its range. For r_0 coordinates any assumptions about other M-type atoms also increase the uncertainties. The dotted line denotes the occasional occurrence of 'wild' substitution coordinates for H atoms.

The occasional reader may be bewildered by the variety of distance definitions discussed above. (To complicate matters further, Watson (1973) has introduced the r_m structure. This should provide a good approximation to the r_e structure, but in practice it is very difficult to obtain.) A summary is given in Chapter 2, Section 2.4. Not all authors state explicitly what kind of distances they have obtained, especially if a mixed r_s/r_0 structure is concerned. Harmony *et al.* (1979) classified geometry parameters into five categories, A through E, depending on the expected deviation from the true r_e geometry. Table 3.2 reflects this author's personal estimate of the general uncertainties in the various structure types. The large spread in these estimates is correlated mainly with the magnitude of the smallest coordinate, according to eqn (3.20). Hopefully this table can provide some guidance when one is trying to assess the significance of geometry comparisons between related molecules.

3.5   Sample calculation on *trans* formic acid

As an illustration the determination of the substitution structure of *trans* formic acid will be outlined. The aim is to demonstrate that a non-specialist can fairly easily follow the steps taken in such a procedure, provided that the determination of the rotational constants is taken for granted. Data published in the literature were converted to planar moments by eqns (3.9) and (3.10) and are collected in Table 3.3.

The molecule is planar, as seen from the inertial defect ($\Delta = -2P_c = 0.076 \, u \, Å^2$) and the very small ΔP_c values. In fact, all c coordinates appear to be imaginary and they are replaced by zero. From eqn (3.13) we calculate the other coordinates (Table 3.4). There are three small coordinates which are italicized. For the other coordinates the relative signs follow unambig-

Table 3.3   Planar moment data ($u \, Å^2$) for *trans* formic acid[a]

	P_a	P_b	P_c
HCOOH	41.96062	6.55820	−0.03821
Isotopomer	ΔP_a	ΔP_b	ΔP_c
$H^{13}COOH$	0.00621	0.16746	−0.00085
$HC^{18}OOH$	2.42088	0.08454	−0.00055
$HCO^{18}OH$	2.47246	0.02646	−0.00020
DCOOH	0.00204	2.24241	−0.00506
HCOOD	1.04416	1.12722	−0.00149
DCOOD	1.05777	3.43055	−0.00537

[a] Calculated from data published by Willemot *et al.* (1978) and by Davis *et al.* (1980).

Table 3.4   Substitution coordinates in *trans* formic acid[a]

Atom	a (Å)	b (Å)
H(1)	(±) *0.044 ± 0.068* → −0.060 ± 0.020	−1.5091 ± 0.0020
C	(±)*0.079 ± 0.019* → −0.0971 ± 0.0030	−0.4124 ± 0.0036
O(1)	−1.1214 ± 0.0013	0.2169 ± 0.0069
O(2)	1.1342 ± 0.0013	*0.121 ± 0.012* → 0.1192 ± 0.0074
H(2)	1.0132 ± 0.0030	1.0856 ± 0.0028

[a] The italicized small coordinates have been improved as described in the text. The error margins in the other coordinates are based on eqn (3.20).

uously from a molecular model (Fig. 3.3). We try to replace the three small coordinates with values calculated from the relations:

$$\sum_i m_i a_i = \sum_i m_i b_i = 0, \qquad (3.23)$$

$$\sum_i m_i a_i b_i = 0. \qquad (3.24)$$

Unfortunately, only $b_{O(2)}$ can be found in this way; the structure of the equations happens to be such that the errors in the a coordinates turn out to be even larger than in the straightforward substitution coordinates. A way around this problem may be found by double substitution. Without going into details, the idea is to use *second differences*

$$\Delta P_a(\text{HCOOD} \to \text{DCOOD}) - \Delta P_a(\text{HCOOH} \to \text{DCOOH}) \qquad (3.25)$$

to determine the a coordinate of the H(1) atom. One hopes that the vibrational effects in the second differences will cancel to even higher order; here we reach the stage where very accurate planar moments are needed. The result is $a_{H(1)} = -0.06 \pm 0.02$ Å (still not very accurate). Then the center of gravity relation (eqn (3.23)) gives a_C, and all coordinates except $a_{H(1)}$ are well determined. The planar moments calculated from the coordinates are lower than the observed ones by 0.3 per cent for P_a and 0.8 per cent for P_b.

The structure thus calculated is given in the column 'r_s structure' of Table 3.5. The structure of the *cis* conformer has been determined in an analogous way and is presented for comparison. Some structural differences are quite outside the estimated error limits, and must be considered physically significant. The other columns will be discussed now.

To obtain the r_z structure, Davis *et al.* (1980) have corrected the rotational constants for the harmonic part of the α constants (eqn (3.17))

Table 3.5 Structure parameters of formic acid[a]

Parameter	r_s structure[b,c]	r_s structure[d,e] (cis conformer)	r_z structure[e,f]	r_α^0 structure[g]	r_e structure[e,f]	r_e structure[c,h]
r(C–H)	1.097 (4)	1.105 (4)	1.097 (5)	1.096	1.091 (5)	1.089 (2)
r(C=O)	1.202 (5)	1.194 (3)	1.205 (5)	1.213	1.201 (5)	1.201 (3)
r(C–O)	1.341 (5)	1.352 (3)	1.347 (5)	1.357	1.340 (5)	1.338 (3)
r(O–H)	0.974 (8)	0.956 (5)	0.966 (5)	0.966	0.969 (5)	0.969 (2)
∠O–C–O	125.1 (5)	122.1 (4)	124.8 (4)	123.6	124.8 (4)	124.9 (4)
∠H–C=O	123.5 (11)	123.2 (6)	123.3 (15)	–	123.3 (15)	123.0 (10)
∠H–C–O	111.4 (11)	114.6 (6)	111.9 (15)	–	111.9 (15)	112.1 (9)
∠C–O–H	106.2 (4)	109.7 (4)	106.6 (4)	–	106.6 (4)	106.6 (2)

[a] Bond distances are given in Å, angles in degrees. Data refer to the *trans* conformer unless otherwise indicated.
[b] Calculated as outlined in the text.
[c] Error margins follow from the least-squares procedure described in Section 3.4.
[d] Bjarnov and Hocking (1978).
[e] For the meaning of the error margins the original literature should be consulted.
[f] Davis et al. (1980).
[g] Calculated by Davis et al. (1980) from the electron diffraction results of Almenningen et al. (1969).
[h] Van Dam and van Eijck, unpublished results.

calculated from a molecular force field. A first result is that all inertial defects are now between -0.0014 and $-0.0024 \, u \, Å^2$, which implies a considerable improvement both with respect to the uncorrected values of about $0.08 \, u \, Å^2$ and with respect to their maximum isotopic variation of about $0.010 \, u \, Å^2$. Accordingly the problems with small coordinates are alleviated. The isotopic changes in the four bond lengths were estimated from eqn (3.22), and incorporated in the geometry determination procedure. The final result is given in the column 'r_z structure', which should be comparable with the r_α^0 structure derived from an electron diffraction experiment (see Chapter 2, Section 2.4.1.3). The discrepancy between the two $O-C-O$ angles seems too large to be acceptable.

Finally Davis *et al.* (1980) estimated the r_e structure by applying eqn (3.21), notwithstanding the very approximate character of this equation. In the last column of Table 3.5 this estimate is compared to another one determined in a different way. *Ab initio* MO calculations were used to find the harmonic and cubic force field from which the α constants were calculated, making a direct determination of the equilibrium structure possible. The correspondence between the two r_e structures is surprisingly good.

References

Almenningen, A., Bastiansen, O., and Motzfeldt, T. (1969). *Acta Chem. Scand.*, **23**, 2848–64.

Bjarnov, E. and Hocking, W. H. (1978). *Z. Naturforsch.*, **33a**, 610–8.

Brown, J. M., Demaison, J., Dubrulle, A., Hüttner, W., and Tiemann, E. (1983). *Molecular constants*, Landolt-Börnstein, Numerical data and functional relationships in science and technology (New series), Group II, Vol. 14b. Springer, Berlin.

Callomon, J. H., Hirota, E., Iijima, T., Kuchitsu, K., and Lafferty, W. J. (1987). *Structure data of free polyatomic molecules*, Landolt-Börnstein, Numerical data and functional relationships in science and technology (New series), Group II, Vol. 15. Springer, Berlin.

Costain, C. C. (1958). *J. Chem. Phys.*, **29**, 864–74.

Costain, C. C. (1966). *Trans. Am. Crystallogr. Assoc.*, **2**, 157–64.

Davis, R. W., Robiette, A. G., Gerry, M. C. L., Bjarnov, E., and Winnewisser, G. (1980). *J. Mol. Spectrosc.*, **81**, 93–109.

Demaison, J., Dubrulle, A., Hüttner, W., and Tiemann, E. (1982). *Molecular constants*, Landolt-Börnstein, Numerical data and functional relationships in science and technology (New series), Group II, Vol. 14a. Springer, Berlin.

Gordy, W. and Cook, R. L. (1984). *Microwave molecular spectra*. Wiley-Interscience, New York.

Harmony, M. D., Laurie, V.W., Kuczkowski, R. L., Schwendeman, R. H., Ramsay, D. A., Lovas, F. J., Lafferty, W. J., and Maki, A. G. (1979). *J. Phys. Chem. Refer. Data*, **8**, 619–721.

Hocking, W. H. (1976). *Z. Naturforsch.*, **31a**, 1113–21.

Kraitchman, J. (1953). *Am. J. Phys.*, **21**, 17–24.

Kroto, H. W. (1974). *Molecular rotation spectra*. Wiley-Interscience, New York.

Kuchitsu, K. and Oyanagi, K. (1977). *Faraday Discuss. Chem. Soc.*, **62**, 20–8.

Kuchitsu, K., Fukuyama, T., and Morino, Y. (1969). *J. Mol. Struct.*, **4**, 41–50.

Kuczkowski, R. L., Gillies, C. W., and Gallaher, K. L. (1976). *J. Mol. Spectrosc.*, **60**, 361–72.

Lide Jr., D. R. (1975). Molecular structure determination by high-resolution spectroscopy. In *Molecular structure and properties*, MTP International Review of Science, Physical Chemistry Ser. 2, Vol. 2 (ed. A. D. Buckingham), pp. 1–25. Butterworths, London.

Lide Jr., D. R. and Paul, M. A. (eds.) (1974). *Critical evaluation of chemical and physical structural information*. National Academy of Sciences, Washington.

Lister, D. G., MacDonald, J. N., and Owen, N. L. (1978). *Internal rotation and inversion: An introduction to large-amplitude motions in molecules*. Academic Press, London.

Mazur, U. and Kuczkowski, R. L. (1977). *J. Mol. Spectrosc.*, **65**, 84–9.

Nemes, L. (1984). Vibrational effects in spectroscopic geometries. In *Vibrational spectra and structure*, Vol. 13 (ed. J. R. Durig), pp. 161–221. Elsevier, Amsterdam.

Penn, R. E. and Curl Jr., R. F. (1971). *J. Chem. Phys.*, **55**, 651–8.

Penn, R. E. and Olsen, R. J. (1976). *J. Mol. Spectrosc.*, **62**, 423–8.

Townes, C. H. and Schawlow, A. L. (1955). *Microwave spectroscopy*. McGraw-Hill, New York.

Van Eijck, B. P. (1981). *J. Mol. Spectrosc.*, **85**, 189–204.

Van Eijck, B. P. (1982). *J. Mol. Spectrosc.*, **91**, 348–62.

Van Eijck, B. P., Maagdenberg, A. A. J., Janssen, G., and van Goethem-Wiersma, T. J. (1983). *J. Mol. Spectrosc.*, **98**, 282–303.

Van Zoeren, E. and van Eijck, B. P. (1983). *J. Mol. Struct.*, **97**, 315–22.

Watson, J. K. G. (1973). *J. Mol. Spectrosc.*, **48**, 479–502.

Willemot, E., Dangoisse, D., and Bellet, J. (1978). *J. Mol. Spectrosc.*, **73**, 96–119.

Wilson, E. B. and Smith, Z. (1981). *J. Mol. Spectrosc.*, **87**, 569–70.

Wollrab, J. E. (1967). *Rotational spectra and molecular structure*. Academic Press, New York.

4

Determination of accurate molecular structure by vibration–rotation spectroscopy

Georges Graner

4.1 Introduction

From a purely quantum mechanical point of view, the concept of molecular structure is by no means a trivial concept. Several scientists have been discussing this point and try to reconcile this classical picture, born in the last century, with a rigorous quantum theory (Claverie and Diner 1980 and references therein). Nevertheless, most chemists and physical chemists agree on a quantum mechanical definition of molecular structure within the framework of the Born–Oppenheimer approximation. Wilson (1979) explains this point of view and points out that, in spite of some drawbacks and difficulties, the *equilibrium* structure of a molecule can be satisfactorily defined in the overwhelming majority of cases.

The present chapter deals with the determination of accurate molecular structure through vibration–rotation spectroscopy. This implies that we are concerned with gas-phase spectroscopy only, since liquids and solids can only provide vibration spectra, from which it is not possible to extract accurate structures.

Although vibration–rotation Raman spectroscopy is an active field (Brodersen 1979), its contribution to structure determinations is still limited so that the bulk of this chapter will be devoted to infrared (IR) spectroscopy. On the other hand, there is such a strong interaction of IR spectroscopy with rotational (i.e. essentially microwave) spectroscopy and to a lesser extent with gas-phase electron diffraction (ED) that there is some overlap here with Chapters 2, 3, and 5 of this book.

As we shall see, IR spectroscopy yields specific structural information completely independent from microwave (MW) spectroscopy in only a few cases. This is clearly the case of molecules having no dipole moment and therefore no pure MW rotation spectrum such as CO_2, BF_3 (plane) or CH_4. In other cases, MW and IR spectroscopies complement each other, the former usually providing better ground-state molecular constants, while the latter seems more appropriate for vibrationally excited states.

4.2   Vibrational effects in molecular structures deduced from spectroscopy

A naive point of view about spectroscopically determined structures would be the following: we are able to determine rotational constants A, B, and C with eight significant figures. Since these constants are inversely proportional to distances squared, one would expect to obtain structural parameters also with eight significant figure accuracy, which means uncertainties in the order of 10^{-8} Å for distances or 10^{-6} degrees for angles. Except for a handful of diatomic molecules, this goal is far from being reached. Most molecular distances are only reliable to within 0.001 Å or even 0.01 Å. The reason is the well-known zero-point vibrational energy, which poses difficulties for the determination of the equilibrium geometry, the so-called r_e geometry.

The rotational constants in the vibrational ground state and excited states are related to the corresponding ones at equilibrium by equations such as

$$B_v = B_e - \sum_s \alpha_s^B \left(v_s + \frac{d_s}{2} \right) + \sum_{r \leqslant s} \sum \gamma_{rs}^B \left(v_r + \frac{d_r}{2} \right) \left(v_s + \frac{d_s}{2} \right) + \dots$$

$$(4.1)$$

and similar ones for A_v and C_v when it is relevant. In eqn (4.1) the summa-
tions are over the vibrational states, each characterized by a quantum
number v_s and degeneracy d_s. The large number of parameters α_s^B and γ_{rs}^B
(and equivalent parameters for A and C) to be determined explains the difficulty in determining the values of A_e, B_e, C_e, and, as a second step, the
equilibrium structure.

Why is it so important to reach this equilibrium structure rather than some
operational structure such as r_0 or r_s? Several reasons can be invoked.

The first one is the internal consistency of the structure determination
method. For example any bent triatomic molecule must be planar and one
should have $I_c = I_a + I_b$ which is equivalent to $1/C = 1/A + 1/B$. It is
easy to check that this condition is far from being fulfilled with the averaged ground-state constants A_0, B_0, C_0, even when they are extremely
accurate. Therefore, for bent symmetrical XY_2 molecules, where only two
geometrical parameters, r and φ, are required, three different structures
would be obtained by choosing A_0 and B_0 or A_0 and C_0 or B_0 and C_0. This
difficulty vanishes with equilibrium constants. Of course, for molecules with
4 or more atoms, which are suspected of being planar, this problem with
ground-state constants prevents us even from ascertaining if a molecule is
planar.

The second reason is the requirement that the structure determined should
be isotopically invariant. This is not a must if we can go *beyond* the Born–
Oppenheimer approximation such as is the case for a few diatomic molecules. But *within* this approximation, two possibilities exist: either the
molecule is so simple that a single isotopomer is sufficient to determine the
structure (e.g. diatomic molecules, $^{12}C^{16}O_2$, H_2O, BF_3, CH_4, ... see Table
4.1); then the other isotopomers each should yield the same structure and
this provides a powerful check. Or if the molecule is more complex, as is
usually the case, one must assume the isotopic invariance and have information on more than one isotopomer just to obtain the structure.

The third reason is the fact that different experimental techniques, such
as spectroscopy and gas-phase electron diffraction, yield different structural
parameters since the zero-point averaging is different. They can best be
reconciled if the comparison is made between the *equilibrium* structures
obtained by the different methods. The same should in principle apply to
theoretical quantum chemical structures which, however, are not at present
accurate enough (except in a few cases) to present a problem.

The last reason is that structural parameters are often used to compare
related molecules. How significant are the differences between *cis* and *trans*

Table 4.1 Polyatomic molecules for which the equilibrium structure can be derived from a single isotopic species

Type of molecule	Example	Structural parameters	Rotational constants available	Observations
Symmetric triatomic linear XY_2	CO_2	r_e	B_e	
Molecule X_n ($n > 2$) represented by a regular planar polygon	H_3	r_e	B_e	The rotational constant C_e should be equal to $B_e/2$.
Molecule X_n ($n > 3$) represented by a regular polyhedron	P_4	r_e	B_e	
Molecule XY_n ($n > 2$) with X at the center of a regular polygon	BF_3, XeF_4	r_e	B_e	The rotational constant C_e should be equal to $B_e/2$.
Spherical tops XY_n ($n > 3$) with X at the center of a regular polyhedron	CH_4, SF_6	r_e	B_e	
Symmetric tops XY_n ($n > 2$) with X at the apex of a regular pyramid	PH_3	r_e, φ_e	A_e and B_e or C_e and B_e	
Symmetric tops XY_{n+2} ($n > 2$) with X at the center of a regular polygon of n Y atoms and two Y atoms on the axis	PF_5	r_e(equat.), r_e(axial)	A_e and B_e or C_e and B_e	
Bent XY_2 molecules with C_{2v} symmetry	H_2O	r_e, φ_e	Two out of three (A_e, B_e, C_e)	Planarity condition $1/C_e = 1/A_e + 1/B_e$

formic acid (see Bjarnov and Hocking 1978 and Chapter 3, Section 3.5)? How significant are the differences in the Si—F bond lengths in the SiH_nF_{4-n} molecules (Davis et al. 1980)? How does the C—H bond distance change with environment? All these questions and many similar ones deal with differences which are of the same size or even smaller than vibrational corrections.

The present chapter will therefore be focused on the determination of *equilibrium structures* by rovibrational and especially IR spectroscopy, complemented of course by MW spectroscopy and sometimes ED. We shall see that for simple molecules, the r_e structure can actually be obtained. For molecules with 4–7 atoms, this is not so easy and one has often to resort to methods for *estimating* the equilibrium structure. The question is then to test the reliability of these methods. For much larger molecules, the task of obtaining equilibrium structures is at present out of reach: at most, the choice is then between purely empirical correlations and structures related to the vibrational ground state (r_0, r_s, r_z, and r_m; as defined in Chapter 2, Section 2.4.1) with all their aforementioned drawbacks. It is then time for IR spectroscopy to quit, since MW spectroscopy is usually more efficient.

4.3 Diatomic molecules, radicals and ions

4.3.1 Simple treatment

For diatomic molecules, eqn (4.1) reduces to a very simple form

$$B_v = B_e - \alpha\left(v + \frac{1}{2}\right) + \gamma\left(v + \frac{1}{2}\right)^2 + \dots \qquad (4.2)$$

Therefore, it is sufficient to obtain a value of the rotational constant B in the vibrational ground state and one or two excited states to derive B_e. From there, it is trivial to derive r_e through the equations

$$I_e = h/8\pi^2 B_e \qquad (4.3)$$

(if B_e is expressed in frequency units) or

$$I_e = h/8\pi^2 c B_e \qquad (4.3')$$

(if B_e is expressed in wavenumber units), and

$$I_e = \mu\, r_e^2. \qquad (4.4)$$

Here μ is the reduced mass of the diatomic molecule A B,

$$\mu = \frac{m_A m_B}{m_A + m_B}. \tag{4.5}$$

The product $I_e B_e$ of eqns (4.3) and (4.3′) has a value of 505379.08 ± 0.61 MHz u Å^2 or 16.857631 ± 0.000020 cm^{-1} u Å^2 according to the most recent values of physical constants (Mills *et al.* 1988). Note that the relatively large uncertainties on Planck and Avogadro constants result in rather large uncertainties on $I_e B_e$. This might well be in some cases the limiting factor if one wishes to get really accurate bond distances. Moreover, older values of this constant such as 505 391 MHz u Å^2 have often been used, even in the recent past (Hoeft and Nair 1987); so the utmost care should be taken in comparing old and new distances.

The procedure described above has been applied to many diatomic molecules, radicals and ions. Huber and Herzberg (1979) report r_e distances for several hundred diatomic molecules in their electronic ground states.

In recent years, this kind of simple approach has been used for many diatomic radicals. One of the earlier examples of radicals studied by I R was CF, the ground electronic state of which is split into two substates $^2\Pi_{1/2}$ and $^2\Pi_{3/2}$. The fundamental vibration band $1 \leftarrow 0$ was recorded with a diode laser spectrometer (Kawaguchi *et al.* 1981). The 31 lines observed yielded ten molecular constants including B_0 and B_1 from which r_e is determined as 1.271977 ± 0.000017 Å. Later, the addition of microwave transitions yielded a revised value of 1.271972 ± 0.000013 Å (Saito *et al.* 1983). In most similar cases the same radical is studied both by M W and I R spectroscopy. The former yields B_0, the latter B_v in one or several excited states, since these transient molecules, produced in discharges, give rather strong vibrational hot bands. In the case of CH, thanks to Fourier-transform spectroscopy, Bernath (1987) was able to observe in *emission* the $1 \rightarrow 0$, $2 \rightarrow 1$ and $3 \rightarrow 2$ bands, about 150 lines altogether from which he deduced $r_e = 1.11983 ± 0.00002$ Å. Note that in these examples, there was no check on the isotopic invariance and no corrections were made for contributions of the electron slipping effect. Similar techniques have been used for ions such as CF$^+$ (Kawaguchi and Hirota 1985), which is simpler to study since it is in a $^1\Sigma^+$ state. Eighteen rovibrational lines of the fundamental band studied by I R diode laser spectroscopy yielded B_0 and B_1 and $r_e = 1.154272 ± 0.000035$ Å. Note the considerable (9 per cent) shortening of distance as compared to the neutral CF molecule.

A handful of negative ions have also been detected, mainly by the technique of velocity-modulation I R spectroscopy (see for example Rosenbaum *et al.* 1986) but also by Fourier-transform spectroscopy (Elhanine *et al.* 1988). Rosenbaum *et al.* find, for the OH$^-$ ion, a bond length $r_e = 0.964317 ± 0.000022$ Å, slightly shorter than in OH, 0.96966 Å according to Huber and Herzberg (1979), but rather different from the distance in OH$^+$, 1.0272 ± 0.0002 Å (Gruebele *et al.* 1986).

4.3.2 Refined treatment

As soon as the number and quality of experimental data increase, the simple treatment presented in the previous paragraph should be improved. First, this simple treatment has certain inadequacies since it is based on a potential function close to the harmonic oscillator potential function, which is clearly insufficient. Then, it does not take full advantage of all the information available from all rovibrational and rotational transitions which are observed. Finally, it is better to treat simultaneously all isotopomers, especially if the amount of information available is not the same for all of them.

The most common procedure is nowadays to represent the rovibrational energies by a power series (Dunham 1932) such as

$$E^i(v, J) = hc \sum_{k=0} \sum_{l=0} Y_{kl}^i \left(v + \frac{1}{2}\right)^k [J(J+1)]^l. \tag{4.6}$$

This formulation is valid in a $^1\Sigma$ electronic state only but similar ones can be used for other cases. The coefficients Y_{kl}^i are closely related to the usual spectroscopic parameters, for instance $Y_{01}^i \cong B_e$, $Y_{11}^i \cong -\alpha_e$, etc., but they are not equal to them. The index 'i' means that they depend on the isotopic species considered. In a first approximation,

$$Y_{kl}^i \cong \mu^{-(k+2l)/2} U_{kl}, \tag{4.7}$$

where U_{kl} is isotopically invariant. Later it was shown (Bunker 1972, 1977; Watson 1973a, 1980) that the U_{kl} coefficients themselves are mass-dependent, partly from the breakdown of the Born–Oppenheimer approximation, and partly from quantum effects in the vibration–rotation calculations. The correct form is therefore

$$Y_{kl}^i = \mu^{-(k+2l)/2} U_{kl} [1 + m_e(\Delta_{kl}^A/m_A + \Delta_{kl}^B/m_B)], \tag{4.8}$$

where m_e is the mass of an electron, m_A and m_B the masses of atoms A and B respectively, and the Δ_{kl} are parameters close to or smaller than 1.

As explained by Ogilvie and Tipping (1983), the determination of the parameters of eqn (4.8) can be made empirically from the wavenumbers of the spectral lines. But it is much better to assume a suitable potential energy function which further imposes relations between Dunham coefficients and reduces the number of free parameters. The minimum of this potential function is taken as the equilibrium distance. The potential function may be expressed as one of many possible analytical functions or determined by the RKR numerical method (see Chapter 2, Section 2.2.1).

This refined method has only been applied to a handful of molecules including HCl (Watson 1973a; Coxon and Ogilvie 1982; Coxon 1986), CO (Watson 1973a, 1980), CS (Bogey *et al.* 1982). For this last molecule r_e is given as 1.5348192(12) Å.

In several other papers, the treatment uses eqns (4.6) and (4.7) only. When several isotopic species are available, the authors can then look for small differences in r_e revealing a breakdown of the Born–Oppenheimer approximation. These differences were not significant for CS (Todd and Olson 1979) until the data were improved by Bogey *et al.* (1982). The recent study of TlF by Hoeft and Nair (1987) is more unusual. By measuring pure rotation lines in several excited states and analyzing them according to a Dunham potential, they find $r_e = 2.08445969(1)$ Å and 2.08446015(2) Å for $^{205}Tl^{19}F$ and $^{203}Tl^{19}F$ respectively, where the exceptionally small uncertainty between parentheses represents one standard deviation, not including huge errors due to fundamental constants.

4.3.3   The case of homonuclear diatomic molecules

Diatomic molecules with two identical atoms have no rovibrational spectrum due to electric dipole transitions. The generally much weaker magnetic dipole and electric quadrupole transitions have been observed, e.g. for O_2, N_2, H_2, either in the laboratory (with high pressures) or, for the first two molecules, in the terrestrial atmosphere.

On the other hand, Raman rotational and rovibrational spectra of these molecules can be recorded (Brodersen 1979) and analyzed to give r_e, as was done for $^{14}N_2$, $^{14}N^{15}N$ and $^{15}N_2$ by Bendtsen (1974).

4.4   Polyatomic molecules

4.4.1   General

More and more equilibrium structures have become available for polyatomic molecules. But the difficulty in determining this structure increases rapidly with molecular complexity. Therefore, it is important to compare, when possible, these equilibrium structures with approximate structures such as r_0, r_s, r_z, r_m ... which often represent the best we can obtain for moderately complex molecules.

The *substitution* or r_s structure, detailed in Chapter 3, is the most used of these approximate structures. Originally, the method was introduced by Kraitchman (1953) as a convenient way to extract the r_e structure from a set of equilibrium moments of inertia for several isotopomers. Later, Costain (1958) suggested the use of this same method with ground-state moments of inertia, I_0, to give the so-called substitution structure. These structures seemed very consistent for different isotopomers and Costain concluded that the r_s distances should not deviate much from the equilibrium values. This point of view was soon criticized by Graner (1963) who claimed that the consistency of r_s distances does not prove that they are close to r_e. The early confidence in the substitution method was somewhat shaken by experience as many problems arose for atoms close to a symmetry element, for atoms without stable isotopes, for hydrogen atoms, for imaginary coordinates, etc. Errors as large as 0.01 Å can indeed be found. Although many interesting results have been found by using the substitution method, it is still somewhat unpredictable.

The approach used to determine the equilibrium structure of a molecule consists of three distinct stages:

First stage: determine all the appropriate ground-state rotational constants for the molecule under study (only B_0 for linear and spherical top molecules, B_0 and A_0 or B_0 and C_0 for symmetric top molecules, A_0, B_0, and C_0 for asymmetric top molecules). MW spectroscopy has been and still is in many cases the best technique for this purpose but Ground-State Combination Differences (GSCDs) from high-quality IR spectra can now be used with profit. These spectra enable us to reach values of J much higher than those measured by MW spectroscopy. In addition, a large number of GSCDs can be gathered from several different rovibrational bands. A recommended practice is to make a least-squares fit of MW transitions and IR GSCDs together, with weights proportional to $(\Delta \nu)^{-2}$, where $\Delta \nu$ is the experimental uncertainty of an observation (e.g. $\Delta \nu = 0.1$ MHz for MW, $\Delta \nu = 3$ MHz for a good IR spectrum).

Sometimes, IR spectroscopy is essentially the only resource. This is the case for molecules with no permanent dipole moment (CH_4) or with a very small one (CH_3D, CH_2D_2). It is also the case for the A_0 (or C_0) constant of symmetric top molecules which can be obtained neither by *normal* MW spectroscopy nor by *normal* IR GSCDs. In a large number of such molecules, the identification of many 'forbidden' or rather 'perturbation-allowed' transitions leads to *abnormal* GSCDs which give access to A_0 (see for example Graner 1976).

The determination of A_0, B_0, and C_0 should of course be made for as many isotopomers as possible.

Second stage: obtain all equilibrium rotational constants A_e, B_e and C_e.

This is not an easy task. Let us look at eqn (4.1) and ignore for the moment all γ parameters (the procedure described for B should of course be applied for A and C also). For each fundamental band ν_s, i.e. a transition between the ground state and $v_s = 1$, eqn (4.1) immediately gives one α_s^B value. Therefore, it may seem sufficient to analyze all fundamental bands of a molecule, extract an α_s value from each $B_s - B_0$ difference and compute B_e through the following equation

$$B_e = B_0 + \frac{1}{2}\sum_s \alpha_s^B d_s. \qquad (4.9)$$

But eqn (4.9) applies only when the effect of resonances among vibrationally excited states has been removed. For instance, if two levels 1 and 2 are mixed by a Fermi-type resonance, the *apparent* B constants are given by

$$B_I = a^2 B_1 + b^2 B_2 \qquad (4.10)$$

$$B_{II} = b^2 B_1 + a^2 B_2,$$

where a and b are the fractional contributions of the unperturbed to the perturbed vibrational eigenfunction. Note that $B_I + B_{II} = B_1 + B_2$. Coriolis-type interactions also contribute, but in a different way, to apparent α_s^B values. These interactions often mix single excited and combination levels. Moreover, the γ_{rs}^B of eqn (4.1) cannot generally be ignored. Therefore we are led to analyze many more IR bands (and/or vibrational satellites in the MW spectrum). The next step is to utilize all available pieces of information to build a potential energy function, first harmonic and later anharmonic (see Chapter 2).

The advantage of deriving such a potential function is that it can predict with a reasonable accuracy the α and γ parameters which have not been actually measured, especially for the less studied isotopomers.

Third stage: from all available A_e, B_e and C_e values, obtain the actual equilibrium structure. In a few cases, detailed in Table 4.1, a single isotopic species brings enough information for deriving the structure. The other isotopomers then give a useful check of this structure, within the Born–Oppenheimer approximation. Note that in many cases, an isotopic substitution destroys the basic symmetry. This is not crucial for the equilibrium structure, but is important when vibrational corrections are necessary.

In most cases, however, more than one isotopomer is needed but it would be an error to think that the amount of information is proportional to the number of isotopomers used. For instance, $^{13}CH_4$ should have exactly the same B_e as $^{12}CH_4$. In a similar way, replacing ^{79}Br by ^{81}Br in CH_3Br changes B_e but not A_e, at least at the present experimental precision, since

A_e depends only on the CH_3 group geometry, which is essentially unaffected by this replacement.

The derivation of the geometrical parameters from all available moments of inertia can be done either by a least-squares fit or by the original Kraitchman method or else by graphical methods in simple cases. A few characteristic examples will be detailed now.

4.4.2 Linear molecules

4.4.2.1 Carbon dioxide
This simple molecule can be used to check the validity of various approximations. From vibrational and rotational constants of 355 bands, Chedin and Teffo (1984) have derived a potential energy function, with 28 parameters up to F_{333333}, which provides all terms needed to extrapolate from B_0 to B_e for almost any isotopic species. Using very accurate B_0 values for six isotopomers and Chedin and Teffo's vibrational corrections, Graner et al. (1986) have obtained precise r_e with a spread of only 3×10^{-6} Å (but not in a random way). The accuracy of these distances is necessarily somewhat worse than this figure due to the uncertainties of the vibrational corrections. The same authors have also computed the approximate structure r_m, as defined by Watson (1973b) with theoretical justifications. In the present case, r_m is within 5×10^{-5} Å of r_e, a rather remarkable result (see Table 4.2). More recently, Harmony et al. (1988) have proposed a modified version of r_m called r_m^ρ. For CO_2, it is within 1×10^{-5} Å of r_e. Note that the r_s value quoted by Harmony et al. differs by 0.00137 Å from r_e (Table 4.2). The high quality of IR data available led Teffo and Ogilvie (1987) to look for a possible breakdown of the Born–Oppenheimer approximation by deriving a formula similar to eqn (4.8).

For similar molecules, r_e distances can be found for CS_2 (Lacy and Whiffen 1981) and for CSe_2 (Bürger and Willner 1988).

Table 4.2 Comparison of C=O distances for CO_2[a]

Type of distance[b]	Value	Range	References
r_e	1.15995884	284×10^{-8}	Graner et al. (1986)
r_m	1.15990735	$13\ 089 \times 10^{-8}$	Graner et al. (1986)
r_m^ρ	1.15995 (2)		Harmony et al. (1988)
r_s	1.16133		Harmony et al. (1988)
r_0	1.16204674	$7\ 033 \times 10^{-8}$	Graner et al. (1986)

[a] All values are given in Å.

[b] Note that r_e, r_m and r_0 are averaged over six isotopic species, namely 16–12–16, 16–13–16, 18–12–18, 18–13–18, 16–12–18, and 16–13–18, whereas r_m^ρ is based on the first three isotopomers.

4.4.2.2 Other linear molecules The equilibrium structures of XYZ molecules have only been published for a few molecules. N_2O, OCS and HCN are the best documented, but results can also be found for HBS, OCSe and XCN (with X = F, Cl, Br, I).

N_2O is an interesting case because the central N atom is very close to the center of mass. Accordingly, its substitution coordinate can vary from 0.11 Å to an imaginary value (Costain 1958, 1966). The corresponding r_s bond lengths found for N—N and N—O are therefore found to differ by at least 0.015 Å from their equilibrium values when calculated by the pure substitution method (Harmony and Taylor 1986). As shown in Table 4.3, most authors are led to extract not $r_e(N—O)$ and $r_e(N—N)$ directly but their sum and their difference. The sum has been well determined for years, with an accuracy of a few 10^{-4} Å at least, whereas the difference changed by several 10^{-3} Å. Griggs *et al.* (1968) analyzed four isotopomers and obtained their B_e through the formula

$$B_e - B_0 = \tfrac{1}{2}(\alpha_1 + 2\alpha_2 + \alpha_3) = \tfrac{1}{2}[B_0 + B(00^01) - B(10^01) - B(02^01)].$$

$$(4.11)$$

This formula is justified by the fact that $\nu_2 \cong 600\,\text{cm}^{-1}$ was not analyzed and that the levels 10^01 and 02^01 are involved in a Fermi resonance, so that the sum of their B constants is unaffected by the resonance, as explained earlier. This formula ignores the γ_{rs}^B and another Fermi resonance. The same approach was used by Amiot (1976) with a larger number of bands and isotopomers. On the other hand, Lacy and Whiffen (1982) were able to determine a sextic force field with 47 parameters to fit data corresponding to 239 different vibrational levels. The vibrational corrections to B_e are therefore supposed to be more accurate. The noticeable differences of bond lengths with Griggs *et al.* (1968) and with Amiot (1976) (Table 4.3) suggest that the neglect of the γ parameters and higher-order terms was not justified. But is Lacy and Whiffen's potential the final one?

OCS seems less difficult because C is further from the center of mass but, again, the sum of bond lengths is more accurately known than each of them individually (Table 4.4). Maki and Johnson (1973) deduced B_e through the following equation which removes the effect of a Fermi resonance between ν_1 and $2\nu_2$

$$B_e = \tfrac{1}{2}[5B_0 - B(10^00) - B(02^00) - B(00^01)].$$

$$(4.12)$$

A first anharmonic force field was derived by Whiffen's group (Foord *et al.* 1975) but this force field, and consequently the distances, were substantially revised later (Whiffen 1980). Comparison of the two values given by Whiffen in 1980 (Table 4.4) suggests that the r_e distances might be known

Table 4.3 Comparison of distances for N_2O[a]

Type of distance	$r_1(N-N)$	$r_2(N-O)$	$r_1 + r_2$	$r_2 - r_1$	References
r_e	1.12665(50)	1.18565(50)	2.31230(3)	0.0590(10)	Griggs et al. (1968)
r_e	1.1282 ± 0.0001	1.1843 ± 0.0001	2.312535 ± 0.000005	0.0561 ± 0.0002	Amiot (1976)
r_e	1.12598(3)	1.18624(3)	2.31222	0.06026	Lacy and Whiffen (1982)
pure r_s	1.1466	1.1695	2.3161	0.0229	Harmony and Taylor (1986)
r_s with first moment	1.12859 ± 0.00022	1.18755 ± 0.00030	2.31614	0.05896	Costain (1958, 1966)
r_m	1.1281(8)	1.1842(8)	2.3123	0.0561(15)	Watson (1973b)
r_m^ρ	1.1277(12)	1.1846(11)	2.3123	0.0569	Harmony and Taylor (1986)

[a] All values are given in Å. The quantities in parentheses represent one standard deviation in units of the last significant figure, whereas ± relates to total uncertainty or total range of distances.

Table 4.4 Comparison of distances for OCS[a]

Type of distance	r_1 (O—C)	r_2 (C—S)	$r_1 + r_2$	References
r_e	1.15431 ± 0.00024	1.56283 ± 0.00037	2.71714 ± 0.00013	Maki and Johnson (1973)
r_e	1.155386(21)	1.562021(17)	2.717407	Foord et al. (1975)
r_e	1.156064(15)	1.561475(13)	2.717539	Whiffen (1980)[b]
r_e	1.15605(3)	1.56127(3)	2.71732	Whiffen (1980)[b]
r_s	1.16021 ± 0.00096	1.56014 ± 0.00100	2.72034 ± 0.00017	Costain (1958)
r_m	1.1587(13)	1.5593(10)	2.7180	Watson (1973b)
r_m^ρ	1.1551(11)	1.5621(9)	2.7172	Harmony and Taylor (1986)

[a] All values are given in Å. The quantities in parentheses represent one standard deviation in units of the last significant figure, whereas ± relates to total uncertainty or total range of distances.

[b] The two values correspond to two different potential functions used by Whiffen (1980).

with an accuracy of a few 10^{-4} Å. The r_m^ρ structure of Harmony and Taylor (1986) seems somewhat better than the pure r_m one.

Among larger linear molecules, acetylene (C_2H_2) is the only molecule for which an equilibrium structure has been derived. IR data alone give $r_e(C—H) = 1.06215 \pm 0.00017$ Å and $r_e(C\equiv C) = 1.20257 \pm 0.00009$ Å (Baldacci *et al.* 1976), whereas the same data, combined with Raman data, give 1.06250 ± 0.00010 Å and 1.20241 ± 0.00009 Å, respectively (Kostyk and Welsh 1980). These values can be compared with older ones (1.0605 and 1.2033 Å, respectively) obtained from a direct fit of an anharmonic force field to the then available spectroscopic constants (Strey and Mills 1976).

4.4.3 Symmetric top molecules

4.4.3.1 Determination of A_0 (or C_0)

As explained above, the first step in the derivation of structural parameters for symmetric top molecules is (after, of course, the determination of B_0, which is not so difficult) the determination of the axial rotational constant, called A_0 for prolate and C_0 for oblate molecules. This is not an easy task. Nemes (1984) recapitulates all the methods which have been advocated or used. The oldest one relies on the sum of $(A\zeta)_t$ quantities for all perpendicular fundamental bands, where ζ_t is the axial Coriolis constant. It is only valid in the harmonic approximation and has led to errors of 1–2 per cent in A_0. A second method was based on the fact that for a perpendicular overtone $2\nu_t^2$, the $(A\zeta)_{eff}$ value is $-2(A\zeta)_t$ but this ignores higher-order terms, resulting in relatively inaccurate values for A_0. In principle, Raman spectroscopy, since it has often different Δk selection rules, can yield the A_0 value but, in practice, its lower resolving power often limits the accuracy of A_0 (Brodersen 1979).

The real breakthrough was due to Olson (1967 and personal communication, 1972). He showed that when an accidental resonance couples two nearby rovibrational levels with different values of $k-l$, the mixing of the wavefunctions sometimes permits observation of normally forbidden lines with *apparent* selection rules different from the usual $\Delta k = 0, \pm 1$. It is therefore possible, with the help of these perturbation-allowed lines, to compute GSCDs in which A_0 is involved. Sometimes, these accidental resonances are only seen for one k value, e.g. in CH_3I (Maki and Hexter 1970; Matsuura *et al.* 1970), so that one can only actually determine a linear combination of A_0 and the centrifugal distortion constant D_K^0. But, in other cases, especially for molecules with A_0/B_0 not too far from 1, many forbidden lines are found from which an accurate value of A_0 can be extracted (see Graner 1976 or Tarrago *et al.* 1987 as examples).

Different methods, based on conceptually related ideas have also been used. We can mention the use of forbidden *rotational* transitions (Chu and

Oka 1974), the technique of avoided crossings in molecular-beam electric-resonance spectroscopy (Ozier and Meerts 1978), IR–MW double-resonance techniques (Yamamoto *et al.* 1986*a* and references therein) and combination differences between a hot band and two related 'cold' bands (Bürger *et al.* 1984; Graner *et al.* 1988; Sakai and Katayama 1988). All these methods are now able to provide good A_0 values for a large number of molecules, mainly belonging to the C_{3v} point group.

Note that even if the final aim is not the equilibrium structure but a substitution structure, the knowledge of A_0 is crucial. Duncan (1974*a*) has shown, for several molecules containing the CH_3 group, that the relations used to determine the off-axis hydrogen coordinates, in the absence of A_0, lead to the prediction of abnormally long $C-H$ bonds (by up to 0.01 Å).

4.4.3.2 Planar XY_3 molecules

The molecules we are concerned with here are SO_3 and BF_3. For SO_3, Dorney *et al.* (1973) adopted an interesting approach. From all the then available information, they built an anharmonic potential function. The inertial defect of this molecule $\Delta = I_c - 2I_b$ is zero for equilibrium but, in the ground state, it is the sum of a large vibrational correction—which can be computed from the potential function—and of a small electronic contribution, that they estimated. From the known B_0, one can obtain in this way B_e and $C_e (= B_e/2)$. From B_e, the distance $r_e(S-O)$ can be calculated as $1.418_4 \pm 0.001$ Å.

Many papers have been devoted to BF_3. As early as 1968, Ginn *et al.* obtained B_e for $^{10}BF_3$ and $^{11}BF_3$ from which they derived $r_e(B-F) = 1.307 \pm 0.002$ Å. Note that the knowledge of C_0 is not essential. More recently Yamamoto and his coworkers used an IR–MW double-resonance technique to obtain GSCDs with $\Delta K = 3$ from which C_0 is well determined (Yamamoto *et al.* 1986*a* and references therein). From B_e alone, a distance $r_e = 1.3070(1)$ Å was found (Yamamoto *et al.* 1986*b*). The constant C_e was not used but it was checked that the ground-state inertial defect agrees with the value predicted from the force field.

4.4.3.3 Pyramidal XY_3 molecules

The molecules we are concerned with here are those with X = N, P, As or Sb and Y = H or D, F, Cl, Br or I but the equilibrium structure is known only for a minority of them. Ammonia NH_3 is by far the most studied of these molecules but, because of the well-known inversion problem, no serious effort has been made to derive an equilibrium structure.

Let us consider PH_3. In the earlier works (Chu and Oka 1974; Helms and Gordy 1977), there was not enough IR data to obtain directly the equilibrium structure. However, forbidden rotational lines gave access to C_0 for PH_3 and PD_3. Chu and Oka computed the zero-point *average* structure (r_z), which gives the average position of nuclei in the ground state, by applying to B_0 and C_0 the *harmonic* part of the vibrational correction

$$B_z = B_0 + \frac{1}{2}\sum_s \alpha_s^{\mathrm{B}}(\mathrm{harm})d_s. \qquad (4.13)$$

This harmonic correction can be accurately calculated provided the harmonic potential function is well known. Note that the r_z structure is not the same for PH_3 and PD_3 (differences are approximately 0.0044 Å for bond length and 0.028° for the $H-P-H$ angle). The difference $r_z(P-H) - r_z(P-D)$ can be well predicted by a semiempirical formula depending on the reduced masses μ_{PH} and μ_{PD} (Chu and Oka 1974). Extrapolation to an infinite reduced mass is supposed to give the equilibrium structure $r_e = 1.4115_9 \pm 0.0006$ Å and $\varphi_e = 93.32_8 \pm 0.02°$ (Helms and Gordy 1977).

Later an IR study of PD_3 (Kijima and Tanaka 1981) gave all α_s for this molecule in spite of the difficulty in obtaining directly the Coriolis interaction constant ζ_{24} between ν_2 and ν_4. Thus, a reasonable value can be given for both B_e and C_e of PD_3 (but, unfortunately, not of PH_3), from which the following values are obtained: $r_e = 1.41175 \pm 0.0005$ Å, $\varphi_e = 93.421 \pm 0.06°$. These results and those of the extrapolative methods are in good agreement. In recent years, many works have been devoted to PH_3, but, as far as we know, no equilibrium structure has appeared in the literature.

4.4.3.4 X_3YZ molecules (C_{3v} point group)

Many molecules corresponding to this general formula have been studied by MW, IR and Raman spectroscopies. An overwhelming majority of these belong to the family of halogenated derivatives of methane (in an extended sense) where Y = C, Si, Ge, Sn and X and Z are different atoms taken in the list H, D, F, Cl, Br, I. A few other molecules exist, such as F_3SN, F_3PO, F_3PS ..., but no equilibrium structure is quoted for them in the literature. The molecules H_3YD and D_3YH (with Y as above) will be treated with the corresponding spherical tops YH_4.

A list of distances of several kinds found in different H_3YZ molecules can be found in a recent paper by Schneider and Thiel (1987) who compare them with values predicted through an *ab initio* MO harmonic force field. Other results concerning H_3CZ molecules can be found in several papers by Duncan and coworkers (Duncan 1970; Duncan *et al.* 1970; Duncan 1974*a*).

For X_3YZ molecules, three geometrical parameters have to be determined in the structure: the $Y-Z$ and $X-Y$ bond lengths and one angle, which can be either $X-Y-X$ or $X-Y-Z$ at choice. Since one isotopic species belonging to the C_{3v} point group yields only two rotational constants A (or C) and B, at least two isotopomers are needed. Such a molecule has six normal modes, three (ν_1, ν_2, ν_3) belonging to the A_1 type of symmetry, three (ν_4, ν_5, ν_6) belonging to the E type of symmetry. Therefore, for each isotopic species, one needs A_0, B_0, six α_s^A and six α_s^B if higher-order terms can be ignored.

Let us examine in some detail the determination of the equilibrium structure for a molecule. For methyl bromide (Graner 1981), sufficient data are available for 4 isotopomers only: $^{12}CH_3{}^{79}Br$, $^{12}CH_3{}^{81}Br$, $^{12}CD_3{}^{79}Br$, and $^{12}CD_3{}^{81}Br$. The first step is to obtain A_0. In the present case, contrary to other methyl halides, it was not possible at that time to find perturbation-allowed transitions, so that one has to rely on a combination of Raman and some IR results. A_0 is given as $5.1800 \pm 0.0010 \, cm^{-1}$ for $^{12}CH_3Br$ and $2.6004 \pm 0.0010 \, cm^{-1}$ for $^{12}CD_3Br$, independently of the bromine isotope. A careful study of all six fundamental bands, taking into account perturbations such as the Coriolis interaction between ν_2 and ν_5, gives the A_e and B_e values reported in Table 4.5. Note the relatively large uncertainties due to the poor knowledge of A_0 and also to the fact that α_3^A for $^{12}CH_3Br$ had to be estimated from the same parameter of $^{12}CD_3Br$, and that α_1^A of $^{12}CD_3Br$ is very inaccurate. At the present level of precision, A_e is essentially insensitive to the bromine isotope. Therefore, we actually have not eight but six constants from which to extract the three geometrical parameters.

Table 4.5 Equilibrium rotational constants for CH_3Br^a

Isotopomer	A_e	B_e
$^{12}CH_3{}^{79}Br$	5.24631 ± 0.00181	0.321920 ± 0.000048
	5.2471 ± 0.0011*	0.321891 ± 0.000005*
$^{12}CH_3{}^{81}Br$	5.24626 ± 0.00181	0.320688 ± 0.000049
	5.2470 ± 0.0009*	0.320667 ± 0.000005*
$^{12}CD_3{}^{79}Br$	2.62533 ± 0.00212	0.259269 ± 0.000030
$^{12}CD_3{}^{81}Br$	2.62545 ± 0.00212	0.258143 ± 0.000030

^{a} All values are in cm^{-1}. The values marked by an asterisk come from Anttila *et al.* (1983), the other ones from Graner (1981).

A first check should be made with the A_e constants. Due to the relation

$$I_a = 4m_H[r(C-H) \cdot \sin(\angle H-C-H/2)]^2 \qquad (4.14)$$

the ratio of A_e for CH_3Br and CD_3Br should be equal to the ratio of the masses $m_D/m_H = 1.99846$. We find here 1.99834 ± 0.00230 and 1.99823 ± 0.00230 for ^{79}Br and ^{81}Br respectively, a much better agreement than expected. Therefore the four A values provide only one piece of information, namely that $r(C-H) \cdot \sin(\angle H-C-H/2)$ is very close to $0.892789 \, Å$. Adopting this value, we have only two parameters to adjust, $r(C-H)$ and

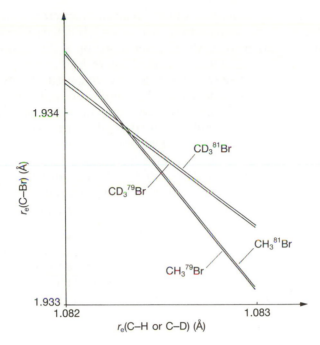

Fig. 4.1 Graphical determination of the equilibrium bond lengths of methyl bromide.

$r(C-Br)$, to reproduce four B_e values. If we plot (Fig. 4.1), for a given isotopomer, all the $r(C-H)$, $r(C-Br)$ pairs which reproduce the corresponding B_e value, we get four different curves. It is found that the curves for $^{12}CH_3^{79}Br$ and $^{12}CH_3^{81}Br$ are nearly identical, and the same for $^{12}CD_3^{79}Br$ and $^{12}CD_3^{81}Br$. The intersection gives: $r_e(C-H$ or $C-D)$ = 1.0823 Å, $r_e(C-Br)$ = 1.9340 Å, $\angle_e H-C-H$ = 111.157°, with uncertainties of the order of 2 or 3×10^{-4} Å and less than 0.1° (Graner 1981). Later, a few perturbation-allowed transitions were found in two different regions of the spectrum of CH_3Br (Lattanzi *et al.* 1982; Anttila *et al.* 1983) so that a more accurate value of A_0 is available for this molecule. We present in Table 4.5 the corresponding values of A_e and B_e given by Anttila *et al.* (1983). Finally, Sakai and Katayama (1988), by combining data from a hot band and two 'cold bands' improved A_0 drastically. They give $A_0 = 5.180632 \pm 0.000021$ and 5.180615 ± 0.000011 cm^{-1} for $^{12}CH_3^{79}Br$ and $^{12}CH_3^{81}Br$, respectively. When similar progress is made for CD_3Br, a much more accurate equilibrium structure can be extracted for methyl bromide.

Methyl fluoride CH_3F provides an interesting example for comparison of ED and spectroscopic determinations of equilibrium structure (Egawa *et al.* 1987). On one hand, ED intensity measurements were analyzed jointly with the rotational constants A_0 and B_0 obtained by spectroscopic methods. On the other hand, an earlier purely spectroscopic work by Duncan (1970), giving a surprisingly large $r_e(C-H)$, but based on $A_0 = 5.081\ \text{cm}^{-1}$, was revised with $A_0 = 5.182009\ \text{cm}^{-1}$; after the revision, this bond length became reasonable. Egawa *et al.* tried also to derive spectroscopic values of A_e and B_e for three isotopomers of methyl fluoride, completing the missing α parameters by predicted values. It is gratifying that ED and spectroscopic equilibrium structures agree well. The bond lengths differ by 0.001 Å and the angle by less than 0.1°, well within their respective estimated error.

When a careful survey is made of other X_3YZ molecules, it must be admitted that the wealth of IR and MW studies of this type of molecules has not resulted in reliable determinations of equilibrium structures. The crop now seems ripe and needs to be harvested.

4.4.3.5 Symmetric top molecules with more than 5 atoms Since increasing the number of atoms means increasing also the number of fundamental bands, it is not surprising to find that real equilibrium structures have not been published for heavier symmetric top molecules and only approximate structures are available. The next candidates seem to be $CH_3-C\equiv N$ and $CH_3-N\equiv C$ (methyl cyanide and isocyanide), propyne $CH_3-C\equiv CH$ and its isomer allene $CH_2=C=CH_2$, and in the more remote future cyclopropane and benzene.

4.4.4 Asymmetric top molecules

These molecules and especially the smallest ones have been the favorites of MW spectroscopists and, to a lesser extent, of IR spectroscopists. Efforts have also been made to correlate spectroscopic information with ED data.

4.4.4.1 XY_2 molecules (C_{2v} point group) Equilibrium structures have been published for about 16 molecules of this type. They can be described as XH_2 (with X = O, S, Se, Te but also C), X_2O (with X = F, Cl), XO_2 (with X = F, Cl, S, Se, N and O itself in ozone) or XF_2 (with X = O, S, Si, Ge). As noted in Table 4.1, there are only two geometrical parameters to be extracted for such a molecule, namely $r_e(X-Y)$ and $\varphi_e = \angle_e Y-X-Y$, whereas a single isotopomer provides three rotational constants, linked by the planarity condition. At first sight, this seems a fairly simple problem, considering that XY_2 molecules have only three fundamental modes. It is tempting, for a MW spectroscopist, to measure a few lines in the vibrational ground state and in three excited states $v_s = 1$, to deduce A, B and C constants in each of these four states, then, by a simple use of eqn (4.9), to obtain the corresponding A, B and C at equilibrium. If the inertial defect $\Delta = I_c - I_a - I_b$ is sufficiently close to zero, r_e and φ_e can be easily computed. At second sight, it is not as simple as that. First, even if low-J rota-

Table 4.6 Equilibrium structure of GeF_2 (after Takeo and Curl 1972)

Parameter	$^{74}GeF_2$	$^{72}GeF_2$	$^{70}GeF_2$
$\Delta_e (u\,Å^2)$	-0.009_4	-0.007_7	-0.006_6
r_e (Å)	1.7321_0	1.7320_9	1.7320_7
φ_e (degrees)	97.1484	97.1475	97.1478

tional lines seem to fit fairly well a rigid-rotor model, centrifugal distortion 'pollutes' the rotational constants obtained. Takeo and Curl (1972) show how to take this problem into account in the case of GeF_2. First the rotational transitions are fitted ignoring centrifugal distortions, the force constants are calculated from the resulting inertial defects and the harmonic frequencies, the centrifugal distortion constants are predicted from these force constants, and then the process is iterated. Table 4.6 shows that Takeo and Curl were quite successful in getting small inertial defects for the three isotopomers considered and that the three equilibrium structures obtained are completely compatible: apparently, the dispersion of the results is a few 10^{-5} Å for r_e and 1×10^{-3} degrees for φ_e. But this apparent accuracy might well be an illusion as shown by the example of NO_2.

For $^{14}N^{16}O_2$, successive equilibrium structures are reported in Table 4.7. Cabana *et al.* (1975) used a large set of IR bands to derive the nine α_s^X and six of the γ_{rs}^X (X = A, B or C). The A_e, B_e and C_e they derive still leave an inertial defect of $-0.0100\,u\,Å^2$, so that the equilibrium structure obtained depends somewhat on the pair of moments of inertia used. Hardwick and Brand (1976) went further by determining the anharmonic potential function within the framework of three different vibrational Hamiltonians. The most sophisticated of these is the so-called non-rigid bender Hamiltonian (the bending vibration has large amplitudes). The corresponding potential function yields $r_e = 1.19464 \pm 0.00015$ Å and $\varphi_e = 133.888 \pm 0.002°$, a tremendous

Table 4.7 Equilibrium structure of NO_2

Inertial defect (u Å²)	r_e (Å)	φ_e (degrees)	References
-0.0100 ± 0.0016	1.1945 ± 0.0005	133.85 ± 0.10	Cabana *et al.* (1975)
–	1.19464 ± 0.00015	133.888 ± 0.002	Hardwick and Brand (1976)
$-0.00216(92)$	1.19389 ± 0.00004	133.857 ± 0.003	Morino *et al.* (1983) (preferred value)
-0.00774	1.19395 ± 0.00012	133.843 ± 0.013	Morino *et al.* (1983) (second choice)
$-0.0046(24)$	$1.19455(3)$	$133.8517(17)$	Morino and Tanimoto (1984)

improvement in accuracy on the previous values. The following years saw many high-resolution I R and M W studies. Morino *et al.* (1983) combined the most recent M W data and some I R results to extract the α and γ parameters. As shown in Table 4.7, the geometrical parameters they derived are surprisingly different from the previous ones. They mention that they ignored a Coriolis interaction between levels $(v_1, v_2 + 2, v_3)$ and $(v_1, v_2, v_3 + 1)$ because when it was taken into account, the inertial defect was degraded. The following year, Morino and Tanimoto (1984) realized that the problem of Coriolis interaction was much more involved. Therefore, a new treatment was made, not using the α and γ parameters but using directly the increments of A, B, and C in seven excited vibrational states. The new equilibrium structure is reported in Table 4.7, last line: the inertial defect is not as good but the r_e value is very close to old values. This example illustrates how dangerous it is to rely heavily on M W measurements, in spite of their accuracy, since they span only a limited range of J and K values.

It is also worth mentioning here the work of Nakata *et al.* (1981) on Cl_2O and also SO_2. From the ground-state rotational constants of four isotopomers, they derive an r_c structure, which cancels small isotopic effects still present in Watson's r_m structure. They claim that in some instances, r_c might even be better than r_e because the electronic contribution in the moments of inertia is cancelled in r_c whereas it has to be taken into account in r_e. The r_c and r_e structures of these molecules are compared in Chapter 2, Table 2.9.

4.4.4.2   Bent XYZ molecules (C_s point group)   In this type are formally included the isotopomers of the previous paragraph with two different X isotopes, such as HDS or $^{35}Cl^{37}ClO$, the *equilibrium* structure of which depends only on two geometrical parameters. For the 'true' bent XYZ molecules, the structure depends on three geometrical parameters, $r(X-Y)$, $r(Y-Z)$ and $\varphi = \angle X-Y-Z$, whereas, because of the planarity condition, a single molecule provides only two independent rotational constants at equilibrium. The use of at least another isotopomer is therefore compulsory.

Among these, one can find molecules such as ONX (X = F, Cl, Br), FSN, HOX (X = F, Cl), radicals like HCO, HNO or HPO but also non-symmetric XXY molecules such as S_2O (Lindenmayer *et al.* 1986), HO_2 or FO_2. Many equilibrium structures for this type of molecules have been determined purely from M W data, especially by the group at Bologna (Degli Esposti *et al.* 1985 and references therein), and are therefore out of the scope of this chapter.

In his book on transient molecules, which also contains a few structures of diatomic radicals and of XY_2 radicals, Hirota (1985) devotes a special paragraph to XYZ bent molecules. He demonstrates that, even when not all vibration–rotation constants are available, the equilibrium structure may still be derived by a clever use of the anharmonic potential function, supplemented by a few *ab initio* MO data.

Very recently, Thiel *et al.* (1988) have used sophisticated *ab initio* MO calculations to derive anharmonic force fields for HOF and DOF (as well as OF$_2$). They used the α values given by this force field and experimental ground-state rotational constants to derive a fairly good equilibrium structure. It is to be noted that the purely *ab initio* MO structures are not as good.

The molecule HOCl provides an interesting comparison between MW and IR results. Anderson *et al.* (1986) used MW lines for six isotopomers. They determined a harmonic force field, which was used to predict the needed centrifugal distortion constants and also to compute an average (r_z) structure and, afterwards, an approximate equilibrium structure based on the changes in r_z on isotopic substitution. They obtained $r_e(O-H) = 0.9636 \pm 0.0025$ Å, $r_e(O-Cl) = 1.6908 \pm 0.0010$ Å and assumed $\varphi_e = \varphi_z = 102.45 \pm 0.42°$. The approach of Deeley (1987) consisted in measuring the three IR fundamentals of H^{16}O^{35}Cl and D^{16}O^{35}Cl and in making use of eqn (4.9) to derive A_e, B_e, and C_e. To account for the neglect of the γ parameters, an error of 1 per cent is assumed on the α values. The inertial defects found are acceptable, $\Delta_e = -0.0026 \pm 0.0040$ and -0.0022 ± 0.0049 u Å^2 for HOCl and DOCl, respectively. The resulting equilibrium structure is $r_e(O-H) = 0.9643 \pm 0.0005$ Å, $r_e(O-Cl) = 1.6891 \pm 0.0002$ Å, and $\varphi_e = 102.96 \pm 0.08°$, apparently five times more accurate than the MW results. But more studies are certainly necessary to substantiate this statement.

4.4.4.3 Tetra-atomic molecules Only a few equilibrium structures have been reported for tetra-atomic asymmetric top molecules. One of the most interesting is phosgene, COCl$_2$, that was extensively studied by Kuchitsu and coworkers (see Chapter 2, Table 2.7). This molecule is planar and has an axis of symmetry so that three geometrical parameters define its structure. In 1980, MW data for eight isotopomers and ED data were combined to extract an r_z, r_m and approximate r_e structures, assuming that the isotopic differences in the average structure can reasonably be approximated (Nakata *et al.* 1980*a*, *b*). Later, the same group undertook an IR study of the main isotopomer of phosgene, resulting in an excellent set of A_e, B_e, and C_e (inertial defect: 0.000 ± 0.020 u Å^2) and a slightly different equilibrium structure (Yamamoto *et al.* 1985). This structure is thus an almost unique combination of MW, ED and IR data. Harmony and Taylor (1986) recapitulate in their Table V all different bond lengths available for this molecule. The substitution structure is quite far from the equilibrium structure (0.008 Å on $r(C=O)$, 0.004 Å on $r(C-Cl)$, and 0.45° on the Cl$-$C$-$Cl angle, relative to the equilibrium structure given by Nakata *et al.* (1980*b*)). Their r_m^ρ structure is within 0.001 Å, 0.001 Å, and 0.1° of

Nakata's equilibrium structure but even closer (0.0001 Å, 0.0006 Å, and 0.02°, respectively) to the latest results obtained by Yamamoto *et al.* (1985).

An interesting point is noted in a paper by Davis and Gerry (1983) concerning the planar molecule HCOCl for which they derive an r_z structure, using MW data from 6 isotopomers containing H and 3 containing D. They find that if only the 6 former are used and no allowance is made for isotopic changes in r_z, a very different structure is obtained (up to 0.018 Å and 6° changes as compared to the 'true' structure), in spite of the fact that the isotopic changes themselves are very small ($\sim 10^{-4}$ Å). The potential user of average structures should therefore be very cautious! Note that a warning of this possibility was given long ago by Kuchitsu and Cyvin (1972). A full discussion of these problems can be found in a recent review by Kuchitsu *et al.* (1988).

4.4.4.4   Penta-atomic molecules   The most studied class of penta-atomic asymmetric top molecules is the family of methane and silane derivatives, especially CH_2X_2 and SiH_2X_2 (X = halogen), but for relatively few of those has the equilibrium structure been derived. We can mention a relatively old paper by Hirota (1978) on CH_2F_2 with an attempt to build an anharmonic potential function. As for CH_2Cl_2, Duncan (1987) used an empirical general harmonic force field, derived from IR and MW data, to convert A_0, B_0, and C_0 into A_z, B_z, and C_z for isotopomers. Then, following the technique already mentioned for $COCl_2$, he predicts the changes in $r_z(C-H)$ and $r_z(C-Cl)$ upon isotopic substitution, which enables him to obtain a set of average structure parameters. An approximate equilibrium structure can thereafter be derived. Similar studies, mostly based on MW data, can be found in papers by Davis and Gerry (1985) for CH_2Br_2, Davis *et al.* (1980) for SiH_2F_2, and Duncan and Munro (1987) for ketene, $CH_2=C=O$.

4.4.4.5   Larger asymmetric top molecules   Structures are more and more difficult to extract for these large molecules, and only a handful of equilibrium structures have been published. Ethylene C_2H_4 might appear as a relatively easy problem since only three geometrical parameters occur at equilibrium. However, in spite of a great deal of IR studies, and a few MW studies of partially deuterated isotopomers, there has been no attempt to derive an equilibrium structure since 1974 (Duncan 1974*b*). For other molecules, one can note a joint study of ED and MW data of 1,1-dichloroethylene $CCl_2=CH_2$ (Nakata and Kuchitsu 1982) and a comparison of IR and ED data for diborane B_2H_6 (Duncan and Harper 1984). In both cases, the equilibrium structures have a stated uncertainty of several 10^{-3} Å and even 0.01 Å for bond lengths and about 0.5° on angles. As already mentioned, for these large molecules, there is no way to escape the use of substitution structures, with their advantages but also their dangers.

4.4.5 Spherical top molecules

The molecules of interest are essentially XY_4 and XY_6 molecules for which a single bond length $r_e(X-Y)$ is enough to completely define the equilibrium structure. But since this structure is supposed to be isotopically invariant, it is natural to consider simultaneously the related symmetric tops XY_3Y' and XYY_3' and the asymmetric top XY_2Y_2' as well as $X'Y_4$, where X' and Y' are isotopes of X and Y (and similar combinations for XY_6).

In spite of the myriad of papers devoted to methane and its isotopic species studied by means of I R, Raman and even, recently, M W spectroscopies, there has been no study of its equilibrium structure since 1979. Hirota (1979) starts from the measured α values of CH_2D_2 to derive an anharmonic potential function with 13 third-order constants. This allows him to convert the ground-state rotational constants into equilibrium constants, also taking into account the 'slipping' effect of the electron (the g factor). His final value is $r_e = 1.087_0 \pm 0.0007$ Å. Gray and Robiette (1979) have a more complete approach since they include all data from $^{12}CH_4$, $^{13}CH_4$, $^{12}CD_4$, $^{12}CH_3D$, $^{12}CHD_3$, and $^{12}CH_2D_2$. They find $r_e = 1.085_8 \pm 0.001$ Å, but according to Hirota himself, Gray's result is somewhat more reliable. Since 1979, many more spectroscopic results are available, e.g. accurate A_0 and C_0 for CH_3D and CHD_3, studies of $^{13}CD_4$, etc., so that a new equilibrium structure determination seems timely. Note that earlier, Bartell and Kuchitsu (1978) derived from E D the following r_e values: $r_e(C-H) = 1.0862(12)$ Å and $r_e(C-D) = 1.0875(13)$ Å.

Ohno and his coworkers have recently studied the structures of MH_4 molecules (M = C, Si, Ge, Sn) essentially from M W but also from I R data (Ohno *et al.* 1985a, b, 1986). We present in Table 4.8 the r_e values they derive by two approximate methods. The first one uses a very simple force field with only harmonic terms plus F_{iii} relative to the M—H stretch. This yields a linear relationship between $A_0 - A_e$ (and similar quantities for B and C) and the mass of the M atom. The second method considers a linear

Table 4.8 Equilibrium distances in MH_4 molecules[a]

	Methane	Silane	Germane	Stannane
Method I[b]	1.0849 ± 0.0031	1.4734 ± 0.0010	1.5158 ± 0.0048	1.6935 ± 0.0084
Method II[b]	1.0835 ± 0.0003	1.4707 ± 0.0006	1.5143 ± 0.0006	1.6909 ± 0.0024
From IR[c]		1.4732 ± 0.0002	1.5163 ± 0.0002	
		1.47395 ± 0.00003[d]		

[a] All values are given in Å.
[b] From Ohno *et al.* (1986). See text for the two methods.
[c] Computed from the data of Bürger and Rahner (1989).
[d] Deduced from A_e of $^{28}SiH_3D$.

relationship between r_0 and the zero-point energy of molecules MH_nD_{4-n}. In a recent review, Bürger and Rahner (1989) obtained A_e directly for $^{28}SiH_4$ and $^{70}GeH_4$ and A_e and B_e for $^{28}SiH_3D$, using experimental IR values of the α parameters. This yields other values of r_e, also reported in Table 4.8.

4.4.6   Molecules with large-amplitude motions

This chapter would not be complete without an allusion to molecules with large-amplitude motions. This includes molecules with internal rotation, such as ethane, and also quasi-linear molecules, such as HCNO, HNCO and many others. The latter are fully discussed in an excellent review by Winnewisser (1985). For well-behaved molecules, the r_0 structure, which is based on ground-state averaged values, is rather close to the equilibrium structure. This is not true for molecules with a large-amplitude mode. It may happen that the molecule is linear in the equilibrium structure but already bent in the ground state and higher. For this type of molecule, a description of the potential function must therefore accompany any representation of the structure.

4.5   Conclusion

This quick survey of the structures — and especially of the equilibrium structures — deduced from spectroscopic data, shows the difficulty of the problem. To quote Lide (1974) in the preface of a very interesting book:

When one tries to interpret *differences* in a certain structural parameter — variations from one molecule to another — the question of accuracy becomes critical. The one point that cannot be disputed is that a realistic assessment of the accuracy of a structural parameter is just as important as the value of that parameter itself.

We have shown that an accuracy for equilibrium bond lengths of the order of 10^{-6} Å is an attainable goal for diatomic molecules and a handful of linear triatomic molecules. For tri-, tetra-, and maybe penta-atomic molecules, provided a serious effort is made and all perturbations are correctly taken into account, an accuracy of a few 10^{-4} Å on distances and of a few 10^{-2} degree on angles seems a not unrealistic goal. For larger molecules, the task of deriving equilibrium structures is at present too formidable. Joint studies of ED, MW and IR data might help to improve approximate structures, which cannot usually be trusted below the 0.01 Å level of accuracy for distances and maybe 0.5° for angles, although some brilliant exceptions do exist.

4.6   Acknowledgements

The author wishes to express his gratitude to his colleagues H. Bürger, J. L. Duncan, K. Kuchitsu, and B. P. van Eijck, as well as to the editor, A. Domenicano, for their helpful comments on the first version of this chapter. He is also grateful to C. Chackerian who tried to improve his English.

References

Amiot, C. (1976). *J. Mol. Spectrosc.*, **59**, 380–95.

Anderson, W. D., Gerry, M. C. L., and Davis, R. W. (1986). *J. Mol. Spectrosc.*, **115**, 117–30.

Anttila, R., Betrencourt-Stirnemann, C., and Dupré, J. (1983). *J. Mol. Spectrosc.*, **100**, 54–74.

Baldacci, A., Ghersetti, S., Hurlock, S. C., and Rao, K. N. (1976). *J. Mol. Spectrosc.*, **59**, 116–25.

Bartell, L. S. and Kuchitsu, K. (1978). *J. Chem. Phys.*, **68**, 1213–5.

Bendtsen, J. (1974). *J. Raman Spectrosc.*, **2**, 133–45.

Bernath, P. F. (1987). *J. Chem. Phys.*, **86**, 4838–42.

Bjarnov, E. and Hocking, W. H. (1978). *Z. Naturforsch.*, **33a**, 610–8.

Bogey, M., Demuynck, C., and Destombes, J. L. (1982). *J. Mol. Spectrosc.*, **95**, 35–42.

Brodersen, S. (1979). High-resolution rotation–vibrational Raman spectroscopy. In *Raman spectroscopy of gases and liquids*, Topics in Current Physics, Vol. 11 (ed. A. Weber), pp. 7–69. Springer, Berlin.

Bunker, P. R. (1972). *J. Mol. Spectrosc.*, **42**, 478–94.

Bunker, P. R. (1977). *J. Mol. Spectrosc.*, **68**, 367–71.

Bürger, H. and Rahner, A. (1989). Vibration and rotation in silane, germane, stannane and their monohalogen derivatives. In *Vibrational spectra and structure*, Vol. 18 (ed. J. R. Durig), pp. 217–370. Elsevier, Amsterdam.

Bürger, H. and Willner, H. (1988). *J. Mol. Spectrosc.*, **128**, 221–35.

Bürger, H., Schippel, G., Ruoff, A., Essig, H., and Cradock, S. (1984). *J. Mol. Spectrosc.*, **106**, 349–61.

Cabana, A., Laurin, M., Lafferty, W. J., and Sams, R. L. (1975). *Can. J. Phys.*, **53**, 1902–26.

Chedin, A. and Teffo, J.-L. (1984). *J. Mol. Spectrosc.*, **107**, 333–42.

Chu, F. Y. and Oka, T. (1974). *J. Chem. Phys.*, **60**, 4612–8.

Claverie, P. and Diner, S. (1980). *Isr. J. Chem.*, **19**, 54–81.

Costain, C. C. (1958). *J. Chem. Phys.*, **29**, 864–74.

Costain, C. C. (1966). *Trans. Am. Crystallogr. Assoc.*, **2**, 157–64.

Coxon, J. A. (1986). *J. Mol. Spectrosc.*, **117**, 361–87.

Coxon, J. A. and Ogilvie, J. F. (1982). *J. Chem. Soc. Faraday Trans. II*, **78**, 1345–62.

Davis, R. W. and Gerry, M. C. L. (1983). *J. Mol. Spectrosc.*, **97**, 117–38.

Davis, R. W. and Gerry, M. C. L. (1985). *J. Mol. Spectrosc.*, **109**, 269–82.

Davis, R. W., Robiette, A. G., and Gerry, M. C. L. (1980). *J. Mol. Spectrosc.*, **83**, 185–201.

Deeley, C. M. (1987). *J. Mol. Spectrosc.*, **122**, 481–9.

Degli Esposti, C., Cazzoli, G., and Favero, P. G. (1985). *J. Mol. Spectrosc.*, **109**, 229–38.

Dorney, A. J., Hoy, A. R., and Mills, I. M. (1973). *J. Mol. Spectrosc.*, **45**, 253–60.

Duncan, J. L. (1970). *J. Mol. Struct.*, **6**, 447–56.

Duncan, J. L. (1974*a*). *J. Mol. Struct.*, **22**, 225–35.

Duncan, J. L. (1974*b*). *Mol. Phys.*, **28**, 1177–91.

Duncan, J. L. (1987). *J. Mol. Struct.*, **158**, 169–77.

Duncan, J. L. and Harper, J. (1984). *Mol. Phys.*, **51**, 371–80.

Duncan, J. L. and Munro, B. (1987). *J. Mol. Struct.*, **161**, 311–9.

Duncan, J. L., Allan, A., and McKean, D. C. (1970). *Mol. Phys.*, **18**, 289–303.

Dunham, J. L. (1932). *Phys. Rev.*, **41**, 721–31.

Egawa, T., Yamamoto, S., Nakata, M., and Kuchitsu, K. (1987). *J. Mol. Struct.*, **156**, 213–28.

Elhanine, M., Farrenq, R., Guelachvili, G., and Morillon-Chapey, M. (1988). *J. Mol. Spectrosc.*, **129**, 240–2.

Foord, A., Smith, J. G., and Whiffen, D. H. (1975). *Mol. Phys.*, **29**, 1685–704.

Ginn, S. G. W., Kenney, J. K., and Overend, J. (1968). *J. Chem. Phys.*, **48**, 1571–9.

Graner, G. (1963). *Spectrochim. Acta*, **19**, 2113–27.

Graner, G. (1976). *Mol. Phys.*, **31**, 1833–43.

Graner, G. (1981). *J. Mol. Spectrosc.*, **90**, 394–438.

Graner, G., Rossetti, C., and Bailly, D. (1986). *Mol. Phys.*, **58**, 627–36.

Graner, G., Demaison, J., Wlodarczak, G., Anttila, R., Hillman, J. J., and Jennings, D. E. (1988). *Mol. Phys.*, **64**, 921–32.

Gray, D. L. and Robiette, A. G. (1979). *Mol. Phys.*, **37**, 1901–20.

Griggs Jr., J. L., Rao, K. N., Jones, L. H., and Potter, R. M. (1968). *J. Mol. Spectrosc.*, **25**, 34–61.

Gruebele, M. H. W., Müller, R. P., and Saykally, R. J. (1986). *J. Chem. Phys.*, **84**, 2489–96.

Hardwick, J. L. and Brand, J. C. D. (1976). *Can. J. Phys.*, **54**, 80–91.

Harmony, M. D. and Taylor, W. H. (1986). *J. Mol. Spectrosc.*, **118**, 163–73.

Harmony, M. D., Berry, R. J., and Taylor, W. H. (1988). *J. Mol. Spectrosc.*, **127**, 324–36.

Helms, D. A. and Gordy, W. (1977). *J. Mol. Spectrosc.*, **66**, 206–18.

Hirota, E. (1978). *J. Mol. Spectrosc.*, **71**, 145–59.

Hirota, E. (1979). *J. Mol. Spectrosc.*, **77**, 213–21.

Hirota, E. (1985). *High-resolution spectroscopy of transient molecules*. Springer, Berlin.

Hoeft, J. and Nair, K. P. R. (1987). *Z. Phys. D: At., Mol. Clusters*, **5**, 345–9.

Huber, K. P. and Herzberg, G. (1979). *Molecular spectra and molecular structure*, Vol. 4, *Constants of diatomic molecules*. Van Nostrand-Reinhold, New York.

Kawaguchi, K. and Hirota, E. (1985). *J. Chem. Phys.*, **83**, 1437–9.

Kawaguchi, K., Yamada, C., Hamada, Y., and Hirota, E. (1981). *J. Mol. Spectrosc.*, **86**, 136–42.

Kijima, K. and Tanaka, T. (1981). *J. Mol. Spectrosc.*, **89**, 62–75.

Kostyk, E. and Welsh, H. L. (1980). *Can. J. Phys.*, **58**, 912–20.

Kraitchman, J. (1953). *Am. J. Phys.*, **21**, 17–24.

Kuchitsu, K. and Cyvin, S. J. (1972). Representation and experimental determination of the geometry of free molecules. In *Molecular structures and vibrations* (ed. S. J. Cyvin), pp. 183–211. Elsevier, Amsterdam.

Kuchitsu, K., Nakata, M., and Yamamoto, S. (1988). Joint use of electron diffraction and high-resolution spectroscopic data for accurate determination of molecular structure. In *Stereochemical applications of gas-phase electron diffraction* (ed. I. Hargittai and M. Hargittai), Part A, pp. 227–63. VCH, New York.

Lacy, M. and Whiffen, D. H. (1981). *Mol. Phys.*, **43**, 1205–17.

Lacy, M. and Whiffen, D. H. (1982). *Mol. Phys.*, **45**, 241–52.

Lattanzi, F., Di Lauro, C., and Guelachvili, G. (1982). *Mol. Phys.*, **45**, 295–307.

Lide Jr., D. R. (1974). Preface. In *Critical evaluation of chemical and physical structural information* (ed. D. R. Lide Jr. and M. A. Paul), pp. iii–vii. National Academy of Sciences, Washington.

Lindenmayer, J., Rudolph, H. D., and Jones, H. (1986). *J. Mol. Spectrosc.*, **119**, 56–67.

Maki, A. G. and Hexter, R. M. (1970). *J. Chem. Phys.*, **53**, 453–4.

Maki, A. G. and Johnson, D. R. (1973). *J. Mol. Spectrosc.*, **47**, 226–33.

Matsuura, H., Nakagawa, T., and Overend, J. (1970). *J. Chem. Phys.*, **53**, 2540–1.

Mills, I., Cvitaš, T., Homann, K., Kallay, N., and Kuchitsu, K. (eds.) (1988). *Quantities, units and symbols in physical chemistry*. Blackwell, Oxford.

Morino, Y. and Tanimoto, M. (1984). *Can. J. Phys.*, **62**, 1315–22.

Morino, Y., Tanimoto, M., Saito, S., Hirota, E., Awata, R., and Tanaka, T. (1983). *J. Mol. Spectrosc.*, **98**, 331–48.

Nakata, M. and Kuchitsu, K. (1982). *J. Mol. Struct.*, **95**, 205–14.

Nakata, M., Kohata, K., Fukuyama, T., and Kuchitsu, K. (1980*a*). *J. Mol. Spectrosc.*, **83**, 105–17.

Nakata, M., Fukuyama, T., Kuchitsu, K., Takeo, H., and Matsumura, C. (1980*b*). *J. Mol. Spectrosc.*, **83**, 118–29.

Nakata, M., Sugie, M., Takeo, H., Matsumura, C., Fukuyama, T., and Kuchitsu, K. (1981). *J. Mol. Spectrosc.*, **86**, 241–9.

Nemes, L. (1984). Vibrational effects in spectroscopic geometries. In *Vibrational spectra and structure*, Vol. 13 (ed. J. R. Durig), pp. 161–221. Elsevier, Amsterdam.

Ogilvie, J. F. and Tipping, R. H. (1983). *Int. Rev. Phys. Chem.*, **3**, 3–38.

Ohno, K., Kawamura, M., and Matsuura, H. (1985*a*). *J. Sci. Hiroshima Univ.*, **49A**, 1–20.

Ohno, K., Matsuura, H., Endo, Y., and Hirota, E. (1985*b*). *J. Mol. Spectrosc.*, **111**, 73–82.

Ohno, K., Matsuura, H., Endo, Y., and Hirota, E. (1986). *J. Mol. Spectrosc.*, **118**, 1–17.

Olson, W. B. (1967). Presentation at the *Symposium on Molecular Structure and Spectroscopy*, Columbus, Ohio, USA. Abstracts, Paper J4.

Olson, W. B. (1972). *J. Mol. Spectrosc.*, **43**, 190–8.

Ozier, I. and Meerts, W. L. (1978). *Phys. Rev. Lett.*, **40**, 226–9.

Rosenbaum, N. H., Owrutsky, J. C., Tack, L. M., and Saykally, R. J. (1986). *J. Chem. Phys.*, **84**, 5308–13.

Saito, S., Endo, Y., Takami, M., and Hirota, E. (1983). *J. Chem. Phys.*, **78**, 116–20.

Sakai, J. and Katayama, M. (1988). *J. Mol. Struct.*, **190**, 113–23.

Schneider, W. and Thiel, W. (1987). *J. Chem. Phys.*, **86**, 923–36.

Strey, G. and Mills, I. M. (1976). *J. Mol. Spectrosc.*, **59**, 103–15.

Takeo, H. and Curl Jr., R. F. (1972). *J. Mol. Spectrosc.*, **43**, 21–30.

Tarrago, G., Delaveau, M., Fusina, L., and Guelachvili, G. (1987). *J. Mol. Spectrosc.*, **126**, 149–58.

Teffo, J.-L. and Ogilvie, J. F. (1987). Presentation at the *Tenth Colloquium on High-Resolution Molecular Spectroscopy*, Dijon, France. Abstracts, Paper D7.

Thiel, W., Scuseria, G., Schaefer III, H. F., and Allen, W. D. (1988). *J. Chem. Phys.*, **89**, 4965–75.

Todd, T. R. and Olson, W. B. (1979). *J. Mol. Spectrosc.*, **74**, 190–202.

Watson, J. K. G. (1973*a*). *J. Mol. Spectrosc.*, **45**, 99–113.

Watson, J. K. G. (1973*b*). *J. Mol. Spectrosc.*, **48**, 479–502.

Watson, J. K. G. (1980). *J. Mol. Spectrosc.*, **80**, 411–21.

Whiffen, D. H. (1980). *Mol. Phys.*, **39**, 391–405.

Wilson, E. B. (1979). *Int. J. Quantum Chem., Quantum Chem. Symp.*, **13**, 5–14.

Winnewisser, B. P. (1985). The spectra, structure and dynamics of quasi-linear molecules with four or more atoms. In *Molecular spectroscopy: Modern research*, Vol. 3 (ed. K. N. Rao), pp. 321–419. Academic Press, Orlando.

Yamamoto, S., Nakata, M., and Kuchitsu, K. (1985). *J. Mol. Spectrosc.*, **112**, 173–82.

Yamamoto, S., Kuchitsu, K., Nakanaga, T., Takeo, H., Matsumura, C., and Takami, M. (1986*a*). *J. Chem. Phys.*, **84**, 6027–33.

Yamamoto, S., Kuwabara, R., Takami, M., and Kuchitsu, K. (1986*b*). *J. Mol. Spectrosc.*, **115**, 333–52.

5

Gas-phase electron diffraction

István Hargittai

5.1 Introduction

The gas electron diffraction technique of molecular structure determination is based on the phenomenon that a beam of fast electrons is scattered by the potential from the charge distribution in the molecule. The resulting interference pattern depends on the molecular geometry, to wit the relative positions of the atomic nuclei and the intramolecular motion.

Gas electron diffraction is one of the few experimental techniques able to determine molecular geometry in the vapor phase. The other principal technique is high-resolution (microwave and modern laser) spectroscopy, see Chapters 3 and 4. There are about 150 papers published per annum on gas electron diffraction structure analysis, theory, and methodological development. The Sektion für Spektren und Strukturdokumentation at the University of Ulm keeps a record of all relevant literature, compiles and disseminates a biannual newsletter and a catalog with structural data, and publishes a virtually complete annotated bibliography (Buck *et al.* 1981; Herde *et al.* 1985). A special value of the Ulm bibliographies is that they feature theoretical and methodological papers as well as reviews, technical reports, etc., in addition to the structural papers. Early structural data have been compiled by Sutton (1958, 1965), followed by two Landolt-Börnstein volumes (Callomon *et al.* 1976, 1987) providing coverage through to 1984.

A recent comprehensive two-volume treatise by a large international group of electron diffractionists provides a detailed description of the stereo-chemical applications of gas-phase electron diffraction (I. Hargittai and M. Hargittai 1988). It includes theory and methodology, the combined application of various techniques (Part A), and gives an account of recent results for selected classes of compounds (Part B).

5.2   Early development

Two important theoretical discoveries preceded the first experimental real-ization of gas electron diffraction. One took place when Debye (1915) examined the diffraction of X rays by randomly oriented rigid systems of electrons, and found that the interference effects do not cancel in the scat-tered intensity distribution. On the contrary, they strongly influence such a distribution as a function of the scattering angle. This distribution depends on the distances between the electrons. Instead of rigid systems of electrons, rigid molecules may also be considered. Such a collection of randomly oriented rigid molecules also shows a strong interference effect in which the intensity distribution depends on the geometry of the molecule.

   The other discovery was that of de Broglie in 1924 (see, for example, Goodman 1981), concerning the wave nature of moving electrons. Ignoring relativistic corrections, the following relationship was established between the wavelength (λ), mass (m) and velocity (v) of electrons

$$\lambda = \frac{h}{mv},$$

where h is Planck's constant, or

$$\lambda = \frac{h}{(2meU)^{1/2}},$$

where U is the accelerating potential and e is the electron charge.

   The first electron diffraction experiments on crystals were reported in 1927 (see, for example, Goodman 1981) followed by a report on the first gas electron diffraction experiment by Mark and Wierl (1930) containing structural data for a series of simple molecules.

   At the beginning, interatomic distances were determined by measuring the positions of maxima and minima on the interference pattern. Soon, however, a more direct method was proposed by Pauling and Brockway (1935): the so-called radial distribution is obtained by Fourier-transforming the intensity data. The radial distribution is related to the probability density distribution of internuclear distances. The position of a maximum in the radial distribution gives the internuclear distance, while its halfwidth provides information on the vibrational amplitude of the internuclear motion.

The influence of molecular vibrations on the interference pattern (of X rays by gas molecules) was first studied in the thirties (James 1932). At the time it was not possible to carry out a quantitative evaluation of the electron scattering intensity distribution. However, the positions of maxima and minima of the molecular interference pattern could be determined surprisingly well against the steeply falling background of atomic scattering intensity, due to the exaggerating ability of the human eye. This is why this technique was called 'visual' electron diffraction.

A breakthrough was due to the introduction of the rotating sector (Finbak 1937; Debye 1939) which is a metallic disc of special shape placed in the path of the scattered electrons in order to compensate for the steeply falling background. Using the rotating sector it was then possible to elucidate quantitatively the intensity distribution for a wide range of scattering angles.

Photometers have been used for the determination of the optical density distribution of the photographic plate which recorded the interference pattern. By the early fifties a technique was developed by the Karles (I. L. Karle and J. Karle 1949, 1950; J. Karle and I. L. Karle 1950) to treat quantitatively the experimental data in order to obtain accurate geometrical and vibrational parameters. The new technique was called the 'sector-microphotometer' method.

From the middle of the fifties new experimental apparatuses have been built which can provide high-precision data. The use of fast electronic computers has also made strong impact. With increasing accuracy of structure determination, the failures of some of the theoretical approximations have become apparent. Thus, for example, attention turned to the failure of the first Born approximation, especially for molecules containing atoms with very different atomic numbers (Glauber and Schomaker 1953).

More detailed interpretation of the results and their critical comparison with those obtained by other physical techniques necessitated a closer look at the physical meaning of the geometrical parameters (Bartell 1955).

The scope of structure determinations has expanded. One of the most important early chemical applications of electron diffraction was the determination of conformational mixtures (Hassel 1943; Bastiansen 1948). In addition to stable free molecules, stable free radicals, unstable free radicals, and ions have also been studied. Extensive high-temperature electron diffraction studies have also begun. More recent developments are characterized by the combined applications of gas electron diffraction with other physical techniques as well as with theoretical calculations.

5.3   Intensity distribution of electron scattering

Some simple expressions for the distribution of elastically scattered electrons
by molecules are given here. The purpose is to show the origin of the expressions used in structure analysis and also to indicate the major approximations involved. The molecules are considered to be built from atoms acting
as independent scattering centers—this is the independent-atom model.
Furthermore, these scattering centers are assumed to possess spherical
symmetry. The electron scattering theory of such scattering centers is not
considered here. For more details of theory and further references, see
Bartell (1988) and I. Hargittai (1988).

In the independent-atom model the amplitude of the scattered wave from
the i-th atom of the molecule at a distance R from the molecule (see Fig. 5.1)
may be given to a good approximation as:

$$\psi_i'(\mathbf{R}) = KA\frac{\exp(ik_0R)}{R}f_i(\theta)\exp[i(\mathbf{s}\cdot\mathbf{r}_i)], \qquad (5.1)$$

where A is a normalizing factor, $f_i(\theta)$ is the electron scattering amplitude
of the i-th atom, $\mathbf{r}_i$ is the position vector of the i-th atom, $K = 2\pi me^2/h^2\epsilon_0$,
ϵ_0 is the vacuum permittivity, m and e are the mass and charge of the
electron, respectively. Further, $s = |\mathbf{s}| = |\mathbf{k}_0 - \mathbf{k}| = 2k_0\sin(\theta/2)$ (cf.
Fig. 5.1), $|\mathbf{k}| = |\mathbf{k}_0| = k_0 = 2\pi/\lambda$, λ is the electron wavelength, and
$s = (4\pi/\lambda)\sin(\theta/2)$.

The amplitude of the scattered wave from an N-atomic molecule may
be obtained as the sum of the amplitudes of the scattered waves by all
atoms:

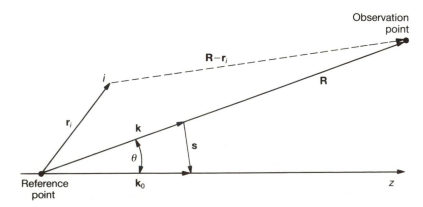

Fig. 5.1 Illustration to the description of electron scattering, see text. The incident beam is
along the z axis.

$$\psi(\mathbf{R}) = \sum_{i=1}^{N} \psi'_i(\mathbf{R}) = KA \frac{\exp(\mathrm{i}k_0 R)}{R} \sum_{i=1}^{N} f_i(\theta) \exp[\mathrm{i}(\mathbf{s} \cdot \mathbf{r}_i)]. \qquad (5.2)$$

5.3.1 Scattering by a rigid molecule

Let us first consider rigid non-vibrating molecules with a fixed orientation. The electron scattering intensity, I, may be expressed as current density,

$$I = \frac{h}{4\pi i m} [\psi'^*(\mathbf{R}) \operatorname{grad} \psi'(\mathbf{R}) - \psi'(\mathbf{R}) \operatorname{grad} \psi'^*(\mathbf{R})]. \qquad (5.3)$$

The intensity I is a function of the scattering angle θ or the s variable, and so is the scattering amplitude f. Using the above expressions we obtain:

$$I(s) = \frac{K^2 I_0}{R^2} \sum_{i=1}^{N} \sum_{j=1}^{N} f_i(s) f_j^*(s) \exp[\mathrm{i}(\mathbf{s} \cdot \mathbf{r}_{ij})], \qquad (5.4)$$

where I_0 is the intensity of the incident beam, and $\mathbf{r}_{ij} = \mathbf{r}_i - \mathbf{r}_j$. Instead of having fixed orientation, the molecules in the gas are randomly oriented, and all orientations appear with equal probability. Considering this, the following expression may be obtained for rigid molecules with random orientations:

$$I(s) = \frac{K^2 I_0}{R^2} \sum_{i=1}^{N} \sum_{j=1}^{N} f_i(s) f_j^*(s) \frac{\sin s r_{ij}}{s r_{ij}}. \qquad (5.5)$$

This expression is important for our physical picture of gas electron diffraction since it shows that the electron scattering of a gas, i.e. of randomly oriented molecules, will contain strong interference effects depending on the internuclear distances.

The intensity $I(s)$ may be considered as the sum of the molecular and atomic contributions

$$I(s) = I_{\mathrm{m}}(s) + I_{\mathrm{a}}(s), \qquad (5.6)$$

$$I_{\mathrm{m}}(s) = \frac{K^2 I_0}{R^2} \sum_{\substack{i=1 \\ i \neq j}}^{N} \sum_{j=1}^{N} f_i(s) f_j^*(s) \frac{\sin s r_{ij}}{s r_{ij}}, \qquad (5.7)$$

$$I_{\mathrm{a}}(s) = \frac{K^2 I_0}{R^2} \sum_{i=1}^{N} |f_i(s)|^2. \qquad (5.8)$$

In addition to the elastically scattered intensity considered so far there is also inelastically scattered intensity. Presuming that the inelastically scattered intensity is completely incoherent, the non-structural part of the intensity distribution is obtained:

$$I_b(s) = I_a(s) + I_{inc}(s).$$ (5.9)

This is usually referred to as the 'background'.

The electron scattering amplitudes of the atoms, $f_i(s)$, may be used in various levels of approximation. A very rough approximation would be to use the atomic number, Z_i, for $f_i(s)$, ignoring scattering by the electron distribution. The scattering amplitudes calculated by the first Born approximation are proportional to $(Z_i - F_i)$ where the second term relates to the electron scattering by the electron density distribution. These amplitudes may not be satisfactory when the molecule contains atoms with large differences in the atomic numbers. The best available scattering amplitudes are obtained by the partial wave method and may be written in the form

$$f_i(s) = |f_i(s)| \exp[i\eta_i(s)],$$ (5.10)

where $|f_i(s)|$ is the absolute value of the atomic electron scattering amplitude and $\eta_i(s)$ is its phase. Accordingly, for a *rigid* molecule,

$$I_m(s) = \frac{K^2 I_0}{R^2} \sum_{\substack{i=1 \\ i \neq j}}^{N} \sum_{j=1}^{N} |f_i(s)| \, |f_j(s)| \cos[\eta_i(s) - \eta_j(s)] \frac{\sin sr_{ij}}{sr_{ij}}.$$ (5.11)

However, to consider the molecules to be rigid is too rough an approximation in any case, since the molecular vibrations strongly influence the electron scattering intensity distribution. The deviations of internuclear distances due to molecular vibrations amount to several hundredths of an ångström, even in the ground vibrational state.

5.3.2 Scattering by a vibrating molecule

According to the definition of the $P(r)$ probability density distribution function, $P_{ij}(r) \, dr$ expresses the probability for a given atomic pair ij that its distance is between r and $r + dr$. The information gained from the gas electron diffraction experiment is closely related to this probability density distribution function. The molecular contribution to the electron scattering intensity distribution may be expressed as follows:

$$I_{\mathrm{m}}(s) = \frac{K^2 I_0}{R^2} \sum_{\substack{i=1 \\ i \neq j}}^{N} \sum_{j=1}^{N} |f_i(s)| \, |f_j(s)| \cos[\eta_i(s) - \eta_j(s)] \int_0^{\infty} P_{ij}(r) \, \frac{\sin sr}{sr} \, \mathrm{d}r.$$

(5.12)

For a diatomic molecule or an atomic pair of a polyatomic molecule, the equilibrium internuclear distance, r_e, corresponds to the position of the minimum of the potential energy function. If r is the instantaneous internuclear distance during vibration and $x = r - r_e$, the system is described by the following Schrödinger equation using $V(x)$ as the vibrational potential of the atomic pair:

$$\frac{\mathrm{d}^2 \psi(x)}{\mathrm{d}x^2} + \frac{8\pi^2 \mu}{h^2} [E - V(x)] \psi(x) = 0,$$

(5.13)

where E is the total energy of the system and μ is the reduced mass. Using the wavefunction $\psi_v(x)$ appearing as the eigenfunction of this equation, the probability distribution function for the vibrational state v is obtained:

$$P_v(x) = |\psi_v(x)|^2.$$

(5.14)

The simplest vibrating system is the harmonic oscillator for which the potential energy is $V(x) = -\frac{1}{2} fx^2$, where the force constant f is $4\pi^2 \mu c^2 \omega_e^2$, c is the speed of light, and ω_e is the vibrational wavenumber. The probability density distribution is a Gaussian function for the harmonic oscillator in the ground ($v = 0$) state:

$$P_0(x) = \frac{1}{l_\alpha \sqrt{2\pi}} \exp\left(-\frac{x^2}{2l_\alpha^2}\right),$$

(5.15)

where $l_\alpha^2 = h/8\pi^2 \mu c \omega_e$ is called the zero-point mean-square vibrational amplitude of the harmonic oscillator.

The normalized probability distribution function for thermal equilibrium at the temperature T may be obtained by taking an average over all states with the Boltzmann weight function,

$$P_T(x) = \frac{\sum_{v=0}^{\infty} |\psi_v(x)|^2 \exp(-E_v/kT)}{\sum_{v=0}^{\infty} \exp(-E_v/kT)}.$$

(5.16)

The thermal-average distribution P_T is also Gaussian:

$$P_T(x) = \frac{1}{l_h \sqrt{2\pi}} \exp\left(-\frac{x^2}{2l_h^2}\right), \tag{5.17}$$

where l_h^2 is the mean-square vibrational amplitude of the harmonic oscillator at the temperature T. The molecular intensity can then be given as

$$I_m(s) = \frac{K^2 I_0}{R^2} \sum_{\substack{i=1 \\ i \neq j}}^{N} \sum_{j=1}^{N} g_{ij}(s) \frac{1}{l_h \sqrt{2\pi}} \int_{-\infty}^{\infty} \exp\left(-\frac{x^2}{2l_h^2}\right) \frac{\sin[s(r_e + x)]}{s(r_e + x)} \, dx, \tag{5.18}$$

where $g_{ij}(s) = |f_i(s)| \, |f_j(s)| \cos[\eta_i(s) - \eta_j(s)]$, r_e is the equilibrium distance between nuclei i and j and l_h is its vibrational amplitude. By expanding $1/(r_e + x)$, and ignoring the terms after the first two, we obtain

$$I_m(s) = \frac{K^2 I_0}{R^2} \sum_{\substack{i=1 \\ i \neq j}}^{N} \sum_{j=1}^{N} g_{ij}(s) \exp\left(-\tfrac{1}{2} l_h^2 s^2\right) \frac{\sin[s(r_e - l_h^2/r_e)]}{sr_e}. \tag{5.19}$$

Equation (5.19) is the temperature-averaged probability density distribution to be used for the molecular intensity since the geometrical parameters from the gas electron diffraction experiment are average parameters corresponding to the thermal equilibrium rather than to a given vibrational state (which is, by contrast, the case in spectroscopy).

In the general case of anharmonic vibrations, the probability density distribution function corresponding to thermal equilibrium, $P_T(r)$, is a somewhat distorted Gaussian function. This may be expressed in the following way:

$$P_T(x) = \frac{A}{l_h \sqrt{2\pi}} \exp\left(-\frac{x^2}{2l_h^2}\right) \left(1 + \sum_{n=1}^{\infty} c_n x^n\right), \tag{5.20}$$

where $x = r - r_e$, A is a normalizing factor (close to unity), and the coefficients c_n depend on the temperature and the anharmonic character of the vibrations. Consequently the following expression may be obtained for the molecular intensities:

$$I_m(s) = \frac{K^2 I_0}{R^2} \sum_{\substack{i=1 \\ i \neq j}}^{N} \sum_{j=1}^{N} g_{ij}(s) \exp\left(-\tfrac{1}{2} l_m^2 s^2\right) \frac{\sin[s(r_a - \kappa s^2)]}{sr_e}. \tag{5.21}$$

There are three effective parameters in this expression: r_a, a certain average internuclear distance parameter; l_m, an effective mean vibrational amplitude which may be considered to be the same as l_h for any practical purpose; and the asymmetry constant κ which is related to the constant a of the Morse potential† to a good approximation by $\kappa = al_h^4/6$. The three kinds of parameters, r_a, l_m, and κ are the quantities determined from the electron diffraction experiment.

A final note concerns some of the approximations used in obtaining the expressions for the molecular intensities. The independent-atom model was applied for the vibrating molecules as well. This means that the atoms are supposed to retain the spherical symmetry of their electron-density distribution during vibration. The influence of chemical bonding on the electron-density distribution is felt in the small-s region of the electron scattering intensities. The complex atomic electron scattering amplitudes used in most up-to-date structure analyses take account of intraatomic multiple scattering, but ignore intramolecular multiple scattering. This problem has been treated in some selected instances. Relativistic effects, polarization of atoms by scattered electrons, and the effect of electron exchange are usually ignored. Tabulated atomic electron scattering amplitudes, both elastic and inelastic, are available.

5.4 Experiment

The general requirements for a gas electron diffraction experiment include a high-energy monochromatic electron beam focussed on the plane of registration, a minimal-optimal scattering volume of the crossing place of the electron beam with the vapor beam, and that as much of the total scattering originates from the sample as possible.

A gas electron diffraction apparatus usually consists of an electron gun, focussing system, sample injecting system, and detector. The latter is most often, though not always, a photographic plate with a rotating sector above it. The scheme of the experiment is given in Fig. 5.2. Commonly two different nozzle-to-plate distances are employed consecutively in an experiment, for example 50 cm and 19 cm. For more details of the experimental techniques, and references, see Tremmel and Hargittai (1988).

A prerequisite for a gas electron diffraction experiment is, of course, that it be possible to obtain an adequate vapor pressure of the compound to be investigated. A typical vapor pressure required in the experiment is about 10–20 torr (1–3 kPa).

The electron diffraction structure determination of high-temperature species is often hindered by the lack of knowledge of the vapor composition or by insufficient concentration of the species to be investigated. Simul-

---

† See Chapter 2, eqn (2.9).

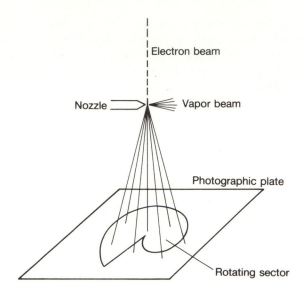

Fig. 5.2 Scheme of the experiment.

taneous mass spectrometric and electron diffraction measurements may solve such problems, as the vapor composition and the optimal experimental conditions are determined by the mass spectrometer prior to and during the diffraction experiment. Such a combined experiment has proved to be especially useful in the investigation of unstable species generated in a reactor nozzle system itself during the diffraction experiment (I. Hargittai *et al*. 1983; Vajda *et al*. 1986).

5.5   Structure analysis

In order to obtain the molecular contribution to the experimental intensity distribution, the optical density distribution of the diffraction pattern is first determined, and then, after introducing various corrections, the atomic contribution is eliminated. The latter is done by an empirical determination of the 'background'. This is illustrated for the experimental intensities of zinc halides in Fig. 5.3. The nozzle temperatures were relatively high (between 580 and 660 K) in the respective experiments and the noise level is somewhat greater than is usual for room-temperature experiments. Another tendency for worsening data quality is observed from the chloride to the iodide as the increasing atomic scattering of the halogen somewhat suppresses the molecular features of the electron scattering. The molecular contribution to the electron diffraction intensities (in short: molecular intensity), already separated from the total intensities, is shown in Fig. 5.4 for the zinc halides.

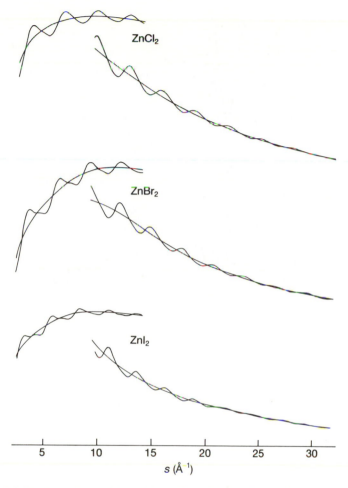

Fig. 5.3 Total experimental intensities and empirical backgrounds, zinc halides.

We have seen that the molecular intensity is in fact a sum of sinusoids and the period and damping of each sinusoid are determined by the inter-nuclear distance and vibrational amplitude, respectively. Thus for a diatomic molecule direct structural information may be gained from the molecular intensity while for a polyatomic molecule this is not possible.

Fourier transformation of the molecular intensity gives the radial distribu-tion, $f(r)$, which is related to the probability density distribution function, $P(r)$,

$$f(r) = \int_0^{s_{max}} sM(s)\exp(-as^2)\sin(sr)\,ds. \qquad (5.22)$$

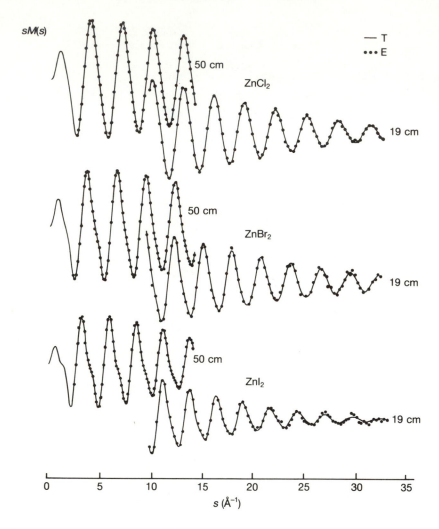

Fig. 5.4 Experimental (E) and theoretical (T) molecular intensities, zinc halides. After M. Hargittai *et al.* (1986).

Here s_{max} corresponds to the largest scattering angle experimentally available. The notation $M(s)$ is used for the molecular intensity rather than $I_m(s)$ as it may be a modified version after the background elimination. The damping factor $\exp(-as^2)$ has an effect as if the temperature of the experiment were higher than in reality and serves to decrease the error wave on the $f(r)$ curve due to the truncation of the experimental data. The coefficient a in the damping factor is chosen to make the term

$$\int_{s_{max}}^{\infty} sM(s)\exp(-as^2)\sin(sr)\,ds \qquad (5.23)$$

negligible. Theoretically the radial distribution is

$$D(r) = \int_0^{\infty} sI_m(s)\sin(sr)\,ds, \qquad (5.24)$$

and can be expressed as

$$D(r) = \frac{\pi}{2}\sum_{\substack{i=1\\i\neq j}}^{N}\sum_{j=1}^{N} c_{ij}\frac{P_{ij}(r)}{r}, \qquad (5.25)$$

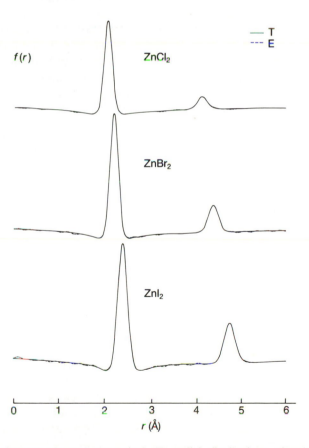

Fig. 5.5 Experimental (E) and theoretical (T) radial distributions, zinc halides. After M. Hargittai *et al.* (1986).

but it has been shown that the $f(r)$ function obtained on the basis of data from a limited s-range usually contains sufficient information to unambiguously determine the molecular geometry. In the above expression of $D(r)$ some constants are omitted and the electron scattering functions are replaced by the coefficients c_{ij}.

Note also that the experimental data never start at $s = 0$. Thus the Fourier transformation is carried out either on the data between s_{min} and s_{max}, or calculated $M(s)$ values are used for the interval between 0 and s_{min}. What is really important is that in comparisons of experimental and calculated theoretical distributions the curves must be obtained in an identical way.

The radial distribution usually yields some direct information on the molecular geometry even for polyatomic molecules. Figure 5.5 shows the radial distributions for zinc halides. Each of these distributions displays two maxima and their assignment to the zinc–halogen bonds and the halogen–halogen non-bonded interactions is straightforward. However, with increasing complexity of the molecules under investigation, the radial distributions become increasingly complex, and the assignment of their various features may become ambiguous. In such assignments, previous knowledge about the molecular structure, spectroscopic and other information on the molecular shape and symmetry are most helpful. An important feature of the $f(r)$ curve is that the area under the maximum corresponding to an interatomic distance r_{ij} is directly proportional to $n_{ij}Z_iZ_j$ (where Z_i and Z_j are the atomic numbers of the two atoms involved, and n_{ij} is the number of times this distance occurs in the molecule), and it is inversely proportional to the distance itself. All this is clearly seen in the series of three curves in Fig. 5.5. The determined parameters are listed in Table 5.1.

Often the experimental radial distribution displays direct evidence about the conformational behavior of a substance. Thus, for example, the radial distribution of acetaldazine in Fig. 5.6 shows the presence of the *anti*

Table 5.1 Molecular parameters of zinc halides[a] (after M. Hargittai *et al.* 1986)

Parameter	$ZnCl_2$	$ZnBr_2$	ZnI_2
$r_g(Zn-X)$ (Å)	2.072 ± 0.004	2.204 ± 0.005	2.401 ± 0.005
$l(Zn-X)$ (Å)	0.062 ± 0.001	0.061 ± 0.002	0.074 ± 0.002
$\kappa(Zn-X)$ (Å$^3 \times 10^6$)	$9.4 \quad \pm 1.8$	$5.5 \quad \pm 2.2$	$20.9 \quad \pm 4.9$
$r_g(X \cdots X)$ (Å)	4.115 ± 0.010	4.370 ± 0.010	4.753 ± 0.010
$l(X \cdots X)$ (Å)	0.088 ± 0.003	0.094 ± 0.003	0.109 ± 0.004
$\kappa(X \cdots X)$ (Å$^3 \times 10^5$)	$2.55 \quad \pm 2.79$	$1.66 \quad \pm 2.66$	$1.83 \quad \pm 3.35$
δ_g (Å)	0.029 ± 0.006	0.029 ± 0.006	0.049 ± 0.006

[a] $D_{\infty h}$ symmetry was assumed for equilibrium geometry. The meaning of parameters is discussed in Section 5.7.

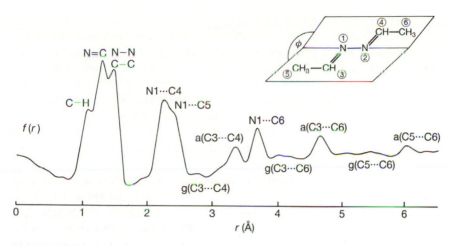

Fig. 5.6 Experimental radial distribution of acetaldazine (after I. Hargittai *et al.* 1976). The nozzle temperature was 368 K. Letters a and g refer to the *anti* and a *gauche* form, respectively, with respect to torsion about the central bond.

torsional isomer unambiguously. It prevails over a *gauche* form which is present as a minor component (I. Hargittai *et al.* 1976). In fact, there are three distinct features in the radial distribution corresponding to three non-bonded C · · · C contributions of the *anti* form. There is a slight indication of the corresponding C · · · C contributions of the *gauche* form as well. Some geometrical parameters are given in Table 5.2 together with parameters of formaldazine, $H_2C=N-N=CH_2$, from another electron diffraction study (Hagen *et al.* 1977). The N=C bond lengths are well determined in both studies and they are the same in the two molecules. The N—N bond length

Table 5.2 Some geometrical parameters (r_a) in acetaldazine and formaldazine[a]

Parameter	Acetaldazine, $CH_3-CH=N-N=CH-CH_3$[b]	Formaldazine, $CH_2=N-N=CH_2$[c]
$r(N=C)$	1.277 ± 0.003	1.277 ± 0.002
$r(N-N)$	1.437 ± 0.013	1.418 ± 0.003
$r(C-C)$	1.486 ± 0.008	–
$r(1 \cdots 4)$[d]	2.231	2.229
$r(3 \cdots 4)$[d]	3.312	3.335
$\angle C=N-N$	110.4 ± 0.9	111.4 ± 0.2

[a] Distances are given in Å, angles in degrees.
[b] I. Hargittai *et al.* (1976).
[c] Hagen *et al.* (1977).
[d] The numbering of atoms is shown in Fig. 5.6.

is poorly determined for acetaldazine as its contribution overlaps with that of the C—C bonds. There is no such C—C bond in formaldazine and its N—N bond is much better determined. Incidentally, for formaldazine it is also the *anti* form that strongly prevails over a *gauche* conformer.

The structure analysis usually starts with establishing a molecular model and comparing experimental and theoretical curves. The refinement of the structural parameters (distances and angles, vibrational amplitudes, asymmetry constants) is usually done by the least-squares technique, in most cases, though not always, based on the molecular intensities (Andersen *et al*. 1969).

The choice of the independent geometrical parameters is worth some consideration. The total number of the independent geometrical parameters is equal to the number of the totally symmetric vibrational modes. This number can be considerably reduced for a symmetrical molecule if the molecular symmetry is known. It is usually convenient to choose bond lengths, bond angles, and torsion angles as independent parameters. Thus, for example, assuming C_{2v} symmetry for five-atomic sulfuryl chloride, SO_2Cl_2, its molecular geometry is described by four parameters (M. Hargittai and I. Hargittai 1981). The radial distribution curve of sulfuryl chloride is shown in Fig. 5.7 with the assignment of internuclear distances. The following bond lengths and bond angles may be convenient to choose as independent parameters: $r(S=O)$, $r(S—Cl)$, $\angle Cl—S—Cl$, and $\angle Cl—S=O$, but any two of the non-equivalent three bond angles may be used. The determined parameters are given in Table 5.3.

Molecular motion, especially large-amplitude motion, strongly influences the electron diffraction pattern. Many kinds of motion may have to be considered, and often from their influence useful additional information may be extracted from the electron diffraction data.

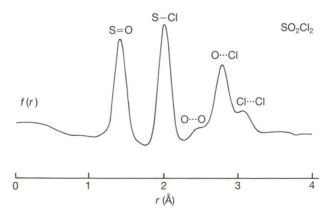

Fig. 5.7 Radial distribution of sulfuryl chloride. After M. Hargittai and I. Hargittai (1981).

Table 5.3 Molecular parameters of sulfuryl chloride, SO_2Cl_2 (after M. Hargittai and I. Hargittai 1981)

Parameter	Value[a]	Parameter	Value (Å)	
			Electron diffraction	Spectroscopic calculations[b]
$r_g(S=O)$	1.418 ± 0.003	$l(S=O)$	0.037 ± 0.001	0.035
$r_g(S-Cl)$	2.012 ± 0.004	$l(S-Cl)$	0.048 ± 0.001	0.048
$\angle_\alpha Cl-S-Cl$	100.3 ± 0.2	$l(Cl\cdots Cl)$	0.082 ± 0.003	0.085
$\angle_\alpha Cl-S=O$	108.0 ± 0.1	$l(Cl\cdots O)$	0.072 ± 0.002	0.070
$\angle_\alpha O=S=O$	123.5 ± 0.2	$l(O\cdots O)$	0.073 ± 0.007	0.064

[a] Distances are given in Å, angles in degrees.
[b] I. Hargittai and Cyvin (1969).

Let us consider, for example, the consequences of torsional motion. One of the consequences is an increase of the effective vibrational amplitudes (l_{eff}) which can be separated into framework and torsional contributions,

$$\langle l_{eff}^2 \rangle = \langle l_f^2 \rangle + \langle l_t^2 \rangle. \tag{5.26}$$

As the framework amplitudes can be calculated from spectroscopic data, the magnitude of the torsional amplitudes characterizes the barrier to torsion. The method of Karle (1981) for the separation of framework and torsional contributions and, accordingly, for the determination of the barrier height implies two assumptions. One is that the shape of the potential function of rotation is known, and the other is that the framework contributions are invariant to the torsion. Another consequence of torsional motion may be the appearance of a distorted average conformation instead of a more symmetric equilibrium conformation. Again, the amount of distortion may be used to estimate the barrier height (Vilkov *et al.* 1978).

It is also possible to utilize the spectroscopically computed framework vibrational amplitudes, l_f, as a function of the torsional angle, τ, in estimating the barrier height (Brunvoll *et al.* 1985). The probability density distribution for a torsion-dependent distance is then expressed as

$$P(r) = k \int_{-\pi/2}^{\pi/2} P(\tau) \frac{1}{l_f(\tau)\sqrt{2\pi}} \exp\left\{ -\frac{[r - r_{ij}(\tau)]^2}{2[l_f(\tau)]^2} \right\} d\tau, \tag{5.27}$$

where k is a normalizing constant. From this $P(r)$, theoretical intensities can be calculated for each torsion-dependent distance at different barrier values. The determination of the barrier thus becomes part of the least-

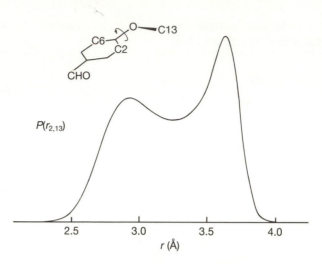

Fig. 5.8 *p*-Anisaldehyde: theoretical probability density distribution of a torsion-dependent distance, C2 · · · C13, assuming free rotation. After Brunvoll *et al.* (1985).

squares refinement. The form of the potential energy function is assumed to be known. Thus, for example, for *p*-anisaldehyde, *p*-CH$_3$O—C$_6$H$_4$—CHO, the assumption was $V = (V_0/2)(1 - \cos 2\tau)$, where V_0 is the barrier to methoxy rotation from the coplanar arrangement. Using the above described procedure the barrier was estimated to be between 3.5 and 12 kJ mol^{-1}. While this is a relatively low barrier, free rotation could be excluded unambiguously. Nonetheless, it may be of interest to show the probability density distribution of a torsion-dependent distance for the free-rotation model (Fig. 5.8). The particular shape of this distribution reflects the changing framework contribution during torsion. This distribution could be modeled by a series of contributions corresponding to an arbitrarily chosen number of steps dividing the range of 0° to 180°.

As the least-squares technique is applied for non-linear relationships in the electron diffraction analysis, it is important that the set of starting parameters be not very far from their real values. Much care must be taken in selecting the initial structures. All available information on the molecular structure has to be considered. It is often desirable to repeat the refinements from different initial structures. The goodness of fit is usually characterized by a so-called R factor,

$$R = \left\{ \frac{\sum_{k=1}^{n} [s_k M_k^{E}(s_k) - s_k M_k^{T}(s_k)]^2 W_k}{\sum_{k=1}^{n} [s_k M_k^{E}(s_k)]^2 W_k} \right\}^{1/2}, \qquad (5.28)$$

where the observations, i.e. molecular intensities, run from $k = 1$ to $k = n$, and the weights of the observations are labelled W_k.

It may happen that more than one model is able to reproduce the experimental distribution equally well. All such models are solutions in a mathematical sense and, accordingly, a decision between them cannot be made on the basis of the electron diffraction data alone. Multiple solutions often occur when there is strong correlation among the parameters in the least-squares refinement.

5.6 Uncertainties

The determination of internuclear distances is possible with a few thousandths of an ångström accuracy in modern gas-phase electron diffraction structure analysis. However, there are many sources of error which may play very different roles in various problems; one may have a major contribution in one kind of study and be negligible in another as compared with the other error sources. It is of great importance to enumerate all errors taken into account in the error estimation when reporting the results of an analysis. With some simplification, we may group all error sources into four classes, viz. theory, experiment, analysis, and physical meaning.

As regards electron scattering theory, Bartell (1985) noted:

If incident electron waves were in reality scattered very weakly by well-separated, spherical scattering centers executing harmonic vibrations, ... today's analysis ... would yield accuracies approaching 0.0001 Å.

However, insufficiencies in up-to-date electron scattering theory hardly influence the accuracy of structure determinations at the current level, and are generally ignored in error estimation, except in some special investigations.

Experimental errors are ignored by some and included by others in the error estimation. The most important experimental error directly influencing the determination of internuclear distances is the experimental systematic error in the determination of the angular variable s, also called scale error (σ_s). It can be excluded by a parallel experiment on a substance with well-determined geometry and corresponding calibration of the experimental scale. Such substance may, e.g., be carbon dioxide or benzene. Typical experimental scale errors may be, say, 0.1 or even 0.2 per cent.

The analysis itself is the origin of various kinds of error. One kind is directly related to the quality of data and correlation among the parameters themselves, and appears as the standard deviation (σ) in the least-squares refinement. The problem of correlation among the measurements is part of the experimental error and can be automatically included in the determination of the least-squares standard deviations, provided a full weight matrix is used. When, however, the off-diagonal elements of the weight matrix are ignored, the standard deviation should be increased in order to compensate for this. Various laboratories determine the corresponding factors (q)

according to their experimental conditions and conditions of data treatment in the range of one to three. For some parameters the standard deviation may be unrealistically low in any case, say, in the order of 0.0001 Å while for some other parameters it may be so high as to render their determination seemingly meaningless. The latter is often the case when strongly correlated parameters, including vibrational amplitudes, are refined simultaneously. Thus reasonable assumptions of some parameters, either known from other sources or having very little relative contribution, may greatly facilitate the determination of the remaining parameters. Insight, intuition, and accumulated experience may play as important a role here as any rigorous criteria. However, all reasonable attempts must be made to keep the arbitrariness of this procedure at a minimum. First of all, incorporation of reliable pieces of information from other sources is preferable to arbitrary assumptions. Secondly, ranges of values rather than single ones are useful to test assumed parameters, and the influence of these assumptions on the determined parameters has to be carefully examined. Finally, the possible consequences of such assumptions can be included into the error estimation by adding a component to the other contributions, reflecting the changes in the parameters (Δp_i) under the impact of various choices of the assumptions.

Taking all the above into account, the estimation of the total experimental uncertainty, σ_t for parameter p_i, may be made according to the following expression,

$$\sigma_t(p_i) = \{[q\sigma(p_i)]^2 + (\sigma_s p_i)^2 + (\Delta p_i)^2\}^{1/2}. \tag{5.29}$$

A fourth error source was mentioned above, relating to the physical meaning of the determined parameters. It is especially important when the molecules under investigation perform large-amplitude motion and also when data from various physical techniques are employed in a combined analysis. Both aspects will get some exposure in the following two sections.

5.7 Average structures†

The r_a distance in the expression for molecular intensities, eqn (5.21), is an operational distance parameter. It gives an effective internuclear distance and is closely related to the thermal average internuclear distance, r_g, referring to temperature T. The following expressions are valid to a good approximation:

$$r_g = r_a + l^2/r_a, \tag{5.30}$$

---

† See also Chapter 2, Section 2.4.1.

$$r_g = r_e + \langle \Delta z \rangle_T + (\langle \Delta x^2 \rangle_T + \langle \Delta y^2 \rangle_T)/2r_e. \tag{5.31}$$

Equation (5.30) relates r_g to r_a which is obtained directly in the refinement. Equation (5.31) relates r_g to the equilibrium distance, r_e, which is the distance between the positions of atomic nuclei in the minimum of the potential energy.

In obtaining this expression for atomic pair ij, the two nuclei are located along the z axis of a Cartesian coordinate system, with one of the two nuclei in its origin. The instantaneous position of the other nucleus will be different from its equilibrium position by displacements Δx, Δy, and Δz. Higher-order terms, which are negligible compared to experimental errors, are ignored. For completeness, centrifugal stretching, δr, should be added to the right-hand side of eqn (5.31).

The following expressions are obtained for the internuclear distance between average nuclear positions, again ignoring terms which are negligible as compared with experimental errors. The distance between average nuclear positions in thermal equilibrium at temperature T is

$$r_\alpha = r_e + \langle \Delta z \rangle_T. \tag{5.32}$$

For the ground vibrational state,

$$r_\alpha^0 = r_e + \langle \Delta z \rangle_0. \tag{5.33}$$

The most unambiguous representation for an internuclear distance is the equilibrium distance which also has the same physical meaning as the distance obtained by quantum chemical calculations. However, it is generally not attainable directly from electron diffraction data alone. The average internuclear distance, r_g, is the most convenient representation for a chemical bond, providing its average length at a given temperature. However, it has no such descriptive geometrical meaning for distances between non-bonded atoms because of the consequences of perpendicular vibrations (cf. the term $(\langle \Delta x^2 \rangle_T + \langle \Delta y^2 \rangle_T)/2r_e$ in eqn (5.31)). The distances between average nuclear positions are free from these effects and the r_α^0 representation is the most suitable for expressing bond angles. On the other hand, it does not correspond to a real average bond distance. A common and useful way to describe molecular geometry from electron diffraction result is by use of r_g bond lengths and r_α bond angles. For relatively rigid molecules, the differences between various representations are usually smaller than the experimental uncertainties. Some examples are listed in Table 5.4. However, for molecules with large-amplitude motion, or with higher accuracy even for relatively rigid molecules, the question of representation, and the physical meaning behind it, gain importance.

Table 5.4 Internuclear distances and vibrational corrections of thionyl chloride, $SOCl_2$ (after I. Hargittai 1974)

(*a*) Internuclear distances[a] (Å)

	r_a	r_g	r_α
S=O	1.442 ± 0.005	1.443	1.442
S—Cl	2.075 ± 0.006	2.077	2.076
Cl · · · O	2.840 ± 0.012	2.842	2.842
Cl · · · Cl	3.087 ± 0.017	3.090	3.090

(*b*) Vibrational corrections (Å $\times 10^4$)

	l^2/r				$(\langle \Delta x^2 \rangle_T + \langle \Delta y^2 \rangle_T)/2r$			
T (K)	0	298	323	373	0	298	323	373
S=O	9	9	9	9	9	14	14	16
S—Cl	14	13	14	15	2	4	4	4
Cl · · · O	14	21	22	25	1	2	2	3
Cl · · · Cl	14	30	32	36	0	0	1	1

[a] From room-temperature experiment (I. Hargittai 1969).

For a symmetric linear triatomic molecule, AB_2, the sum of the two bond lengths, $2r(A—B)$, is rigorously equal to the non-bonded distance $B \cdot \cdot \cdot B$. This is true, however, only for the equilibrium distances. Bending vibrations will make the average distance $r_g(B \cdot \cdot \cdot B)$ shorter than $2r_g(A—B)$, by a difference δ_g which is called shrinkage. Illustration of the origin of shrinkage for a linear AB_2 molecule is given in Fig. 5.9. For zinc

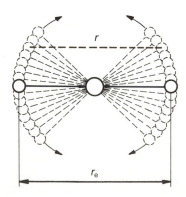

Fig. 5.9 Consequences of bending vibrations of a linear symmetric triatomic molecule AB_2; r_e is the equilibrium $B \cdot \cdot \cdot B$ distance and r is the instantaneous distance.

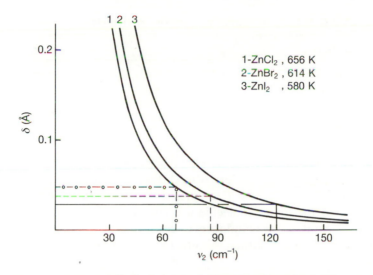

Fig. 5.10 Calculated shrinkage (δ) versus the bending vibrational frequency (ν_2) of zinc halides. The experimental shrinkages and the corresponding estimated ν_2 values are indicated. After M. Hargittai *et al.* (1986).

halides (cf. Fig. 5.5 for their radial distributions) this shrinkage amounts to 0.03 to 0.05 Å and makes the apparent bond angle much smaller (by, say, 15°) than the 180° corresponding to the linear configuration. Independent information on the linearity of these molecules makes it possible to estimate their bending vibrational frequency on the basis of the experimentally determined shrinkage, thereby turning this problem into an advantage. This is illustrated for zinc halides in Fig. 5.10. The relevant parameters of zinc halides are collected in Table 5.1.

Conversion of the r_g into r_α parameters means indeed the introduction of shrinkage corrections and can be done by calculating the terms $(\langle \Delta x^2 \rangle_T + \langle \Delta y^2 \rangle_T)/2r_e$ in eqn (5.31). The temperature-dependence of structural parameters as determined from electron diffraction was clearly stated in the definition of the distances r_g and r_α referring to thermal equilibrium

Table 5.5 Molecular parameters of I_2 at three temperatures (after Ukaji and Kuchitsu 1966)

T (K)	353	573	773
r_g (Å)	2.672 ± 0.003	2.679 ± 0.002	2.687 ± 0.002
l_m (Å)	0.061 ± 0.005	0.070 ± 0.004	0.077 ± 0.005
κ (Å^3 × 10^6)	3.9 ± 8.4	10.0 ± 11.2	13.0 ± 10.2

Table 5.6 Vibrational effects on internuclear distances

Effects	r_g	r_α	r_α^0	r_e
Shrinkage	+	−	−	−
Temperature	+	+	−	−
Isotope	+	+	+	−

at temperature T. The average parallel displacement, $\langle \Delta z \rangle$, the mean amplitudes of vibration, l, and the centrifugal distortion, δr, are especially sensitive to temperature changes. Examples of the temperature-dependence of parameters of the iodine molecule are collected in Table 5.5 (Ukaji and Kuchitsu 1966). Relative abundances of conformers is another kind of highly temperature-dependent structural information obtainable from electron diffraction.

Whereas representations r_g and r_α are temperature-dependent, r_α^0 is not. However, even this distance between average nuclear positions for the ground vibrational state contains the consequences of anharmonicity in molecular vibrations, as is seen in eqn (5.33). This effect is seen for bonds differing in isotopic composition only, due to mass-dependent vibrational effects. It is only the equilibrium distance, r_e, which is totally free from all of the vibrational effects enumerated above (cf. Table 5.6).

5.8 Interaction with other techniques

The presentation of structural parameters in terms of r_α^0, i.e. distances between average nuclear positions in the ground vibrational state, has special importance. It is attainable by both principal techniques of gas-phase molecular structure determination, i.e. electron diffraction and high-resolution spectroscopy. This representation is labelled r_z in the microwave spectroscopy literature but has exactly the same physical meaning. If the electron diffraction structural parameters are converted into r_α^0 distances and the rotational constants from the microwave spectra (A_0, B_0, C_0: see Chapter 3) into A_z, B_z, C_z rotational constants, referring to the distances between average nuclear positions in the ground vibrational state, then the measurements from the two techniques can be used in a combined way in the structure refinement. The conversion can be carried out by means of a harmonic force field and normal-coordinate analysis (see Kuchitsu *et al.* 1988). Note that while the conversion is done on the geometrical parameters for electron diffraction, it is done on the rotational constants for microwave spectroscopy. In a combined refinement the quantity to be minimized may be, for example,

$$\sum_{k=1}^{n} [s_k M_k^E(s_k) - s_k M_k^T(s_k)]^2 W_k + \sum_{j=a,b,c} (I_j^E - I_j^T)^2 W_j, \qquad (5.34)$$

where the first term is identical with the numerator in eqn (5.28) and I_a, I_b, and I_c are the principal moments of inertia calculated from the A_z, B_z, C_z rotational constants, respectively. The choice of the relative weights W_k and W_j is a delicate matter for which no general formula is available. Depending on the data and the structural problem under investigation, various schemes have to be tested. Whereas the electron diffraction intensity measurements may number over a hundred, the introduction of three, or sometimes even fewer, pieces of data from high-resolution spectroscopy may eliminate the ambiguity of a structure determination and result in noticeable enhancement of accuracy. This is because the pattern of correlation among the parameters may be strikingly different for the two sets of observables. Thus, for example, for sulfone molecules, such as SO_2Cl_2 and SO_2F_2, the tetrahedral sulfur bond configuration results in somewhat closely spaced non-bonded distances, which hinders their accurate determination from electron diffraction, while their bond lengths can be determined with high accuracy. On the other hand, the distances $O \cdots O$ as well as $Cl \cdots Cl$ for SO_2Cl_2, or $F \cdots F$ for SO_2F_2 can be determined directly from the principal moments of inertia with no similar information on the bond lengths of these molecules from the microwave spectra. Even without vibrational corrections, the utilization of complementary information from the two techniques has great importance. In such a case, however, the final refinements should be done by removing data from the other technique lest systematic errors be introduced by the uncorrected measurements.

The above example illustrates the point that a general scheme for combined application of different techniques could hardly be useful. Indeed, a very broad range of approaches may be followed when incorporating additional information in an electron diffraction structure analysis. The procedure may range from the use of some chemical experience in constructing models in the initial stages of the analysis, to built-in calculated vibrational amplitudes from normal-coordinate analysis in the least-squares refinements, to assuming differences between the lengths of similar bonds from quantum chemical calculations. The following listing gives an impression of some possibilities in the combined applications of selected techniques from the point of view of electron diffraction, covering experiment, analysis and the interpretation of results.

Mass spectrometry (cf. M. Hargittai *et al.* 1979 and I. Hargittai *et al.* 1980)
Vapor composition
Optimization of experimental conditions
Combined experiment
Relative bond strengths

Vibrational spectroscopy (cf. Cyvin 1972)
  Vapor composition
  Molecular symmetry
  Normal-coordinate analysis
  Geometry
  Mean vibrational amplitudes
  Force field
X-ray crystallography (see, for example, Chapters 18 and 19; M. Hargittai and I. Hargittai 1987)
  Geometry
  Consequence of intermolecular interactions
High-resolution spectroscopy (cf. Chapters 2 and 3; Kuchitsu and Cyvin 1972; Kuchitsu 1981; Kuchitsu *et al.* 1988)
  Geometry
  Rotational constants
  Combined analysis of electron scattering intensities and rotational
    constants
Liquid-crystal NMR spectroscopy (cf. Chapter 12 and Rankin 1988)
  Geometry, complementary data
Quantum chemical calculations (cf. Chapter 13; Boggs 1988; Schäfer *et al.* 1988)
  Geometry
  Small structural differences
  Correlation of structure and bonding
Molecular mechanics calculations (cf. Chapter 14 and Geise and Pyckhout 1988)
  Relative stability of conformers

It should also be mentioned that the input of electron diffraction data into other techniques has gradually increased during recent years. Suffice it to mention the extended versions of molecular intensity expressions used to provide a considerable amount of information on molecular vibrations and force fields from the diffraction data (see, for example, Spiridonov 1988).

5.9 Perspectives

Some past failures of gas-phase electron diffraction in structure elucidation have resulted from overextension of its real capabilities. The combined application of various techniques is probably the best safeguard against misinterpretation of the data. The progress of gas-phase electron diffraction has not only made its application more reliable but has also consolidated the expanded scope of its applicability. The fruits of this recent progress are summarized in the following passage quoted from Jerome Karle (1988):

... As a result of the dedicated efforts in a relatively small number of laboratories, gas electron diffraction has served as a valuable tool in the investigation of molecular structure. Much information has been obtained concerning molecular configuration, bond distances and angles, internal motion (including hindered internal rotation and barrier heights), preferred orientation in conformers, and conjugation and aromaticity. Investigations have also concerned mixtures in equilibrium, including evaluations of thermodynamic quantities, free radicals, a wealth of high-temperature studies, clusters, isotope effects, and the joint use of other techniques such as laser excitation, microwave spectroscopy and mass spectrometry. We thus have the view of gas electron diffraction as a technique of wide application to many aspects of molecular structure, and when it is combined with various spectroscopic techniques, the value of each may be considerably enhanced. ...

These words eloquently express how far gas-phase electron diffraction has moved from its beginnings when its single, though fundamental, importance was to be a source of interatomic distances.

Herman Mark, the initiator of the technique, recently looked back to the pioneering days and stated the following (1988; italics by H. M.):

This method permits the direct experimental determination of the distance between atoms which are joined together by a covalent bond. In the late 1920's such 'interatomic distances' were of considerable interest to arrive at *quantitative* models of organic molecules; ...

The potential information from electron diffraction has recently been gradually upgraded for the other physical techniques as well. Its complementary nature for the other techniques and its uniqueness in general is widely recognized.

Structural chemistry itself is undergoing some transition in that emphasis is being shifted from ground-state, stable structures towards structural variations in chemical processes. Gas-phase electron diffraction is one of the few experimental tools applicable *to the following of structural changes accompanying chemical changes*. By mentioning this, it is also implied that the meaning of chemical changes has recently broadened and goes beyond the formation and destruction of a compound. All processes involving changes in reactivity belong to chemical changes. Thus, for example, transition of a molecule from one stable conformer into another stable conformer is a process in which the reactivity of the molecule changes. Incidentally, these two torsional isomers are not separable based on common physical properties, although they may differ considerably in other physical properties, such as electric dipole moment. During a conformational transition, there are changes in the bond configurations in addition to the generally anticipated torsional angle changes (Scharfenberg and Hargittai 1984; Wiberg and Murcko 1987). Another example occurs when molecules enter, or leave, crystal structures, a process which is not merely a physical change since there are conformational changes and even changes in the bond configurations. The greater these changes and the more sensitive the analysis, the more of these changes can be detected.

In gas/crystal comparisons, the conformational changes are the largest in magnitude because they require the least amount of energy. Changes in the bond angles and bond lengths are usually relatively very small because of the large energy requirements for such changes. Conversely, however, even small angle and bond length changes are important energetically. The changes in their entirety then are a compromise among all kinds of change and it may be a crude approximation to consider the very visible conformational changes only, and ignore the rest.

The accuracy of the determination of structural changes accompanying chemical changes depends very much on the particular problem under investigation. Consider, for example, the geometries of two conformers, which have probably both been determined in the same phase, by the same technique, and probably even in the same experiment. In this case the *differences* will be largely independent of operational effects and most experimental errors, and the accuracy of their determination may be considerably higher than the accuracy of the individual parameters.

The determination of gas/crystal structural changes is an entirely different case. All possible operational effects may enter the picture. Most probably, as different physical techniques are used with different matter/radiation interaction, the averaging over molecular motion will be different, and the error estimation will be different. Great care must be exercised in eliminating all possible consequences of operational effects before anything can be said about genuine changes in molecular structure accompanying the molecule entering, or leaving, a crystal.

Further structural changes accompanying chemical changes can be investigated in bond formation, e.g. for coordination molecules, and, most significantly, in the process of molecules getting ready to enter a reaction, or just emerging, even as intermediate products, from a reaction. Electron diffraction should be employed in very close interaction with other techniques, such as X-ray crystallography and quantum chemical calculations, in such research.

References

Andersen, B., Seip, H. M., Strand, T. G., and Stølevik, R. (1969). *Acta Chem. Scand.*, **23**, 3224–34.

Bartell, L. S. (1955). *J. Chem. Phys.*, **23**, 1219–22.

Bartell, L. S. (1985). *J. Mol. Struct.*, **126**, 331–44.

Bartell, L. S. (1988). Status of electron scattering theory with respect to accuracy in structure analyses. In: I. Hargittai and M. Hargittai (1988), Part A, pp. 55–83.

Bastiansen, O. (1948). Om noen av de forhold som hindrer den fri dreibarhet om en enkeltbinding. Dissertation, University of Trondheim. Garnaes' boktrykkeri, Bergen.

Boggs, J. E. (1988). Interaction of theoretical chemistry with gas-phase electron diffraction. In: I. Hargittai and M. Hargittai (1988), Part B, pp. 455-75.

Brunvoll, J., Bohn, R. K., and Hargittai, I. (1985). *J. Mol. Struct.*, **129**, 81-91.

Buck, I., Maier, E., Mutter, R., Seiter, U., Spreter, C., Starck, B., Hargittai, I., Kennard, O., Watson, D. G., Lohr, A., Pirzadeh, T., Schirdewahn, H. G., and Majer, Z. (1981). *Bibliography of gas phase electron diffraction 1930-1979.* Physics Data, No. 21-1. Fachinformationszentrum, Karlsruhe.

Callomon, J. H., Hirota, E., Kuchitsu, K., Lafferty, W. J., Maki, A. G., and Pote, C. S. (1976). *Structure data of free polyatomic molecules*, Landolt-Börnstein, Numerical data and functional relationships in science and technology (New series), Group II, Vol. 7. Springer, Berlin.

Callomon, J. H., Hirota, E., Iijima, T., Kuchitsu, K., and Lafferty, W. J. (1987). *Structure data of free polyatomic molecules*, Landolt-Börnstein, Numerical data and functional relationships in science and technology (New series), Group II, Vol. 15. Springer, Berlin.

Cyvin, S. J. (ed.) (1972). *Molecular structures and vibrations.* Elsevier, Amsterdam.

Debye, P. (1915). *Ann. Phys.*, **46**, 809-23.

Debye, P. P. (1939). *Phys. Z.*, **40**, 66 and 404-6.

Finbak, C. (1937). *Avhandl. Norske Videnskaps-Akad. Oslo, I, Mat.-Naturv. Klasse*, No. 13, pp. 3-20.

Geise, H. J. and Pyckhout, W. (1988). Self-consistent molecular models from a combination of electron diffraction, microwave, and infrared data together with high-quality theoretical calculations. In: I. Hargittai and M. Hargittai (1988), Part A, pp. 321-46.

Glauber, R. and Schomaker, V. (1953). *Phys. Rev.*, **89**, 667-71.

Goodman, P. (ed.) (1981). *Fifty years of electron diffraction.* Reidel, Dordrecht.

Hagen, K., Bondybey, V., and Hedberg, K. (1977). *J. Am. Chem. Soc.*, **99**, 1365-8.

Hargittai, I. (1969). *Acta Chim. Acad. Sci. Hung.*, **60**, 231-6.

Hargittai, I. (1974). Az elektrondiffrakciós atomtávolság. In *A kémia újabb eredményei*, Vol. 21 (ed. B. Csákvári), pp. 7-173. Akadémiai Kiadó, Budapest.

Hargittai, I. (1988). A survey: The gas-phase electron diffraction technique of molecular structure determination. In: I. Hargittai and M. Hargittai (1988), Part A, pp. 1-54.

Hargittai, I. and Cyvin, S. J. (1969). *Acta Chim. Acad. Sci. Hung.*, **61**, 51-6.

Hargittai, I. and Hargittai, M. (eds.) (1988). *Stereochemical applications of gas-phase electron diffraction.* Part A, *The electron diffraction technique.* Part B, *Structural information for selected classes of compounds.* VCH, New York.

Hargittai, I., Schultz, Gy., Naumov, V. A., and Kitaev, Yu. P. (1976). *Acta Chim. Acad. Sci. Hung.*, **90**, 165-78.

Hargittai, I., Bohátka, S., Tremmel, J., and Berecz, I. (1980). *Hung. Sci. Instrum.*, **50**, 51-6.

Hargittai, I., Schultz, Gy., Tremmel, J., Kagramanov, N. D., Maltsev, A. K., and Nefedov, O. M. (1983). *J. Am. Chem. Soc.*, **105**, 2895-6.

Hargittai, M. and Hargittai, I. (1981). *J. Mol. Struct.*, **73**, 253-5.

Hargittai, M. and Hargittai, I. (1987). *Phys. Chem. Minerals*, **14**, 413-25.

Hargittai, M., Tamás, J., Bihari, M., and Hargittai, I. (1979). *Acta Chim. Acad. Sci. Hung.*, **99**, 127–35.

Hargittai, M., Tremmel, J., and Hargittai, I. (1986). *Inorg. Chem.*, **25**, 3163–6.

Hassel, O. (1943). *Tidsskr. Kjemi, Bergvesen, Metallurgi*, **3**, 32–4. English translation (1970): *Kjemi*, **30**(5A), 25–7.

Herde, E., Maier, E., Mez-Starck, B., Mutter, R., Seiter, U., Spreter, C., Hargittai, I., Watson, D. G., Lohr, A., and Selz, G. J. (1985). *Bibliography of gas phase electron diffraction, Supplement 1980–82*. Physics Data, No. 21–2. Fachinformationszentrum, Karlsruhe.

James, R. W. (1932). *Phys. Z.*, **33**, 737–54.

Karle, I. L. and Karle, J. (1949). *J. Chem. Phys.*, **17**, 1052–8.

Karle, I. L. and Karle, J. (1950). *J. Chem. Phys.*, **18**, 963–71.

Karle, J. (1981). Internal rotation and electron diffraction. In *Diffraction studies on non-crystalline substances* (ed. I. Hargittai and W. J. Orville-Thomas), pp. 243–67. Elsevier, Amsterdam, and Akadémiai Kiadó, Budapest.

Karle, J. (1988). Foreword. In: I. Hargittai and M. Hargittai (1988), Part A, pp. ix–xii.

Karle, J. and Karle, I. L. (1950). *J. Chem. Phys.*, **18**, 957–62.

Kuchitsu, K. (1981). Geometrical parameters of free molecules: Their definitions and determination by gas electron diffraction. In *Diffraction studies on non-crystalline substances* (ed. I. Hargittai and W. J. Orville-Thomas), pp. 63–116. Elsevier, Amsterdam, and Akadémiai Kiadó, Budapest.

Kuchitsu, K. and Cyvin, S. J. (1972). Representation and experimental determination of the geometry of free molecules. In: Cyvin (1972), pp. 183–211.

Kuchitsu, K., Nakata, M., and Yamamoto, S. (1988). Joint use of electron diffraction and high-resolution spectroscopic data for accurate determination of molecular structure. In: I. Hargittai and M. Hargittai (1988), Part A, pp. 227–63.

Mark, H. F. (1988). Introduction. In: I. Hargittai and M. Hargittai (1988), Part A, pp. xv–xviii.

Mark, H. and Wierl, R. (1930). *Z. Phys.*, **60**, 741–53.

Pauling, L. and Brockway, L. O. (1935). *J. Am. Chem. Soc.*, **57**, 2684–92.

Rankin, D. W. H. (1988). Combined application of electron diffraction and liquid crystal NMR spectroscopy. In: I. Hargittai and M. Hargittai (1988), Part A, pp. 451–82.

Schäfer, L., Ewbank, J. D., Siam, K., Chiu, N.-S., and Sellers, H. L. (1988). Molecular orbital constrained electron diffraction (MOCED) studies: The concerted use of electron diffraction and quantum chemical calculations. In: I. Hargittai and M. Hargittai (1988), Part A, pp. 301–19.

Scharfenberg, P. and Hargittai, I. (1984). *J. Mol. Struct.*, **112**, 65–70.

Spiridonov, V. P. (1988). Spectroscopic information from electron diffraction. In: I. Hargittai and M. Hargittai (1988), Part A, pp. 265–99.

Sutton, L. E. (ed.) (1958). *Tables of interatomic distances and configuration in molecules and ions*. Spec. Publ. No. 11. The Chemical Society, London.

Sutton, L. E. (ed.) (1965). *Tables of interatomic distances and configuration in molecules and ions* (Suppl.). Spec. Publ. No. 18. The Chemical Society, London.

Tremmel, J. and Hargittai, I. (1988). Gas electron diffraction experiment. In: I. Hargittai and M. Hargittai (1988), Part A, pp. 191–225.

Ukaji, T. and Kuchitsu, K. (1966). *Bull. Chem. Soc. Jpn.*, **39**, 2153–6.

Vajda, E., Tremmel, J., Rozsondai, B., Hargittai, I., Maltsev, A. K., Kagramanov, N. D., and Nefedov, O. M. (1986). *J. Am. Chem. Soc.*, **108**, 4352–3.

Vilkov, L. V., Penionzhkevich, N. P., Brunvoll, J., and Hargittai, I. (1978). *J. Mol. Struct.*, **43**, 109–15.

Wiberg, K. B. and Murcko, M. A. (1987). *J. Phys. Chem.*, **91**, 3616–20.

6

X-ray crystallography: an introduction

Jenny P. Glusker and Aldo Domenicano

6.1   The visualization of atoms and molecules

The chemist would like to be able to visualize molecules and the atoms in them and so obtain a direct three-dimensional view. For example, when considering reaction mechanisms it is difficult to determine the extent of steric hindrance or the accessibility of functional groups if one does not have a three-dimensional view of the system. The electron microscope can, as was shown, for example, by Labaw and Wyckoff (1958), be used to view macromolecules such as those of viruses. It can also be used to image the lattice defects in certain stable inorganic compounds (Iijima 1973). But when one needs precise geometrical data on molecules, an excellent method to choose is an analysis of the X-ray or neutron diffraction patterns obtained from single crystals of the substance of interest. The result of such an analysis is a vast wealth of data on the relative positions of atoms, ions or molecules in a crystal, and a measure of the variability of these positions. From these data, chemical formulae, stereochemistry, bond types, steric strain, displacement parameters and many other features of the arrangement of atoms can be calculated. No other method to date gives this amount of detailed information on the internal (atomic) geometry of both molecular and non-molecular crystals.

The present chapter contains an elementary introduction to X-ray crystallography. This is followed by a glossary of common crystallographic terms. Taken together, the chapter and glossary should enable the reader who is unfamiliar with X-ray crystallography to become acquainted with the principles on which the technique is based and to gain some familiarity with basic concepts and terms that will be used in the next chapters. Since there is not space here for a detailed description of X-ray crystallographic methods, the reader is referred to the many texts available on the subject (e.g. James 1962; Dunitz 1979; Luger 1980; Vainshtein *et al.* 1982; Glusker and Trueblood 1985; Ladd and Palmer 1985; Stout and Jensen 1989).

The possibilities of viewing objects are limited. Two objects cannot be distinguished from each other unless they are separated by at least half the wavelength of the radiation being used to study them. The distances between atoms in molecules are of the order of 1 Å. X rays of the order of 1 Å wavelength must be used to 'view' atoms, implying that it is necessary to use an X ray microscope. Unfortunately, this is not possible because no material has yet been found that is appropriate for focusing X rays in the way that a glass lens focuses visible light. As a result, if one wishes to view atoms, some other technique must be devised.

The experimental strategy is to carry out part of the process for which an optical microscope is used, by measuring the scattered (diffracted) beams obtained when a beam of X rays passes through a crystal. Then the action of the lens of the optical microscope, which recombines all these beams to form a magnified image, is *simulated*, in the X-ray experiment, by an appropriate calculation. The result is a picture of the electron density in a crystal; in this electron-density map atomic positions appear as peaks on a fairly flat background. Studies with long-wavelength X radiation (Sayre *et al.* 1977) or with tunneling electron microscopy (Binnig *et al.* 1982) give us hope that, in the future, appropriate microscopic methods may be efficient for viewing molecules. However, at present, the best general method for obtaining precise measurements of molecular dimensions in the solid state involves analyses of the X-ray or neutron diffraction patterns obtained from single crystals.

The subject of this chapter mainly concerns how the mathematical simulation of the action of a lens is performed when X rays are scattered by the electrons in atoms. Since the use of a microscope involves a recombination of radiation waves, the concept of the phase of a wave becomes important.

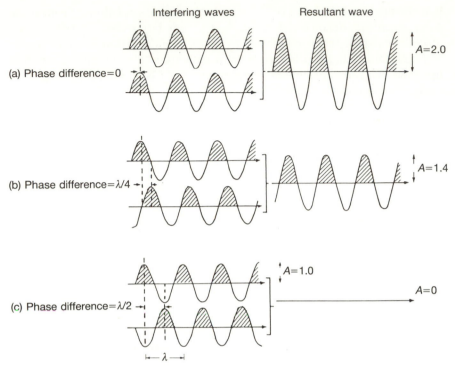

Fig. 6.1 Interference of two waves of the same wavelength (λ) and amplitude (A) travelling in the same direction (after Glusker and Trueblood 1985). Analogous principles apply when there are more than two waves. (a) In phase. (b) Partially in phase (phase difference $\lambda/4 \equiv \pi/2$). (c) Out of phase (phase difference $\lambda/2 \equiv \pi$).

When waves of the same wavelength travel in the same direction they are considered in phase when their wave crests reinforce each other and out of phase when they cancel each other; this is shown in Fig. 6.1. It is the *relative phases* that are important in a consideration of how two waves will interfere with each other. Intermediate relative phases between two waves cause varying amounts of constructive or destructive interference and these relative phases are determined by the path differences of the two waves. In an optical microscope the light flows through the object and beyond through the lens. This flow is continuous and the phase relationships of all the scattered waves are maintained when they are recombined by the lens. In the X-ray diffraction experiment, where atoms scatter X rays, the relative phases of the waves have to be deduced in order for them to be correctly recombined. The methods for making such deductions constitute attempts to solve the 'phase problem' (see Section 6.10).

6.2 Diffraction

When radiation passes through a small aperture or past the edge of an opaque object, the light will appear to spread into the region that would be expected to be in the area of the geometric shadow giving fringes of light and dark bands. This phenomenon is called diffraction and may be explained by use of the proposal of Huygens (1690) that each point on a wave front may be regarded as a new source of waves (secondary waves). Interference between such secondary waves causes the diffraction effect. Diffraction is particularly noticeable when the slit width and radiation wavelength are of the same order of magnitude (see, for example, Jenkins and White 1976). When the single slit is replaced by a series of slits, spaced regularly at distances similar to the wavelength of the light being used, a

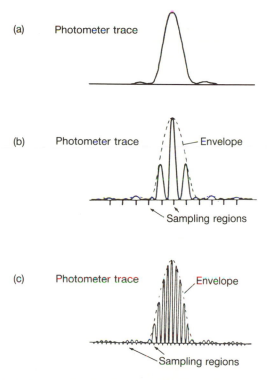

Fig. 6.2 Diffraction by one and by two slits (after Glusker and Trueblood 1985). The effect of changing the distance between the two slits is shown. Note that the shape of the 'envelope' remains that of a single slit but, when the spacing of two slits is changed, the distance between sampling regions is also changed. (a) Diffraction pattern of one slit of width a. (b) Diffraction pattern of two slits (narrow spacing between slits ($d = 2a$), wide spacing between sampling regions). (c) Diffraction pattern of two slits (wide spacing between slits ($d = 6a$), narrow spacing between sampling regions).

pattern of fringes is obtained on photographic film used to intercept these diffracted beams. The spacings between the fringes are inversely proportional to the separation of the slits and directly proportional to the slit-to-film distance and to the wavelength of the radiation.

The overall intensity distribution of the diffraction pattern of a series of equidistant slits is similar to that of the diffraction pattern of a single slit shown in Fig. 6.2(a). The intensity distribution of a multiple-slit diffraction pattern is a sampling, at points whose locations depend on the separations of the slits, of the diffraction pattern of a single slit (Fig. 6.2(b) and (c)). In other words the diffraction pattern of a single slit has been broken up, as a result of interference from adjacent slits, into a series of 'sampling' areas, and a series of fringes is obtained when photographic film is used as a detector. The intensities of these fringes are related to the intensity of the diffraction pattern of a single slit.

6.3   Crystals and their internal regularity

Crystals provide excellent diffraction gratings for X rays. Studies by crystallographers, mineralogists, and mathematicians in the eighteenth and nineteenth centuries (and earlier) already formulated the idea that crystals should have a regularly repeating internal structure. In defining a crystal, the *internal regularity*, not the development of faces, is considered to be fundamental; a crystal is a solid with a regularly repeating internal arrangement of atoms. It was the internal periodicity of crystals that von Laue and his co-workers (Friedrich *et al.* 1912) used in a successful attempt to diffract X rays to show that these rays have a wave-like nature; the crystal provided an ideal diffraction grating since it was not then readily possible to obtain precisely ruled diffraction gratings with the necessarily small spacings. This successful experiment also confirmed the validity of the ideas about the periodic structure of crystals.

The internal periodicity of crystals may be described by a *unit cell* which

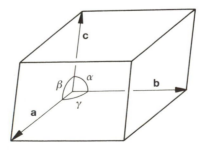

Fig. 6.3 The use of vectors to describe the shape and size of a unit cell.

contains all the atoms in a specific unit that is repeated translationally to give the crystal. In other words, the unit cell is the small building block from which a crystal may be considered to be derived. Each unit cell is characterized by three vectors **a**, **b** and **c** that define its edges (Fig. 6.3). The symbols α, β and γ are used to denote the angles between them. Several early observers (Kepler 1611; Hooke 1665; Huygens 1690; Haüy 1784) drew diagrams of crystals from such a buildup of basic units, sometimes spheres, sometimes unit cells. There are seven basic shapes for unit cells depending on whether or not their edges a, b, c are equal and whether or not the angles are 90°; they correspond to the *crystal systems*, and are listed in Table 6.1. Often, and particularly in triclinic and some monoclinic crystals, there are several possible choices of unit cell, but when this problem arises the choice is made of the unit cell with angles nearest 90° and with edge lengths that are as similar as possible. Precise optical measurements of macroscopic crystal forms will give only relative values for the lengths of a, b and c but they will give absolute values for α, β and γ. It must, however, be remembered that the unit cell is primarily a geometrical construction, useful to describe the internal periodicity of the crystal. A crystal is composed of atoms, molecules, or ions and the manner by which they pack ultimately determines the physical properties of the crystal, for example, in which directions it may be cleaved and how anisotropic these properties are.

The axial and angular relationships mentioned above for seven unit-cell shapes are not enough to define the crystal systems because axial lengths may, by chance, be equal and angles may, by chance, be 90°. A knowledge of the symmetry of a crystal is essential for defining which one of the seven crystal systems a crystal belongs to.

The symmetry of the internal structures of crystals has been the subject of study by mineralogists throughout the nineteenth century, much before the actual structures were known. There are two types of symmetry operations to consider. *Proper symmetry operations* (also called symmetry operations of the first kind) retain the chirality of an object; that is, they could, on a macroscopic scale, convert a right hand to another right hand, generally in a different position. Such symmetry operations are translations and rotations. A *translation* is simply a motion of the entire body in one direction. An object has an n-fold *rotation axis* if it can be rotated about an axis passing through it and then appears indistinguishable from the starting view at each rotation of $2\pi/n$. The effect of a fourfold rotation axis on a chiral object is shown in Fig. 6.4(a). When a rotation is coupled with a translation, the result is a *screw axis*, designated as n_r, which corresponds to an n-fold rotation axis and a translation in a direction parallel to that axis by r/n fractions of the unit-cell repeat (Fig. 6.4(b)). *Improper symmetry operations* (also called symmetry operations of the second kind) are those that convert an object into its mirror image, that is, they convert a right-handed object

Table 6.1 The seven crystal systems, 14 Bravais lattices, 32 crystallographic point groups (crystal classes), and 230 space groups

Crystal systems (7)	Axial and angular relationships	Characteristic symmetry	Lattice symmetry	Bravais Lattices (14)[a]	Crystallographic point groups (32)[b]	Common and/or representative space groups (out of 230)[c]
Triclinic	$a \neq b \neq c$ $\alpha \neq \beta \neq \gamma$	Identity or inversion (onefold rotation or rotatory inversion axis in any direction)	$\bar{1}$	P	$1, \bar{1}$ (C_1, C_i)	$P\bar{1}$
Monoclinic	$a \neq b \neq c$ $\alpha = \gamma = 90°$, $\beta \neq 90°$	Twofold rotation or rotatory inversion axis (along **b**)	$2/m$	P, C	$2, m, 2/m$ (C_2, C_s, C_{2h})	$P2_1$ $C2/c$ $P2_1/c$
Orthorhombic	$a \neq b \neq c$ $\alpha = \beta = \gamma = 90°$	Three twofold rotation or rotatory inversion axes (along **a**, **b** and **c**)	mmm	P, C, I, F	$222, mm2,$ mmm (D_2, C_{2v}, D_{2h})	$P2_12_12_1$ $Pbca$
Tetragonal	$a = b \neq c$ $\alpha = \beta = \gamma = 90°$	Fourfold rotation or rotatory inversion axis (along **c**)	$4/mmm$	P, I	$4, \bar{4}, 4/m, 422,$ $4mm, \bar{4}2m,$ $4/mmm$ (C_4, S_4, C_{4h}, D_4, C_{4v}, D_{2d}, D_{4h})	$I4/mmm$

System	Axes/angles	Symmetry		Lattice	Point groups	Space groups
Cubic	$a = b = c$ $\alpha = \beta = \gamma = 90°$	Four threefold axes (along $\mathbf{a} + \mathbf{b} + \mathbf{c}$, $-\mathbf{a} + \mathbf{b} + \mathbf{c}$, $\mathbf{a} - \mathbf{b} + \mathbf{c}$, $-\mathbf{a} - \mathbf{b} + \mathbf{c}$)	$m3m$	P, I, F	$23, m3, 432$ $\bar{4}3m, m3m$ (T, T_h, O, T_d, O_h)	$Pm3m$ $Fm3m$ $Fd3m$
Trigonal	$a = b = c$ $\alpha = \beta = \gamma \neq 90°$	Threefold rotation or rotatory inversion axis (along $\mathbf{a} + \mathbf{b} + \mathbf{c}$)	$\bar{3}m$	R	$3, \bar{3}, 32, 3m, \bar{3}m$ ($C_3, S_6, D_3, C_{3v}, D_{3d}$)	$R\bar{3}m$
Hexagonal	$a = b \neq c$ $\alpha = \beta = 90°, \gamma = 120°$	Sixfold rotation or rotatory inversion axis (along $\mathbf{c}$)	$6/mmm$	P	$6, \bar{6}, 6/m, 622,$ $6mm, \bar{6}m2,$ $6/mmm$ ($C_6,$ $C_{3h}, C_{6h}, D_6,$ C_{6v}, D_{3h}, D_{6h})	$P6_3/mmc$

[a] P = primitive; C = double-face centered; F = all-face centered; I = body centered. The (primitive) trigonal lattice is usually denoted as R.
[b] Hermann–Mauguin system followed by Schoenflies system of notation.
[c] More than 60% of organic compounds crystallize in one of the first six space groups listed (Nowacki et al. 1967).

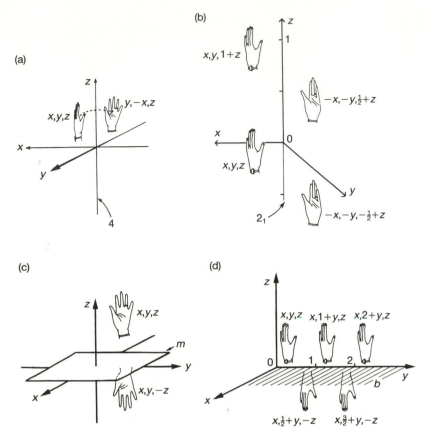

Fig. 6.4 Some symmetry operations (after Glusker and Trueblood 1985). (a) Fourfold rotation axis. (b) 2_1 screw axis. (c) Mirror plane. (d) b-type glide plane.

into a left-handed object. Those improper symmetry operations that do not involve translation may be defined as *rotatory inversion* (or *alternating*) axes, $\bar{n}$. These correspond to a rotation by $2\pi/n$, followed by inversion through a point lying on the axis. The first of these is a center of symmetry or inversion center, $\bar{1}$ (or i); if situated at the origin it converts a point at x, y, z into a point at $-x$, $-y$, $-z$. Reflection across a mirror plane, $\bar{2}$ (or m), is another improper symmetry operation (see Fig. 6.4(c)). When a mirror plane is coupled with a translation, the result is a *glide plane*. This consists of reflection across a plane combined with translation in a direction parallel to the plane, see Fig. 6.4(d). Glide planes are designated as a, b, c, etc., depending on the direction of translation. It can be shown that, as a result of the internal periodicity in a crystal, only those symmetry operations for which $n = 1, 2, 3, 4, 6$ are possible in crystals.

To describe the internal structure of crystals it is advantageous to introduce the concept of *space lattice* or *crystal lattice*. This is a set of infinite, evenly spaced equivalent points related to each other by translational symmetry. There are, in all, 14 types of three-dimensional space lattices, called the Bravais lattices. They are listed in Table 6.1 and are all of the possible periodic three-dimensional arrangements of points such that the environment of any point is identical in every respect to that of every other point. The 14 Bravais lattices can be classified according to symmetry into seven groups, corresponding to the seven crystal systems. Note that Bravais lattices are not always primitive (type P), that is with just one point in the unit cell; additional points can be found at the center of the cell (type I), at the centers of two opposite faces (types A, B, C), or at the centers of all faces (type F). We stress that the actual crystal structure is a 'convolution' of the crystal lattice and the contents of the unit cell. This means that the contents of a unit cell are laid down systematically ('convoluted') on each point of a (primitive) crystal lattice.

It is now appropriate to consider the symmetries of structures (such as arrays of molecules) arranged on these crystal lattices. The crystallographically admissible operations that do not involve translation can be combined in three dimensions to give 32 *crystallographic point groups*. These can be applied to the shapes of crystals and, indeed, the point group of a crystal may often be determined by a detailed examination of the development of faces (the *habit* of a crystal). Point-group symmetries are denoted by crystallographers using the Hermann–Mauguin system (see, for example, Hahn 1987, p. 795), but spectroscopists and theoretical chemists may be more familiar with the Schoenflies system (see, for example, Cotton 1971); both are listed in Table 6.1.

But the internal periodicity of atoms in crystal structures must necessarily be represented by symmetry operations involving translations. Thus *point-group* symmetry operations may be combined with translations to give *space-group* symmetry operations. An enumeration of all possible combinations of symmetry elements that are compatible with three-dimensional periodicity leads to a total of 230 space groups. These combinations are catalogued in the *International Tables for Crystallography* (Hahn 1987) for each of these space groups in terms of an *asymmetric unit*, that is, that fractional part of the unit cell from which the entire crystal structure may be generated by the space-group symmetry operations, including translations. Sometimes there is additional, approximate symmetry within the asymmetric unit, termed 'non-crystallographic symmetry', but, since the two or more entities are not related by crystallographic symmetry, they have different environments. The importance of the asymmetric unit is that it is the only portion of the crystal structure that needs to be determined by the X-ray diffraction experiment. The rest of the crystal structure may then be derived straightforwardly. Excellent simple introductions to the subjects of symmetry have been written by Steadman (1982) and I. Hargittai and M. Hargittai (1986).

In view of the many early measurements of crystals, particularly by mineralogists, a specific way of describing the crystal habit was needed. The 'law of rational indices' (Haüy 1784) states that each face of a crystal may be described, with reference to three non-coplanar axes, by three small integers; these numbers are customarily designated h, k and l where the three non-coplanar axes are the directions of the three axes of the unit cell, as shown in Fig. 6.5. This law follows from the concept that crystals are built up from basic building blocks and that edges may be represented by 'steps' in this construction. The use of the indices hkl was introduced by Whewell, and popularized by Miller (1839); they are generally referred to as 'Miller indices'. The importance of these indices lies in the fact that not only crystal faces, but also sets of parallel *lattice planes* (defined as planes through three non-colinear lattice points) may be denoted through them. In addition, as will be seen later, the diffracted beams from a crystal are also denoted as hkl.

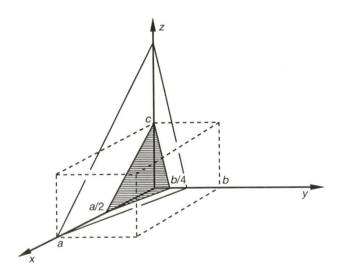

Fig. 6.5 Miller indices for a set of parallel lattice planes. The unit cell is outlined by broken lines and the set of lattice planes (or possible crystal face) is shown by solid lines. The indices are defined as the smallest integers corresponding to the ratios between the unit-cell edges (a, b, c) and the intercepts that the first plane from the origin makes with the crystal axes (e.g. $a/2$, $b/4$, c). Thus the set of lattice planes shown here has Miller indices 241.

6.4 Diffraction by crystals

X rays are scattered by atoms in crystals by an amount that depends on the number of electrons, the scattering angle, the amount of atomic vibration, and, possibly, disorder of atoms in the crystal. The atomic scattering factors of a non-vibrating atom may be computed from theoretical wavefunctions for free atoms. On the other hand, neutrons are scattered by nuclei and the amount of scattering is not a regularly decreasing function of atomic number; as a result atoms may be located or differentiated when the X-ray experiment gives problems because of very weak scattering (hydrogen atoms) or similarity of atomic numbers (see Chapter 11). A major disadvantage of neutron diffraction methods is that, for most sources, larger crystals are required than for X-ray studies.

The X-ray diffraction pattern of a crystalline solid consists — when recorded on a photographic film — of a series of spots; these represent a sampling, at points whose positions depend on the size of the repeat unit, of the diffraction pattern of the contents of a single unit cell. This result is analogous to the results obtained by diffraction by slits, described earlier (Section 6.2). The larger the repeating unit (unit cell) the closer together the spots of the diffraction pattern. This is well illustrated by the few spots seen in the X-ray diffraction pattern of crystals of sodium chloride and the many spots seen in the diffraction pattern of crystalline hemoglobin for the same wavelength of radiation. A feature that is evident on an X-ray diffraction photograph is that there is considerable variation in the intensities of the diffraction spots. Some are intense, while others are absent or only faintly detectable. While the dimensions of the unit cell may be derived from the scattering angles of the diffracted beams, the positions of the atoms can only be determined if the intensities of these beams are also measured. An experimental setup for such measurements will be described in Section 6.5.

Just as the crystal may be considered in terms of a crystal lattice, so the diffraction pattern may be considered in terms of a 'reciprocal lattice' (Ewald 1913, 921) which may be geometrically constructed with points (one for each diffracted beam) designated *hkl* and with distances of each reciprocal-lattice point from the origin inversely proportional to the spacings between equivalent crystal-lattice planes perpendicular to that direction (see Fig. 6.6(a) for a two-dimensional example). An important property of the reciprocal lattice is clearly seen from Fig. 6.6(a): the three vectors defining its unit cell, $\mathbf{a}^*$, $\mathbf{b}^*$, and $\mathbf{c}^*$, satisfy the following conditions:

$$\mathbf{a}^* \cdot \mathbf{a} = \mathbf{b}^* \cdot \mathbf{b} = \mathbf{c}^* \cdot \mathbf{c} = 1 \tag{6.1}$$

$$\mathbf{a}^* \cdot \mathbf{b} = \mathbf{a}^* \cdot \mathbf{c} = \mathbf{b}^* \cdot \mathbf{a} = \mathbf{b}^* \cdot \mathbf{c} = \mathbf{c}^* \cdot \mathbf{a} = \mathbf{c}^* \cdot \mathbf{b} = 0, \tag{6.2}$$

where $\mathbf{a}$, $\mathbf{b}$, and $\mathbf{c}$ are the three vectors of the direct cell.

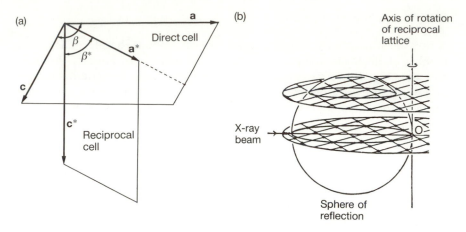

Fig. 6.6 The reciprocal lattice. (a) The relationship between direct and reciprocal unit cells (a two-dimensional example). (b) The reciprocal lattice passing through the sphere of reflection (radius $1/\lambda$). Reflection occurs when a reciprocal-lattice point touches the surface of the sphere of reflection (after Bunn 1961).

The reciprocal lattice is very useful in deducing conditions for diffraction. For example, Ewald (1923) showed that if a sphere of radius $1/\lambda$ is placed with the incident beam direction along a diameter and the origin of the reciprocal lattice at the point where the incident beam emerges from the sphere, as shown in Fig. 6.6(b), it is found that whenever a reciprocal-lattice point P touches the surface of this 'sphere of reflection', the conditions for diffraction are satisfied. The direction of the diffracted beam is given by the line connecting the center of the sphere with P. The use of the sphere of reflection (Ewald sphere) makes it possible to predict which reciprocal-lattice points will be in a reflecting position for a given crystal orientation.

When an X-ray diffraction study is carried out, the unit-cell dimensions and space group of the crystal are first ascertained. Partial information on the symmetry of the crystal (the 'Laue class') is deduced from the symmetry of the diffraction pattern. Additional information may sometimes be obtained by other methods, e.g. from the crystal habit. The possible space groups corresponding to the Laue class of the crystal may be distinguished, though not always uniquely, from systematic absences in the diffracted beams. Some beams, expected from the unit-cell dimensions, may be systematically absent (for example, $h00$ when h is odd); this indicates the presence of symmetry operations involving translation (in this case a 2_1 screw axis along $\mathbf{a}$). The systematic absences allow screw axes, glide planes and lattice centerings to be detected; a detailed documentation is given in the *International Tables for Crystallography* (Hahn 1987).

At this point the investigator will know the unit-cell dimensions and the space group (or a set of possible space groups) for the crystal. If the crystal density is then measured, the unit-cell atomic contents can be deduced if the chemical composition of the material is known.

6.5 Methods of measuring the X-ray diffraction pattern

The experimental setup used today for recording the diffraction pattern, although now much more sophisticated, is basically the same as that used in the first experiment by von Laue and his co-workers (Friedrich *et al.* 1912). So-called characteristic X rays (which are almost monochromatic) obtained with appropriate metal targets in the X ray tube, are used; wavelengths of 0.71 Å (Mo Kα radiation) or 1.54 Å (Cu Kα radiation) are those most commonly employed. A beam of such X rays, finely collimated by passage through a narrow metal tube, is directed at a small crystal mounted on a fiber (or, if it is unstable in the open air, in a thin-walled capillary) which is held in a goniometer head (see Fig. 6.7). Most of the X-ray beam passes directly through the crystal and is intercepted by a beam catch. However, some of the incident beam is diffracted in various directions by the crystal; these diffracted beams are intercepted and detected by a photographic film or by an electronic detecting device (a scintillation counter, for example). The direction (2θ, see Fig. 6.7) and relative intensity (I) of each diffracted beam is then recorded. Measurements are made on thousands to hundreds of thousands of diffracted beams for a given crystal, depending on the size of the unit cell. If the unit cell is large the spacing between diffraction spots on a film is smaller; as a result the number of diffracted beams accessible with a particular wavelength is larger than for a crystal with a small unit cell.

The best crystals for study in this way are equidimensional, 0.1–0.3 mm on an edge; they should have what is called 'mosaic spread', that is, a slight irregularity of orientation of small blocks of unit cells in the crystal. Because of the slight imprecision of alignment there is less likely to be further diffraction within the crystal, an effect called extinction which modifies especially the most intense diffracted beams (see Chapter 7, Section 7.2.3). There are two theories of diffraction by crystals. In the *kinematic theory* any mutual interaction between incident and scattered beams is neglected, as, for example, in a weakly diffracting mosaic crystal. This is the situation normally encountered in X-ray crystallography. In more perfect crystals the interaction between the incident and the scattered beams becomes important but is difficult to correct for; this *dynamical theory* of diffraction is more of a problem in electron diffraction by crystals than in X-ray diffraction.

Bragg (1912, 1913) showed that the angular distribution of scattered radiation from a crystal behaves as if each diffracted beam were reflected from a set of parallel lattice planes. The angular deviation of the diffracted beam from the incident X-ray beam is 2θ as shown in Fig. 6.7. If the distance between successive lattice planes hkl is d_{hkl}, and the angle between the n-th order of diffraction of X rays (wavelength λ) and the normal to the set of lattice planes is $\pi/2 - \theta_{hkl}$, then the following equation holds:

$$2d_{hkl}\sin \theta_{hkl} = n\lambda. \tag{6.3}$$

This is known as Bragg's law. X rays are diffracted by a crystal when, and only when, eqn (6.3) is satisfied. The diffracted beam then appears to have been reflected from the hkl lattice planes.

Each diffracted beam is identified by the Miller indices of the set of parallel planes causing the reflection, hkl. If these are known, then d_{hkl} (and θ_{hkl}) may be calculated from the unit-cell dimensions. The diffracted beam is usually referred to as a 'reflection' or a 'Bragg reflection'; the latter term will be used here to emphasize its origin. The order of diffraction, n, is usually incorporated in the indices as a multiplying factor; thus the third-order Bragg reflection from the 213 set of lattice planes is indexed as 639. The reciprocal-lattice vector $\mathbf{H} = h\mathbf{a}^* + k\mathbf{b}^* + l\mathbf{c}^*$, associated with the hkl Bragg reflection, is called the *scattering vector*. By the definition of reciprocal lattice, the scattering vector is perpendicular to the set of reflecting direct-lattice planes. Its magnitude is $H = 1/d_{hkl} = 2\sin \theta_{hkl}/\lambda$, and its end is on the surface of the sphere of reflection.

The methods for measuring the diffraction of X rays by a crystal are many, although for most single-crystal studies a computer-controlled diffractometer is most commonly used. It automatically measures and records the intensities of diffracted beams. The crystal is first aligned so that it is always at the geometric center of the instrument and fully bathed in the X-ray beam. The orientation of the crystal with respect to the geometry of the diffractometer is initially determined by measuring the angles at which a few diffracted beams are detected, using measurements from a photograph of the diffraction pattern. The beams selected from the photographic film are automatically centered, thereby giving orientation angles for each of them. These orientation angles may then be used to derive the crystal orientation (orientation matrix) which allows the diffractometer to be programmed to scan all accessible Bragg reflections.

Most diffractometers have the 'four-circle geometry', shown in Fig. 6.7. The spike of the characteristic radiation is selected by reflection from a graphite crystal to ensure a single wavelength. The detector may be rotated about the 2θ circle parallel to the base of the instrument to intercept the diffracted beam. The other three circles orient the crystal so that it will be in the diffracting position for the Bragg reflection under study. The ϕ circle rotates the spindle axis of the goniometer head. The χ circle is perpendicular to the ϕ circle and allows reflections to be brought into the equatorial plane

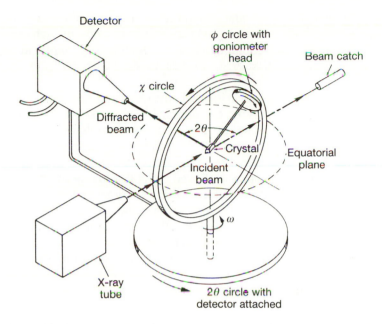

Fig. 6.7 A four-circle diffractometer (after Glusker and Trueblood 1985). The four circles can be used to adjust the orientation of the crystal and detector so as to bring any set of lattice planes into diffracting position and detect the reflection. ϕ is the angle of rotation about the spindle axis of the goniometer head. χ is the angle between the spindle axis of the goniometer head and the diffractometer axis (normal to the equatorial plane). ω is the angle that the χ circle makes with the incident beam. 2θ is the angle between the directions of incident and diffracted beams.

of the diffractometer. The ω circle lies in the plane of the 2θ circle and brings the set of lattice planes into the correct position for reflection.

There are two generally used methods of scanning a peak in order to measure its intensity. They are called *scan modes*. In the ω scan mode the ω circle is driven through a small angular range (about 1° for low-order reflections) while the counter is kept in a fixed position on the 2θ circle. In the $\omega/2\theta$ scan mode the crystal and counter are moved simultaneously so that $\omega = 2\theta/2$ throughout the entire scan. The latter technique gives higher-quality data, especially at high 2θ values. The total peak intensity, recorded by the detector, is measured by summing intensities for a number of individual angular steps that cover the entire peak area. At specified angular distances from the peak center the background intensities are measured (at both sides of the peak) and the value is appropriately subtracted from the peak intensity. (A more detailed treatment of these points will be found in Chapter 7, Section 7.2.4). If possible, the temperature of the crystal is lowered by blowing a stream of cold gas over it, so that thermal motion of atoms in the crystal may be reduced and therefore their positions determined more accurately. The estimated standard deviation (e.s.d.) of each intensity, $\sigma(I_{hkl})$, is assessed from counting statistics, modified by empirical adjustments that take care of those errors that do not derive strictly from counting statistics.

When measuring the intensities of the X-ray diffraction pattern a multitude of problems may be encountered. They may be categorized as instrumental problems, crystal problems and user problems. These include random errors such as misalignment of the instrument, fluctuations in the power of the X-ray beam, failure of the detecting system, non-uniformity of a low-temperature gas stream, if used, crystals moving on their mounts and intermittent X-ray shutter failures. There are other errors, such as those derived from crystal decay in the X-ray beam, which are systematic and may be corrected for by measuring periodically a few standard reflections to check on the constancy of the experimental setup. There are also other types of errors that are more difficult to correct for unless more information is available; these include multiple reflection, extinction, and thermal diffuse scattering. A detailed account of these sources of error will be found in Chapter 7.

Once the intensity, I_{hkl}, of a Bragg reflection has been measured, it must be corrected by some factors that take into account the relative length of time the crystal was in the diffracting position, the extent of polarization of the X-ray beam and the absorption of X rays by the crystal. This is done by the following equation:

$$F_{hkl}^2 = I_{hkl}(KLpA)^{-1}, \tag{6.4}$$

where F_{hkl} is the *structure amplitude*, i.e. the amplitude of the diffracted wave measured relative to the amplitude of scattering of a single electron which is taken as unity at $\sin\theta = 0$, and K is a scale factor. The Lorentz factor, L, takes into consideration how long each set of lattice planes was in the reflecting position; it depends on the technique used for data collection. The polarization factor, p, originates from the fact that the variation of the reflection efficiency of X rays with the scattering angle depends on the polarization status of the incident beam. The transmission factor, A, takes into account the reduction of the intensity due to absorption of X rays by the crystal. It may be computed geometrically from path lengths through a crystal of known shape and size or estimated by measuring the variation in intensity of a reflection as the crystal is rotated about the scattering vector. As a result of the measurements described above and the data reduction, we now have h, k, l, F_{hkl} (and $\sigma(F_{hkl})$) for all the experimentally accessible diffracted beams.

6.6   The phase problem

The measurement of the diffraction pattern represents an analogy to the first stage in the study of an object by use of a microscope. We now proceed to a discussion of methods of simulating the action of a lens. This involves the determination of the relative phase angle of each Bragg reflection. The phase of a Bragg reflection is the phase relative to a wave scattered by an atom at the origin in the same direction (Fig. 6.8). Values of the phase angle for each Bragg reflection are essential for calculating the electron-density map (see Section 6.9) which requires the measured amplitudes (from the intensities of the diffracted beams, eqn (6.4)) *and* the phases so that they may be correctly combined.

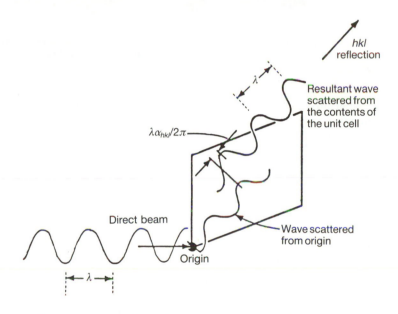

Fig. 6.8 The physical meaning of phase angle α_{hkl} (after Nyburg 1961).

6.7   The structure factor

The amount that an atom in a crystal can scatter X rays depends on the number of electrons, the diffracting angle, and the amount of thermal motion (or disorder) since this latter will smear out the electron-density cloud around an atom and make it appear larger and more diffuse. To compute the structure amplitudes it is necessary to know how each atom in the structure scatters X rays and how the X rays scattered from the various atoms interfere with each other.

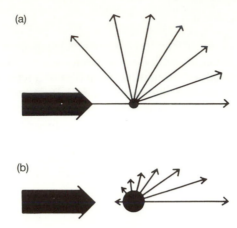

Fig. 6.9 Effect of particle size on diffraction fall-off. (a) Scattering of radiation by a particle small relative to the wavelength. (b) Scattering of radiation by a particle large relative to the wavelength. From Glusker and Trueblood (1985).

When radiation is scattered by particles that are very small relative to the wavelength of the radiation, the scattered radiation has the same intensity in all directions. When it is scattered by larger particles, the radiation scattered from different regions of the particle will still be in phase in the forward direction, but at higher scattering angles there is interference between radiation scattered from various parts of the particle. The intensity of radiation scattered at higher angles is thus less than that scattered in the forward direction (see Fig. 6.9). This effect is greater if the particle size is larger relative to the wavelength. Thus, if an atom is vibrating, its effective size, as viewed over the millions of unit cells as an average, is larger, and this effect will increase as the temperature is raised and the vibration amplitudes are increased. Therefore the fall-off in the atomic *scattering factors*, f, with $\sin \theta/\lambda$ is increased (an example of the reciprocal relationship between real space (the atom which is smeared out by temperature effects) and the diffraction space (the diffraction pattern which is less extensive, narrower, as the temperature is raised)). If the vibration is large enough then f will fall more steeply to negligible values with increasing $\sin \theta/\lambda$, as shown in Fig. 6.10. The peak for that atom in the electron-density map is, as a result, more diffuse and the atom is less precisely defined in position. This type of effect is also seen when disorder is present, that is, when there is a random variation in the contents of different unit cells, often only in one region of the unit cell. The average of all the unit cells in the crystal gives a highly smeared electron density in the region of disorder.

If the motion is isotropic the scattering factor for an atom at rest, f^0, is modified by an exponential factor $\exp(-B\sin^2 \theta/\lambda^2)$, called *temperature*

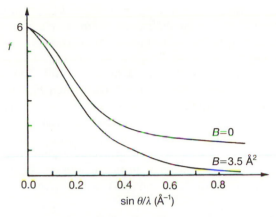

Fig. 6.10 Variation of the scattering factor of carbon, $f = f^0 \exp(-B\sin^2\theta/\lambda^2)$, as a function of scattering angle. After Glusker and Trueblood (1985).

factor (or Debye–Waller factor), where B is related to the mean-square amplitude of atomic vibration, $B = 8\pi^2\langle u^2 \rangle$. Thus the scattering factor is now $f = f^0 \exp(-B\sin^2\theta/\lambda^2)$. In minerals and some salts, B values are low (in the range 1–3 Å^2 corresponding to 0.01–0.04 Å^2 in $\langle u^2 \rangle$). In organic compounds they are higher in value.

However, thermal vibration in molecules and crystals is usually far from isotropic and so a better way to represent atomic motion is by an ellipsoid; this means that the vibration may be described by six parameters (three to define the lengths of the principal axes of the ellipsoid and three to define their orientation relative to the crystal axes). More details will be given in Chapters 8 and 9, which describe the analysis of molecular motions in crystals.

The wave scattered by the contents of the unit cell in the direction of the *hkl* reflection is described in amplitude and phase by the *structure factor*, $\mathbf{F}_{hkl}$. This may be derived by noting that the radiation scattered by one atom will interfere constructively or destructively with radiation scattered by other atoms in the unit cell and therefore *the scattering from the entire crystal will depend on where each atom lies in the unit cell*. The structure factor is then given by:

$$\mathbf{F}_{hkl} = F_{hkl}\exp(\mathrm{i}\alpha_{hkl}) = \sum_j f_j \exp[2\pi\mathrm{i}(hx_j + ky_j + lz_j)] =$$

$$\sum_j f_j \cos[2\pi(hx_j + ky_j + lz_j)] + \mathrm{i}\sum_j f_j \sin[2\pi(hx_j + ky_j + lz_j)] =$$

$$A_{hkl} + \mathrm{i}B_{hkl}, \tag{6.5}$$

where $\alpha_{hkl} = \tan^{-1}(B_{hkl}/A_{hkl})$ and $F_{hkl} = (A_{hkl}^2 + B_{hkl}^2)^{1/2}$ are the phase and amplitude, respectively, of the diffracted beam. The sum is over all atoms in the unit cell and f_j is the scattering factor of the j-th atom at the value of $\sin \theta / \lambda$ corresponding to the Bragg reflection hkl. The positional coordinates of the atoms, x_j, y_j and z_j, are expressed as fractions of the unit-cell edges. Thus if the atomic positions are known, then F_{hkl} can be calculated and so can the phase angle α_{hkl}. Note that if the structure is centro-symmetric and the origin of the reference system is at a center of symmetry then B_{hkl} vanishes and the structure factor is no longer expressed by a complex number. This means that the phase angle, α_{hkl}, can only be 0 or π, according to the positive or negative sign of A_{hkl}.

It is often convenient to write eqn (6.5) using vector notation:

$$\mathbf{F_H} = \sum_j f_j \exp(2\pi i \mathbf{H} \cdot \mathbf{r}_j),\tag{6.6}$$

where $\mathbf{H} = h\mathbf{a^*} + k\mathbf{b^*} + l\mathbf{c^*}$ is a *reciprocal-lattice vector* (denoting a Bragg reflection) and $\mathbf{r}_j = x_j\mathbf{a} + y_j\mathbf{b} + z_j\mathbf{c}$ is a *direct-lattice vector* (denoting an atomic position).

It is important to realize that *each* atom in the unit cell (and hence in the crystal) contributes to the intensity of *each* reflection by an amount that depends on its position in the unit cell, its scattering power and, to a lesser extent, the scattering angle. Thus each F_{hkl} value contains information about the entire crystal structure.

6.8 Anomalous scattering

The absorption of X rays by an element slowly increases as the wavelength increases, but shows discontinuities where the absorption drops suddenly and then starts to rise again as shown in Fig. 6.11. These absorption edges occur at wavelengths which represent the energy necessary to excite a bound electron to a vacant higher-energy level or to eject it altogether. At wavelengths slightly below this value (higher-energy radiation) the absorption is higher and the scattering factor f becomes a complex number

$$\hat{f} = f + \Delta f' + i\Delta f'',\tag{6.7}$$

where $\Delta f'$ and $\Delta f''$ vary with the wavelength of the incident radiation (and, to a much lesser extent, with $\sin \theta / \lambda$). Thus, if an atom absorbs X rays strongly, there will be a phase change for the X rays scattered by that atom relative to the phases of the X rays scattered by other atoms, and this is called *anomalous scattering*.

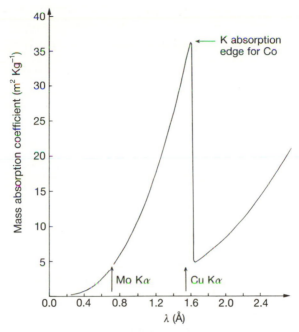

Fig. 6.11 Absorption curve for cobalt as a function of wavelength.

From eqn (6.5) it is easily seen that the structure amplitudes of two centro-symmetrically related reflections, F_{hkl} and $F_{\bar{h}\bar{k}\bar{l}}$, and hence their intensities I_{hkl} and $I_{\bar{h}\bar{k}\bar{l}}$, are equal, even for a non-centrosymmetric crystal. This is known as Friedel's law. However, Friedel's law is no longer valid if the crystal is non-centrosymmetric and there is an anomalous scatterer in the structure. Bijvoet *et al.* (1951) used the differences in intensity resulting from anomalous scattering to determine the chirality (absolute configuration) of the tartrate ion. This method is now used routinely for absolute determinations of chirality.

6.9   The electron-density function

The electron density in a crystal is normally represented by means of a three-dimensional map of the electron-density function, $\rho(xyz)$, calculated by Fourier synthesis at various points in the unit cell. This function is usually computed at spacings of about 0.3 Å. Fourier summation methods are used because the electron density within a crystal is a triply periodic function and hence can be expressed by the triple summation of appropriately phased waves. Thus it is the simulation of the action of a lens and is a summation of waves of known amplitude F_{hkl}, frequency hkl, and phase α_{hkl}.

The electron-density function is given by the following equation:

$$\rho(xyz) = \frac{1}{V}\sum_{hkl} F_{hkl}\exp[-2\pi i(hx + ky + lz)];\qquad (6.8)$$

or, in vector notation:

$$\rho(\mathbf{r}) = \frac{1}{V}\sum_{\mathbf{H}} F_{\mathbf{H}}\exp(-2\pi i\mathbf{H}\cdot\mathbf{r}),\qquad (6.9)$$

where V is the volume of the unit cell. Equation (6.8) can also be written in the form:

$$\rho(xyz) = \frac{1}{V}\sum_{hkl} F_{hkl}\cos[2\pi(hx + ky + lz) - \alpha_{hkl}],\qquad (6.10)$$

where the amplitudes, F_{hkl}, and phases, α_{hkl}, of the diffracted beams appear explicitly. Note that while the structure amplitudes are directly accessible to experiment, the phase angles are not; they have to be obtained in some way if we wish to calculate the electron-density map. If the structure is centrosymmetric the phase angles, α_{hkl}, are all either 0 or π. Therefore if there are N different diffracted beams, hkl, there are 2^N possible electron-density functions and only one of these is the correct one. If the structure is non-centrosymmetric the phase angles, α_{hkl}, may have any value between 0 and 2π and the true electron-density function is only one in an infinite number of such functions. However, it can be seen from eqn (6.10) that the largest F_{hkl} values contribute most to the electron density and therefore some simplification is possible if only the largest terms are used. The importance of having the correct phase angles in order to compute a correct electron-density map is stressed in Fig. 6.12.

Since the electron-density function may be expressed as a Fourier series with the structure factors as coefficients, the structure factors may also be expressed in terms of the electron density. Thus the structure factor, F_{hkl}, is the *Fourier transform* of the electron density $\rho(xyz)$:

$$\mathbf{F}_{hkl} = \int_V \rho(xyz)\exp(i\phi)\,\mathrm{d}V\qquad (6.11)$$

(where ϕ is $2\pi(hx + ky + lz)$), and the electron density is the *inverse Fourier transform* of the structure factor:

$$\rho(xyz) = \frac{1}{V}\sum_{hkl} F_{hkl}\exp(-i\phi).\qquad (6.12)$$

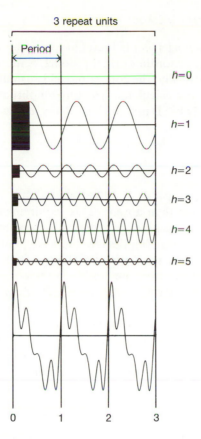

Fig. 6.12 Summation of waves to give a one-dimensional electron-density map. The solid bars mark the distance to peak from origin, which corresponds to the relative phase angle, $\alpha_h/2\pi h$.

A triple summation has been used in eqn (6.12) because F_{hkl} is only observed at discrete points. Thus the intensity of a Bragg reflection is proportional to the value at that reciprocal-lattice point, hkl, of the square of the Fourier transform of the electron density. This concept highlights the relationship between an object and its diffraction pattern.

6.10   Estimating the phase angles

The relative phase angles for each diffracted beam may be found in one of several ways. Here we outline only those methods that are of importance in small molecules crystallography.

6.10.1 Trial-and-error methods

A tentative atomic arrangement is found in some way and the phase angles are calculated from the coordinates in the model. A number of arrangements can be tested systematically; the correct model is usually identified by comparing the observed and calculated structure amplitudes. This method was the one used originally but is not much used any longer, unless the structure is extremely simple.

6.10.2 Direct methods

The electron density must be positive or zero but never negative since electrons are never less than absent. This requirement, together with the information that there are peaks at atomic positions but that elsewhere the map is fairly flat (near zero), implies that there are mathematical constraints on possible phase angles (see Karle and Hauptman 1950). The best set of phases may be found by 'direct methods'. In these methods the phases are chosen to give the least negative electron-density map. The simplest case is that of centrosymmetric structures, where — as we have seen — the phase problem reduces to that of determining the sign of each A_{hkl}. For example, for three intense reflections with indices hkl, $h'k'l'$, and $h + h'$, $k + k'$, $l + l'$, the signs of the respective A terms of the structure factor expression (eqn (6.5)) are related by the following equation

$$s(hkl) \times s(h'k'l') \approx s(h + h', k + k', l + l'), \qquad (6.13)$$

where s means 'sign of' ($+$ or $-$) and $\approx$ means 'probably equal to' and stresses the importance of the probability aspects of these equations. If two of these signs (phases) are known, it is possible to derive the third. An example is given in Fig. 6.13. From such relationships it is often possible to derive phases for almost all intense Bragg reflections, and, since it is the intense Bragg reflections that dominate the electron-density map, to obtain a good approximation to that map. Analogous methods are used for non-centrosymmetric structures.

For such analyses the structure amplitudes are normalized by removing the effects of fall-off of scattering with angle (scattering factor and temperature factor) so that only the geometric component, E_{hkl}, is left. Usually only the highest E values are used; any intense high-θ reflections are thereby given high weight in the calculations that follow. An analysis of the statistical distribution of E values will indicate whether the structure is centrosymmetric or non-centrosymmetric. The triple products among high E values, that is, any triplets with indices hkl, $h'k'l'$, and $h + h'$, $k + k'$, $l + l'$, are found and, after fixing the phases of some origin-

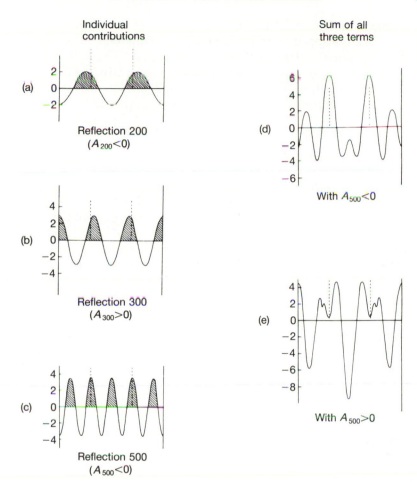

Fig. 6.13 An example of direct phase determination. If the reflections 200, 300, and 500 are all intense, they will give a significant contribution to the electron-density map. Suppose that, by some method, it is found that A_{200} has a negative sign (phase π) and A_{300} has a positive sign (phase 0). The areas in which these two terms contribute in a positive manner to the electron-density map are shaded. The regions in which these areas overlap, near $x = \pm 0.3$, correspond to regions to which A_{500} contributes positively only if it is negative (phase π). On summation of these three terms the least negative background is obtained when A_{500} has a negative sign; if A_{500} is positive the map is less satisfactory. The relation among these signs may then be written as $s(500) \approx s(200) \times s(300)$, where $\approx$ means 'probably equal to'. After Glusker and Trueblood (1985). (a) Contribution of reflection 200, with amplitude 2.0 and phase π (A_{200} negative). (b) Contribution of reflection 300, with amplitude 3.0 and phase 0 (A_{300} positive). (c) Contribution of reflection 500, with amplitude 3.5 and phase π (A_{500} negative). (d) Sum of all three terms (with A_{500} negative). (e) Sum of all three terms (with A_{500} positive).

determining reflections[†], the signs or phases of as many E values as possible are developed from triple products of the type listed above and probability formulae. Usually there are several sets of phases that satisfy the truncated data set so used. Often it is possible to identify the correct set on the basis of the probability associated with sign relationships. Otherwise several electron-density maps (computed with E values) may have to be calculated and searched for reasonable geometry of peak positions.

6.10.3   Heavy-atom and Patterson methods

If one or a few atoms in the structure have high atomic numbers and therefore dominate the scattering, the structure determination may be simplified by first finding the positions of these few atoms. The method for doing this is to calculate a *Patterson function* (Patterson 1934, 1935), which uses the squares of the amplitudes of the diffracted beams and a cosine function:

$$P(uvw) = \frac{1}{V} \sum_{hkl} F_{hkl}^2 \cos\left[2\pi(hu + kv + lw)\right]. \qquad (6.14)$$

This equation should be compared with that for the electron-density function, eqn (6.10). There are two main differences for the Patterson function; the values of F_{hkl} are squared and there are no phase terms. Thus this function can be calculated straightforwardly from the experimental data. Peaks in the Patterson map can be shown to represent vectors between atoms. The height of a peak will be proportional to the product of the atomic numbers of the two atoms at the ends of the vector. So, while a Patterson map for a unit cell containing N atoms will contain nearly N^2 peaks, if one or two atoms have high atomic numbers the vectors between them will be very high and will dominate the map. Therefore, if the space group is known, the positions of the heavy atoms can be found.

There are then two methods of proceeding. One can further analyze the Patterson map since the next highest peaks will be between the heavy atom(s) and lighter atoms. This can be done by moving the origin of the Patterson map onto each heavy atom in turn, using the space-group symmetry, and noting where peaks superpose in all origin positions of the map (Beevers and Robertson 1950). This is known as the vector superposition or minimum function method. The other method is to compute phases for the heavy atom(s) and then calculate the electron-density map using measured structure amplitudes, F_{hkl}^{obs}, and phase angles α_{hkl} calculated from the position(s) of the heavy atom(s). This map will, of course, favor the heavy atoms because information on their positions is contained in the phases used but

---

[†] These phases may be chosen arbitrarily but they then determine the location of the origin.

the structure amplitudes contain information on the entire structure. The combination of these two effects will give weak images of some if not all of the positions of the lighter atoms in addition to the high peaks at the heavy-atom positions. The phases are then recomputed, including the newly-found atomic positions, and the electron density searched for more peaks until the entire structure is found. With small, symmetrical molecules, the entire structure may often be derived directly from the Patterson map, even if heavy atoms are not present.

6.10.4 The method of isomorphous replacement

Isomorphous compounds are isostructural and capable of forming solid solutions. In the method of isomorphous replacement, differences in intensities from crystals of isostructural compounds differing only in the identity of one atom are measured. The varying atom — usually a heavy atom — is located by Patterson methods and from this the phases may be derived if the structure is centrosymmetric. Otherwise more isomorphs must be studied. This method of solving the phase problem is of great value in protein crystallography (see Blundell and Johnson 1976), but it is only occasionally applied to small molecules.

6.11 The use of electron-density maps

Once the relative phases of all the reflections have been determined by one of the methods listed above, an electron-density map may be computed. If part of the structure was omitted or slightly wrong this information will be revealed in this map. However, it is usually easier to interpret a map from which the known part of the structure has been removed. This is known as a *difference map* and uses as coefficients (with reference to eqn (6.8)) the differences $\Delta \mathbf{F}_{hkl} = \mathbf{F}_{hkl}^{obs} - \mathbf{F}_{hkl}^{calc}$, where $\mathbf{F}_{hkl}^{calc}$ is the structure factor calculated for that part of the structure that has been determined, and $\mathbf{F}_{hkl}^{obs}$ is the 'observed' structure factor (consisting of the observed structure amplitude and the calculated phase).

The electron-density distribution of a molecule contains information on its bonding properties and its potential reactivity. The *valence-electron distribution* may be analyzed through very accurate X-ray diffraction studies, as discussed in Chapter 10.

At this stage a reasonably good model of the structure has been obtained. Some methods for improving this model will now be described.

6.12 Refinement of the model

The molecular model that has been derived may still be improved by making slight adjustments to any or all of the positional and thermal (displacement) parameters. While inspection of a difference map may reveal the locations of hydrogen atoms and incorrect positions of certain atoms, the method generally used for refining atomic parameters is the method of least squares. This method finds the best fit of a *particular assumed model* to a set of experimental observations when there are more experimental observations than parameters to be determined. This latter condition is well realized in the X-ray diffraction experiment. The method involves minimization of the sum of the squares of the deviations between the experimental quantities and their values calculated with the derived parameters. In general, each atom has three positional parameters and one (isotropic) to six (anisotropic) thermal parameters, so that if there are N anisotropic atoms in a structure there are $9N + 1$ parameters to refine (the 1 is for an overall scale factor, that is needed to put observed and calculated structure amplitudes on the same scale). Sometimes a fractional occupancy factor is assigned to an atom and refined as an additional parameter. Ideally there should be at least 10 independent observations per parameter to be determined, so that there should be about 100 Bragg reflections measured per atom in the asymmetric unit. Individual measurements of F_{hkl} are weighted in the least-squares procedure by the reciprocal of the square of their estimated standard deviations, $\sigma(F_{hkl})$; but empirical weighting schemes are also used. According to the least-squares method, a set of normal equations is obtained from the observational equations that relate the observed structure amplitudes to the atomic parameters. Note that the trigonometric functions in the structure factor expression (see eqn (6.5)) must be linearized by Taylor expansion, followed by truncation at the first-order term. Due to this approximation the least-squares method is applied iteratively; several cycles may be necessary to achieve convergence. Further problems may arise because some of the parameters may be correlated so that changes in one result in changes of others. Additional information on the application of the least-squares method to the refinement of atomic parameters will be found in Chapter 10, Section 10.6.

The final coordinates of each atom in the structure, combined with the unit-cell dimensions, allow one to derive the geometry of the molecule. Estimated standard deviations in all these quantities should be used to calculate the standard deviations in the derived geometrical parameters.

6.13   Assessing the accuracy of X-ray crystallographic results

The determination of atomic coordinates by X-ray crystallographic techniques requires the measurements of intensities, the estimation of phase angles, the computation of an electron-density map and the refinement of a model that includes thermal vibration parameters. The intensity measurements are affected by errors which depend on the construction of the instrument used and the manner in which the experimenter uses it. Since there are many more data than parameters, one may hope that errors will tend to cancel out—that is, that they are random. Unfortunately, this is not generally true, because the diffracted intensities suffer from a number of *systematic* effects (absorption, extinction, multiple reflection, thermal diffuse scattering, crystal decay in the X-ray beam, scan-truncation errors) which may prove hard to eliminate (see Chapter 7). Additional systematic errors originate from inadequacies in the model used in the refinement. The scattering model that is normally adopted in X-ray crystallography is based on an assemblage of independent, spherical atoms; this is clearly inadequate to describe the contribution of valence electrons in molecules (see Chapter 10). The usual treatment of thermal vibration is based on the harmonic approximation; it may prove inadequate if the vibrational motion is highly anharmonic or curvilinear (see Chapter 8, Section 8.8, and Chapter 11, Section 11.3.2).

The following criteria should first of all be considered when assessing the correctness of an X-ray crystallographic study. The agreement of individual structure amplitudes with those calculated for the refined model should be consistent with the estimated precision of the experimental measurements. A difference electron-density map, with phases computed from the final refined parameters, should show no electron-density fluctuations greater than those expected on the basis of the estimated precision of this map. Isotropic and/or anisotropic thermal parameters should have reasonable values. Finally any anomalies in molecular geometry or molecular packing should be scrutinized with great care and regarded with skepticism.

A word about the R factor. This is a popular indicator of the correctness of a structure and is a measure of how well the observed structure amplitudes are matched by the values calculated for a proposed model.† The R factor is defined as:

$$R = \sum_{hkl} \left| F_{hkl}^{\text{obs}} - F_{hkl}^{\text{calc}} \right| \Big/ \sum_{hkl} F_{hkl}^{\text{obs}}. \tag{6.15}$$

Final R values less than about 0.06 are reported for reliable crystal structures. However, if one atom dominates the scattering by virtue of a high atomic number, the other atoms may not be well determined even if the R value is reasonably low. Therefore R values are not always definitive in assessing the precision of a structure. The correctness of the model used in the refinement must also be investigated. It is also important to check whether there are sufficient data per parameter refined (Jones 1984).

---

† Another indicator of the quality of the agreement is the *goodness-of-fit* (see Chapter 10, Section 10.6.1).

In the calculation of molecular geometry the estimated standard deviations from the least-squares refinement are used. These may be optimistically low because values of $\sigma(F_{hkl})$ (derived from $\sigma(I_{hkl})$) are often underestimated. Moreover, systematic errors arising from inadequate treatment of absorption, extinction, etc., are nearly always present. It is also important that estimated errors in cell dimensions should be taken into account (but they are only rarely). Of the final parameters the most questionable are those of the hydrogen atoms. This is because hydrogen atoms have only one electron; their apparent position is displaced towards the atom they are bonded to; they undergo high-amplitude thermal vibrations and are often disordered (e.g. in methyl groups linked to aromatic molecules and in a number of hydrogen-bonded systems). Sometimes hydrogen atoms are introduced in fixed, idealized positions in the refinement and the reader of a crystallographic paper should be aware of this. On the other hand, neutron diffraction gives better hydrogen positions and thermal parameters because hydrogen scatters neutrons to an extent comparable to other elements (see Chapter 11).

In spite of all these problems today's crystal structure analysis, particularly when the experiment is carefully carried out at low temperature, may reach a high level of accuracy and the resulting molecular geometries merit detailed analysis, as shown in Chapters 15 to 21.

6.14 Acknowledgements

We are pleased to acknowledge the valuable critical comments of our colleagues Carlo Maria Gramaccioli and Silvio Cerrini on early versions of this chapter. This work was supported by NIH grants CA-10925, CA-06927 and by an appropriation from the Commonwealth of Pennsylvania.

References

Beevers, C. A. and Robertson, J. H. (1950). *Acta Crystallogr.*, **3**, 164.

Bijvoet, J. M., Peerdeman, A. F., and van Bommel, A. J. (1951). *Nature*, **168**, 271–2.

Binnig, G., Rohrer, H., Gerber, Ch., and Weibel, E. (1982). *Appl. Phys. Lett.*, **40**, 178–80.

Blundell, T. L. and Johnson, L. N. (1976). *Protein crystallography*. Academic Press, New York.

Bragg, W. L. (1912). *Nature*, **90**, 410.

Bragg, W. L. (1913). *Proc. Cambridge Phil. Soc.*, **17**, 43–57.

Bunn, C. W. (1961). *Chemical crystallography: An introduction to optical and X-ray methods* (2nd edn). Clarendon Press, Oxford.

Cotton, F. A. (1971). *Chemical applications of group theory* (2nd edn), Chapter 3. Wiley-Interscience, New York.

Dunitz, J. D. (1979). *X-ray analysis and the structure of organic molecules.* Cornell University Press, Ithaca.

Ewald, P. P. (1913). *Phys. Z.*, **14**, 465–72.

Ewald, P. P. (1921). *Z. Kristallogr.*, **56**, 129–56.

Ewald, P. P. (1923). *Kristalle und Röntgenstrahlen.* Springer, Berlin.

Friedrich, W., Knipping, P., and Laue, M. (1912). *Sitzungsber. math. phys. Klasse kgl. Bayer. Akad. Wiss.*, pp. 303–22. English translation by J. J. Stezowski (1981) in *Structural crystallography in chemistry and biology* (ed. J. P. Glusker), pp. 23–39. Hutchinson and Ross, Stroudsburg.

Glusker, J. P. and Trueblood, K. N. (1985). *Crystal structure analysis: A primer* (2nd edn). Oxford University Press, New York.

Hahn, T. (ed.) (1987). *International tables for crystallography*, Vol. A, *Space-group symmetry* (2nd edn). Reidel, Dordrecht.

Hargittai, I. and Hargittai, M. (1986). *Symmetry through the eyes of a chemist.* VCH, Weinheim.

Haüy, R. J. (1784). *Essai d'une théorie sur la structure des crystaux, appliquée a plusieurs genres de substances crystallisées.* Gogué and Née de la Rochelle, Paris. Additional treatises were published in 1801 and 1822.

Hooke, R. (1665). *Micrographia: Or some physiological descriptions of minute bodies made by magnifying glasses. With observations and inquiries thereupon.* Martyn and Allestry (for the Royal Society), London. Reprinted as: Gunther, R. W. T. (1938). *The life and work of Robert Hooke. Part V. Micrographia, 1665.* Early Science in Oxford, Vol. 13. Oxford University Press.

Huygens, C. (1690). *Traité de la lumière.* Van der Aa, Leiden.

Iijima, S. (1973). *Acta Crystallogr.*, **A29**, 18–24.

James, R. W. (1962). *The optical principles of the diffraction of X-rays* (4th reprint, with additions). Bell, London.

Jenkins, F. A. and White, H. E. (1976). *Fundamentals of optics* (4th edn). McGraw-Hill, New York.

Jones, P. G. (1984). *Chem. Soc. Rev.*, **13**, 157–72.

Karle, J. and Hauptman, H. (1950). *Acta Crystallogr.*, **3**, 181–7.

Kepler, J. (1611). *Strena seu de nive sexangula.* Tampach, Francofurti ad Moenum. English translation of excerpts by J. S. Silverman (1977): A New Year's gift, or on a hexagonal snowflake. In *Crystal form and structure* (ed. C. J. Schneer), pp. 16–7. Dowden, Hutchinson, and Ross, Stroudsburg.

Labaw, L. W. and Wyckoff, R. W. G. (1958). *J. Ultrastruct. Res.*, **2**, 8–15.

Ladd, M. F. C. and Palmer, R. A. (1985). *Structure determination by X-ray crystallography* (2nd edn). Plenum Press, New York.

Luger, P. (1980). *Modern X-ray analysis on single crystals.* De Gruyter, Berlin.

Miller, W. H. (1839). *A treatise on crystallography.* Cambridge.

Nowacki, W., Matsumoto, T., and Edenharter, A. (1967). *Acta Crystallogr.*, **22**, 935–40.

Nyburg, S. C. (1961). *X-ray analysis of organic structures.* Academic Press, New York.

Patterson, A. L. (1934). *Phys. Rev.*, **46**, 372–6.

Patterson, A. L. (1935). *Z. Kristallogr.*, **90**, 517–42.
Sayre, D., Kirz, J., Feder, R., Kim, D. M., and Spiller, E. (1977). *Science*, **196**, 1339–40.
Steadman, R. (1982). *Crystallography*. Van Nostrand-Reinhold, New York.
Stout, G. H. and Jensen, L. H. (1989). *X-ray structure determination: A practical guide* (2nd edn). Wiley-Interscience, New York.
Vainshtein, B. K., Fridkin, V. M., and Indenbom, V. L. (1982). *Modern crystallography*, Vol. 2, *Structure of crystals*. Springer, Berlin.

Appendix

A glossary of selected crystallographic terms

Compiled by Jenny P. Glusker, Fred L. Hirshfeld, Kenneth N. Trueblood, and Aldo Domenicano.

Absolute configuration, absolute structure: The structure of a chiral crystal or molecule expressed in an absolute frame of reference. In 1951 Bijvoet and co-workers demonstrated that the absolute structure can be determined using the differences between $(F_{hkl})^2$ and $(F_{\bar{h}\bar{k}\bar{l}})^2$ that are found when there is anomalous scattering. (See also *Friedel's law*.)

Absolute scale: Values of, for example, structure amplitudes, relative to the amplitude of scattering by a classical point electron, are said to be on an *absolute scale*. Experimentally, it is much easier to measure the *relative* intensities of the reflections from a particular crystal than to determine their absolute values. Hence, the experimentally derived relative structure amplitudes are often placed approximately on an absolute scale by scaling against values calculated from a model structure.

Absorption edges: The absorption of X rays by an element increases as the wavelength increases. However, there are discontinuities at which the absorption coefficient abruptly drops to a low value and then starts to rise again. These are called *absorption edges* and they occur at wavelengths below which the incident X-ray quantum has just sufficient energy to excite a bound electron to a vacant higher orbital or to eject it altogether.

Absorption effects: As an X-ray beam passes through a crystal, its intensity is reduced by scattering and by absorption. The intensity, I, of a beam after passing through a thickness t of an absorbing crystal is: $I = I_0 \exp(-\mu t)$, where I_0 is the intensity of the incident beam and μ the linear absorption coefficient; μ is a function of wavelength and atomic composition.

Anomalous scattering, anomalous dispersion: The occurrence of a phase change on scattering of X rays that are strongly absorbed by one or more kinds of atoms present in the crystal.

Asymmetric unit: The unique portion of a crystal structure from which the entire structure can be generated by the space-group symmetry operations. The asymmetric unit is usually a fraction of the unit cell; it coincides with the unit cell in the triclinic space group $P1$. For molecular crystals, the asymmetric unit may contain part of a molecule, a whole molecule, or all or part of several molecules not related by crystallographic symmetry.

Atomic coordinates: A set of coordinates that specifies the position of an atom with respect to a specified reference system, usually the crystal axes. Atomic coordinates are generally expressed as fractions of unit-cell edges.

Atomic parameters: A set of quantities relative to each crystallographically independent atom in the crystal structure. In most cases they are the three *atomic coordinates* (q.v.) and the *thermal* (or, better, *displacement*) *parameters* (q.v.). Only one displacement parameter is needed to describe the isotropic thermal vibration of an atom, while six are required to define (harmonic) anisotropic motion. With disordered structures, an additional atomic parameter is the *occupancy factor* (q.v.).

Bragg equation: Each diffracted beam may be considered as a 'reflection' from a family of parallel lattice planes *hkl*. If the perpendicular spacing between successive lattice planes is d_{hkl}, and the angle between the diffracted beam (wavelength λ, *n*-th order of diffraction) and the normal to the set of lattice planes is $\pi/2 - \theta_{hkl}$, then the following equation holds: $2d_{hkl}\sin\theta_{hkl} = n\lambda$. If X rays strike a crystal they will be diffracted when, and only when, this equation is satisfied.

Bragg reflection: See *Reflection*.

Centrosymmetric structure: A crystal structure whose space group, and therefore arrangement of atoms, contains a center of symmetry. When the origin is chosen at the center of symmetry, the phase angle for each reflection is either 0 or π.

Constrained geometry: A molecular fragment of known or assumed geometry, such as a benzene ring, can be treated in the least-squares refinement as a group that requires only six coordinates to define it, i.e. three defining position and three defining orientation in the unit cell. In this way, by constraining the geometry of the fragment, the number of parameters to be refined is reduced.

Convolution: Given two functions $f(xyz)$ and $g(xyz)$, their *convolution* at the point uvw is defined as:

$$c(uvw) = \int\!\!\int\!\!\int_{-\infty}^{+\infty} f(xyz)\cdot g(u-x, v-y, w-z)\,\mathrm{d}x\,\mathrm{d}y\,\mathrm{d}z$$

(where the roles of the two functions are interchangeable). Note that a crystal structure is a convolution of the crystal lattice and the contents of the unit cell, and that the Patterson function is a self-convolution of the electron density.

Crystal classes: See *Crystallographic point groups*.

Crystal lattice: Crystals are composed of atoms or groups of atoms repeated at regular intervals in three dimensions with the same orientation. Each such group of atoms may be replaced by a representative point to give the space lattice or *crystal lattice*. The representative points may be centers of atoms or any point within the unit cell, provided they are all completely equivalent to each other. Thus the crystal lattice may be defined as a set of infinite, evenly spaced equivalent points related to each other by translational symmetry. The meaning is specific and *the term should not be used to denote the entire atomic arrangement*.

Crystallographic point groups, crystal classes: As a general definition, a *point group* is a group of symmetry operations that leave unmoved at least one point within the object to which they apply. There is, of course, an infinite number of such groups. The 32 *crystallographic point groups*, also called *crystal classes*, are those that contain only the symmetry operations that are allowed in a crystal, i.e. rotation (n) and rotatory inversion ($\bar{n}$) axes with $n = 1, 2, 3, 4, 6$. The crystallographic point groups characterize the external symmetry of (well-formed) crystals.

Debye–Waller factor: The (isotropic) *temperature factor* (q.v.).

Deformation density: The difference between the true electron density of a molecule or crystal and the corresponding *promolecule* or *procrystal* density. Both could be either static (non-vibrating) or dynamic (smeared by identical atomic vibrations).

Difference synthesis: A Fourier synthesis (or map) for which the Fourier coefficients are the differences between the observed and calculated structure factors for each reflection. Such a map will have peaks where not enough electron density was included in the trial structure and troughs where too much was included. It is used both for locating missing atoms and for correcting the positions of atoms already included in the model.

Diffuse scattering: Halos or streaks that appear around intense reflections and indicate the presence of disorder in the structure (static) or high thermal motion of atoms (dynamic). In principle, for any crystal structure there is diffuse scattering due to thermal motion (*thermal diffuse scattering* or TDS). This scattering has a maximum coinciding with the Bragg peak and in most cases it is evaluated as a part of the peak intensity and of the background. It is very difficult to correct diffraction data for this effect, and its neglect results in an underevaluation of the atomic displacement parameters.

Direct methods: Methods of deriving relative phases of Bragg reflections by a consideration of probability relationships among the phases of the more prominent reflections. These relationships come from the conditions that the structure is composed of atoms (giving independent, isolated peaks in the electron-density map) and that the electron density is nowhere negative. Only certain values for the phases are consistent with these conditions.

Discrepancy index: See *R factor*.

Disordered structure: A crystal structure in which atoms or molecules pack randomly (non-periodically) in alternative ways in different unit cells. Such disorder may be limited to a portion of the unit cell; it is revealed by the presence of diffuse scattering around intense reflections.

Displacement parameters: Atomic vibrations are displacements from mean positions with periods that are typically smaller than 10^{-12} s. Because *static* displacements of a given atom in a structure that vary in an essentially random fashion from one unit cell to another will simulate vibrations of that atom, the vibration parameters of an atomic temperature factor are better referred to as *displacement parameters* unless there is evidence that they truly denote vibrational displacements uncontaminated by static disorder. See also: *Temperature factor*.

Double reflection (also called *Renninger effect* after its discoverer): X rays, reflected by one set of lattice planes, may then be reflected by another set of planes which, by chance or design, are in exactly the right orientation. The twice-reflected beam emerges in a direction that corresponds to single reflection from a third set of planes, whose Miller indices are the sums of the indices of the two sets of planes producing the reflection. It thus adds spurious intensity to a reflection, an effect that can be serious if the true intensity of that reflection is weak or zero. This may cause an ambiguity in the space-group determination if a systematically absent reflection gains intensity by it. It can generally be eliminated by reorienting the crystal, or by changing the wavelength of the X rays.

Dynamical diffraction: A theory of diffraction that takes into account the attenuation of the primary beam on passage through the crystal, as well as the mutual interactions between primary and scattered beams. These effects are important for perfect crystals, and in the case of electron diffraction by crystals.

Electron-density map: A contoured representation of electron density at various points in a crystal structure. Electron density is highest near atomic centers. The map may be calculated by Fourier summation from the experimental structure amplitudes, F_{hkl}^{obs}, and an appropriate set of phases, α_{hkl}:

$$\rho(xyz) = \frac{1}{V}\sum_{hkl} F_{hkl}^{obs} \cos\left[2\pi\left(hx + ky + lz\right) - \alpha_{hkl}\right],$$

where V is the volume of the unit cell.

E-map: A Fourier map, comparable to an electron-density map, with phases derived by direct methods and normalized structure amplitudes, E_{hkl}, replacing F_{hkl} in the Fourier summation. Since the E values correspond to point atoms at rest, the peaks on the resulting map are sharper than those computed with F values.

Equivalent positions: The complete set of positions produced by the operation of the symmetry elements of the space group upon any general position.

Equivalent reflections: A set of reflections whose intensities are required by the crystal class to be equal. In the absence of anomalous scattering, any two reflections related by inversion through the reciprocal-lattice origin have equal intensity whatever the crystal class (see *Friedel's law*). Combination of these two kinds of equivalence defines the *Laue symmetry* of the diffraction pattern. According to this symmetry, the reflections hkl, $\bar{h}k\bar{l}$, $h\bar{k}l$, $\bar{h}\bar{k}\bar{l}$ are, for example, equivalent for any monoclinic crystal. But if anomalous scattering is appreciable this set of reflections will, for a non-centrosymmetric monoclinic crystal, split into two pairs of true symmetry equivalents (hkl, $h\bar{k}l$ and $\bar{h}k\bar{l}$, $\bar{h}\bar{k}\bar{l}$). In high-symmetry crystals other sets of reflections may be equivalent, e.g. hkl, klh, and lhk for cubic crystals.

Ewald sphere: See *Sphere of reflection*.

Extinction: A specific type of systematic weakening of the incident beam as it passes through a crystal. Part of the incident beam in a single perfect block may be multiply reflected so that it returns to its original direction but is out of phase with the main beam, thus reducing the intensity of the latter. Also the intensity of the diffracted beam is decreased by this mechanism (*primary extinction*). When the crystal is mosaic, part of the beam will be diffracted by one mosaic block and therefore is not available for diffraction by a following block that is accurately aligned with the first. Thus the second block contributes less than expected to the diffracted beam (*secondary extinction*). Both effects are evidenced by a tendency for F^{obs} to be systematically smaller than F^{calc} for very intense reflections. Extinction can be reduced by dipping the crystal in liquid nitrogen, thereby increasing its mosaicity.

Form factor: See *Scattering factor*.

Fourier map: A map computed for a periodic function by addition of waves of known frequency, amplitude and phase. The term is generally used by crystallographers for an electron-density or difference electron-density map.

Fourier synthesis: A function $f(t)$ that is periodic with period T may be represented by a Fourier series:

$$f(t) = a_0/2 + \sum_{n=1}^{\infty} a_n \cos(2\pi nt/T) + \sum_{n=1}^{\infty} b_n \sin(2\pi nt/T);$$

or, using the exponential form:

$$f(t) = \sum_{n=-\infty}^{\infty} c_n \exp(2\pi int/T).$$

The Fourier theorem states that any periodic function may be resolved into cosine and sine terms. Since a crystal has a periodic three-dimensional structure, its electron density may be represented by a three-dimensional Fourier summation or *Fourier synthesis*.

Fourier transform: In the pair of equations

$$f(x) = \int_{-\infty}^{\infty} g(y) \exp(2\pi ixy) \, dy$$

$$g(y) = \int_{-\infty}^{\infty} f(x) \exp(-2\pi ixy) \, dx$$

$f(x)$ is the Fourier transform of $g(y)$, and $g(y)$ is the inverse Fourier transform of $f(x)$. In X-ray diffraction the structure factor, $\mathbf{F}_{hkl}$, is related to the electron density, $\rho(xyz)$, by the equation

$$\mathbf{F}_{hkl} = \int_V \rho(xyz) \exp(i\phi) \, dV,$$

and conversely

$$\rho(xyz) = \frac{1}{V} \sum_{hkl} \mathbf{F}_{hkl} \exp(-i\phi),$$

where $\phi = 2\pi(hx + ky + lz)$ and V is the volume of the unit cell. Thus the structure factor is the Fourier transform of the electron density and the electron density is the inverse Fourier transform of the structure factor. Summation replaces integration in the latter equation because the diffraction pattern of a crystal is observed only at discrete points. The intensity at each of these points is proportional to F_{hkl}^2 and thus to the value at that point of the square of the Fourier transform of the electron density.

Friedel's law: This law states that the intensities of centrosymmetrically related reflections are equal, $I_{hkl} = I_{\bar{h}\bar{k}\bar{l}}$. This law does not hold with non-centrosymmetric structures that contain anomalous scatterers.

Glide plane: A *glide plane* is a symmetry element for which the symmetry operation is reflection across the plane combined with translation in a direction parallel to the plane. The translation component leads to the generation of an infinitely repeating periodic pattern in the direction of the translation.

Habit: The usual appearence of a crystal especially in terms of the presence of sets of symmetry-related faces. The faces of a crystal on which growth is slowest are the largest and best developed.

Heavy-atom method: A method of structure determination in which the phases calculated from the position of a heavy atom are used to compute the first approximate electron-density map, from which further portions of the structure are recognizable as additional peaks. If necessary, successive approximate electron-density maps are computed to give the entire structure.

Integrated intensity: The total intensity of radiation (i.e. the total number of photons) reflected from a family of lattice planes as the crystal is rotated at constant speed with respect to the incident beam through a small angular range about the Bragg angle.

Intensity distribution: Intensities of Bragg reflections from a non-centrosymmetric crystal tend to be clustered more tightly around their mean value than are those from a centrosymmetric one. This effect can be used to test for the presence or absence of a center of symmetry in the crystal.

Isomorphism: Similarity of crystal shape, unit-cell dimensions, and structure between substances of similar chemical composition. Ideally, the substances are so closely similar that they can form a continuous series of solid solutions.

Isomorphous replacement method: A method of deriving phases from the differences in intensity between corresponding reflections from two or more isomorphous crystals.

Kinematic diffraction: A theory of diffraction in which attenuation of the primary beam on passage through the crystal and coherent interference between incident and scattered beams are neglected, as in a weakly diffracting mosaic crystal.

Laue classes, Laue groups: The eleven possible symmetry arrangements of X-ray diffraction intensities (in the absence of anomalous scattering). They correspond to the 11 crystallographic point groups having a center of symmetry. See also: *Equivalent reflections*.

Libration: A form of rigid-body vibrational motion in which each atom in a molecule moves along an arc rather than along a straight line. See also: *Thermal-motion corrections*.

Lorentz factor: A correcting factor in the reduction of intensity data. It takes into account the time that it takes for a given reflection (represented as a reciprocal-lattice point with finite size) to pass through the surface of the sphere of reflection. The value depends on the scattering angle and on the geometry of the measurement of the reflection.

Miller indices: A set of three integers used to identify a face of a crystal, a set of lattice planes, or a particular order of reflection from these planes. Of the set of planes with Miller indices h, k, and l, the first plane from the origin makes intercepts

a/h, b/k, and c/l with the unit-cell edges a, b, and c. The *law of rational indices* states that the indices of the natural faces of a crystal are *small* integers, seldom greater than three.

Mosaic spread: Although Bragg's law implies that X-ray diffraction is possible only when the orientation of the crystal with respect to the incident beam satisfies the Bragg equation exactly, in fact the diffracted intensity is usually appreciable over a range of several tenths of a degree around the Bragg angle. To explain this behavior, we may imagine the crystal to be built as a mosaic of many tiny blocks differing slightly in orientation. The *mosaic spread* is a measure of the degree of orientational inhomogeneity in a particular crystal. Extinction (q.v.) is much weaker in a mosaic than in a perfect crystal, allowing the diffracted intensities to be predicted by the kinematic theory of diffraction.

Non-centrosymmetric structure: A crystal structure with no center of symmetry in the atomic arrangement. The phase angle of each reflection may have any value between 0 and 2π.

Non-crystallographic symmetry: Local symmetry within the asymmetric unit of a crystal structure. For example, the asymmetric unit may contain two molecules having identical structure but, since they are not related by crystallographic symmetry, they have different environments. Non-crystallographic symmetry can be used as an aid in structure determination.

Normalized structure factor: The ratio of the value of the structure amplitude to its root-mean-square expectation value. It is given by $E_{hkl} = F_{hkl}/(\epsilon \Sigma_j f_j^2)^{1/2}$, where the f_j are the atomic scattering factors (corrected for an overall temperature factor) and ϵ is a factor that takes into account the fact that groups of reflections in particular areas of the reciprocal lattice may have an average intensity greater than that for the general reflections. Values of ϵ for the various groups of reflections depend on the space group of the crystal.

Occupancy factor: A parameter that defines the fractional occupancy of a given site by a particular atom in a disordered structure.

Order of diffraction: An integer associated with a given interference fringe of a diffraction pattern. It is first order if it arises as a result of a radiation path difference of one wavelength. Similarly, the n-th order corresponds to a path difference of n wavelengths.

Patterson synthesis: A summation of a Fourier series that has the squares of the structure amplitudes as coefficients:

$$P(uvw) = \frac{1}{V}\sum_{hkl} F_{hkl}^2 \cos\left[2\pi(hu + kv + lw)\right].$$

Because the values of F_{hkl}^2 can be calculated from the diffraction intensities, the Patterson map can be computed directly with no phase information needed. Ideally, the positions of the maxima in the map represent the end points of vectors between atoms, all referred to a common origin.

Phase: The difference in position of the crests of two waves of the same wavelength traveling in the same direction. Also, the point to which the crest of a given wave has advanced in relation to a standard position, for example, the starting point. The phase is usually expressed as a fraction of the wavelength in angular measure, with one cycle or period being 2π; that is, if the crests are displaced by Δx for a wavelength λ, the phase difference is $2\pi\Delta x/\lambda$. The *phase of a structure factor* is expressed relative to the phase of a wave scattered at the origin of the unit cell in the same direction. It is given by the expression $\alpha_{hkl} = \tan^{-1}(B_{hkl}/A_{hkl})$, where A_{hkl} and B_{hkl} are the components of the structure factor (q.v.). If the structure is centrosymmetric and the origin is set at a center of symmetry then B_{hkl} vanishes and the phase angle can only be 0 or π, according to the positive or negative sign of A_{hkl}.

Phase problem: The problem of determining the phase angle to be associated with each structure amplitude, so that an electron-density map may be calculated from a Fourier series with structure factors (including both amplitude and phase) as coefficients. The measured intensities of diffracted beams give only the squares of the amplitudes; the phases cannot normally be determined experimentally. Phases may, however, be calculated for any postulated structure and combined with the experimentally determined amplitudes to give a tentative electron-density map. They may also, in special cases, be deduced by various methods, e.g. Patterson synthesis (q.v.), heavy-atom method (q.v.), isomorphous replacement method (q.v.), and direct methods (q.v.).

Polarization factor: A correcting factor in the reduction of intensity data. It takes into account the fact that the variation of the reflection efficiency of X rays with the scattering angle depends on the polarization status of the incident beam. Unlike the Lorentz factor (q.v.), the polarization factor does not depend on the method used for data collection, except when a crystal monochromator is used.

Polymorphism: A substance may adopt different crystal structures, depending on the conditions of crystallization. This is a widespread phenomenon and is called *polymorphism*.

Probability density function (pdf): A function, $P(\mathbf{u})$, giving the probability density of a particle in space. The probability of finding the particle in a volume element dV at position $\mathbf{u}$, measured from some origin, is $P(\mathbf{u})dV$.

Promolecule (procrystal): A hypothetical electron density defined by summation of the densities of independent, ground-state atoms at the same positions as the atoms of a given molecule (or crystal). Usually the atomic densities are spherically averaged, but alternative definitions are sometimes preferred.

R factor, discrepancy index: An index that gives a measure of the disagreement between observed and calculated structure amplitudes and hence a crude measure of the correctness of the derived model and the quality of the data. It is defined as:

$$R = \sum_{hkl} |F_{hkl}^{obs} - F_{hkl}^{calc}| / \sum_{hkl} F_{hkl}^{obs}.$$

Values less than about 0.06 are considered good for present-day structure determinations. However, some partially incorrect structures have had R values below 0.10, and many basically correct but imprecise structures have higher R values.

Reciprocal lattice: The lattice defined by axes $\mathbf{a^*}$, $\mathbf{b^*}$, $\mathbf{c^*}$; it is related to the crystal lattice or direct lattice (with axes $\mathbf{a}$, $\mathbf{b}$, $\mathbf{c}$) in such a way that $\mathbf{a^*}$ is perpendicular to $\mathbf{b}$ and $\mathbf{c}$ and its length is inversely proportional to the thickness of the direct cell in this direction; $\mathbf{b}^*$ is similarly related to $\mathbf{a}$ and $\mathbf{c}$; and $\mathbf{c}^*$ to $\mathbf{a}$ and $\mathbf{b}$. Therefore, for instance,

$$\mathbf{a}^* = (\mathbf{b} \times \mathbf{c})/(\mathbf{a} \cdot \mathbf{b} \times \mathbf{c}) = (\mathbf{b} \times \mathbf{c})/V, \text{ etc.,}$$

where V is the volume of the direct cell. From this definition, it follows that $\mathbf{a} \cdot \mathbf{a}^* = 1$, $\mathbf{a}^* \cdot \mathbf{b} = \mathbf{a}^* \cdot \mathbf{c} = 0$, etc. Rows of points in the direct lattice are normal to planes of the reciprocal lattice, and vice versa. The repeat distance between points in a particular row of the reciprocal lattice is inversely proportional to the interplanar spacing between the crystal-lattice planes that are normal to this row of points. The same relation holds between the distances between points of the crystal lattice and the spacings of planes of the reciprocal lattice. Using the reciprocal lattice greatly facilitates the interpretation of X-ray diffraction patterns of crystals.

Refinement: A process of improving the parameters of an approximate (trial) structure until the best fit of calculated structure amplitudes to those observed is obtained. This is usually done by the method of least squares. Since the dependence of the structure amplitudes on atomic parameters is not linear, the process involves a series of iterations until convergence is reached. To avoid falling into physically meaningless minima, it is absolutely necessary to start with a sufficiently good set of initial parameters. Ideally, there should be at least 10 different F_{hkl} measured per parameter to be determined.

Reflection, Bragg reflection: Since diffraction by a crystal may be considered as reflection from a set of lattice planes (a view suggested by W. L. Bragg), this term has come to be used to denote a diffracted beam.

Renninger effect: See *Double reflection*.

Resolution: The ability to distinguish adjacent parts of an object when examining it with radiation. Most X-ray structures of small molecules are determined to a resolution of 0.8 Å or better. At this resolution each atom is fairly distinct. The resolution improves with an increase in the maximum value of $\sin \theta / \lambda$ at which reflections are measured. Thermal vibration and crystal quality generally limit the resolution that is ultimately attainable.

Rigid-body model: A model of vibrational motion that assumes a molecule to be rigid, so that all its interatomic distances are constant.

Scattering factor, form factor: X rays set the electrons of atoms into vibration, causing them to act as sources of secondary radiation. The scattering power of a single atom depends on its electronic structure and the angle of scattering. This scattering is reduced as the vibration of the atom increases (see *Temperature factor*). These effects are expressed in *atomic scattering factors*, which represent the scattering power of an atom measured relative to the scattering by a single point electron at the atomic center; the scattering power falls off as the scattering angle increases. Atomic scattering factors are computed from theoretical wavefunctions for free atoms (neutral or charged). They are modified by anomalous scattering if the incident wavelength is near an absorption edge of the scattering element.

Scattering vector: The reciprocal-lattice vector associated with (and perpendicular to) a set of reflecting crystal-lattice planes *hkl*:

$$\mathbf{H} = h\mathbf{a}^* + k\mathbf{b}^* + l\mathbf{c}^*.$$

Its magnitude is given by $H = 1/d_{hkl} = 2\sin\theta_{hkl}/\lambda$, where d_{hkl} is the interplanar spacing. The order of diffraction, n, is contained in the Miller indices hkl as a multiplying factor.

Screw axis: A *screw axis*, designated n_r, is a symmetry element for which the symmetry operation is a rotation about the axis by $2\pi/n$ coupled with a translation parallel to the axis by r/n of the unit-cell length in that direction. The translation component leads to the generation of an infinitely repeating periodic pattern in the direction of the translation.

Series-termination error: The error in a (periodic) function arising from trunca-tion of the (Fourier) series by which it is calculated. In an X-ray diffraction experi-ment, the number of Fourier coefficients available for evaluation of the electron density will depend on the maximum $\sin\theta/\lambda$ value to which the data are collected. Because of truncation of the Fourier series, peaks in the resulting Fourier synthesis are surrounded by series of ripples. These are especially noticeable around a heavy atom. The use of difference synthesis (q.v.) obviates most of the effects of series-termination errors.

Sharpened Patterson synthesis: A Patterson synthesis computed with values of F_{hkl}^2 modified by an appropriate function that enhances those coefficients with high values of $\sin\theta/\lambda$. The resulting interatomic vectors appear as sharper peaks and the Patterson map may therefore be simpler to interpret.

Space groups: Groups of symmetry operations consistent with an infinitely extended pattern repeated periodically along three dimensions. There are 230 such groups; they can be derived by addition of mutually consistent translational sym-metry to the 32 crystallographic point groups. Allowed translational elements are simple translations, screw axes, and glide planes. Each space group may be con-sidered as the group of operations that converts the unique portion of the structure, the asymmetric unit, into an infinitely extending pattern. The space group of a crystal can be identified (although sometimes with some ambiguity) from the Laue symmetry of, and systematic absences in, the diffraction pattern.

Sphere of reflection, Ewald sphere: A construction for deducing conditions for diffraction in terms of the reciprocal rather than the direct lattice. It is a sphere, of radius $1/\lambda$, with the incident beam along a diameter. The origin of the reciprocal lattice is positioned at the point where the incident beam emerges from the sphere. Whenever a reciprocal-lattice point, P, touches the surface of the sphere, the condi-tions for a beam to be diffracted (or reflected, in the sense used by Bragg) are satisfied. The direction of the diffracted beam is given by the line connecting the center of the sphere with P. Thus, for any orientation of the crystal relative to the incident beam, it is possible to predict which reciprocal-lattice points, and thus which planes in the crystal, will be in a reflecting position.

Structure amplitude: The magnitude of the *structure factor* (q.v.).

Structure factor: The wave scattered by the contents of the unit cell in the direction of the hkl reflection is described in amplitude and phase by the *structure factor*, $\mathbf{F}_{hkl}$. The *magnitude of the structure factor*, F_{hkl}, is the ratio of the amplitude of the radiation scattered by the contents of the unit cell to that scattered by a classical point electron; it is also called *structure amplitude*. In the approximation usually employed for crystal structure calculations, the structure factor is expressed as a sum of single-atom contributions, extended to all atoms in the unit cell:

$$\mathbf{F}_{hkl} = \sum_j f_j \exp\left[2\pi i(hx_j + ky_j + lz_j)\right]$$

$$= \sum_j f_j \cos\left[2\pi(hx_j + ky_j + lz_j)\right] + i\sum_j f_j \sin\left[2\pi(hx_j + ky_j + lz_j)\right]$$

$$= A_{hkl} + iB_{hkl},$$

where f_j is the atomic *scattering factor* (q.v.), corrected for the temperature factor (q.v.). The atomic coordinates x_j, y_j, z_j are expressed as fractions of the unit-cell edges. If the structure is centrosymmetric and the origin is set at a center of symmetry then B_{hkl} vanishes and the structure factor is no longer a complex quantity. The *phase of the structure factor*, $\alpha_{hkl} = \tan^{-1}(B_{hkl}/A_{hkl})$, is expressed relative to the phase of a wave scattered at the origin of the unit cell in the same direction. From the intensity of a reflection, I_{hkl}, the magnitude but not the phase of $\mathbf{F}_{hkl}$ can be derived directly. The experimentally determined structure amplitudes are denoted by F_{hkl}^{obs}; those calculated for a model of the structure are designated as F_{hkl}^{calc}. For the relation between structure factor and electron density, see *Fourier transform*.

Superposition method: A method of analyzing the Patterson map that involves setting the origin of the Patterson map in turn on the known positions of certain atoms and suitably combining the superposed maps.

Temperature factor: An exponential expression by which the scattering of an atom is reduced as a consequence of vibration (or a simulated vibration resulting from static disorder). For isotropic harmonic motion the exponential factor is $\exp(-B\sin^2\theta/\lambda^2)$ (Debye–Waller factor), with B called, loosely but commonly, the 'isotropic temperature factor'. It equals $8\pi^2\langle u^2\rangle$, where $\langle u^2\rangle$ is the mean-square displacement of the atom from its mean position. For anisotropic motion the exponential expression most commonly used contains six parameters, the anisotropic vibration or displacement parameters, which describe ellipsoidal rather than isotropic motion or average static displacements.

Thermal diffuse scattering: See *Diffuse scattering*.

Thermal-motion corrections: Adjustments to intramolecular dimensions from a crystal structure determination for distortions arising from atomic vibrations, especially libration (q.v.). The appropriate corrections depend on a model for specifying the correlations between the motions of the several atoms. For librational motion, the apparent distances are often shorter, never longer, than the true distances. For example, in a rigid molecule undergoing libration, all atoms appear displaced toward the libration axis. The 'librational motion' considered may be genuine or largely apparent, a consequence of static disorder in the positions of the atoms.

Thermal parameters: See *Displacement parameters*, *Temperature factor*.

Translation: The word *translation* has two different meanings in crystallography. Generally it indicates the symmetry element that is typical of lattices, but it is also used to denote a type of rigid-body thermal motion, in which all atoms of a molecule move the same distance in the same direction, that is, along the same or parallel lines.

Twin: A composite crystal built from two or more individuals that have grown together in a specific relative orientation.

Unit cell: The basic building block of a crystal, repeated infinitely by translation in three dimensions. It is characterized by three vectors, **a**, **b**, and **c**, that form the edges of a parallelepiped. The angles between these vectors are denoted as α (between **b** and **c**), β (between **a** and **c**), and γ (between **a** and **b**).

Wilson plot: A plot of the logarithm of the average ratio of observed Bragg intensities to the theoretical values expected for a random arrangement of the same (stationary) atoms in the unit cell, in successive ranges of $\sin^2 \theta / \lambda^2$. Such a plot typically approaches a straight line, whose intercept yields the factor needed to place the observed intensities on an absolute scale (q.v.) and whose slope yields an average isotropic displacement parameter for the whole structure.

X rays, characteristic: X rays of definite wavelengths, characteristic of an element. They are emitted, for instance, together with a continuous background, if a substance (a metal in the case of X-ray tubes) is bombarded by fast electrons. In this process an electron that has been displaced from an inner shell is replaced by another electron that falls in from an outer shell, the energy lost by this second electron being emitted as a quantum of X radiation.

7

Measurement of accurate Bragg intensities

Paul Seiler

7.1 Introduction

The accuracy of atomic parameters obtained by X-ray crystal structure analysis depends on many factors: the measuring procedure and the treatment of the measured intensities, the scattering power of the crystal at high Bragg angles, the methods used to refine the atomic parameters, and others. This chapter will deal mainly with the problems of measuring accurate integrated intensities on a standard four-circle diffractometer and of correcting these intensities for various factors to derive the kinematic Bragg intensities.

The collection of a complete set of accurate X-ray intensities to high resolution with a standard diffractometer is very time consuming, even for small unit cells. For example for the structure analysis of Li_2BeF_4 at 81 K (Seiler and Dunitz 1986) more than 40 000 X-ray intensities were measured over a period of about three months. In most cases, however, the available diffractometer time is limited. Also crystal specimens may suffer from X-radiation damage, and thus not all reflections can be measured with the desired accuracy. So, one may ask: which reflections should one measure accurately for optimal determination of the atomic parameters? Some aspects of this problem are discussed in Section 7.3.

One formula that has been proposed for the total integrated diffracted power $I_{\mathbf{H}}^{\text{meas}}$ (Rees 1977) for a crystal in a homogeneous beam is

$$I_{\mathbf{H}}^{\text{meas}} = I_{\mathbf{H}}^{\text{Bragg}} A(1 + \alpha) + \sum_m p_m I_{\mathbf{H}_m}^{\text{Bragg}} + \text{background} \qquad (7.1)$$

$$I_{\mathbf{H}}^{\text{Bragg}} = I_0 v Q_{\mathbf{H}} y \qquad (7.2)$$

$$Q_{\mathbf{H}} = \frac{a^2 \lambda^3}{V^2} Lp\, F_{\mathbf{H}}^2, \qquad (7.3)$$

where $\mathbf{H}$ is the scattering vector, I_0 the power per unit area of the incident beam, $I_{\mathbf{H}}^{\text{Bragg}}$ the integrated diffracted power for reflection $\mathbf{H}$ (elastic scattering only), $I_{\mathbf{H}}^{\text{meas}}$ the integrated power effectively measured, A the transmission factor ($1/A$: absorption factor), α the thermal diffuse scattering (TDS) correction (mostly first-order TDS due to the acoustic lattice vibrations), $p_m I_{\mathbf{H}_m}^{\text{Bragg}}$ the contribution of reflection $\mathbf{H}_m$ through multiple scattering (p_m may be positive or negative), $Q_{\mathbf{H}}$ the kinematic integrated reflectivity per volume of the crystal, v the irradiated crystal volume, y the extinction factor, $a = e^2/mc^2 = 2.818 \times 10^{-13}$ cm is the classical radius of an electron, λ the wavelength, V the volume of the unit cell, $p = (1 + \cos^2 2\theta)/2$ the polarization factor for an unpolarized incident beam, θ the Bragg angle, L the Lorentz factor, $F_{\mathbf{H}}$ the magnitude of the structure factor.

In order to determine the *kinematic* Bragg reflection power $I_{\mathbf{H}}^{\text{Bragg}(k)} = I_0 v Q_{\mathbf{H}}$ we have to subtract the diffuse background from $I_{\mathbf{H}}^{\text{meas}}$, avoid multiple scattering as far as possible, and then correct for the remaining systematic errors, such as absorption, extinction, thermal diffuse scattering (eqns (7.1) to (7.3)), as well as for scan truncation, radiation damage, etc. After all these corrections one can write:

$$I_{\mathbf{H}}^{\text{Bragg}(k)} = K Lp\, F_{\mathbf{H}}^2 \qquad (7.4)$$

or $\qquad\qquad F_{\mathbf{H}}^2 = K^{-1} (Lp)^{-1} I_{\mathbf{H}}^{\text{Bragg}(k)}. \qquad (7.5)$

The terms in eqn (7.5) that vary from one reflection to another are $I_{\mathbf{H}}^{\text{Bragg}(k)}$, L, p, and $F_{\mathbf{H}}^2$. L and p can easily be calculated and $F_{\mathbf{H}}^2$ is the quantity we wish to determine for each reflection; K is a scale factor. For routine measurements, the correction of systematic errors is often neglected and instead of the kinematic, elastic Bragg intensity $I_H^{Bragg(k)}$, only the net intensity $I_H^{net} = I_H^{obs} -$ background is evaluated, where I_H^{obs} is the total peak count (see Section 7.2.4).

7.2   Sources of error in integrated X-ray intensities

The basis of every accurate X-ray crystal structure analysis is a set of integrated intensities as free as possible of systematic and random errors, and extending so far in reciprocal space that high resolution of the atomic parameters is guaranteed. The sources of error that can bias the observed intensities are manifold (see, for example, Young 1974; Feil 1977; Rees 1977; Lehmann 1980a, b; Flack 1985; Gabe 1985; Blessing 1987; Seiler 1987). Some of these errors depend on experimental and instrumental details, such as the quality of the crystal, the alignment and the stability of the X-ray diffractometer used and the measurement procedure. On the other hand, some systematic errors are really a consequence of the inadequacy of the kinematic theory, which forms the basis for nearly all X-ray diffraction experiments. In this theory it is assumed that only a single collision between photon and crystal occurs and that there is no loss of energy as the radiation is propagated through the crystal. In other words, all the energy contained in the diffracted beam is additional to that of the unmodified primary beam. In principle, this theory is only valid for the limiting case of a very small crystal, and, in modified form, for an ideally imperfect crystal. In contrast, the dynamical theory accounts for multiple collisions between photon and crystal, i.e. for a dynamic exchange of energy between primary and diffracted beams. The range of application of this theory, however, is in general limited to ideally perfect crystals. Systematic errors like absorption, extinction or multiple scattering are essentially the result of inadequacy of the kinematic theory.

7.2.1   Selection of crystals

Careful selection of single crystals is vital for attaining high quality X-ray diffraction data. The ideal crystal for data collection should be as pure as possible at the molecular level, untwinned, optimal in size and shape, and should show a small, uniform mosaic spread.

A spherical crystal shape helps to minimize errors caused by inhomogeneity in the cross-section of the primary beam. Moreover, it is straightforward to calculate absorption corrections for such a crystal of known radius.[†]

[†] The main effect of neglecting absorption correction for a spherical, absorbing crystal is to produce an overall displacement parameter (temperature factor) that is too low. For a crystal

With respect to the kinematic theory approximation, the ideal crystal size should be as small as possible to reduce multiple scattering, extinction and absorption errors. On the other hand, when high-order data are required, the crystal must be large enough to diffract significantly at high Bragg angles. For most soft molecular crystals the Debye–Waller factor is high at room temperature. Thus, to obtain accurate high-order data (say, beyond a reciprocal radius $H = 2\sin\theta/\lambda = 1.6\ \text{Å}^{-1}$) from crystals of typical size (linear dimensions about 0.3 to 0.5 mm), intensity measurements have to be carried out at low temperature. However, such low-temperature experiments must be done with special care to avoid introducing additional sources of error. In particular, too rapid cooling may produce a large and inhomogeneous mosaic spread.

7.2.2 Simultaneous diffraction

Simultaneous diffraction (or multiple reflection, Renninger 1937) can be a serious source of error in accurate intensity measurements, especially for the weak low-order reflections. When simultaneous diffraction occurs, the integrated intensity of a reflection may be reduced, if power is removed from the incident or the (primary) diffracted beam ('Aufhellung'; p_m in eqn (7.1) is negative), or increased by a rescattering process (see below) in the direction of the (primary) diffracted beam ('Umweganregung', p_m is positive). Aufhellung is the dominant effect for strong reflections (see, for example, Coppens 1968); the resulting error in the observed intensities is relatively small compared to Umweganregung so only the latter is considered here.

When two reciprocal-lattice points, P_1 and P_2, intersect the Ewald sphere simultaneously, some intensity of the stronger reflection can be diffracted in the direction of the weaker one, as indicated schematically in Fig. 7.1. We assume that the planes associated with the reciprocal-lattice vectors $\mathbf{H}_1$ and $\mathbf{H}_2$ are oriented with respect to the primary beam so as to satisfy Bragg's law for both planes:

$$\mathbf{H}_1 = \mathbf{S}_1 - \mathbf{S}_0 \qquad (7.6)$$

$$\mathbf{H}_2 = \mathbf{S}_2 - \mathbf{S}_0, \qquad (7.7)$$

where $\mathbf{S}_1$ and $\mathbf{S}_2$ are vectors of magnitude $1/\lambda$ in the directions of the two diffracted beams, $\mathbf{S}_0$ being along the incident beam. Subtraction of eqn (7.7) from eqn (7.6) gives

$$\mathbf{H}_1 - \mathbf{H}_2 = \mathbf{S}_1 - \mathbf{S}_2. \qquad (7.8)$$

---

of irregular shape (e.g. a plate-like crystal) absorption errors can also lead to an incorrect degree of anisotropy of the atomic displacement tensors $\mathbf{U}$. (For a review of absorption correction methods see Walker and Stuart (1983).)

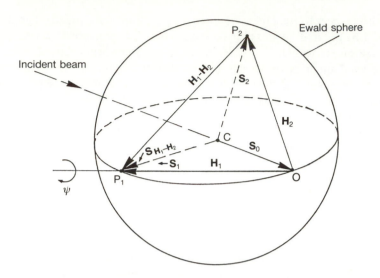

Fig. 7.1 Simultaneous three-beam diffraction. In addition to the primary point O, the scattering vectors $\mathbf{H}_1$ and $\mathbf{H}_2$ intersect the Ewald sphere. The diffracted beam $\mathbf{S}_2$ can act as 'incident beam' to produce a reflection $(\mathbf{H}_1 - \mathbf{H}_2)$ in the $\mathbf{S}_1$ direction.

The diffracted beam from $\mathbf{H}_2(\mathbf{S}_2)$ now acts as an incident beam that is diffracted by the plane $\mathbf{H}_1 - \mathbf{H}_2$ in the direction of $\mathbf{S}_1$. Thus, the intensity measured at the detector contains in addition to the $\mathbf{H}_1$ reflection the intensity of the superimposed $\mathbf{H}_1 - \mathbf{H}_2$ reflection, produced by double reflection of $\mathbf{S}_2$. Similarly, the reflection $\mathbf{H}_2 - \mathbf{H}_1$ produced by double reflection of $\mathbf{S}_1$ may be superimposed on the $\mathbf{H}_2$ reflection.

The probability of multiple reflection increases with the radius of the Ewald sphere, with the 'size' of the reciprocal-lattice point, i.e. with the width of the reflection profile (see Section 7.2.4), and also with the volume of the unit cell. The additional intensity produced by double reflection depends on the reflecting power of the planes involved, the crystal perfection and other factors (see Zachariasen 1965).

Multiple reflection can be detected by analysing the azimuthal intensity profile of a reflection, i.e. by rotating the crystal about the diffraction vector $\mathbf{H}$ and measuring the intensity at different values of the azimuthal angle ψ (see Fig. 7.1). In order to obtain reliable intensity profiles it is advantageous to use a spherical crystal specimen with a small, uniform mosaic spread. Also, it is vital to have a well-aligned diffractometer (for accurate alignment of a four-circle diffractometer see, for example, Samson and Schuelke (1967)). Fig. 7.2 shows the azimuthal profile for the weakest low-order reflection from a crystal of Li_2BeF_4 (space group $R\bar{3}$, Mo Kα radiation,

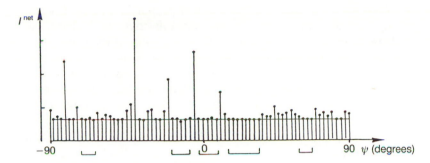

Fig. 7.2 Azimuthal intensity profile showing simultaneous diffraction for the weakest low-order reflection $4\bar{1}2$ from a crystal of Li_2BeF_4 (Mo Kα radiation, $T = 81$ K). After Seiler and Dunitz (1986).

$T = 81$ K; Seiler and Dunitz 1986). The intensity due to multiple scattering occurs as sharp spikes, varying between a small amount and several hundred per cent. When the averaging of the individual intensity measurements is limited to the indicated flat regions of the profile, the internal agreement for symmetry-equivalent reflections is better than 1 per cent. Nevertheless, the averaged intensity may still contain small contributions from multiple scattering. Moreover, this averaged intensity may also be contaminated by harmonic contributions from the incident white radiation (for a discussion see, for example, Young (1974)). We may note, in passing, that for this crystal the ratio of the diffracted intensity between the strongest and weakest low-order reflections is about 12 000:1.

To reduce multiple reflection the crystal should be mounted in a random orientation (i.e. no symmetry axis approximately parallel to a rotation axis of the diffractometer; see Burbank (1965)). Each reflection should be measured at several ψ angles or in several symmetry-equivalent positions. By inspection of the measured data, the individual intensity measurements suffering from severe multiple diffraction errors can be detected and eliminated.

Methods for calculating appropriate ψ settings to avoid large contributions from double reflections have been described (e.g. Coppens 1968; Soejima *et al.* 1985).

7.2.3 Extinction

Extinction is essentially the result of inadequacy of the kinematic theory (Darwin 1914) and reduces mainly the intensities of the strong low-order reflections. Within the kinematic approximation, the diffracted intensity of an ideally imperfect crystal is proportional to F_H^2. For ideally perfect crystals, however, the diffracted intensity is proportional to F_H, according to the dynamical theory, and for real crystals the observed intensity is somewhere between the two extremes. The deviation of the measured from the kinematically calculated reflection power of a Bragg reflection is called extinction. Darwin's (1914, 1922) first treatment of extinction was based on a mosaic crystal model and led to a distinction between 'primary' and 'secondary' extinction. Darwin's theory has been significantly improved, e.g. by Hamilton (1957), Zachariasen (1967), and especially by Becker and Coppens (1974), and adapted for use in the least-squares refinement of crystal structures.

Primary extinction leads to attenuation of the incident and diffracted beams as they pass through a coherently diffracting domain ('mosaic block'). The attenuation is the result of dynamic interactions between the beams rescattered by successive lattice planes and the incident beam, and thus the corresponding corrections should be based on the dynamical scattering theory. Primary extinction becomes significant only if the size (r) of the perfect domains is larger than the primary extinction distance, a distance inversely proportional to λ and F_H. If these domains are small enough (say $r < 10^{-7}$ m) then primary extinction becomes negligible and the crystal may be regarded as ideally imperfect. However, even with such a crystal the diffracted intensity can suffer from *secondary extinction*. This phenomenon can be described as follows: When a particular ray of the primary beam propagates through the crystal, it will be attenuated when several perfect domains are in diffracting orientation. On the other hand, part of the diffracted power is restored when beams diffracted by some domains are rediffracted via other domains in the direction of the primary beam. This exchange of energy between the primary beam and the diffracted beams of different blocks is considered to be incoherent, and thus the corrections are based on the kinematic theory. The larger λ and F_H, the thicker the crystal, and the thicker and better aligned the blocks, the more severe the secondary extinction will be.

The distinction between primary and secondary extinction (or between coherent and incoherent scattering) is somewhat arbitrary with reference to real imperfect crystals. Kato (1976) considers that if a mean spatial correlation length within the crystal is larger than a critical distance (comparable to the primary extinction distance or Pendellösung fringe spacing) then a wave-optical treatment of the problem is to be preferred to the kinematic one.

7.2.3.1 Detection of anisotropic extinction

Detection of anisotropic extinction   Anisotropic extinction, similarly to anisotropic absorption, can produce slow variation of the azimuthal intensity profile and lead to differences among the intensities of (Laue) symmetry-equivalent reflections. However, in contrast to photoelectric

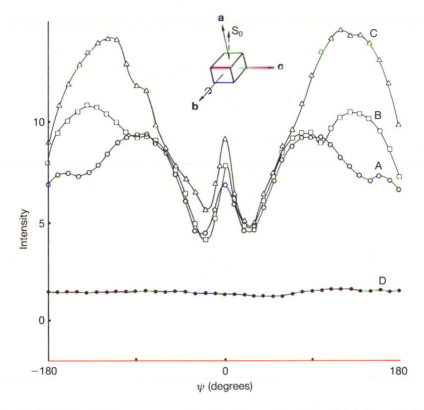

Fig. 7.3 Variation of the integrated intensity for different azimuthal angles ψ, in a crystal of the complex $C_{12}H_{24}O_6$.KNCS. A: 040 reflection with $\Delta T = 0$; B: 040 reflection with $\Delta T \cong 7$ K; C: 040 reflection with $\Delta T \cong 20$ K; D: 080 reflection with $\Delta T = 0$. After Seiler and Dunitz (1978).

absorption, the variations are most marked for very strong reflections (see, for example, Denne 1972; Thornley and Nelmes 1974).

Figure 7.3 shows an example of anisotropic extinction for a crystal of the complex of 1,4,7,10,13,16-hexaoxacyclooctadecane(18-crown-6) with potassium thiocyanate, $C_{12}H_{24}O_6$.KNCS (Seiler and Dunitz 1978). The variation of the integrated intensity of the strong 040 reflection ($F^{calc} = 74.7$, Mo Kα radiation) measured at different azimuthal angles ψ is as large as 40 per cent (curve A), in contrast to the weaker 080 reflection ($F^{calc} = 40$) which shows a nearly constant intensity with respect to ψ (curve D). Experiments with several crystal specimens of the same compound showed that the azimuthal intensity profiles of strong reflections are very sensitive to mechanical strains produced, for example, by cutting and mounting the crystal. By avoiding mechanical treatment of any kind,

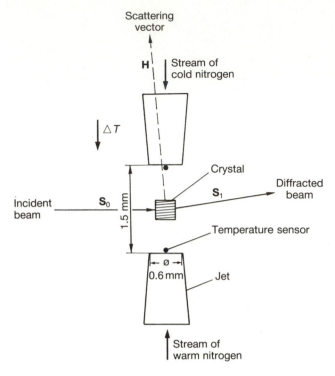

Fig. 7.4 Schematic representation of the device for producing a temperature gradient. In most of our experiments the gradient was in the plane of S_1 and S_0 and perpendicular to S_0. The measured temperature difference ΔT is the difference between the temperatures of the two streams of nitrogen as they left the jets. The mean temperature of the crystal was held constant at approximately 293 K. After Seiler and Dunitz (1978).

a practically flat azimuthal intensity profile could be obtained, but although the extinction was almost isotropic, it was also more severe. Elastic strain can also be produced by applying a temperature difference (ΔT) at the crystal (Fig. 7.4). The resulting change in the diffracted intensity for the 040 reflection is shown in Fig. 7.3 (curves B and C) and in more detail for the 100 reflection in the next section.

7.2.3.2 Experimental estimation of extinction   The extinction corrections obtained by least-squares refinement are often based on physically unreasonable extinction models or even on no model at all. In some least-squares programmes (e.g. in XTAL88 (Stewart *et al.* 1988)), the extinction corrections are based on the kinematic-theory approximation described by Becker and Coppens (1974). In most cases, the calculated extinction corrections lead to a significant improvement between observed and calculated structure factors of the affected reflections, and thus to lower R values. In some cases, however, e.g. when extinction is very strong or if primary extinction is significant, these corrections may be inaccurate. Moreover, the corrections become erratic if the observed intensity loss of the strongest reflections is caused not only by extinction but also by inaccurate intensity measurements (e.g. by non-linear counting chains, improperly calibrated beam attenuators etc.). Other problems may arise, if the refined crystal-structure model is incorrect or when several models are equally compatible with the X-ray data. We encountered this problem during an accurate structure analysis in which we hoped to determine whether the Li_2BeF_4 crystal is built from neutral atoms or conventional ions (Seiler and Dunitz 1986). The least-squares refinements based on neutral and ionic form factors, including an isotropic extinction correction, led to essentially the same low R value for the five strongest, extinction affected, low-order reflections, in spite of considerable differences in the calculated structure factors. In other words, the calculated extinction corrections are clearly unreliable in such a case, and the kinematic intensity for the strongest reflections should be estimated experimentally. Several approaches have been proposed; for example, by altering the crystal thickness and measuring the intensity for different path lengths of the X-ray beam within the crystal and extrapolating to zero path length (Bragg et al. 1921; Lawrence and Mathieson 1977), or by making intensity measurements at several wavelengths and extrapolating to zero wavelength (Lawrence 1977). Both methods can lead to satisfactory results, if primary extinction is negligible. A specially notable method involves the use of a plane-polarized incident beam (Chandrasekhar 1956; Chandrasekhar et al. 1969).

In another method, strain is produced in the crystal (i.e. the crystal perfection is reduced) by applying a temperature gradient in the plane of S_1 and S_0 and perpendicular to S_0 during the diffraction process (see Fig. 7.4). The relation between a gradual, elastic change in crystal perfection and a corresponding change in the extinction behaviour was investigated for a crystal of the $C_{12}H_{24}O_6$.KNCS complex (Seiler and Dunitz 1978). From Fig. 7.5 we see that increasing the temperature gradient at the crystal leads to a striking increase in the diffracted intensity for the strong 100 reflection ($F^{calc} = 63.7$), that is, to a reduction in the extinction. In fact, with a temperature difference (ΔT) of about 50 K measured near the two sides of the crystal, the diffracted intensity becomes practically equal to the expected kinematic value $I_H^{Bragg(k)}$ for Mo Kα radiation (curve A) and for Cu Kα radiation (curve B). The intensity increase caused by a temperature gradient is also accompanied by a broadening of the diffraction peak. Up to a certain limiting value, both effects are reversible, i.e. on removal of the gradient, both the integrated intensity and the peak profile are indistinguishable from those of the original state ($\Delta T = 0$). Similar observations were made many years ago by Sakisaka and Sumoto (1931) for a quartz plate. The increase

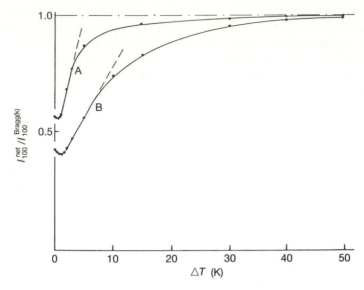

Fig. 7.5 Approach of the integrated intensity I_{100}^{net} to the expected kinematic intensity $I_{100}^{Bragg(k)}$ with increasing ΔT, in a crystal of the complex $C_{12}H_{24}O_6 \cdot KNCS$. A: for Mo K$\alpha$ radiation; B: for Cu Kα radiation. After Seiler and Dunitz (1978).

in intensity can be interpreted as follows: when the crystal is subjected to a temperature gradient, lattice nets perpendicular to the gradient become curved because of the unequal thermal expansion in different parts of the crystal. We may assume that this lattice deformation destroys the exact regularity in phase of the radiation scattered at different parts of the crystal (which is necessary for primary extinction). In terms of the more general formalism of Kato (1976), the intensity increase would be connected with a decrease of the mean spatial correlation length within the crystal.

7.2.4   Effective integration of intensities

7.2.4.1   Choice of reflection scan mode and counter aperture   The shape of the intensity profile obtained by an ω scan (moving crystal, stationary counter) or an $\omega/2\theta$ scan (moving crystal and counter in a ratio of 1/2, see Fig. 7.6) depends mainly on four factors: wavelength dispersion of the X-ray beam (τ_λ), crystal cross section (τ_c), mosaic spread (τ_m), and divergence of the X-ray source (τ_x). In fact, the intensity profile can be regarded as the convolution of these four components (Alexander and Smith 1962).

The total diffracted power (I_H^{meas}) is obtained by summing the individual intensities produced by rotating the reciprocal-lattice point through the surface of the Ewald sphere. The minimum scan range ($\Delta\omega$) should include

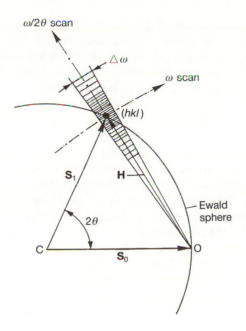

Fig. 7.6 Schematic representation of the ω and $\omega/2\theta$ scan. After Alexander and Smith (1962).

the entire Bragg peak, and is normally estimated as $\Delta\omega = A + B \tan\theta$, where A depends on τ_c, τ_m and τ_x and B depends on τ_λ.

The background scattering must be subtracted from the peak I_H^{meas}. We assume that the background consists of a more or less flat part due to air scattering, fluorescence, Compton scattering, electronic noise, 'optic' mode contributions of the thermal diffuse scattering (TDS), and a non-linear part due to white radiation and the 'acoustic' mode contributions of TDS (see Section 7.2.5, Fig. 7.8). The flat portion can be estimated by measurements at the scan limits. The non-linear contribution due to white radiation can be drastically reduced by use of a high-quality crystal monochromator. (Advantages and disadvantages of crystal monochromators for background control are discussed by Young (1974)). The non-linear TDS contribution due to acoustic lattice vibrations, however, cannot be subtracted out completely by this method (see Section 7.2.5).

When the X-ray intensities are measured with a scintillation counter and when the counting losses are negligible the integrated peak count (I_H^{obs}) is proportional to the integrated reflection power I_H^{meas}. The net, scaled peak intensity is then estimated as

$$I_H^{net} = K[I_H^{obs} - (BG_L + BG_R)t_P/t_{L+R}],\qquad(7.9)$$

where I_H^{obs} is the integrated peak count, including the background under the peak, t_P is the scan time to measure the peak, $BG_L + BG_R$ is the sum of the background counts measured at the left and right scan limits over the time t_{L+R}. If the proportionality factor K is constant during the measurement, the standard deviation of I_H^{net} can be estimated as

$$\sigma(I_H^{net}) = \{K^2[I_H^{obs} + (BG_L + BG_R)(t_P/t_{L+R})^2] + P^2(I_H^{obs})^2\}^{1/2}. \quad (7.10)$$

The last term (where the factor P is called 'instability constant' or 'ignorance factor') is supposed to account for instabilities of the experimental set up (see McCandlish *et al.* 1975). As Burbank (1964) has pointed out, the scan mode to be preferred is the one which yields the true integrated Bragg intensity by scanning the smallest volume in reciprocal space, to reduce background and TDS contributions to a minimum. For high-order reflections the $\omega/2\theta$ scan is normally preferred because of the increasing effect of wavelength dispersion with increasing Bragg angle. For low-order reflections an ω scan may sometimes be more useful, especially when the crystal mosaicity is large. In principle ω, ω/θ, and $\omega/2\theta$ scan modes can be used if an appropriate counter aperture is chosen and the background subtracted correctly.

   An experimental approach to the choice of scan type, scan angle, and counter aperture is shown in Fig. 7.7 for a nearly spherical crystal of ammonium hydrogen 2S-malate (Seiler, unpublished results). The integrated intensities for the 120 and the 16,1,6 reflections ($H = 0.28$ and 2.20 Å^{-1}, respectively) are plotted as a function of scan angle (a, b) and counter aperture (c, d). For the low-order reflection a $1°$ ω or $\omega/2\theta$ scan yields practically the entire integrated intensity (Fig. 7.7(a)); an aperture of about $0.8°$ (Fig. 7.7(c)) would suffice for both scans. For the high-order reflection the intensity differences between the two scan modes are far more marked. Figure 7.7(d) shows that for the ω scan we need an aperture at least double that of the $\omega/2\theta$ scan. Figure 7.7(b) shows that for both types of scan a very large scan angle (about $7°$) is needed. It is clear, however, that such large scan angles would lead to a poor signal-to-noise ratio for weak reflections and also possibly to an overestimate of the background scattering because of overlap with a neighbouring reflection (see Fig. 7.7(a), curve A).

   In order to improve the accuracy of data collection, alternative methods have been proposed to optimize some of these experimental parameters. The profile analysis described by Lehmann and Larsen (1974), for example, assigns an optimal peak and background region to each reflection profile (see also Blessing *et al.* 1974; Blessing 1987). A different procedure for obtaining the integrated intensity I_H^{net} is to fit a peak and a background function to the observed reflection profile (for details see, for example, Diamond (1969); an improvement of this procedure and a summary of

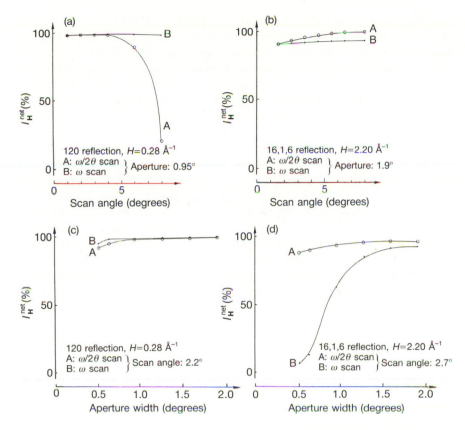

Fig. 7.7 Experimental estimation of optimal scan angle $\Delta\omega$ (a, b) and counter aperture (c, d) for a spherical crystal (diameter 0.5 mm) of ammonium hydrogen 2S-malate (CAD4 diffractometer equipped with graphite monochromator, Mo Kα radiation, $T = 81$ K). In curve A of (a), the increasing intensity loss with increasing scan angle is due to overlap with a neighbouring reflection, i.e. the background is overestimated (Seiler, unpublished results).

similar methods are described by Clegg (1981)). The main result of these methods is to improve $I_H^{net}/\sigma(I_H^{net})$ or to increase the speed of data collection without significant loss in the relative precision, compared to the traditional estimation of peak and background intensity. In practice, the criteria for 'optimal' assignment of peak and background region for all the procedures mentioned remain more or less subjective, especially for high-order reflections.

In a more recent study, Mathieson (1982, 1983, 1984) has proposed that the identification of peak and background region, as well as the measurement of the integrated intensity of a Bragg reflection, should be based on a two-dimensional $\Delta\omega$, $\Delta 2\theta$ intensity distribution, rather than on a one-dimensional $\Delta\omega$ reflection profile. At present, however, this procedure would be far too time-consuming for evaluating the net intensities for routine structure analysis but it may be appropriate for very small unit cells, where highest accuracy of the observed structure factors is needed.

The two-dimensional intensity distribution of a Bragg reflection can be produced automatically with some modern diffractometers (see, for example, the option OTPLOT in the CAD4 software), so at least scan parameters and counter aperture can be optimized routinely by this method.

7.2.4.2 Scan-truncation errors

The main difficulty in assigning a 'true' reflection scan angle, especially at high θ, is that the Bragg intensity does not fall to zero within a feasible scan angle $\Delta\omega$. According to Alexander and Smith (1962), Kheiker (1969), Denne (1977), Destro and Marsh (1987), a typical $\omega/2\theta$ scan range ($2°$ plus the $\alpha_1 - \alpha_2$ splitting) leads to an increasing underestimation of the true net intensity with larger scattering angle. As seen in Fig. 7.7(a, b) the scan angle needed for a low- and high-order reflection may differ by about $6°$, due mainly to the spectral dispersion of the primary beam. It is usually assumed that τ_λ can be approximated by a Cauchy function, while τ_m is represented by a Gaussian. Since the wavelength dispersion $d\theta/d\lambda$ for a Bragg reflection is $\tan \theta/\lambda$, the effective scan range $\Delta\omega$ should increase faster with Bragg angle than is usually done, i.e. by adding the $\alpha_1 - \alpha_2$ splitting to a constant scan range. Scan truncation is a substantial source of error. For example, from Alexander and Smith (1962) we obtain a correction of about 8 per cent to be applied to a typical reflection intensity at $H = 2.4$ Å^{-1} (Mo Kα radiation, $3°$ $\omega/2\theta$ scan), whereas at $H = 1.4$ Å^{-1} the error is less than 1 per cent. Denne (1977) has derived a correction for scan-truncation errors, based on a wavelength spectrum taken from a single-crystal diffractometer and on the assumption that 'the effect of finite source size, finite crystal size and mosaic spread will tend to average out' in a reflection profile. He showed that after the scan-truncation correction the atomic displacement parameters decrease.

More recently, Destro and Marsh (1987) found that the equation given by Denne underestimates scan-truncation errors when common diffractometers and data-collection techniques are used. These authors proposed an empirical correction method and showed that with an additional scattering-angle dependent function (called 'aberration' function), a better estimate of the truncation losses could be obtained. In practice, the change of a reflection profile with increasing scattering angle depends not only on the wavelength dispersion, but also on the amount of thermal diffuse intensity (and possibly also static diffuse intensity) included in the profile (see below). In addition, the crystal properties, i.e. size, shape, and perfection, and instrumental factors such as the collimation of the primary beam, the quality of the crystal monochromator, the width of the receiving counter aperture, etc., can bias the truncation losses. Thus, the observed scan-truncation errors are in general different for different crystals and for different diffractometers, even when the intensities are measured with the same procedure. Because it is difficult to consider all experimental factors in a theoretical approach, the experimental correction method may in general produce better agreement between observed and calculated intensities for high-order reflections.

In our laboratory the scan-truncation error has been estimated for several low-temperature data sets. With our diffractometers and procedures, the intensity losses due to this error appear to be much smaller than those reported by other authors (see, for example, Hirshfeld and Hope 1980; Destro and Marsh 1987). For example, after the least-squares refinements of the structure of Li_2BeF_4 at 81 K (Seiler and Dunitz 1986), we analyzed the ratio of F^{obs}/F^{calc} for all measured reflections as a function of scattering angle, by dividing the reflection sphere into 11 shells (from $0 < H < 0.2$ Å^{-1} up to $2.4 < H < 2.7$ Å^{-1}). The value of $\Sigma F^{obs}/\Sigma F^{calc}$ in each shell is almost unity, even for that containing the highest-order reflections. As it is difficult to distinguish between the tails of a Bragg peak and the non-linear thermal diffuse intensity, we do not know to what extent the scan-truncation losses were compensated by the TDS contributions. We can assume, however, that TDS must be quite small for this crystal at 81 K, as the melting point is about 800 K.

7.2.5 Thermal diffuse scattering (TDS)

Thermal motion not only diminishes the elastic Bragg intensity, it also produces an inelastic thermal diffuse intensity (TDS) which is not uniform throughout reciprocal space. When the incident radiation interacts with the lattice vibrations, it exchanges momentum and energy with the latter, raising or lowering its energy by one phonon (quantum of lattice vibrational energy) or in some circumstances by two or more phonons. The corresponding change of wavelength is small for X rays because the energy of phonons ($\sim 10^{-2}$ eV) is only a tiny fraction of the X-ray energy ($\sim 10^4$ eV).

Lattice vibrations can be distinguished as low-frequency 'acoustic' modes, and relatively high-frequency 'optic' modes (see, for example, Cochran 1969). The inelastic, coherent, one-phonon scattering due to the acoustic lattice vibrations contributes most to the phonon scattering and peaks at the reciprocal-lattice points, but less sharply than the Bragg reflections (indicated in Fig. 7.8). The TDS due to optic lattice vibrations and multiphonon processes, however, varies much more slowly in reciprocal space. Thus, the remaining TDS in a reflection profile (see eqn (7.1)), after the usual background correction, is mainly first-order TDS from the acoustic modes. The resulting overestimation of the true Bragg intensity depends on the physical

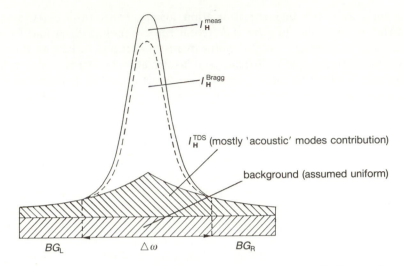

I_H^{meas}

I_H^{Bragg}

I_H^{TDS} (mostly 'acoustic' modes contribution)

background (assumed uniform)

BG_L $\triangle \omega$ BG_R

Fig. 7.8 Idealized reflection profile, showing the various intensity contributions to the total scattering. It is assumed that the background scattering is uniform over the scan range and includes air and Compton scattering, the 'optic' mode contributions of TDS, fluorescence, electronic noise, etc. After Lucas (1969).

properties of the crystal (elastic constants, temperature, mosaic spread), the instrumental profile, as well as on the measuring procedure (type of scan, scan range, counter aperture; see, for example, Göttlicher 1968; Cochran 1969; Willis and Pryor 1975). The TDS is in general anisotropic; it is large for soft molecular compounds (with relatively low elastic constants) and increases with temperature and scattering angle. The relative amount of TDS observable near a Bragg peak increases with the volume of reciprocal space scanned and with the size of the receiving counter aperture.

With a standard X-ray diffractometer we can not distinguish between Bragg intensity and TDS measured inside the (main) reflection peak. On the other hand, the presence of TDS can be detected by analysing the slope of the background of a peak (see, for example, Young 1974).

7.2.5.1 Detection of TDS in an actual low-temperature experiment The experiment described in Fig. 7.7 was made for several high-order reflections with similar scattering angles and peak intensities, using the same crystal specimen. The proportional increase in intensity with increasing scan angle or counter aperture varies considerably from reflection to reflection. From a comparison of the resolved $\alpha_1 - \alpha_2$ peak profiles, we concluded that the variations were too large to be explained by scan-truncation error alone (Seiler, unpublished results). On the other hand, it seemed likely that the peaks must contain varying amounts of TDS intensity, which is not subtracted out by the background correction. The atomic displacement tensors for the crystal in question are quite anisotropic, the direction of largest displacement ($\langle U^{22} \rangle \cong 0.018 \text{ Å}^2$) being roughly parallel to the **b** axis, the direction showing also largest thermal expansion. In contrast, the displacement along **a** is small ($\langle U^{11} \rangle \cong 0.006 \text{ Å}^2$) and the corresponding thermal expansion is almost zero.

For example, using the same experimental conditions as in Fig. 7.7(b), i.e. at 81 K, the intensity increase for the 1,17,1 reflection ($H = 2.12 \text{ Å}^{-1}$, $F^{\text{calc}} = 4.1$) is about 25 per cent, whereas for the 16,0,0 reflection ($H = 2.10 \text{ Å}^{-1}$, $F^{\text{calc}} = 4.5$) we gain only about 6 per cent (Fig. 7.9, curves A, C; intensities are averaged over 8 symmetry-equivalent reflections). At 123 K the relative gains for the same two reflections are about 35 per cent and 10 per cent, respectively (Fig. 7.9, curves B, D). This shows that the tails of the peak profiles extend out further at the higher temperature owing to the larger TDS contribution.

With this simple experiment, we cannot determine the total TDS under a Bragg peak. However, these observations show that TDS is anisotropic for this crystal and that, for a typical $\omega/2\theta$ scan range, say 2.5°, the 1,17,1 Bragg peak must include considerably more TDS than the 16,0,0 one. To

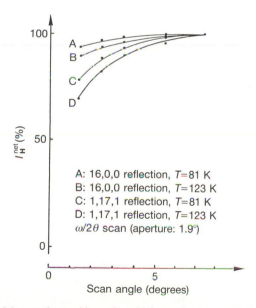

Fig. 7.9 Increase of the net observed intensity with increasing scan angle $\Delta\omega$ for two high-order reflections of ammonium hydrogen 2S-malate, measured at 81 K and 123 K. The different slopes of the curves (A–D) indicate that the TDS contribution in the observed reflection profiles of the 16,0,0 reflection is much smaller than for the 1,17,1 reflection (Seiler, unpublished results).

estimate the true Bragg intensity we would have to measure the TDS contribution for each reflection separately, which would require special equipment and be very time consuming. For example, using a highly perfect crystal of silicon and an extremely narrow instrumental profile, obtainable from a double-crystal spectrometer, Renninger (1967) was able to separate most of the (one-phonon) TDS from the Bragg intensity.

7.2.5.2   Influence of TDS on atomic parameters   When TDS corrections are applied, it is usually not by experiment but by calculating a correction factor $\alpha_1 = I_H^{TDS}/I_H^{Bragg}$ where I_H^{TDS} is the integrated (one-phonon) TDS intensity and I_H^{Bragg} the Bragg intensity (see, for example, Cochran 1969; Harada and Sakata 1974; Stevens 1974; Willis and Pryor 1975). This calculation requires the elastic constants to be known, which is not the case for most molecular compounds.

For α-oxalic acid dihydrate (Dam *et al.* 1983), α_1 values have been calculated at 100 K for one-phonon TDS. The correction factors amount to 50 per cent at high Bragg angles. Least-squares refinement using corrected data did not alter the positional parameters but increased the U^{ij} values by about 12 per cent. The α_1 values for axial reflections show that TDS is anisotropic and largest along **b**. As in our observations, this corresponds to the direction of largest atomic displacement.

7.2.6   Long-term intensity fluctuations of standard reflections, when crystal suffers X-radiation damage

Especially during long intensity measurements it is essential to monitor the stability of the experimental system (X-ray source, crystal, counting device, low-temperature device, etc.). This is usually done by remeasuring the intensity of one or a few chosen Bragg reflections at regular intervals during the data collection (see, for example, McCandlish *et al.* 1975). In order to obtain a uniform scale factor for all observed intensities, a correction must be applied to compensate for any slow drift in the intensity of the standard reflections. To save counting time, crystallographers like to choose strong reflections as standards, and these are mostly low-order reflections. However, the assumption that the intensity loss for low- and high-order reflections will be the same, is by no means assured, especially for crystals suffering X-radiation damage. In fact, protein crystallographers have been aware for years that the effects of radiation damage are more marked at higher scattering angle (see, for example, Blake and Phillips 1962; Blundell and Johnson 1976). We encountered this problem in a low-temperature charge-density study of **1** (Dunitz and Seiler 1983; Seiler and Dunitz 1985).

At the beginning of the experiment the intensities of 23 standard reflections distributed between $H = 0.32$ Å^{-1} and $H = 2.30$ Å^{-1} were measured

1

as a function of temperature. Figure 7.10 shows that the intensities of the high-order reflections decrease faster with increasing temperature. Then, over a period of six weeks the same standard reflections were remeasured at 96 K (Fig. 7.11). It is clear that the intensity fall-off with increasing exposure time also depends on the scattering angle. For example, while the 110 reflection at $H = 0.32$ Å^{-1} decreases only by about 4 per cent, the $\bar{9}$,12,4 reflection at $H = 2.30$ Å^{-1} loses as much as 40 per cent (Fig. 7.11).

Radiation damage also produced an increase in the unit-cell volume ΔV amounting to about 0.6 per cent after six weeks. This is about the same ΔV as would have been produced by warming the fresh undamaged crystal from 96 to about 136 K. We may therefore regard ΔV as a function of exposure time $\Delta V = f(t)$ or of temperature $\Delta V = g(T)$. The abscissas of Figs. 7.10 and 7.11 are scaled so that corresponding time intervals and temperature rises produce the same ΔV. The comparison shows that for all the standards there is a striking similarity between the intensity fall-off with increasing exposure time and with increasing temperature. These observations suggest that the main initial effect of radiation damage is to increase the defects and the static disorder in the crystal. The resulting change in the intensities after the six-weeks exposure time can be simulated by introducing an additional overall displacement parameter amounting to about $\Delta B \cong 0.4$ Å^2 or $\Delta U \cong 0.005$ Å^2 for the crystal studied.

A correction factor for radiation damage which is based only on the intensity fall-off of low-order reflections can lead to serious errors in the displacement parameters. Also, for accurate data collection of radiation-sensitive crystals it is advantageous to begin the measurements at high θ and work inwards in small shells (Seiler and Dunitz 1985). In our experiment we applied a scaling correction (Fig. 7.12) accounting for the exposure time and scattering-angle dependent intensity fall-off. After this correction, the scale and overall displacement parameter, estimated from three high-order least-squares refinements ($H > 1.6$ Å^{-1}, $H > 1.8$ Å^{-1}, $H > 2.0$ Å^{-1}) using all reflections with $I > 3\sigma(I)$, agree within about one standard deviation ($R = 0.013$–0.014). In addition, from a high-order refinement, against the reflections with $H > 1.4$ Å^{-1} and $I > 3\sigma(I)$, all six crystallographically independent C—H distances could be estimated rather accurately; they vary between 1.07(2) Å and 1.10(2) Å and thus agree well with typical values obtainable from other methods.

In a recent, accurate low-temperature X-ray structure analysis of a highly

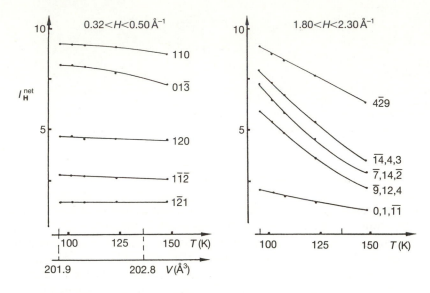

Fig. 7.10 Temperature dependence of intensities of 10 standard reflections of a crystal of **1** measured at the start of the experiment (after Seiler and Dunitz 1985). For comparison with Fig. 7.11, the unit-cell volumes at two temperatures (96 K and 136 K) are indicated.

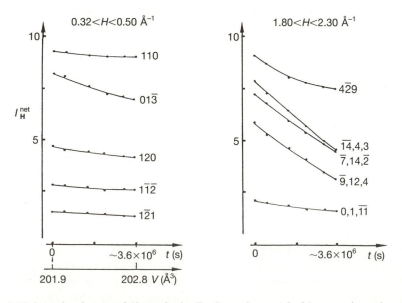

Fig. 7.11 Intensity changes of 10 standard reflections of a crystal of **1** over a six-week period at 96 K (after Seiler and Dunitz 1985). The intervals along the abscissa are exposure times required to produce the same ΔV as produced by the temperatures indicated in Fig. 7.10.

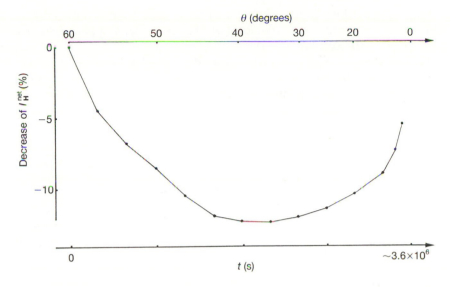

Fig. 7.12 Proportional intensity decrease of observed reflections as function of scattering angle or exposure time (after Seiler and Dunitz 1985). Scaling of reflection intensities based on this curve for the structure analysis of **1** gave satisfactory results.

reactive [1.1.1]propellane derivative (Seiler *et al.* 1988), we have observed a similar scattering-angle dependent intensity loss of several standard reflections. In this experiment, the crystal (enclosed in a glass capillary, melting point ~ 208 K) was irradiated over a period of about four weeks at 81 K. At the end of the measurement, the intensity loss due to radiation damage for reflections at $H = 0.33$ Å^{-1} and $H = 1.67$ Å^{-1} was 6.5 and 12.5 per cent, respectively. The intensity loss was accompanied by a small but significant increase of the unit-cell volume of about 0.13 per cent and a relatively marked increase of the monoclinic β angle of 0.11°.

7.3   Which reflections should be measured in order to obtain accurate atomic parameters?

Most routine X-ray measurements are made at room temperature, and the observed intensities hardly pass $H = 1.3$ Å^{-1} ($\theta = 27°$ for Mo Kα, $\theta = 90°$ for Cu Kα radiation). Even with very accurate X-ray measurements at such limited resolution, the positional and displacement parameters obtained by least-squares refinements may be considerably in error because of the inadequacy of the free-atom scattering factors for low-order reflections. However, when sufficiently extensive and accurate high-order measurements (say $1.6 < H < 2.8$ Å^{-1}) are available, atomic parameters can be derived that are closer to the 'true' values and, in general, also closer to those obtainable from neutron diffraction (see, for example, Coppens 1977; see also Chapter 10, Section 10.8.1).

For tetrafluoroterephthalonitrile, **2**, studied at 98 K (Seiler *et al.* 1984), we have investigated the effect on the positional and displacement parameters of including only low-order $(H < 1.3 \text{ Å}^{-1})$ or high-order $(H > 1.7 \text{ Å}^{-1})$ or all reflections $(0 < H < 2.3 \text{ Å}^{-1})$ in the least-squares refinements.

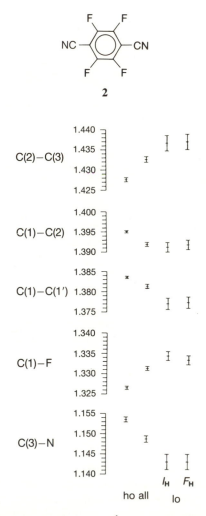

Fig. 7.13 Comparison of interatomic distances (Å) for tetrafluoroterephthalonitrile obtained by high-order refinement (ho; $H > 1.7 \text{ Å}^{-1}$), all-reflections refinement (all), and low-order refinements (lo; $H < 1.3 \text{ Å}^{-1}$) based on I_H and F_H values. After Seiler *et al.* (1984).

Figure 7.13 shows systematic differences in interatomic distances of more than 0.01 Å (or several standard deviations) among the three refinements. In the low-order refinement the C(1)—F and C(2)—C(3) bonds are too long and the C(3)—N bond is too short compared with the high-order values. With the low-order data, the C(3) and N atoms are drawn towards the deformation density of the triple bond, which is the strongest feature of the charge-density difference map (Fig. 7.14). The refinement including all observed reflections out to $H = 2.3$ Å^{-1} leads to atomic positions and bond lengths about half way between low- and high-order values. Subsequent calculations (Hirshfeld 1984) using the same data in a multipole refinement (static deformation-density map) yield atomic parameters that are very close to our high-order values; see Chapter 10, Section 10.11.1.

For α-oxalic acid dihydrate ($T \cong 100$ K) positional parameters obtained from five neutron and four high-order X-ray data sets are reproducible to a precision of 0.001 Å or better (Coppens *et al.* 1984). The discrepancies among the displacement tensor elements U^{ij} for the same studies, however, are large. This is not surprising, because the U^{ij} values are more sensitive to systematic errors in the observations. As discussed before, neglect of TDS (and absorption correction) can lead to underestimation of the overall displacement parameter, whereas truncation errors and neglect of radiation

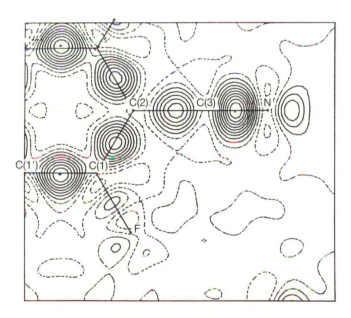

Fig. 7.14 Charge-density difference map for tetrafluoroterephthalonitrile in the molecular plane. Contours are drawn at intervals of 0.075 eÅ^{-3}. After Seiler *et al.* (1984).

damage can produce the opposite effect. Additional errors may occur by incorrect calibration of the crystal temperature or by instability of this temperature during data collection, etc.

7.3.1   Selective exclusion of weak intensities

Another source of systematic error of the derived atomic parameters may be introduced by omitting the weak, often inaccurately measured reflections from the least-squares refinement. Crystallographers frequently specify an 'arbitrary' $I/\sigma(I)$ threshold (say $I > 3\sigma(I)$) for a reflection to be included in the structure refinement. However, it has been argued by Hirshfeld and Rabinovich (1973) that selective exclusion of weak (or negative) intensities leads to a bias in the remaining observations towards too high F^2 (or I) values and thus to systematic errors in the refined parameters. Although the arguments against rejection of weak reflections are accepted in principle, there is still the question of how serious the resulting errors caused by such malpractices actually are.

For tetrafluoroterephthalonitrile (Seiler *et al.* 1984) we have analysed the results of several high-order refinements with different criteria for including weak reflections (see Table 7.1). Starting with all 1406 reflections having $H > 1.7$ Å^{-1} (refinement A2), we successively excluded the weaker ones until only 436 with $I > 20\sigma(I)$ were retained (F2). The atomic parameters (U^{ij} values not listed) and the scale factor from the different refinements (A2–F2) are practically the same; the effect of excluding more and more weak reflections lies well below the threshold of physical significance. In contrast, refinements based only on low-order data (H2) or even on all observations (G2) lead to substantial changes in all atomic parameters, including the scale factor.

The effect of selectively excluding weak reflections from the least-squares refinements was also examined in detail during the structure analysis of Li_2BeF_4 (Seiler and Dunitz 1986). For this low-temperature study (81 K), all reflections out to $H = 2.7$ Å^{-1} were measured, mostly in all symmetry-equivalent orientations, a total of about 40 000 measurements (6345 unique reflections). Analogously to the previous study, we have carried out a series of high-order refinements, beginning with all 1913 reflections at $H > 2.4$ Å^{-1} ($R_1 = 0.019$);   then   the   reflections   with $I < 10\sigma(I)$, $I < 20\sigma(I)$ up to $I < 60\sigma(I)$ were successively omitted, until only 499 reflections remained. The 63 atomic parameters and the scale factor obtained from these refinements did not change more than about a standard deviation. Besides, no systematic bias of the atomic parameters was observed when more and more weak reflections were rejected, not even for the refinement using only 26 per cent of the observed high-order reflections.

In this connection it should be mentioned that the accurate measurement

Table 7.1 Results of various refinements for tetrafluoroterephthalonitrile ($T = 98$ K). A2, B2, C2, F2 are high-order refinements; G2 includes all accessible reflections out to $H = 2.3$ Å^{-1}; H2 is a low-order refinement. After Seiler et al. (1984)

	A2	B2	C2	F2	G2	H2
Number of reflections	1406	1238	1238	436	2387	431
Inclusion criteria { on I	None	$I > 0$	$I > 0$	$I > 20\sigma(I)$	None	$I > 0$
{ on H	$H > 1.7$ Å^{-1}	$H > 1.7$ Å^{-1}	$H > 1.7$ Å^{-1}	$H > 1.7$ Å^{-1}	None	$H < 1.3$ Å^{-1}
Refinement based on	$I_\mathbf{H}$	$I_\mathbf{H}$	$F_\mathbf{H}$	$F_\mathbf{H}$	$I_\mathbf{H}$	$I_\mathbf{H}$
Scale factor	0.2200(6)	0.2200(6)	0.2202(6)	0.2202(7)	0.2225(3)	0.2181(9)
Discrepancy indexes[a] { R_F	0.081	0.060	0.060	0.010	0.055	0.028
{ R_I	0.052	0.042	0.042	0.017	0.045	0.041
{ R_{wF}	0.016	0.016	0.016	0.016	0.037	0.035
Goodness-of-fit[b]	1.106	1.084	1.120	1.307	4.764	8.483
Bond distances (Å) { C(1)—C(2)	1.3950(3)	1.3951(3)	1.3951(3)	1.3950(3)	1.3918(5)	1.3911(12)
{ C(1)—F	1.3264(3)	1.3264(3)	1.3262(3)	1.3264(4)	1.3313(5)	1.3345(11)
{ C(1)—C(1')	1.3841(2)	1.3841(3)	1.3841(3)	1.3841(3)	1.3814(5)	1.3768(14)
{ C(2)—C(3)	1.4270(4)	1.4270(4)	1.4270(4)	1.4270(5)	1.4322(7)	1.4366(19)
{ C(3)—N	1.1538(4)	1.1538(5)	1.1537(5)	1.1537(6)	1.1489(8)	1.1432(19)

[a] Defined as $R_F = \Sigma_\mathbf{H}(F_\mathbf{H}^{obs} - F_\mathbf{H}^{calc})/\Sigma_\mathbf{H} F_\mathbf{H}^{obs}$; $R_I = \Sigma_\mathbf{H}(I_\mathbf{H}^{obs} - I_\mathbf{H}^{calc})/\Sigma_\mathbf{H} I_\mathbf{H}^{obs}$; $R_{wF} = [\Sigma_\mathbf{H} w_\mathbf{H}(F_\mathbf{H}^{obs} - F_\mathbf{H}^{calc})^2/\Sigma_\mathbf{H} w_\mathbf{H}(F_\mathbf{H}^{obs})^2]^{1/2}$.

[b] Defined as $S = [\Sigma_\mathbf{H} w_\mathbf{H}(F_\mathbf{H}^{obs} - F_\mathbf{H}^{calc})^2/(N_{ref} - N_{par})]^{1/2}$; see Chapter 10, Section 10.6.1.

of the weakest intensities (those with a net intensity of only a few counts) is not a trivial matter. First, it is clear that the signal-to-noise ratio $I_H^{net}/\sigma(I_H^{net})$ for these reflections is very poor. Second, the weakest intensities are especially sensitive to systematic errors such as multiple scattering, higher harmonic contributions and faulty background estimates (due to overlap with neighbouring reflections or non-random electronic noise). Third, a substantial systematic bias (towards too low I values) can be introduced by improper treatment of pre-scan measurements (see Seiler *et al.* 1984). So, when sufficient, accurate observations are available (i.e. when the crystal model is clearly overdetermined) and when the weak, inaccurately measured reflections suffer from large systematic errors, it may even be advisable to exclude the latter from the least-squares process.

7.4   Acknowledgements

I am indebted to Professor Jack D. Dunitz for his critical and constructive corrections on several versions of the manuscript, and to Professors Fred Hirshfeld and Riccardo Destro for helpful comments and criticisms.

References

Alexander, L. E. and Smith, G. S. (1962). *Acta Crystallogr.*, **15**, 983–1004.
Becker, P. J. and Coppens, P. (1974). *Acta Crystallogr.*, **A30**, 129–47.
Blake, C. C. F. and Phillips, D. C. (1962). Effects of X-irradiation on single crystals of myoglobin. In *Biological effects of ionizing radiation at the molecular level*, pp. 183–91. Symposium, International Atomic Energy Agency, Vienna.
Blessing, R. H. (1987). *Crystallogr. Rev.*, **1**, 3–58.
Blessing, R. H., Coppens, P., and Becker, P. (1974). *J. Appl. Crystallogr.*, **7**, 488–92.
Blundell, T. L. and Johnson, L. N. (1976). *Protein crystallography*, pp. 330–1. Academic Press, New York.
Bragg, W. L., James, R. W., and Bosanquet, C. H. (1921). *Phil. Mag.*, **42**, 1–17.
Burbank, R. D. (1964). *Acta Crystallogr.*, **17**, 434–42.
Burbank, R. D. (1965). *Acta Crystallogr.*, **19**, 957–62.
Chandrasekhar, S. (1956). *Acta Crystallogr.*, **9**, 954–6.
Chandrasekhar, S., Ramaseshan, S., and Singh, A. K. (1969). *Acta Crystallogr.*, **A25**, 140–2.
Clegg, W. (1981). *Acta Crystallogr.*, **A37**, 22–8.
Cochran, W. (1969). *Acta Crystallogr.*, **A25**, 95–101.
Coppens, P. (1968). *Acta Crystallogr.*, **A24**, 253–7.
Coppens, P. (1977). *Isr. J. Chem.*, **16**, 159–62.
Coppens, P., Dam, J., Harkema, S., Feil, D., Feld, R., Lehmann, M. S., Goddard, R., Krüger, C., Hellner, E., Johansen, H., Larsen, F. K., Koetzle, T. F., McMullan, R. K., Maslen, E. N., and Stevens, E. D. (1984). *Acta Crystallogr.*, **A40**, 184–95.

Dam, J., Harkema, S., and Feil, D. (1983). *Acta Crystallogr.*, **B39**, 760–8.

Darwin, C. G. (1914). *Phil. Mag.*, **27**, 315–33 and 675–90.

Darwin, C. G. (1922). *Phil. Mag.*, **43**, 800–29.

Denne, W. A. (1972). *Acta Crystallogr.*, **A28**, 192–201.

Denne, W. A. (1977). *Acta Crystallogr.*, **A33**, 438–40.

Destro, R. and Marsh, R. E. (1987). *Acta Crystallogr.*, **A43**, 711–8.

Diamond, R. (1969). *Acta Crystallogr.*, **A25**, 43–55.

Dunitz, J. D. and Seiler, P. (1983). *J. Am. Chem. Soc.*, **105**, 7056–8.

Feil, D. (1977). *Isr. J. Chem.*, **16**, 149–53.

Flack, H. D. (1985). Avoidance, detection and correction of systematic errors in intensity data. In *Crystallographic computing 3: Data collection, structure determination, proteins, and databases* (ed. G. M. Sheldrick, C. Krüger, and R. Goddard), pp. 18–27. Clarendon Press, Oxford.

Gabe, E. J. (1985). Reducing random errors in intensity data collection. In *Crystallographic computing 3: Data collection, structure determination, proteins, and databases* (ed. G. M. Sheldrick, C. Krüger, and R. Goddard), pp. 3–17. Clarendon Press, Oxford.

Göttlicher, S. (1968). *Acta Crystallogr.*, **B24**, 122–9.

Hamilton, W. C. (1957). *Acta Crystallogr.*, **10**, 629–34.

Harada, J. and Sakata, M. (1974). *Acta Crystallogr.*, **A30**, 77–82.

Hirshfeld, F. L. (1984). *Acta Crystallogr.*, **B40**, 484–92.

Hirshfeld, F. L. and Hope, H. (1980). *Acta Crystallogr.*, **B36**, 406–15.

Hirshfeld, F. L. and Rabinovich, D. (1973). *Acta Crystallogr.*, **A29**, 510–3.

Kato, N. (1976). *Acta Crystallogr.*, **A32**, 453–7 and 458–66.

Kheiker, D. M. (1969). *Acta Crystallogr.*, **A25**, 82–8.

Lawrence, J. L. (1977). *Acta Crystallogr.*, **A33**, 232–4.

Lawrence, J. L. and Mathieson, A. McL. (1977). *Acta Crystallogr.*, **A33**, 288–93.

Lehmann, M. S. (1980*a*). Optimal experimental conditions for charge density determination. In *Electron and magnetization densities in molecules and crystals*, NATO ASI Series B, Physics, Vol. 48 (ed. P. Becker), pp. 287–322. Plenum Press, New York.

Lehmann, M. S. (1980*b*). Error analysis in experimental density determination. In *Electron and magnetization densities in molecules and crystals*, NATO ASI Series B, Physics, Vol. 48 (ed. P. Becker), pp. 355–72. Plenum Press, New York.

Lehmann, M. S. and Larsen, F. K. (1974). *Acta Crystallogr.*, **A30**, 580–4.

Lucas, B. W. (1969). *Acta Crystallogr.*, **A25**, 627–31.

McCandlish, L. E., Stout, G. H., and Andrews, L. C. (1975). *Acta Crystallogr.*, **A31**, 245–9.

Mathieson, A. McL. (1982). *Acta Crystallogr.*, **A38**, 378–87.

Mathieson, A. McL. (1983). *Acta Crystallogr.*, **A39**, 79–83.

Mathieson, A. McL. (1984). *Acta Crystallogr.*, **A40**, 355–63.

Rees, B. (1977). *Isr. J. Chem.*, **16**, 154–8.

Renninger, M. (1937). *Z. Phys.*, **106**, 141–76.

Renninger, M. (1967). Experimental determination of the integrated contribution of temperature diffuse scattering in X-ray reflections. In *Advances in X-ray analysis*, Vol. 10 (ed. J. B. Newkirk and G. R. Mallett), pp. 42–5. Plenum Press, New York.

Sakisaka, Y. and Sumoto, I. (1931). *Proc. Phys. Math. Soc. Jpn.*, **13**, 211–7.

Samson, S. and Schuelke, W. W. (1967). *Rev. Sci. Instrum.*, **38**, 1273–83.

Seiler, P. (1987). *Chimia*, **41**, 104–16.

Seiler, P. and Dunitz, J. D. (1978). *Acta Crystallogr.*, **A34**, 329–36.

Seiler, P. and Dunitz, J. D. (1985). *Austr. J. Phys.*, **38**, 405–11.

Seiler, P. and Dunitz, J. D. (1986). *Helv. Chim. Acta*, **69**, 1107–12.

Seiler, P., Schweizer, W. B., and Dunitz, J. D. (1984). *Acta Crystallogr.*, **B40**, 319–27.

Seiler, P., Belzner, J., Bunz, U., and Szeimies, G. (1988). *Helv. Chim. Acta*, **71**, 2100–10.

Soejima, Y., Okazaki, A., and Matsumoto, T. (1985). *Acta Crystallogr.*, **A41**, 128–33.

Stevens, E. D. (1974). *Acta Crystallogr.*, **A30**, 184–9.

Stewart, J. M., Hall, S. R., Alden, R. A., Olthof-Hazekamp, R., and Doherty, R. M. (1988). *The XTAL system of crystallographic programs*. Computer Science Center, University of Maryland, College Park, MD, USA.

Thornley, F. R. and Nelmes, R. J. (1974). *Acta Crystallogr.*, **A30**, 748–57.

Walker, N. and Stuart, D. (1983). *Acta Crystallogr.*, **A39**, 158–66.

Willis, B. T. M. and Pryor, A. W. (1975). *Thermal vibrations in crystallography*. Cambridge University Press.

Young, R. A. (1974). Single-crystal intensities. In *X-ray diffraction* (L. V. Azaroff, R. Kaplow, N. Kato, R. J. Weiss, A. J. C. Wilson, and R. A. Young), pp. 500–80. McGraw-Hill, New York.

Zachariasen, W. H. (1965). *Acta Crystallogr.*, **18**, 705–10.

Zachariasen, W. H. (1967). *Acta Crystallogr.*, **23**, 558–64.

8

Diffraction studies of molecular motion in crystals

Kenneth N. Trueblood

8.1 Introduction

Debye (1913) recognized almost immediately after the discovery of X-ray diffraction that diffraction experiments could make possible determination of the mean-square vibration amplitudes of atoms in crystals. Until the advent of high-speed computing in the 1950s, however, it was not possible to determine atomic vibration amplitudes sufficiently well to permit analysis of the motion of molecules in crystals. Then, in a remarkable series of papers, Cruickshank showed how to determine parameters to describe the anisotropic motion of atoms (Cruickshank 1956*a*), how to use these parameters to describe rigid-body motion of molecules (Cruickshank 1956*b*), and how to correlate this motion with the thermodynamic and spectroscopic properties of molecular crystals (Cruickshank 1956*d*, *e*, *f*). Furthermore he showed (Cruickshank 1956*c*, 1961) that librational molecular motion (i.e. rotational oscillation) could lead to apparent foreshortening of intramolecular distances, an important conclusion reached independently by Busing and Levy (1957, 1964).

In this account of the analysis of molecular motion in crystals, after some preliminary comments to provide background, we focus on two topics: the kinds of information about the motion of molecules in crystals that can be learned from diffraction experiments, and, more briefly, the implications of that motion for the determination of precise molecular geometry from crystal diffraction data. Further details can be found in numerous reviews of various aspects of the analysis of molecular motion in crystals (e.g.

Johnson 1970a, b, c, 1980; Johnson and Levy 1974; Willis and Pryor 1975; Dunitz 1979; Dunitz et al. 1988a, b).

8.2 Background

Mean-square displacements of individual atoms derived from X-ray or neutron diffraction data represent averages over *time* and over *space*. In other words, because the diffraction experiment normally extends at least over minutes or hours, and more often days, the atomic positions obtained are averages over countless vibration periods. Furthermore, because the radiation is being scattered simultaneously from millions of unit cells, *assumed* to be identical throughout the crystal lattice, the unit cell actually imaged represents an average over any minor 'static disorder', i.e. variations in atomic positions that may exist from one unit cell to another throughout the diffraction experiment. (If there are major *ordered* variations, photographs will quickly show that the lattice being used is not correct.)

The problem of distinguishing static disorder from vibrations of atoms can be approached in various ways, *e.g.* by using other techniques for detecting atomic and molecular motion (such as solid-state NMR spectroscopy), or by carrying out the diffraction experiment over a range of temperatures for which the crystal structure is stable. Although comparatively few careful studies of the same structure have been made at widely differing temperatures, there have been a number of such investigations and more are under way. Among those which have clearly demonstrated that atomic displacements do increase significantly with increasing temperature, and thus do really represent (at least in significant part) atomic motion, have been those of Becka and Cruickshank (1963) on hexamethylenetetramine, of Kvick *et al.* (1980) on γ-glycine and of Brock and Dunitz (1982) on naphthalene. Parts (a) and (b) of Fig. 8.1 are ORTEP (Johnson 1965) illustrations of γ-glycine (studied by neutron diffraction, so that the H atoms could be refined anisotropically) at 83 K and 298 K and of naphthalene (by X-ray diffraction, so the H atoms were not refined anisotropically and are not represented here) at 92 K and 239 K, respectively. The principal axes of the ellipsoids drawn at each atomic position represent the mean-square displacement amplitudes (MSDAs) along each of three mutually perpendicular directions. These ellipsoids have been drawn to represent 50 per cent probability; that is, there is an equal chance that the atom is inside the ellipsoid or outside it. It is clear that the average displacement increases markedly as the temperature is increased. Furthermore, the orientation of the ellipsoids is, in general, quite in accord with patterns of motion to be expected.

A contrasting example is that of monoclinic ferrocene (Seiler and Dunitz 1979a). Between room temperature and 173 K, the MSDAs of the cyclopentadienyl ring C atoms decrease very little in the tangential direction

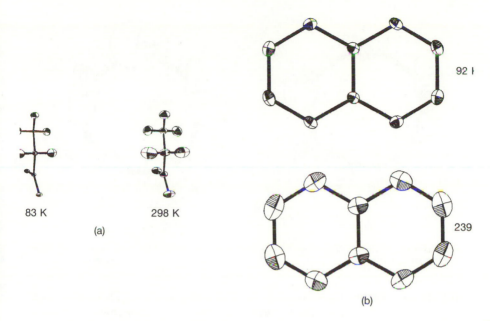

Fig. 8.1 Atomic displacement ellipsoids at different temperatures, drawn with ORTEP (Johnson 1965). (a) For γ-glycine, studied by neutron diffraction at 83 K and 298 K (Kvick *et al*. 1980). (b) For naphthalene, studied by X-ray diffraction at 92 K and 239 K (Brock and Dunitz 1982).

(Fig. 8.2), implying mainly static disorder among at least two ring orientations. Below 164 K, the temperature of the phase transition to the triclinic structure, the tangential MSDAs are approximately proportional to T, consistent with libration of the rings about local fivefold axes (Seiler and Dunitz 1979*b*).

Typical mean-square atomic displacements, $\langle u^2 \rangle$, for carbon atoms in a representative organic crystal without significant disorder vary from around 0.015 Å^2 at 100 K to 0.045 Å^2 or more at 300 K. For atoms in a particularly flexible portion of a molecule, the values might be twice as large. (Note that if $\langle u^2 \rangle = 0.04$ Å^2, then the r.m.s. displacement is 0.2 Å.) In protein and nucleic acid molecules, typical r.m.s. displacements range up to about 1.5 Å (Yu *et al*. 1985), and although some of this effect may be due to static disorder, it has been estimated that in representative macromolecular crystals half or more of the mean-square displacement arises from atomic motion, usually quite anisotropic and often highly anharmonic.

The *probability density function* (pdf) for atom k, $P_k(\mathbf{u})$, is defined so that $P_k(\mathbf{u})\, dV$ denotes the probability of finding atom k in the volume element dV when the atom is at a distance $\mathbf{u}$ from its mean position. Nothing is implied about whether the displacement is dynamic or static, so that

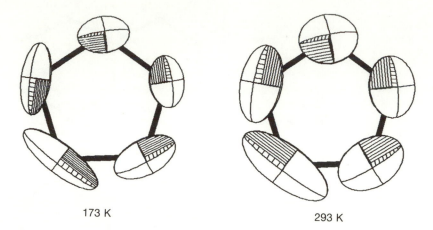

173 K   293 K

Fig. 8.2 Atomic displacement ellipsoids for monoclinic ferrocene at 173 K and 293 K (Seiler and Dunitz 1979a).

throughout the ensuing discussion we shall refer to quantities proportional to $\langle u^2 \rangle$ as *atomic displacement parameters* (ADPs) rather than using heretofore more common terms such as 'vibration parameters' or 'thermal parameters'.

In X-ray and neutron diffraction studies of crystals containing small molecules, the atomic pdf is usually taken as a Gaussian function:

$$D(\mathbf{x}) = [(\det \mathbf{U}^{-1})^{1/2})/(2\pi)^{3/2}] \exp(- \mathbf{x}^T\mathbf{U}^{-1}\mathbf{x}/2), \qquad (8.1)$$

where $\mathbf{U}^{-1}$ is the inverse of the second-moment matrix $\mathbf{U} = \langle \mathbf{uu}^T \rangle$ of the pdf. If an atomic pdf actually is Gaussian, corresponding to the motion of the atom in an effectively harmonic potential, the pdf is completely defined by $\mathbf{U}$. More generally, provided that sufficient high-angle diffraction data are available, the higher cumulants (Johnson 1969) of non-Gaussian pdfs can also be determined. In practice, however, this is uncommon.

If $\mathbf{U}$ is anisotropic and positive definite, the surfaces corresponding to the set of equations

$$\mathbf{x}^T\mathbf{U}^{-1}\mathbf{x} = \text{constant}$$

are ellipsoids; if $\mathbf{U}$ is isotropic, they are spherical. The Fourier transform of the pdf given by eqn (8.1) is the corresponding 'temperature factor', which we designate as $\tau(\mathbf{H})$:

Isotropic: $\tau(\mathbf{H}) = \exp(- 2\pi^2\langle u^2 \rangle H^2)$

$\qquad\qquad\qquad = \exp(- 8\pi^2\langle u^2 \rangle \sin^2 \theta/\lambda^2)$

Anisotropic: $\tau(\mathbf{H}) = \exp(- 2\pi^2 \mathbf{H}^T \mathbf{U} \mathbf{H})$ in matrix notation, or

$$= \exp(- 2\pi^2 h_i a^i U^{ij} h_j a^j)$$ in tensor notation,

where $\mathbf{H} = h_i a^i = h\mathbf{a}^* + k\mathbf{b}^* + l\mathbf{c}^*$ is a reciprocal-lattice vector; $\mathbf{U} = \langle \mathbf{u}\mathbf{u}^T \rangle$ a symmetric matrix or tensor; and a superscript T means 'transpose'.

For convenience in calculation, the anisotropic temperature factor is often written as $\tau(\mathbf{H}) = \exp(- h_i \beta^{ij} h_j)$ (summation implied), with $\mathbf{H}$ the scattering vector $h_i a^i$ and then $\beta^{ij} = 2\pi^2 a^i a^j U^{ij}$ (no summation). The most commonly used forms of the anisotropic displacement parameters (ADPs) in the temperature factor expression are the U^{ij} defined above, those here called β^{ij} and a type that is variously symbolized as b^{ij} or B^{ij}, with $b^{ii} = \beta^{ii}$ and $b^{ij} = 2\beta^{ij}$ ($i \neq j$). Unfortunately, the writers of crystallographic papers often say their parameters are of one type when they are not, and although such errors can sometimes be detected by inspection or by tests (Hirshfeld 1976; Trueblood 1978; Chandrasekhar and Bürgi 1984), it is occasionally not possible to be sure what form has, in fact, been reported.

If n_i represents a component of a unit vector in its covariant representation (Patterson 1959), then the *mean-square displacement amplitude, MSDA, of a given atom in the direction given by* $\mathbf{n}$ is:

$$\langle u^2 \rangle_{\mathbf{n}} = n_i (\beta^{ij}/2\pi^2) n_j \quad \text{or} \quad \mathbf{n}^T (\boldsymbol{\beta}/2\pi^2) \mathbf{n},$$

where $\boldsymbol{\beta}$ is the tensor defined in the preceding paragraph.

8.3   Rigid-bond and rigid-molecule tests

For a perfectly rigid molecule, by definition the interatomic distances do not change no matter how the molecule may move, and all the atoms must move in phase. Therefore, the mean-square displacement amplitudes (MSDAs) for every pair of atoms A,B must be equal along their interatomic vector:

$$\Delta_{A,B} = (MSDA)_A - (MSDA)_B = \mathbf{n}^T (\boldsymbol{\beta}/2\pi^2)_A \mathbf{n} - \mathbf{n}^T (\boldsymbol{\beta}/2\pi^2)_B \mathbf{n} = 0.$$

Although no molecule is perfectly rigid, Hirshfeld (1976) pointed out that, since the amplitudes of bond-stretching vibrations for atoms other than H and D are normally much smaller than the amplitudes resulting from other vibrations, the MSDAs of a pair of bonded atoms (not including H or D) in a typical organic molecule should be almost the same *along the bond direction*, even though they may be widely different in other directions. This condition, which has been termed the *rigid-bond criterion*, has, indeed, been found to hold for well-refined organic structures based on good data: the differences in MSDAs for bonded pairs, when neither atom is H or D, are smaller than 10×10^{-4} Å^2. Rosenfield *et al.* (1978) used this criterion more generally to detect non-rigidity in molecules. If a molecule *is* rigid, then *all* intramolecular distances, not just bond distances, remain invariant, so the differences in MSDAs for *each* pair of atoms should be close to zero along the direction of the line between the atoms. The converse does not necessarily hold true; for example, for planar or linear molecules in which there are small vibrations of atoms normal to the plane or line, the differences of MSDAs in the interatomic direction will remain essentially zero despite the perpendicular motions of the atoms. Furthermore, two atoms might move along their interatomic line with equal amplitudes but quite out of phase, so that there is appreciable motion of one atom relative to the other even though their MSDAs are equal. This example illustrates one of the major shortcomings of analyses based on Bragg diffraction data: *no information is obtainable about the correlations between the motions of different atoms*. Such information is implicit in many *models* used to interpret ADPs, but it must always be borne in mind that it is a property of the model, not of the data or, necessarily, of the crystal.

Nonetheless, when these cautions are heeded, inspection of a table of MSDA values is often very revealing. If the data satisfy the rigid-bond test, we may assume that the displacement parameters probably do represent genuine atomic vibrations. In such a case, gross violation of the condition of equality of the MSDAs for particular pairs of atoms along their interatomic lines indicates that the molecule is *not* rigid, and, more specifically, that there is appreciable motion of one of these atoms relative to the other. This test has been used convincingly to provide information about the pattern of internal motions in many organic structures. Examples include triphenylphosphine oxide (Brock *et al.* 1985), a [3.1.1]propellane derivative (Chakrabarti *et al.* 1981) for which this test provided evidence of incipient inversion at a nitrogen atom (Dunitz *et al.* 1988*a*), and a rigid spherand (Trueblood 1985). In the first two examples mentioned, the intensity data were measured with considerable care near 100 K; for the spherand structure, data were measured at room temperature with no special efforts taken, but they were nonetheless remarkably good. This is not common; for detecting subtle effects, reliable data collected at low temperature are normally much to be preferred. (It is worth noting that, with good neutron diffraction data, the difference in MSDA values for C and H (or D) atoms in a C—H (or C—D) bond is consistently close to 50×10^{-4} Å^2, with H showing the larger amplitude. Similar results are found for N—H and O—H bonds.)

In coordination complexes, where the stretching force constants are much weaker than those of ordinary covalent bonds and where the ligand atoms usually have an appreciably smaller mass than the metal atom to which they are bonded, the rigid-bond postulate takes a modified form: for such ligand–metal bonds, the differences in MSDA values may be as much as

30×10^{-4} Å^2 (Bürgi 1984), with the lighter ligand atom having the larger MSDA value. Gross violations of even this less stringent criterion have been found and have been correlated convincingly with spin-crossover effects in crystalline Fe^{III} complexes (Chandrasekhar and Bürgi 1984), with dynamic Jahn–Teller and pseudo Jahn–Teller distortions in crystals of Cu^{II} complexes (Ammeter et al. 1979; Stebler and Bürgi 1987), and with valence disorder in a binuclear Mn^{III}–Mn^{IV} complex (Stebler et al. 1986).

8.4   Rigid-body model

Cruickshank (1956b) showed that the anisotropic displacement parameters of the individual atoms in a molecule that might be assumed to behave as a reasonably rigid body could be analyzed by a straightforward linear least-squares process to provide tensors describing the overall motion of the molecule. His analysis was generally appropriate only for a centrosymmetric molecule at a crystallographic center of inversion. Schomaker and Trueblood (1968) generalized the treatment to molecules at positions of any site symmetry. In the general case, the motion of a 'rigid' molecule can be described, in the harmonic approximation, in terms of three tensors:

T Pure translation (the same motion for every atom in the molecule)
L Pure rotation (motion of a particular atom depends on its position in the molecule, as well as on the amplitude of the rotation)
S Coupled rotation and translation (screw motion)

These tensors represent quadratic averages over all manner of instantaneous rotations and translations (and over lattice position as well in many structures). **T** and **L** are symmetric second-rank tensors. **S** is not, in general, symmetric, but one of its diagonal elements is indeterminate from diffraction data (often the sum of the diagonal elements is arbitrarily constrained to zero). **S** vanishes for crystals with a center of symmetry. There are thus at most 20 determinable tensor elements (fewer if the molecule lies at a position in the crystal with any special symmetry).

Two general approaches have been used in analyzing rigid-body motion in terms of these tensors. The first, that originally suggested by Cruickshank (1956b), is to refine the individual atomic displacement parameters independently and then determine the molecular tensor components by a linear least-squares fit to these ADPs. In this least-squares fit, there are usually many more observations (the individual atomic β^{ij} or U^{ij}) than quantities to be determined. The second approach, suggested by Pawley (1966), is to determine **T**, **L** and **S** directly during the least-squares refinement based on the diffraction data; the displacement parameters of the atoms of the rigid body are then dependent variables. This procedure is especially valuable when the ratio of the number of independent diffraction intensities to the number of parameters to be determined (if one tried to refine many atoms aniso-tropically) is low. For the purposes of the present discussion, the first approach is the preferred one, because we wish to examine and use the individual atomic displacement parameters to detect and analyze overall and internal molecular motion, rather than to assume in advance a particular kind of motion.

During the past three decades, the rigid-body model has been applied (sometimes rather indiscriminately) to the analysis of ADPs from many hundreds, if not thousands, of crystal structures. When the assumption of molecular rigidity seems reasonably justified, e.g. with carbon skeletons of polycyclic molecules, the model fits well and seems quite plausible. Examples of almost perfect agreement between observed ADPs and those calculated after a rigid-body fit include the centrosymmetric isomer of tetramethyltetraasterane (Chesick *et al.* 1973) and a tricyclic ketal (Brown *et al.* 1982).

The physical significance of the translational, librational and screw tensors (**T, L,** and **S**) derived from a rigid-body analysis of a molecule in a particular crystal structure can also be assessed by comparing them with the same quantities obtained by a lattice-dynamical calculation (see Chapter 9). For such a calculation, one needs the equilibrium positions of the atoms in the crystal and an appropriate set of atom–atom potentials. For crystals of several aromatic hydrocarbons (e.g. benzene, naphthalene, anthracene, pyrene: Filippini *et al.* 1973, 1974; Gramaccioli *et al.* 1982; see also Chapter 9, Table 9.4), the **T, L,** and **S** tensors calculated by lattice dynamics are in at least fair agreement with those obtained by analysis of the experimental ADPs. This gives confidence that these quantities do indeed have some physical significance, despite the various uncertainties in their interpretation.

In some of the most careful studies of molecular motion in crystals, correction has been made for internal vibrations. Following Higgs (1953, 1955), it is assumed that the internal vibrations are not correlated with the overall motion and may be estimated from spectroscopic or force-field calculations. Becka and Cruickshank (1963) first applied this correction to the C and N atoms in hexamethylenetetramine in their study of the motion of this molecule as a function of temperature. More recently, Brock and Dunitz (1982) corrected for the internal vibrations of C atoms in studying the temperature-dependence of molecular motion in crystalline naphthalene. The contributions of these vibrations to MSDA differences along the bonds were only 0.0002 Å^2 even at the highest temperature used (239 K), although their effects on **T** and **L** were more significant.

Johnson (1970*c*) showed that plausible estimates of C—H and C—D vibration amplitudes can be obtained from ADPs measured by neutron dif-fraction on organic crystals, and thus demonstrated that the 'observed'

ADPs derived by least-squares refinement can indeed be physically mean-ingful. He used neutron diffraction data for several phenylnorbornyl deriva-tives to calculate (segmented) rigid-body parameters (**T**, **L**, **S** and torsional amplitudes) *using only the heavy-atom skeleton* (all but the H or D atoms). Then he calculated the displacement parameters U^{ij} that these motions implied for the H(D) atoms and subtracted these calculated values from the observed U^{ij} for the H(D) atoms. The differences for the H(D) atoms were invariably positive, and those for the atoms on the phenyl groups agreed remarkably well in amplitude and direction with those implied by a normal-mode calculation for the H(D) atoms in benzene.

8.5 Corrections to intramolecular distances

Cruickshank (1956c, 1961) and Busing and Levy (1957, 1964) recognized early that a careful distinction must be made between the *distance between the mean positions* found for a pair of atoms in a structure and the aver-age of the *instantaneous separation* of those atoms. The observed distance between the mean positions of a pair of atoms in a molecule in a crystal (as found, e.g. by Fourier or least-squares methods) is *usually shorter (never longer) than the actual average distance between the instantaneous positions* of the atoms. It is easy to appreciate that libration of a *rigid* group of atoms will lead to an apparent contraction of all distances that are not parallel to the libration axis. In essence, the centroid of the pdf for a librating atom is somewhat displaced toward the axis of libration, away from the midpoint of the arc along which the atom moves. In a strictly rigid molecule, all atoms move in phase, but in general we cannot assume strict rigidity, and any esti-mate of the correction for atomic and molecular motion must thus depend on assumptions about the relative phases of the motions of the different atoms.

Thus, corrections can only be made when the correlation between the motions of the atoms concerned is known, i.e. when the phase relationship (if any) between their motions is known. The rigid-body model implies one such correlation, and correction of intramolecular distances is straight-forward (Cruickshank 1956c, 1961; Schomaker and Trueblood 1968). Any model for internal motion, such as that of Dunitz and White (1973) or the segmented-body model of Johnson (1970c), also implies a correlation and thus a way of correcting distances within the group moving internally. Busing and Levy (1964) considered the limits that could be placed on such corrections for extreme cases of correlated motion as well as for some other plausible models (see also Johnson 1970b). In room-temperature struc-tures, the foreshortening may often amount to 0.03 Å or more, considerably greater than all other errors combined in a well-determined structure. Cor-rections are therefore essential, even though they may be very uncertain if the rigid-body model is grossly inadequate and no convincing model of internal motion can be postulated. In such situations, even the corrected distances must be regarded as unreliable. One possible recourse might be to a suitable lattice-dynamical calculation if such is available or can be made; such calculations yield the generalized atomic mean-square displacement matrices, which include the tensors corresponding to *correlated* mean-square displacements of *all* pairs of atoms in the structure. They can thus be used for calculating corrections to the interatomic distances (Scheringer 1972; Johnson 1980; Gramaccioli and Filippini 1985; see also Chapter 9, Section 9.4). However, such lattice-dynamical calculations are lengthy; comparatively few have been made for molecules of general interest, and especially few for molecules with internal degrees of freedom.

It is important to note that, in the absence of a lattice-dynamical calculation, it is never possible to correct *inter*molecular distances, because the correlation between the motion of different molecules is unknown. Johnson (1970*b*) gives a valuable summary of the effect of molecular motion on interatomic distances and angles.

Temperature-dependent studies of thermal motion in several crystal structures containing molecules or polyatomic ions have shown that *apparent* bond distances shrink as the temperature increases if there is significant molecular motion. When rigid-body corrections to the distances are taken into account, however, bond distances do not, in fact,

Table 8.1 C—N and C—C bond distances at various temperatures for hexamethylenetetramine, $(CH_2)_6N_4$ (Becka and Cruickshank 1963), and *t*-butylammonium chloride, $Me_3C-NH_3^+$, Cl^- (Trueblood 1987), before and after correction for the effects of molecular libration

Compound	T (K)	Bond	Distance (Å)	
			Uncorrected	Corrected
Hexamethylene-	34	C—N	1.474(4)	1.476(4)
tetramine	100		1.474(7)	1.479(7)
	298		1.464(4)	1.478(4)
t-Butylammonium	115	C(1)—N	1.513(4)	1.517(5)
ion	298		1.500(5)	1.515(6)
	115	C(1)—C(2)	1.520(5)	1.527(6)
	298		1.504(8)	1.523(9)
	115	C(1)—C(3)	1.515(4)	1.521(5)
	298		1.506(7)	1.526(8)
	115	C(1)—C(4)	1.523(4)	1.528(5)
	298		1.512(7)	1.531(8)

vary significantly with temperature (Tables 8.1 and 8.2). The data in Table 8.1, for hexamethylenetetramine (Becka and Cruickshank 1963) and *t*-butylammonium ion in *t*-butylammonium chloride (Trueblood 1987), show quite clearly the decrease of *apparent* distances when the temperature increases sufficiently that the libration amplitude is appreciable. The r.m.s. libration amplitudes in the former structure vary from 2.4° at 34 K to 6.6° at 298 K; the largest r.m.s. amplitudes in the latter range from 3.8° at 115 K to 7.7° at 298 K. With naphthalene (Table 8.2) the data are less clear-cut, at least in part because there is less motion, but for those bonds that showed the greatest change in apparent length from 92 K to 239 K, the librational correction eliminates about half of this deviation; the maximum libration amplitudes vary from 2.6° at 92 K to 4.4° at 239 K.

Table 8.2 C—C bond distances at various temperatures for naphthalene (Brock and Dunitz 1982), before and after correction for the effects of molecular libration

T (K)	Bond[a]	Distance[b] (Å)	
		Uncorrected	Corrected
92	C(1)—C(2)	1.376	1.377
143		1.374	1.377
239		1.368	1.373
92	C(4)—C(5)	1.376	1.378
143		1.376	1.378
239		1.374	1.380
92	C(2)—C(3)	1.425	1.426
143		1.425	1.428
239		1.425	1.431
92	C(3)—C(4)	1.422	1.424
143		1.421	1.423
239		1.419	1.424
92	C(1)—C(5′)	1.415	1.417
143		1.412	1.415
239		1.407	1.413
92	C(3)—C(3′)	1.422	1.424
143		1.423	1.426
239		1.420	1.426

[a] The C atoms of the naphthalene nucleus are numbered as follows:
[b] Least-squares standard deviations are about 0.002 Å.

With groups that may librate appreciably even at low temperature, or be statistically disordered so as to simulate libration — for example, a nearly spherical ion like ClO_4^- — the effects on apparent bond distances may be significantly larger than those evident in Tables 8.1 and 8.2. Nonetheless, when the ADPs are subjected to rigid-body motion analysis (i.e. it is assumed that the observed ADPs represent either real or simulated rigid motion) and the resulting values of L are used to correct the apparent distances, the originally disparate measurements at different temperatures and even in different structures often become quite consistent. When the r.m.s. libration amplitude is so large (greater than *ca.* 0.3 radian, or 17°) that the harmonic approximation for the atomic displacements is seriously violated, however, the corrections become unreliable.

When bond distances are to be determined with the highest possible accuracy, it is best to reduce motional corrections by measuring the intensities at as low a temperature as feasible. However, even bond distances obtained by low-temperature X-ray analysis and corrected for residual libration may be in error by more than 0.01 Å (several times the nominal error of a typical precise experiment) if the refined atomic positions are displaced from the centroids of the nuclear pdfs because of contamination with non-spherical bonding electron density. Appropriate deformation terms (see Chapter 10, Section 10.9) must be included in the least-squares refinement, or this refinement must be based on sufficiently high-order data that the contributions of the valence electrons to the atomic form factors become negligible. For typical organic crystals, low-temperature measurements and wavelengths appreciably shorter than that provided by Cu Kα radiation (e.g. Mo Kα) are required to obtain the high-order data (to $H = 1.4$ Å^{-1} or beyond) necessary to provide molecular dimensions comparable with those obtainable with neutron diffraction or other physical methods, or (for relatively small molecules) by calculation. The effects of intermolecular interactions in the solid state on molecular geometry must be considered when such comparisons are made, but in some careful studies, excellent agreement has been found, e.g. for the molecular dimensions of 2-aminopropenenitrile determined by X-ray analysis at 97 K (Seiler and Dunitz 1985) and by microwave spectroscopy, and calculated with a 6–31G* basis set. Examples of such comparisons will be found in Chapters 11 and 18.

One of the most popular models for correction of bond distances when a light atom is bonded to a relatively heavy one is the 'riding model' of Busing and Levy (1957, 1964), clearly discussed by Johnson (1970*b*). In this model, it is assumed that the lighter atom has all of the translational motion of the atom to which it is bonded, plus an additional uncorrelated motion. However, a *word of caution* is in order: corrections for anharmonic effects in X—H bonds (Ibers 1959; Craven and Swaminathan 1984; Jeffrey and Ruble 1984; see also Chapter 11, Section 11.3.2.2) are usually of opposite sign and may offset much of a riding-model correction in just those situations when the riding model is most often invoked (H on C, N, or O).

8.6 Barriers to internal and overall rotation in crystals

Studies of intramolecular torsional motion in crystals were pioneered by Johnson (1967, 1970c), with his segmented-body approach, and by Hamilton and his collaborators in their neutron diffraction studies of amino acids and other molecules containing methyl groups (Hamilton 1969; Hamilton *et al.* 1969; Schlemper *et al.* 1971; Lehmann *et al.* 1972), although occasional estimates of torsional amplitudes and frequencies had been made almost a decade earlier (Trueblood *et al.* 1961). The recent studies of transition-metal complexes by Bürgi and his coworkers that were alluded to earlier illustrate a different approach to the analysis of internal dynamic effects in crystals; they have been reviewed by Bürgi (1984) and will not be further discussed here, but this approach obviously has considerable potential.

Trueblood and Dunitz (1983) demonstrated the feasibility of using MSDA values from crystal diffraction data in the estimation of force constants and barriers to internal motion (chiefly torsional) for many different kinds of groups. The model used was the appealingly simple one-parameter torsional model of Dunitz and White (1973), which had been applied extensively (Trueblood 1978) in the estimation of torsional amplitudes of a variety of groups in different crystal structures. In applying this model, a least-squares fit to the atomic displacement tensors is made not only of T, L and S but also of a mean-square torsional amplitude for each group in the molecule that is suspected of having appreciable torsional motion. This mean-square amplitude about an assumed axis of libration is the only additional parameter in the Dunitz–White model. The group undergoing torsion is assumed to be itself internally rigid; the computer program used (Trueblood 1978, and unpublished work; the current version of the program is THMA11) allows up to seven different groups to undergo torsional motions within a given molecule.

In the Dunitz–White model, all correlations between the internal motion and the overall motion are ignored; however, this approach is oversimplified. Just as, in the rigid-body model, the components of the S tensor are needed to allow for the quadratic correlation between the pure translational and pure librational motion, so here analogous terms are needed to allow for the correlation between the internal torsional motion and the overall molecular translation and libration (Schomaker and Trueblood 1984). In general, there are six quadratic correlation terms, three of the internal torsion, ϕ, with overall rotation and three of ϕ with overall translation. It is not possible to determine separately $\langle \phi^2 \rangle$ about the internal torsional axis and the correlation of ϕ with the overall rotation about this axis (or any parallel axis), which we designate $\langle \phi \lambda^{\|} \rangle$. Only the sum $\langle \phi^2 \rangle + 2 \langle \phi \lambda^{\|} \rangle$ can be determined. It is apparently true, however, that when the value of $\langle \phi^2 \rangle$ calculated with the simple Dunitz–White model is large relative to $\mathbf{L}$ about the internal torsional axis, than $\langle \phi \lambda^{\|} \rangle$ is likely to be quite small relative to $\langle \phi^2 \rangle$, so that $\langle \phi^2 \rangle + 2 \langle \phi \lambda^{\|} \rangle$ is a good approximation to $\langle \phi^2 \rangle$. It is clear that in some structures correlations of internal and overall motion are important in improving the agreement between observed and calculated U^{ij}; in one study cited by Dunitz *et al.* (1988*a*), the effect of including these correlations improved the 'agreement index' (between observed and calculated U^{ij}) dramatically, by a factor of four.

In our earlier study (Trueblood 1978), we had demonstrated that, without including these correlations, this simple model gave results not significantly different from those of more elaborate models that included anharmonicity corrections, even for r.m.s. amplitudes as high as 0.4 radian (23°). Although not all of those systems have been reinvestigated including correlations, for many of them in which the torsional amplitudes were large, the results are not greatly different when correlations are included.

Most of the data for the study of frequencies, force constants and barriers to internal torsion were taken from the literature, although a few were from our own unpublished work. We began with careful searches of the Cambridge Structural Database (see Chapter 15) to locate structures of interest, eliminating those that were of low precision, those based on X-ray data that contained any atom of atomic number greater than 17 (chlorine), and those that were disordered. A library check was almost always necessary, if only to ascertain whether the displacement parameters had been published and, if not, how they might be obtained.

In all, calculations were made for around 300 librating groups in more than 125 crystal structures (Trueblood and Dunitz 1983). When there were many structures containing the group of interest, we selected on the basis of the ease of obtaining the data, the reported precision, and the general interest and possible significance of the results (e.g. the presence of several identical groups, some of which might be chemically but not crystallographically equivalent). Most of the data used were from diffractometer studies made since 1970, although even rather imprecise displacement parameters can sometimes reveal quite clearly torsional and other motions of sufficiently high amplitude. Some structural studies that seemed *a priori* likely candidates for inclusion had to be rejected because the displacement parameters obviously contained gross errors (the rigid-bond test was routinely applied); others could be used only after corrections had been made, the most common error being incorrect specification by authors of the form in which their anisotropic displacement parameters were reported.

After a mean-square torsional amplitude was obtained for each group, the quadratic force constant for the torsional motion was estimated, from either the classical expression,

$$f = kT/\langle \phi^2 \rangle,$$

or the quantum mechanical expression for the amplitude,

$$\langle \phi^2 \rangle = [h/(8\pi^2 I\nu)] \coth (h\nu/2kT)$$

with $f = 4\pi^2 I\nu^2$; where k is Boltzmann's constant, h is Planck's constant, ν is the frequency, and I is the moment of inertia of the librating group. Estimation of the barrier from the force constant depends upon the assumption of a potential function. For a cosine potential of the form:

$$V = (V_n/2)(1 - \cos n\phi)$$

(where V_n is the n-fold torsional barrier) expansion of $\cos n\phi$ ignoring powers of $n\phi$ higher than the second, i.e. assuming that $n\phi \ll 1$, leads to

$$V = (V_n/2)[1 - (1 - n^2\phi^2/2 + \ldots)] \cong V_n n^2 \phi^2/4.$$

Equating this last expression with that for a quadratic potential, $V = f\phi^2/2$, we get

$$V_n \cong 2f/n^2.$$

Table 8.3 summarizes some of the results for a few common groups. All of the studies of methyl derivatives were, of course, based on neutron-diffraction structure analyses. The agreement between the methyl barriers estimated by the present method and those found by other methods is encouragingly good. Although there are no reliable literature values to compare with the results for the barriers to torsion of the NH_3^+ group, the values found are quite plausible, being larger than those for the isoelectronic methyl group and well correlated with the hydrogen-bonding environment (Lehmann *et al.* 1972). The barriers for the bulkier groups, CF_3 and *t*-butyl, extend to higher ranges than those found in other studies, which were made chiefly in the gas phase; this discrepancy may well be caused by intermolecular interactions arising from the comparative inflexibility of the packing around the molecule. Packing energy and related calculations for the structural regions around molecules containing bulky groups are needed, however, before such an interpretation can be confirmed.

There are several obvious limitations to this approach to the estimation of torsional barriers, among which are:

Table 8.3 Some threefold torsional potential barriers[a] derived from atomic displacement parameters, ADPs (Trueblood and Dunitz 1983)

Group	Attached atom	Range of barriers (kJ mol^{-1})	
		From ADPs	From other methods[b]
CH_3	C, N (trigonal)	1.5–9.6	1.2–8.5 (s)
	C, N, P (tetrahedral)	4–26	12–22 (g)
	O	1–9	4.5 (CH_3OH, g)
			15–19 (other, s)
NH_3^+	C	14–62^c	–
CF_3	C (trigonal)	4–10	4–7 (g)
	C (tetrahedral)	9–66	13–25 (g)
$C(CH_3)_3$	C, N, P	7–112	2–27 (g)

[a] These are equivalent cosine-function barriers derived from force constants. In the classical approximation $V_3 \cong 2RT/(9\langle\phi^2\rangle)$.
[b] s, solid-state measurement; g, gas-phase measurement.
[c] Well correlated with the hydrogen-bonding environment.

1. The actual physical process occurring in the crystal is by no means clear. Is the motion in (or of) a single molecule, or possibly a concerted rotational motion in (or of) a small, or perhaps large, group of molecules? What is the relationship between the time- and space-averaged rotational MSDA and the actual physical process? This relationship would be clear in a situation in which there is an energetically highly-degenerate phonon corresponding to the hypothesized internal rotation, and, although this is admittedly a rather artificial situation, it is implicitly assumed here. Some measure of the cooperativity involved in the rotation would be desirable. It is also assumed, as mentioned, that $\langle\phi^2\rangle + 2\langle\phi\lambda^{\parallel}\rangle$ is a good approximation to $\langle\phi^2\rangle$ in the cases examined here.

2. There is no unique way to resolve the total vibrational motion into different modes, and other modes than those singled out may, in fact, be important. He and Craven (1985) have adopted a quite different approach to the analysis of internal modes using crystal diffraction data. They perform a 'quasi normal-mode' analysis, fitting force constants for a number of (presumably) low-frequency modes, as well as **T**, **L** and **S**, to the atomic U^{ij} by a least-squares method. Their approach is in some ways more flexible, but it too depends on chemical intuition in the recognition of the likely important low-frequency internal motions.

3. The harmonic assumption and the assumption that the potential is periodic, with only a single dominant term of known n, are both over-

simplifications in real situations. In fact, however, the former does not seem (Trueblood 1978) to have a significant effect except in the most extreme conditions (beyond those considered in the study referred to), and the latter is plausible for most of the threefold symmetrical groups investigated.

4. Systematic errors in the data might be expected to affect the results significantly. However, experience suggests that they will affect primarily the translational tensor rather than **L** or the individual torsional amplitudes, which depend on *differences* in the displacement parameters of individual atoms, such differences being far less subject to systematic errors than are the absolute values. It has been shown in several studies that quite imprecise (and, presumably, inaccurate) data can yield quite plausible patterns of motion.

5. Intermolecular effects must clearly be taken into account in interpreting the results of any analysis of barriers to internal motion in crystals. By attempting to correlate the results for groups that are chemically but not crystallographically identical (e.g. the methyl groups in durene, or the *t*-butyl groups in 1,2,4,5-tetra-*t*-butylbenzene) with the environment of these groups in the crystal, one should be able to assess the reliability of this approach. A similar comparison of results for crystals containing several chemically identical molecules in the asymmetric unit should serve the same purpose. We are now undertaking such studies.

Maverick and Dunitz (1987) have recently applied the approach used by Trueblood and Dunitz (1983) to the estimation of barriers to rotation of cyclopentadienyl rings in 22 different environments in crystalline ferrocene, nickelocene and ruthenocene, at temperatures varying from 98 to 293 K. They had to extend the method to cover situations in which $n^2\langle\phi^2\rangle$ was no longer much smaller than 1, by numerically integrating the classical Boltzmann expression. The rotation barriers they found, which for different rings in different crystals varied from 2.7 to more than 50 kJ mol^{-1}, compared remarkably well in both magnitude and precision with independent measurements on the same substances made using various spectroscopic techniques. These barriers in the solid state are considerably larger than those found for the free molecules by various methods; the increase is due to intermolecular packing effects and the trend is reproduced nicely by force-field calculations (Brock, Maverick, and Dunitz, unpublished work).

Despite its limitations, the ADP approach to measuring rotation barriers in crystals gives reasonable agreement with other techniques (all of which themselves have limitations). It is a valuable complement to other methods, especially since it often permits resolving many different librational motions in a complex molecule.

8.7   Macromolecules of biological import

In recent years it has become clear that diffraction data provide a way of assessing some of the dynamic properties of macromolecules. Initially, this was done through analysis of isotropic displacement parameters (Frauenfelder *et al.* 1979; Artymiuk *et al.* 1979; Singh *et al.* 1980). More recently, anisotropic parameters have been obtained for some atoms in macromolecules, although the resolution is normally so limited that they cannot be obtained for all. However, for some small proteins (rubredoxin: Watenpaugh *et al.* 1980; avian pancreatic polypeptide: Glover *et al.* 1983), ADPs have been obtained for atoms in certain residues, and analysis of these parameters provided evidence for concerted motion in distinct regions of each molecule. A different approach, involving the determination during refinement not of *individual* ADPs but rather of 'group rigid-body' parameters, as if different portions of the molecule were able to move independently, has recently been tried, with some (though limited) success (Moss *et al.* 1984; Holbrook *et al.* 1985). Advancements in techniques of data collection and restrained refinement will doubtless soon make it possible to determine ADPs for some of the atoms of even more large biologically-important molecules, and hence provide further clues to the dynamic properties of these molecules in crystals containing them.

8.8   Higher-order models for thermal motion

Although the Gaussian pdf has proved adequate for thermal-motion analysis in most circumstances, including some structures in which there is demonstrably high-amplitude internal motion, there are situations in which significant anharmonicity makes the quadratic approximation inadequate. Various approaches have been used, including addition of terms to the Gaussian expression (notably, the cumulant expansion model introduced by Johnson (1969, 1970c); see also Prince and Finger (1973)) and introduction of models based on curvilinear density functions (on a circle or a sphere). These have been discussed well by Johnson and Levy (1974). Recently there has been increased interest in anharmonic 'temperature factors' (e.g. Schulz 1984; Schulz *et al.* 1985; Scheringer 1985, 1987), especially in relation to the onset of phase transitions (Yoshiasa *et al.* 1987). A versatile computer program for investigating them has been described (PROMETHEUS: Zucker *et al.* 1983).

8.9   Acknowledgements

It is a pleasure to acknowledge the valuable critical comments of Professors Hans-Beat Bürgi, Carlo Maria Gramaccioli, Fred Hirshfeld, and Emily

Maverick on an early version of this chapter, and my long-time colleagues, Professors Jack Dunitz and Verner Schomaker. Some of this work has been supported by the National Science Foundation.

References

Ammeter, J. H., Bürgi, H.-B., Gamp, E., Meyer-Sandrin, V., and Jensen, W. P. (1979). *Inorg. Chem.*, **18**, 733–50.

Artymiuk, P. J., Blake, C. C. F., Grace, D. E. P., Oatley, S. J., Phillips, D. C., and Sternberg, M. J. E. (1979). *Nature*, **280**, 563–8.

Becka, L. N. and Cruickshank, D. W. J. (1963). *Proc. Roy. Soc.*, **A273**, 435–54 and 455–65.

Brock, C. P. and Dunitz, J. D. (1982). *Acta Crystallogr.*, **B38**, 2218–28.

Brock, C. P., Schweizer, W. B., and Dunitz, J. D. (1985). *J. Am. Chem. Soc.*, **107**, 6964–70.

Brown, K. L., Down, G. J., Dunitz, J. D., and Seiler, P. (1982). *Acta Crystallogr.*, **B38**, 1241–5.

Bürgi, H.-B. (1984). *Trans. Am. Crystallogr. Assoc.*, **20**, 61–71.

Busing, W. R. and Levy, H. A. (1957). *J. Chem. Phys.*, **26**, 563–8.

Busing, W. R. and Levy, H. A. (1964). *Acta Crystallogr.*, **17**, 142–6.

Chakrabarti, P., Seiler, P., Dunitz, J. D., Schlüter, A.-D., and Szeimies, G. (1981). *J. Am. Chem. Soc.*, **103**, 7378–80.

Chandrasekhar, K. and Bürgi, H.-B. (1984). *Acta Crystallogr.*, **B40**, 387–97.

Chesick, J. P., Dunitz, J. D., Gizycki, U. V., and Musso, H. (1973). *Chem. Ber.*, **106**, 150–6.

Craven, B. M. and Swaminathan, S. (1984). *Trans. Am. Crystallogr. Assoc.*, **20**, 133–5.

Cruickshank, D. W. J. (1956a). *Acta Crystallogr.*, **9**, 747–53.

Cruickshank, D. W. J. (1956b). *Acta Crystallogr.*, **9**, 754–6.

Cruickshank, D. W. J. (1956c). *Acta Crystallogr.*, **9**, 757–8.

Cruickshank, D. W. J. (1956d). *Acta Crystallogr.*, **9**, 915–23.

Cruickshank, D. W. J. (1956e). *Acta Crystallogr.*, **9**, 1005–9.

Cruickshank, D. W. J. (1956f). *Acta Crystallogr.*, **9**, 1010–1.

Cruickshank, D. W. J. (1961). *Acta Crystallogr.*, **14**, 896–7.

Debye, P. (1913). *Verhand. Deutsch. Phys. Gesell.*, **15**, 738–52.

Dunitz, J. D. (1979). *X-ray analysis and the structure of organic molecules*, pp. 244–61. Cornell University Press, Ithaca.

Dunitz, J. D. and White, D. N. J. (1973). *Acta Crystallogr.*, **A29**, 93–4.

Dunitz, J. D., Schomaker, V., and Trueblood, K. N. (1988a). *J. Phys. Chem.*, **92**, 856–67.

Dunitz, J. D., Maverick, E. F., and Trueblood, K. N. (1988b). *Angew. Chem. Int. Ed. Engl.*, **27**, 880–95.

Filippini, G., Gramaccioli, C. M., Simonetta, M., and Suffritti, G. B. (1973). *J. Chem. Phys.*, **59**, 5088–101.

Filippini, G., Gramaccioli, C. M., Simonetta, M., and Suffritti, G. B. (1974). *Acta Crystallogr.*, **A30**, 189–96.

Frauenfelder, H., Petsko, G. A., and Tsernoglou, D. (1979). *Nature*, **280**, 558–63.

Glover, I., Haneef, I., Pitts, J., Wood, S., Moss, D., Tickle, I., and Blundell, T. (1983). *Biopolymers*, **22**, 293–304.

Gramaccioli, C. M. and Filippini, G. (1985). *Acta Crystallogr.*, **A41**, 356–61.

Gramaccioli, C. M., Filippini, G., and Simonetta, M. (1982). *Acta Crystallogr.*, **A38**, 350–6.

Hamilton, W. C. (1969). *Mol. Cryst. Liq. Cryst.*, **9**, 11–24.

Hamilton, W. C., Edmonds, J. W., Tippe, A., and Rush, J. J. (1969). *Discuss. Faraday Soc.*, **48**, 192–204.

He, X. M. and Craven, B. M. (1985). *Acta Crystallogr.*, **A41**, 244–51.

Higgs, P. W. (1953). *Acta Crystallogr.*, **6**, 232–41.

Higgs, P. W. (1955). *Acta Crystallogr.*, **8**, 99–104 and 619–20.

Hirshfeld, F. L. (1976). *Acta Crystallogr.*, **A32**, 239–44.

Holbrook, S. R., Dickerson, R. E., and Kim, S.-H. (1985). *Acta Crystallogr.*, **B41**, 255–62.

Ibers, J. A. (1959). *Acta Crystallogr.*, **12**, 251–2.

Jeffrey, G. A. and Ruble, J. R. (1984). *Trans. Am. Crystallogr. Assoc.*, **20**, 129–32.

Johnson, C. K. (1965). *ORTEP*. Report ORNL-3794, Oak Ridge National Laboratory, Oak Ridge, Tennessee, USA.

Johnson, C. K. (1967). Presentation at the *American Crystallographic Association Summer Meeting*, Minneapolis, Minnesota, USA. Abstracts, p. 82.

Johnson, C. K. (1969). *Acta Crystallogr.*, **A25**, 187–94.

Johnson, C. K. (1970*a*). An introduction to thermal-motion analysis. In *Crystallographic computing* (ed. F. R. Ahmed), pp. 207–19. Munksgaard, Copenhagen.

Johnson, C. K. (1970*b*). The effect of thermal motion on interatomic distances and angles. In *Crystallographic computing* (ed. F. R. Ahmed), pp. 220–6. Munksgaard, Copenhagen.

Johnson, C. K. (1970*c*). Generalized treatments for thermal motion. In *Thermal neutron diffraction* (ed. B. T. M. Willis), pp. 132–60. Oxford University Press.

Johnson, C. K. (1980). Thermal motion analysis. In *Computing in crystallography* (ed. R. Diamond, S. Ramaseshan, and K. Venkatesan), pp. 14.01–19. Indian Academy of Sciences, Bangalore.

Johnson, C. K. and Levy, H. A. (1974). Thermal-motion analysis using Bragg diffraction data. In *International tables for X-ray crystallography*, Vol. 4, *Revised and supplementary tables* (ed. J. A. Ibers and W. C. Hamilton), pp. 311–36. Kynoch Press, Birmingham.

Kvick, Å., Canning, W. M., Koetzle, T. F., and Williams, G. J. B. (1980). *Acta Crystallogr.*, **B36**, 115–20.

Lehmann, M. S., Koetzle, T. F., and Hamilton, W. C. (1972). *J. Am. Chem. Soc.*, **94**, 2657–60.

Maverick, E. and Dunitz, J. D. (1987). *Mol. Phys.*, **62**, 451–9.

Moss, D. S., Haneef, I., and Howlin, B. (1984). *Trans. Am. Crystallogr. Assoc.*, **20**, 123–7.

Patterson, A. L. (1959). Fundamental mathematics. In *International tables for X-ray crystallography*, Vol. 2, *Mathematical tables* (ed. J. S. Kasper and K. Lonsdale), pp. 3–98 (see especially pages 11–22 and 52–64). Kynoch Press, Birmingham.

Pawley, G. S. (1966). *Acta Crystallogr.*, **20**, 631–8.

Prince, E. and Finger, L. W. (1973). *Acta Crystallogr.*, **B29**, 179–83.

Rosenfield Jr., R. E., Trueblood, K. N., and Dunitz, J. D. (1978). *Acta Crystallogr.*, **A34**, 828–9.

Scheringer, C. (1972). *Acta Crystallogr.*, **A28**, 512–5 and 616–9.

Scheringer, C. (1985). *Acta Crystallogr.*, **A41**, 73–9 and 79–81.

Scheringer, C. (1987). *Acta Crystallogr.*, **A43**, 703–6.

Schlemper, E. O., Hamilton, W. C., and La Placa, S. J. (1971). *J. Chem. Phys.*, **54**, 3990–4000.

Schomaker, V. and Trueblood, K. N. (1968). *Acta Crystallogr.*, **B24**, 63–76.

Schomaker, V. and Trueblood, K. N. (1984). *Acta Crystallogr.*, **A40** (suppl.), p. C339.

Schulz, H. (1984). Presentation at the *American Crystallographic Association Spring Meeting*, Abstracts, p. 15.

Schulz, H., Zucker, U., and Frech, R. (1985). *Acta Crystallogr.*, **B41**, 21–6.

Seiler, P. and Dunitz, J. D. (1979*a*). *Acta Crystallogr.*, **B35**, 1068–74.

Seiler, P. and Dunitz, J. D. (1979*b*). *Acta Crystallogr.*, **B35**, 2020–32.

Seiler, P. and Dunitz, J. D. (1985). *Helv. Chim. Acta*, **68**, 2093–9.

Singh, T. P., Bode, W., and Huber, R. (1980). *Acta Crystallogr.*, **B36**, 621–7.

Stebler, M. and Bürgi, H.-B. (1987). *J. Am. Chem. Soc.*, **109**, 1395–401.

Stebler, M., Ludi, A., and Bürgi, H.-B. (1986). *Inorg. Chem.*, **25**, 4743–50.

Trueblood, K. N. (1978). *Acta Crystallogr.*, **A34**, 950–4.

Trueblood, K. N. (1985). *J. Mol. Struct.*, **130**, 103–15.

Trueblood, K. N. (1987). *Acta Crystallogr.*, **C43**, 711–3.

Trueblood, K. N. and Dunitz, J. D. (1983). *Acta Crystallogr.*, **B39**, 120–33.

Trueblood, K. N., Goldish, E., and Donohue, J. (1961). *Acta Crystallogr.*, **14**, 1009–17.

Watenpaugh, K. D., Sieker, L. C., and Jensen, L. H. (1980). *J. Mol. Biol.*, **138**, 615–33.

Willis, B. T. M. and Pryor, A. W. (1975). *Thermal vibrations in crystallography*. Cambridge University Press.

Yoshiasa, A., Koto, K., Kanamaru, F., Emura, S., and Horiuchi, H. (1987). *Acta Crystallogr.*, **B43**, 434–40.

Yu, H., Karplus, M., and Hendrickson, W. A. (1985). *Acta Crystallogr.*, **B41**, 191–201.

Zucker, U. H., Perenthaler, E., Kuhs, W. F., Bachmann, R., and Schulz, H. (1983). *J. Appl. Crystallogr.*, **16**, 358.

9

Lattice-dynamical interpretation of crystallographic thermal parameters

Carlo M. Gramaccioli

9.1 Introduction

In usual crystallographic calculations, the effect of atomic motion on the structure factor $\mathbf{F_H}$ is considered in the so-called *temperature factor*, $\tau_j(\mathbf{H})$:

$$\mathbf{F_H} = \sum_j f_j \tau_j(\mathbf{H}) \exp(2\pi i \mathbf{H} \cdot \mathbf{r}_j), \qquad (9.1)$$

where $\mathbf{H}$ is a reciprocal-lattice vector, denoting a reflection, and $\mathbf{r}_j$ a direct-lattice vector, denoting the position of the j-th atom. The temperature factor is the Fourier transform of the probability density function (pdf) of an atom around its mean position. If the pdf is isotropic and the distribution Gaussian, then (see Chapter 6, Section 6.7, and Chapter 8, Section 8.2):

$$\tau_j(\mathbf{H}) = \tau_j(H) = \exp(-B_j \sin^2\theta/\lambda^2), \qquad (9.2)$$

where $B_j = 8\pi^2 \langle u_j^2 \rangle$.

Such isotropic temperature factors were used in early crystal structure determinations. Here the most critical approximation is the assumption that the pdf is the same in all directions. This is actually the case for some of the simplest structures, where the atoms are located in special positions of high symmetry, and which were the first to be determined (for instance, NaCl-type compounds, or simple metals, oxides, etc.). However, already in the first electron-density maps of structures where atoms are not in highly symmetrical positions, the peaks typically appeared as flat or elongated, suggesting strongly anisotropic thermal motion. This was more and more apparent as the accuracy of measurements was improved, and it became standard practice to collect complete sets of three-dimensional intensity data.

Anisotropic temperature factors were first introduced in an essentially up-to-date formulation by Cruickshank (1956a). They can be written as:

$$\tau_j(\mathbf{H}) = \exp(-2\pi^2 \mathbf{H}^{\mathrm{T}} \mathbf{U}_j \mathbf{H}). \tag{9.3}$$

The reciprocal-lattice vector $\mathbf{H}$ is here referred to Cartesian axes; the matrix $\mathbf{U}_j = \langle \mathbf{u}_j \mathbf{u}_j^{\mathrm{T}} \rangle$ is the second-moment matrix of the pdf (see again Chapter 8, Section 8.2), referred to the same Cartesian set; and the superscript T means transpose. An alternative expression can be given if $\mathbf{H}$ is referred to the reciprocal axes. In this case $\mathbf{H} = \mathbf{C}^{\mathrm{T}}\mathbf{h}$, where $\mathbf{C}$ is the transformation matrix from a Cartesian set of coordinates to a crystallographic unit-cell reference; the elements of $\mathbf{h}$ are the components of the reciprocal-lattice vector with respect to the reciprocal axes, i.e. they are the Miller indices hkl. It will be: $\tau_j(\mathbf{H}) = \tau_j(\mathbf{h}) = \exp(-\mathbf{h}^{\mathrm{T}}\boldsymbol{\beta}_j\mathbf{h})$, where:

$$\boldsymbol{\beta}_j = 2\pi^2 \mathbf{C} \mathbf{U}_j \mathbf{C}^{\mathrm{T}}. \tag{9.4}$$

This transformation is typical of a second-rank tensor.† As $\mathbf{U}_j$, also $\boldsymbol{\beta}_j$ is a symmetric 3×3 matrix.

The elements of the $\boldsymbol{\beta}$ and $\mathbf{U}$ tensors—β^{ij} and U^{ij}, respectively—are called 'thermal parameters', or (better) *atomic displacement parameters*, ADPs. Needless to say, these different ways for expressing the ADPs (there are still more: see Chapter 8, Section 8.2) may generate confusion. This is especially the case when authors of crystallographic papers fail to mention which temperature-factor expression they actually used in the refinement. More than the $\boldsymbol{\beta}$ tensors, the $\mathbf{U}$ tensors are particularly appropriate for immediate comparison between different crystal structures, since their values are independent of the unit-cell dimensions. Thus in the final set of parameters resulting from a crystal structure refinement, the ADPs should preferably be reported as U^{ij}.

The considerable increase in the number of parameters with respect to the isotropic case—nine instead of four for each atom, in general—has too often been considered only as a tolerable way for improving the agreement between observed and calculated structure amplitudes (the final discrepancy index R may decrease substantially on introducing anisotropic tempera-ture factors). Little attention has generally been paid to the $\boldsymbol{\beta}$ or $\mathbf{U}$ tensors,

†Equation (9.4) coincides with the relationship given in Chapter 8 only when the crystal axes are orthogonal. In other cases, the reference system of Chapter 8 (which is the most widely used at present) implies a *non-Cartesian set* coinciding in direction with the crystal axes, but with unit lengths of 1 Å. If this reference system is chosen, then the transformation matrix $\mathbf{C}$ in eqn (9.4) is diagonal, with $C_{ii} = 1/a_i$ (where a_i is the corresponding direct-cell parameter).

as if they were of little significance. This point of view is reflected in the recommendation by most qualified scientific journals to omit publication of ADPs, or to have them deposited. At the same time, more elaborate expressions of the temperature factor have been proposed by various authors, especially Johnson (1969, 1970a, b). These expressions are particularly appropriate with manifestly anharmonic motion, and derive from including higher cumulants in the Edgeworth expansion of the pdf, which is no longer necessarily Gaussian. If only the next terms beyond second moments are introduced, this involves an additional third-rank tensor for each atom, with ten independent elements. Use of such models requires an appropriate number of particularly accurate intensity measurements.

9.2 The importance of crystallographic thermal parameters

The determination of atomic displacement parameters is becoming a matter of increasing attention to scientists because:

1. They are essential in order to correct bond lengths for thermal-motion effects.

2. They provide information for deconvoluting thermal effects from diffraction data which are used for electron-density studies (see Chapter 10).

3. They are useful for checking consistency of crystallographic data with other physicochemical data (spectroscopic, thermodynamic, etc.).

The problem of correcting interatomic distances for thermal-motion effects was considered for the first time by Cruickshank (1956c, 1961) and Busing and Levy (1957, 1964); then by several other authors, especially Schomaker and Trueblood (1968), Johnson (1970a, c, 1980), Scheringer (1972c). In the general case, the corrected distance r_c between atoms j and k is related to the uncorrected distance r_o by the equation:

$$r_c = r_o + \left[\text{tr}\,(\mathbf{Z}) - \mathbf{r}_o^T \mathbf{Z} \mathbf{r}_o / r_o^2 \right] / 2 r_o. \tag{9.5}$$

Here $\mathbf{r}_o$ is the vector corresponding to the distance r_o and the matrix $\mathbf{Z}$ is defined as:

$$\mathbf{Z} = \mathbf{U}_j + \mathbf{U}_k - \mathbf{U}_{jk} - \mathbf{U}_{jk}^T. \tag{9.6}$$

In eqn (9.6), $\mathbf{U}_j$ and $\mathbf{U}_k$ are the thermal-motion tensors of atoms j and k, respectively ($\mathbf{U}_j = \langle \mathbf{u}_j \mathbf{u}_j^T \rangle$, $\mathbf{U}_k = \langle \mathbf{u}_k \mathbf{u}_k^T \rangle$). $\mathbf{U}_{jk}$ is a probability distribution tensor correlating the motions of atoms j and k, $\mathbf{U}_{jk} = \langle \mathbf{u}_j \mathbf{u}_k^T \rangle$; *this coupling tensor cannot be evaluated from Bragg diffraction measurements.*

Such a situation would apparently preclude the possibility of relying on crystallographic data alone to obtain accurate bond-length estimates. When a model correlating the thermal motions of different atoms can be assumed to hold, then it is possible to overcome this difficulty, since the information corresponding to the U_{jk} tensors can be deduced. For instance, in a number of molecular crystals, the whole molecule or at least part of it behaves as a rigid body. In such a case, the displacement u_j of the j-th atom from its mean position can be expressed as:

$$u_j = t + \lambda \times r_j = t + A_j \lambda, \tag{9.7}$$

where t represents a translational displacement of the whole molecule, r_j the distance of the mean position of the atom from an arbitrary reference point (usually the centre of mass of the molecule), and λ is an axial vector which represents rotation. From eqn (9.7) the matrix A_j is easily shown to be (see, for example, Willis and Pryor 1975, p. 181):

$$A_j = \begin{bmatrix} 0 & r_3 & -r_2 \\ -r_3 & 0 & r_1 \\ r_2 & -r_1 & 0 \end{bmatrix}$$

where r_1, r_2, and r_3 are the components of r_j in the reference system (here Cartesian). Accordingly, the U_j tensors can be written in terms of three molecular thermal-motion tensors, $T = \langle tt^T \rangle$, $L = \langle \lambda\lambda^T \rangle$, $S = \langle \lambda t^T \rangle$:

$$U_j = T + A_j L A_j^T + A_j S + S^T A_j^T. \tag{9.8}$$

This is the basis for Cruickshank's (1956b) and especially Schomaker and Trueblood's (1968) interpretation of the ADPs in terms of rigid-body motion. According to their procedure, the elements of the T, L, and S tensors (with the exception of the trace of S) are obtained from a least-squares fit to the individual ADPs. If the molecule is indeed a rigid body, the correction of bond distances can be shown to depend on L only, and in most cases no additional data besides the final parameters of a crystal structure determination are necessary. In other cases, however, even with a rigid body, L cannot be uniquely determined from Bragg diffraction data: this happens, for instance, when the molecule is too small, or when all atoms lie on a conic section (see Johnson 1970b).

The degree of fit of the individual U_j tensors to T, L, and S is often considered as a good indication of the rigid-body behaviour of a molecule; sometimes, however, this fit seems to indicate substantial rigidity even for molecules where easy deformation would be expected. In several instances this happens because packing forces in the crystal tend to 'squeeze' the molecule, which then assumes a comparatively inflexible conformation. In other instances the internal motions contribute to the parameters that are *supposed* to describe the rigid-body motion — T, L, and S — leading to questionable results (Filippini *et al*. 1974; Gramaccioli and Filippini, 1985*b*).

The rigid-body behaviour is persuasive evidence for the physical significance of individual atomic thermal parameters, as obtained from crystallographic measurements. For classic rigid molecules, there are some striking examples, which leave little doubt wih respect to this point: see, for example, Chapter 8, Section 8.4, or Willis and Pryor (1975, Chapter 6). Additional evidence comes from the 'rigid-bond' test (Hirshfeld 1976; Rosenfield *et al*. 1978; see also Chapter 8, Section 8.3): in view of the much more difficult deformation along a covalent bond with respect to other degrees of freedom, the mean-square displacement amplitudes along the bond of atoms which are directly linked to each other should be essentially the same. This rule is closely followed in the majority of crystal structure results. Violations of it for structures of good to moderate accuracy can be accounted for by specific effects, see the extensive literature referred to in Chapter 8, Section 8.3.

Besides the rigid body, other models are used sometimes for carrying out thermal-motion corrections. One of these is the so-called 'riding model' (Busing and Levy 1964), where a peripheral group is expected to 'ride', i.e. to move independently in addition to the movement of the atomic skeleton to which it is attached; such a model seems to be especially appropriate for hydrogen atoms (but see Chapter 11, Section 11.3.2.2, for a more detailed discussion). Another scheme of thermal motion is the 'segmented-body model' (Johnson 1970*a*); it should be used when the molecule is made of several rigid parts joined together. It is not always easy, however, to define how these parts are moving relative to each other. In many instances these models are able to interpret fairly well the individual experimental ADPs.

9.3 Lattice-dynamical interpretation

The physical significance of crystallographic thermal parameters is substantially confirmed by direct calculation. For this purpose, the classic dynamical theory of crystal lattices or *lattice dynamics* (Born and Huang 1954) can be used, following a number of different models of varying complexity.

The earliest attempts of this kind were based on Debye's models (see, for example, James *et al*. (1929); Zener and Bilinsky (1936); Nicklow and Young (1966); an extensive list of references can be found in James (1962)). Such models, however, cannot be applied to structures which are even moderately complex.

The Born–von Kármán formalism in lattice-dynamical calculations (see Born and Huang (1954) for a detailed discussion) has several advantages:

1. The theory is considerably more elaborate than for Einstein's or Debye's models, and a much better fit to reality is obtained, particularly for the so called 'density-of-states' function. The complexity of structures might be a practical limit of application, but it is not a theoretical limit.

2. The formalism is not far from common calculations in molecular spectroscopy, or from procedures for evaluating the minimum-energy conformation, the packing energy, or other physical properties in crystals. Besides some computing techniques, also the results obtained from the fit to experimental data of parameters such as those of force fields, van der Waals' constants, etc., can usually be transferred and used as such, with surprisingly good agreement of the calculated frequencies with the corresponding measurements.

3. In Debye's or Einstein's models, a characteristic frequency (or temperature) is introduced; this choice is made *a posteriori* for each particular structure, by fitting the calculated physical properties to the corresponding experimental results. Instead, the empirical force constants and van der Waals' parameters seem to be quite easily transferable within wide groups of chemically similar substances or functional groups. This means that, whenever such empirical fields are known, the Born–von Kármán treatment leads directly to estimates of the ADPs starting from unit-cell parameters, atomic coordinates, and space-group operations *only*, i.e. from usual crystallographic information (plus temperature).

4. In addition to the ADPs, further information is obtained as a spinoff. Besides thermodynamic functions (a common feature with the other models), such information includes the U_{jk} tensors, Raman- or IR-active frequencies, phonon dispersion curves, etc.; these data are often more important than the ADPs themselves.

The detailed mathematical procedure may be found, for instance, in Willis and Pryor (1975, Chapter 3). In spite of the interest and advantages, such calculations have so far been restricted to relatively few examples. There are, in fact, considerable difficulties:

1. The order of the dynamical matrices **D** to be diagonalized is equal to $3N$, where N is the number of atoms in the primitive unit cell. Since in general the **D** matrices are complex, the actual order is $6N$. Many molecular crystals have four molecules in the unit cell; this implies handling matrices

whose order is eight times larger than that of the corresponding dynamical matrices for vibrational frequency calculations in the isolated molecule.

2. The diagonalization of the dynamical matrices must be repeated for a considerable number of points in the Brillouin zone; in the first works of this kind, this number was 30 000 to 60 000.

3. Force fields are not always available. This problem is especially important for charged atoms or groups, where the charge itself and its distribution are not well known. Furthermore, polarization effects might require more sophisticated models.

However, in spite of all these difficulties, a number of considerable simplifications have been worked out, so that — at least with some types of molecules — these calculations appear to be quite feasible nowadays, being often of the same order of magnitude as for the corresponding crystal structure refinement.

Of these simplifications, the rigid-body model is of fundamental importance for molecular crystals. It leads to equations of motion which are similar to the general case, but for each molecule only six (or five) degrees of freedom are considered, three translational and three (or two) rotational. Instead of the masses of the individual atoms, the mass of the whole molecule and the three principal moments of inertia are used. The rigid-body model leads directly to estimates of the molecular-motion tensors T, L, and S; these are obtained in the same way as the U_j tensors in the general case. From T, L, and S, estimates of the individual U_j tensors are readily obtained through eqn (9.8). With the rigid-body approximation the average order of the dynamical matrices is 48, a reasonable number as compared to nearly 300, which is already required for benzene in the general case.

The first applications of this technique are due to Cochran and Pawley (1964), and Pawley (1967). The relationships between the lattice-dynamical procedure and the crystallographic formulation have been clearly pointed out by several authors, especially Pawley (1967, 1968), Scheringer (1972a, b, c), and Willis and Pryor (1975).

The difficulty concerning the excessive number of points in the Brillouin zone can be overcome (at least in part) by using a convenient sampling, of increasing density near the origin (Filippini et al. 1976). An alternative procedure has been proposed by Kroon and Vos (1978, 1979). By such techniques, the number of points can be reduced to about 200 for usual crystal structures.

Extension of this simplified procedure to non-rigid molecules is straightforward. This is possible because in most cases the internal degrees of freedom corresponding to low vibrational frequencies (and which can therefore interfere with lattice modes) are comparatively few in number relative to the total; therefore the dynamical matrices remain substantially small. The determination of the low-frequency degrees of freedom, including the rotational and translational ones, is a routine operation in vibrational spectroscopy, since these degrees of freedom are characterized by the corresponding eigenvectors of the dynamical matrix for the isolated molecule (see, for example, Wilson *et al.* 1955, or Gwinn 1971).

If we build a rectangular matrix $\mathbf{V}$, whose columns are these eigenvectors, reduced force-constant matrices $\mathbf{\Phi}' = \mathbf{V}^T \mathbf{\Phi} \mathbf{V}$ are easily obtained, and introduced into blocks in the dynamical matrix. Here $\mathbf{\Phi}$ is the *complete* force-constant matrix, of order $3N$ (N being the number of atoms in the molecule), whose elements are the second derivatives of the potential energy with respect to the atomic shifts. If only the eigenvectors corresponding to zero frequency are used, this procedure coincides with the rigid-body model; introduction of the internal degrees of freedom is possible without altering the general scheme, up to convergence of the results with a complete treatment. Because of the degeneracy at zero frequency, suitable linear combinations of the eigenvectors are taken so that they coincide with translational or rotational motion along (or about) the principal axes of inertia; this facilitates comparison with Schomaker–Trueblood's (1968) rigid-body model. The contribution of the higher-frequency modes can be considered separately as a constant throughout the Brillouin zone, and added to the total. An account of mathematical details is given, for instance, by Bonadeo and Burgos (1982), who used a simplified version for crystalline biphenyl, and by Gramaccioli and Filippini (1983, 1985*a*).

From this procedure, instead of the individual atomic $\mathbf{U}_j$ tensors, or the rigid-body tensors $\mathbf{T}$, $\mathbf{L}$, and $\mathbf{S}$, a mean-square displacement tensor relative to the whole molecule, $\mathbf{W} = \langle \mathbf{q}_i \mathbf{q}_j^T \rangle$ is obtained directly, where $\mathbf{q}_i$ and $\mathbf{q}_j$ are molecular displacements in terms of the normal modes of the isolated molecule. The first two 3×3 blocks along the main diagonal of the $\mathbf{W}$ matrix are associated with the $\mathbf{T}$ and $\mathbf{L}$ tensors, respectively, whereas the adjacent block corresponds to $\mathbf{S}$. From $\mathbf{W}$, by applying the transformation:

$$\mathbf{\Omega} = \mathbf{V} \mathbf{m}^{-1/2} \mathbf{W} \mathbf{m}^{-1/2} \mathbf{V}^T \tag{9.9}$$

(where $\mathbf{m}$ is a diagonal matrix composed of the atomic masses) the matrix $\mathbf{\Omega}$ is obtained, whose 3×3 blocks along the main diagonal are the atomic $\mathbf{U}_j$ tensors, and the off-diagonal blocks are the $\mathbf{U}_{jk}$ coupling tensors defined in Section 9.2. Thus the lattice-dynamical procedure provides essential information for correcting the molecular geometry *in the most general case*.

9.4 Results

Examples of ADPs from lattice-dynamical calculations are reported in Tables 9.1 to 9.3 together with the corresponding experimental values. Table

Table 9.1 Anisotropic displacement parameters ($\text{Å}^2 \times 10^4$) for selected atoms of pyrene at room temperature

Atom[a]	U^{11}	U^{22}	U^{33}	U^{12}	U^{13}	U^{23}
C(1)	777	872	964	94	−102	−254
	822	938	982	132	−91	−246
	832	929	1017	82	−119	−242
	841	933	1027	76	−125	−246
	841(15)	768(14)	824(16)	93(12)	−15(13)	−149(12)
C(2)	640	703	1126	−44	28	−215
	676	755	1144	−13	34	−211
	685	751	1182	−50	23	−215
	694	755	1196	−63	17	−215
	571(12)	647(12)	1010(17)	−38(10)	22(12)	−230(12)
C(3)	557	525	915	0	136	−78
	585	560	932	13	142	−74
	603	569	957	0	147	−82
	612	577	971	−19	147	−86
	498(10)	452(10)	803(15)	−12(8)	183(10)	−64(9)
H(2)	749	838	1534	−189	57	−289
	978	1115	1608	−94	85	−258
	987	1094	1657	−183	62	−262
	996	1089	1682	−220	51	−258
	859(33)	980(34)	1538(49)	−327(27)	85(32)	−300(32)
H(4)	941	729	1450	−195	437	51
	1151	985	1527	−113	459	82
	1188	998	1566	−164	482	70
	1215	998	1580	−239	482	59
	942(34)	875(31)	1625(50)	−359(26)	409(33)	122(31)

For each atom the values in the first four lines are from lattice-dynamical calculations (Gramaccioli and Filippini 1983), using different models as specified in the text. Values in the fifth line are from a single-crystal neutron diffraction study (Hazell *et al*. 1972).

[a] C atoms are numbered as shown in the following formula. H atoms are numbered according to the C atom to which they are bonded.

9.1 refers to pyrene, a classic rigid-body molecule (Gramaccioli and Filippini 1983). For each atom the first line reports values obtained from the rigid-body model, without any contribution of internal modes. The second line includes internal mode contribution, but no mixing with lattice modes is considered. In the next two lines, the internal modes of the isolated molecule with frequencies up to 200, or 300 cm^{-1} (two and five in number, respectively) are allowed to mix with the lattice modes; convergence is virtually reached in the last case. The last line reports the corresponding experimental values, taken from a neutron diffraction study by Hazell *et al.* (1972). The agreement is quite good, and the rigid character of the molecule is evident.

A more detailed discussion, with a series of examples regarding several aromatic hydrocarbons, is given by Gramaccioli and Filippini (1983). On the whole, the agreement with the experimental data is reasonably good, between 15 and 35 per cent in most cases. There are a few exceptions, however, such as benzene at 138 K, where the calculated values exceed the corresponding experimental data by about 50 per cent, and deuterated anthracene at 17 K, where almost the opposite occurs.

The disagreement for benzene is most probably connected with the large amplitude of motion, which implies inadequacy of a harmonic model; this is especially evident in the rotational component, and is likely to arise from the regular hexagonal shape of the molecule, which behaves as a wheel. Recent measurements at 15 K (Jeffrey *et al.* 1987) and at 4.2 K (David and Ibberson 1989), where the amplitude of motion is appreciably reduced, and new lattice-dynamical calculations (Filippini and Gramaccioli 1989) give results that are in almost perfect agreement: see Chapter 11, Table 11.12.

As for some simple ionic crystals (NaCl, KCl: see Reid and Smith 1970), most of the observed ADPs are systematically smaller than the corresponding calculated values: this suggests a possible effect of the so-called 'thermal diffuse scattering', TDS. The amount of influence of TDS on the experimental data is difficult to establish in each case, since it depends on the data-collecting strategy, and especially on how the background is estimated for each reflection (see Chapter 7, Section 7.2.5, and Willis and Pryor (1975, Chapter 9)). The agreement with the observed frequencies is good to nearly excellent in all the cases so far examined; this confirms the view that any disagreement in the ADPs should depend on anharmonic behaviour for large amplitude of vibration, or lack of appropriate corrections in the experimental data (TDS, absorption).

The greatest majority of substances for which such calculations have been performed are uncharged molecules, like hydrocarbons (see above) or octasulphur (Gramaccioli and Filippini 1984): here reliable empirical force fields and van der Waals' parameters are available. A few calculations have been performed on molecules with heteroatoms, although no charge effects have been considered as such (Filippini *et al.* 1981; Criado *et al.* 1984); the agreement with the experimental ADPs is good, although in other cases there are substantial difficulties. When atoms strongly differing in electro-negativity are present, it is awkward to neglect Coulombic effects, and attempts are presently being made to solve the problem for these systems.

With non-rigid molecules, the agreement between calculated and observed ADPs is as good as with rigid molecules: two examples—o-terphenyl (Gramaccioli and Filippini 1985b) and tetraphenylmethane (Filippini and Gramaccioli 1986)—are reported in Tables 9.2 and 9.3, respectively. Espe-cially for the latter molecule, which is clearly non-rigid, general bond-length correction for thermal-motion effects can be made, using the calculated coupling tensors $\mathbf{U}_{jk}$, as specified in Section 9.2. Using the corrected values, the mean C—C bond length in the phenyl ring becomes quite close to the corresponding r_g value for toluene, 1.399 ± 0.002 Å, as obtained by gas-phase electron diffraction (Seip et $al.$ 1977). The effect of bond-length correction for a structure of this kind at room temperature is evident when different models are compared: the mean C—C bond length in the phenyl ring is 1.389 Å (uncorrected), 1.392 Å (rigid-body motion, from improper application to the whole molecule of Schomaker–Trueblood's fit), 1.396 Å

Table 9.2 Anisotropic displacement parameters (Å$^2 \times 10^4$) for selected atoms of o-terphenyl at room temperature

Atom[a]	U^{11}	U^{22}	U^{33}	U^{12}	U^{13}	U^{23}
C(1)	577	533	453	−62	−11	−4
	517(15)	509(18)	382(12)	−57(14)	−25(12)	14(10)
C(2)	560	643	585	11	66	−72
	487(16)	595(21)	542(15)	−7(16)	27(13)	−71(10)
C(2′)	402	575	481	−6	0	−14
	386(13)	601(19)	407(12)	−32(14)	15(10)	12(10)
C(3′)	455	711	557	74	33	−36
	521(17)	731(24)	510(16)	156(18)	82(15)	51(20)
C(4′)	437	943	641	136	44	22
	498(17)	1040(30)	651(20)	206(22)	103(16)	52(20)

For each atom the values in the first line are from lattice-dynamical calculations (Gramaccioli and Filippini 1985b), those in the second line are from a single-crystal neutron diffraction study (Brown and Levy 1979).

[a] Atoms are numbered as shown in the adjacent formula:

Table 9.3 Anisotropic displacement parameters ($\text{Å}^2 \times 10^4$) for the C atoms of tetraphenylmethane at room temperature

Atom[a]	U^{11}	U^{22}	U^{33}	U^{12}	U^{13}	U^{23}
C(0)	319	319	301	0	0	0
	307(6)	307(6)	309(8)	0	0	0
C(1)	331	349	328	−12	0	−20
	337(6)	362(6)	299(8)	−30(6)	40(8)	0(8)
C(2)	373	439	360	−36	−32	−28
	427(6)	497(10)	347(10)	−30(6)	−16(8)	−24(8)
C(3)	487	517	416	−114	−32	−92
	496(10)	664(10)	360(10)	−150(10)	4(8)	−81(8)
C(4)	626	463	521	−90	4	−161
	664(10)	514(10)	471(10)	−155(10)	92(4)	−168(10)
C(5)	620	379	553	30	4	−125
	640(10)	401(10)	525(10)	18(10)	92(10)	−85(8)
C(6)	445	355	424	42	−8	−56
	455(10)	391(6)	387(10)	18(6)	16(8)	−44(8)

For each atom the values in the first line are from lattice-dynamical calculations (Filippini and Gramaccioli 1986), those in the second line are from an X-ray diffraction study (Robbins *et al.* 1975).

[a] C(0) is the central atom of the molecule, which has $\bar{4}$ (S_4) crystallographic symmetry. The ring carbons are numbered sequentially from C(1) (linked to C(0)) to C(6).

(riding motion), 1.398 Å (general case, using the U_j and U_{jk} tensors derived from lattice dynamics). Therefore, no accuracy beyond 0.01 Å can be claimed for light-atom structures at room temperature if corrections for thermal-motion effects are not applied, and for accuracy of the order of 0.001 Å a very good model should be used. For hydrogen atoms the situation is still more difficult, because anharmonicity does not seem to be negligible (Craven and Swaminathan 1984; Jeffrey and Ruble 1984; see also Chapter 11, Section 11.3.2.2).

Table 9.4 reports some examples of estimates of the **T**, **L**, and **S** tensors from Schomaker–Trueblood's fit and from lattice-dynamical calculations. For a rigid molecule such as pyrene the agreement is good, especially for **L**, which is the only quantity bearing on the bond-length correction. For anthracene there is discrepancy, owing to the fact that the carbon atoms nearly lie on an ellipse (here no H-atom data have been considered), and the matrix of the normal equations is unstable (see Filippini *et al.* 1974). For *o*-terphenyl, since the molecule is not rigid, a rigid-body fit to the whole molecule interprets most of the internal motion in terms of **T**, **L**, and **S**: for this reason the **T** and **L** tensors are overestimated.

Table 9.4 Examples of estimates of the **T**, **L**, and **S** rigid-body tensors: (1) from Schomaker–Trueblood's fit to observed ADPs; (2) from non-rigid lattice-dynamical model[a]

Molecule		**T** ($\text{Å}^2 \times 10^4$)			**L** ($\text{rad}^2 \times 10^4$)			**S** ($\text{Å rad} \times 10^4$)		
Pyrene[b]	(1)	413	47	189	53	−18	4	(10)	10	−3
			354	33		47	14	9	(0)	9
				618			49	−7	−6	(−10)
	(2)	504	81	−12	41	−14	5	2	−5	3
			454	−23		50	8	−6	−1	−10
				727			44	−4	1	−11
Deuterated anthracene[c]	(1)	313	163	−235	50	13	−34			
			309	15		36	4			
				355			67			
	(2)	420	23	104	32	0	−5			
			395	12		26	−12			
				484			52			
o-Terphenyl[d]	(1)	427	−54	22	46	−15	−1	(1)	−12	9
			452	70		46	16	10	(28)	−37
				543			77	51	3	(−29)
	(2)	400	−14	7	34	−14	−3	1	−21	20
			462	1		30	9	1	29	−17
				389			48	50	2	−25

[a] All studies are at room temperature. The reference system is Cartesian, with the origin at the centre of mass of the molecule, and the axes oriented along $\mathbf{a^*}$, $\mathbf{b}$, $\mathbf{a^*} \times \mathbf{b}$.
[b] Gramaccioli and Filippini (1983). The experimental data are from Hazell *et al.* (1972).
[c] Gramaccioli and Filippini (1983). The experimental data are from Lehmann and Pawley (1972); the molecule has $\bar{1}$ (C_i) crystallographic symmetry.
[d] Gramaccioli and Filippini (1985b). The experimental data are from Brown and Levy (1979).

The agreement between the lattice-dynamical and Schomaker–Trueblood's estimates of **S** is generally less good than for **T** and **L**, except in some cases when the molecule is really rigid, and quite good experimental data are available. Since tr(**S**) is set arbitrarily to zero, due to its indeterminacy in Schomaker–Trueblood's least-squares treatment, the diagonal elements are especially liable to disagree; there is in fact no physical reason (except for symmetry requirements, when the molecule is located in special positions) for tr(**S**) to be zero.

For a non-rigid molecule, an example of coupling between the various degrees of freedom (in terms of modes of vibration of the isolated molecule, including rotations and translations: see Section 9.3) is shown in Table 9.5. This table reports the first 12×12 block of the tensor $\mathbf{W} = \langle \mathbf{q}_i \mathbf{q}_j^T \rangle$ for

Table 9.5 The first 12×12 block of the tensor $\mathbf{W} = \langle \mathbf{q}_i \mathbf{q}_j^{\mathrm{T}} \rangle$ for o-terphenyl.[a] From Gramaccioli and Filippini (1985b)

Translational modes			Rotational modes			Internal modes (frequency in cm^{-1})					
q_1	q_2	q_3	q_4	q_5	q_6	q_7 (56.1)	q_8 (58.1)	q_9 (68.6)	q_{10} (102.1)	q_{11} (115.9)	q_{12} (153.1)
9.549	0.543	−0.702	−0.626	1.627	−2.107	−0.130	−0.026	−0.166	−0.095	−0.128	0.032
	9.466	−0.363	0.268	0.123	0.105	−0.060	−0.045	−0.154	0.179	0.061	−0.025
		9.778	−2.246	1.478	0.593	0.255	0.149	0.213	−0.006	0.114	0.011
			4.813	−1.772	−0.033	−0.233	0.363	0.219	−0.101	−0.103	−0.016
				4.187	−1.071	0.333	0.053	−0.241	0.216	−0.044	0.049
					5.236	0.159	−0.727	0.561	−0.044	0.197	0.008
						1.020	0.040	0.010	0.046	−0.013	0.005
							1.301	−0.004	0.003	−0.051	−0.036
								0.937	−0.043	0.006	−0.019
									0.540	0.001	0.007
										0.428	0.000
											0.281

The rows and columns refer, in sequence, to translational, rotational, and internal coordinates in order of increasing frequency.

[a] All values are given in u Å^2. The reference system is defined by the principal axes of inertia.

o-terphenyl. The coupling between the five internal modes with the lowest frequencies and the translational or rotational modes is not negligible, although the most extensive coupling occurs between the translational and rotational modes; the importance of **S** in a general rigid-body case is therefore emphasized.

Since the A D Ps can be accurately obtained from calculations of this kind, one might wonder whether such crystallographic information is useful, especially in view of its low accuracy with respect to spectroscopic data. However, at present these calculations can be carried out routinely only for a few systems, i.e. small, essentially neutral molecules entirely made of light atoms, for which the amplitude of motion is not too large.

Even in such favourable cases, although methods such as those described in this chapter might be expected to deal successfully with a wider variety of molecules, a check with experiment is always necessary. For instance, our confidence about the substantial correctness of the lattice-dynamical U_{jk} tensors is founded on the good agreement between the U_j tensors (which are calculated in essentially the same manner) and their experimental counterparts. Moreover, although the experimental A D Ps are not very accurately determined, they provide useful information from a wider point of view than routine crystal structure analysis, because they reflect the vibrational behaviour of the crystal in the whole Brillouin zone, including frequencies not observable by I R or Raman spectroscopy.

9.5 Acknowledgements

I am deeply grateful to Professors Fred L. Hirshfeld and Kenneth N. Trueblood for their critical comments on an early version of this chapter.

References

Bonadeo, H. and Burgos, E. (1982). *Acta Crystallogr.*, **A38**, 29–33.
Born, M. and Huang, K. (1954). *Dynamical theory of crystal lattices*. Clarendon Press, Oxford.
Brown, G. M. and Levy, H. A. (1979). *Acta Crystallogr.*, **B35**, 785–8.
Busing, W. R. and Levy, H. A. (1957). *J. Chem. Phys.*, **26**, 563–8.
Busing, W. R. and Levy, H. A. (1964). *Acta Crystallogr.*, **17**, 142–6.
Cochran, W. and Pawley, G. S. (1964). *Proc. Roy. Soc.*, **A280**, 1–22.
Craven, B. M. and Swaminathan, S. (1984). *Trans. Am. Crystallogr. Assoc.*, **20**, 133–5.
Criado, A., Conde, A., and Márquez, R. (1984). *Acta Crystallogr.*, **A40**, 696–701.
Cruickshank, D. W. J. (1956*a*). *Acta Crystallogr.*, **9**, 747–53.
Cruickshank, D. W. J. (1956*b*). *Acta Crystallogr.*, **9**, 754–6.
Cruickshank, D. W. J. (1956*c*). *Acta Crystallogr.*, **9**, 757–8.
Cruickshank, D. W. J. (1961). *Acta Crystallogr.*, **14**, 896–7.

David, W. I. F. and Ibberson, R. M. (1989). *ISIS Annual report*, pp. 43–6. Swindon Press, Swindon.

Filippini, G. and Gramaccioli, C. M. (1986). *Acta Crystallogr.*, **B42**, 605–9.

Filippini, G. and Gramaccioli, C. M. (1989). *Acta Crystallogr.*, **A45**, 261–3.

Filippini, G., Gramaccioli, C. M., Simonetta, M., and Suffritti, G. B. (1974). *Acta Crystallogr.*, **A30**, 189–96.

Filippini, G., Gramaccioli, C. M., Simonetta, M., and Suffritti, G. B. (1976). *Acta Crystallogr.*, **A32**, 259–64.

Filippini, G., Gramaccioli, C. M., and Simonetta, M. (1981). *Chem. Phys. Lett.*, **79**, 470–5.

Gramaccioli, C. M. and Filippini, G. (1983). *Acta Crystallogr.*, **A39**, 784–91.

Gramaccioli, C. M. and Filippini, G. (1984). *Chem. Phys. Lett.*, **108**, 585–8.

Gramaccioli, C. M. and Filippini, G. (1985a). *Acta Crystallogr.*, **A41**, 356–61.

Gramaccioli, C. M. and Filippini, G. (1985b). *Acta Crystallogr.*, **A41**, 361–5.

Gwinn, W. D. (1971). *J. Chem. Phys.*, **55**, 477–81.

Hazell, A. C., Larsen, F. K., and Lehmann, M. S. (1972). *Acta Crystallogr.*, **B28**, 2977–84.

Hirshfeld, F. L. (1976). *Acta Crystallogr.*, **A32**, 239–44.

James, R. W. (1962). *The optical principles of the diffraction of X-rays* (4th reprint, with additions). Bell, London.

James, R. W., Brindley, G. W., and Wood, R. G. (1929). *Proc. Roy. Soc.*, **A125**, 401–19.

Jeffrey, G. A. and Ruble, J. R. (1984). *Trans. Am. Crystallogr. Assoc.*, **20**, 129–32.

Jeffrey, G. A., Ruble, J. R., McMullan, R. K., and Pople, J. A. (1987). *Proc. Roy. Soc.*, **A414**, 47–57.

Johnson, C. K. (1969). *Acta Crystallogr.*, **A25**, 187–94.

Johnson, C. K. (1970a). Generalized treatments for thermal motion. In *Thermal neutron diffraction* (ed. B. T. M. Willis), pp. 132–60. Oxford University Press.

Johnson, C. K. (1970b). An introduction to thermal-motion analysis. In *Crystallographic computing* (ed. F. R. Ahmed), pp. 207–19. Munksgaard, Copenhagen.

Johnson, C. K. (1970c). The effect of thermal motion on interatomic distances and angles. In *Crystallographic computing* (ed. F. R. Ahmed), pp. 220–6. Munksgaard, Copenhagen.

Johnson, C. K. (1980). Thermal motion analysis. In *Computing in crystallography* (ed. R. Diamond, S. Ramaseshan, and K. Venkatesan), pp. 14.01–19. Indian Academy of Sciences, Bangalore.

Kroon, P. A. and Vos, A. (1978). *Acta Crystallogr.*, **A34**, 823–4.

Kroon, P. A. and Vos, A. (1979). *Acta Crystallogr.*, **A35**, 675–84.

Lehmann, M. S. and Pawley, G. S. (1972). *Acta Chem. Scand.*, **26**, 1996–2004.

Nicklow, R. M. and Young, R. A. (1966). *Phys. Rev.*, **152**, 591–6.

Pawley, G. S. (1967). *Phys. Status Solidi*, **20**, 347–60.

Pawley, G. S. (1968). *Acta Crystallogr.*, **B24**, 485–6.

Reid, J. S. and Smith, T. (1970). *J. Phys. Chem. Solids*, **31**, 2689–97.

Robbins, A., Jeffrey, G. A., Chesick, J. P., Donohue, J., Cotton, F. A., Frenz, B. A., and Murillo, C. A. (1975). *Acta Crystallogr.*, **B31**, 2395–9.

Rosenfield Jr., R. E., Trueblood, K. N., and Dunitz, J. D. (1978). *Acta Crystallogr.*, **A34**, 828–9.

Scheringer, C. (1972*a*). *Acta Crystallogr.*, **A28**, 512–5.

Scheringer, C. (1972*b*). *Acta Crystallogr.*, **A28**, 516–22.

Scheringer, C. (1972*c*). *Acta Crystallogr.*, **A28**, 616–9.

Schomaker, V. and Trueblood, K. N. (1968). *Acta Crystallogr.*, **B24**, 63–76.

Seip, R., Schultz, Gy., Hargittai, I., and Szabó, Z. G. (1977). *Z. Naturforsch.*, **32a**, 1178–83.

Willis, B. T. M. and Pryor, A. W. (1975). *Thermal vibrations in crystallography.* Cambridge University Press.

Wilson, E. B., Decius, J. C., and Cross, P. C. (1955). *Molecular vibrations.* McGraw-Hill, New York.

Zener, C. and Bilinsky, S. (1936). *Phys. Rev.*, **50**, 101–4.

10

The role of electron density in X-ray crystallography

Fred L. Hirshfeld†

10.1   Structural *vs.* electron-density studies

Most X-ray crystallographic studies are aimed at structural information and pay little attention, if any, to the details of the electron distribution. But the two are, in practice, inseparably linked. It is only when we are content

† Deceased May 20, 1991.

with a routinely approximate structure that we can afford to disregard the electron density. When better than everyday accuracy is required, the relation between the nuclear and the electronic distributions becomes more symmetric. An accurate set of nuclear coordinates and a detailed map of the electron density can be obtained, via X-ray diffraction, only jointly and simultaneously, never separately or independently.

If nothing at all is known about the electron distribution $\rho(\mathbf{r})$ in a particular molecule or crystal, we can always approximate it, in terms of the nuclear coordinates, by summing the spherically averaged free-atom densities $\rho_a^{\mathrm{at}}(r)$ about the positions $\mathbf{R}_a$ of the several nuclei. So for our purpose this independent-atom model, known as the *promolecule* or *procrystal*

$$\rho^{\mathrm{pro}}(\mathbf{r}) = \sum_a \rho_a^{\mathrm{at}}(|\mathbf{r} - \mathbf{R}_a|), \tag{10.1}$$

is tantamount to complete ignorance. Any information provided by X-ray diffraction must, to be of any relevance for the electron density, be sufficiently precise and detailed to permit significant conclusions about the *deformation density*

$$\delta\rho(\mathbf{r}) = \rho(\mathbf{r}) - \rho^{\mathrm{pro}}(\mathbf{r}),$$

i.e. it must tell us how the true density differs from the hypothetical promolecule or procrystal.

10.2 The electron-density function

The electron density in a crystal, being a triply periodic function, can be represented by a triply infinite Fourier sum (see Chapter 6, Section 6.9). This is conventionally written as

$$\rho(\mathbf{r}) = (1/V) \sum_{\mathbf{H}} \mathbf{F}_{\mathbf{H}} \exp(-2\pi i \mathbf{H} \cdot \mathbf{r}). \tag{10.2}$$

The vector $\mathbf{r}$ denotes an arbitrary position in real space, with fractional coordinates $x^i = x, y, z$ measured from a chosen origin, i.e.

$$\mathbf{r} = x^i \mathbf{a}_i = x\mathbf{a} + y\mathbf{b} + z\mathbf{c}.$$

V is the volume of the unit cell and the sum is over all reciprocal-lattice vectors

$$\mathbf{H} = h_i \mathbf{a}^i = h\mathbf{a}^* + k\mathbf{b}^* + l\mathbf{c}^*$$

with integral indices h, k, l. The scalar product $\mathbf{H} \cdot \mathbf{r}$ is thus given by

$$\mathbf{H} \cdot \mathbf{r} = h_i x^i = hx + ky + lz.$$

The structure factor $\mathbf{F_H} = \mathbf{F}_{hkl}$ provides the connection with the X-ray diffraction experiment. If the X-ray scattering is weak, i.e. if the crystal is small and not too perfect and/or short-wavelength radiation is used, the diffracted intensities will be adequately described by the kinematic approximation. This implies that the integrated intensity of the hkl Bragg reflection is proportional (if we neglect anomalous scattering; see below) to the square of the magnitude F_H of the structure factor $\mathbf{F_H}$. So our first and major task is to measure these Bragg intensities. If we are aiming for the most reliable results, we should measure as many of them as we can, as accurately as we know how.

10.3 The X-ray experiment

The crucial stage of any crystallographic study, then, is the diffraction experiment. The quality and quantity of our intensity data ultimately determine what information can be extracted from them. This information may be degraded by improper treatment at later stages of the investigation, but it can never be enhanced, however careful or elegant our refinement procedures. It also pays to be lucky in our choice of crystal specimen; even the most accurate and extensive X-ray data will yield disappointing results if our crystal turns out to be twinned, disordered, or otherwise recalcitrant.

10.3.1 Experimental accuracy

Accuracy in X-ray measurements poses a highly demanding experimental challenge. This has been pursued diligently and imaginatively in a number of crystallographic laboratories, with occasionally spectacular success. Some of the principal difficulties, and the methods adopted to assure optimal results, are authoritatively discussed in Chapter 7. Problems of data reduction have been considered in some detail by Blessing (1987). In general, success depends on taking the greatest pains to identify and eliminate all possible kinds of systematic error, and then measuring each reflection as often and as long as necessary to reduce the random errors to within acceptable tolerances. In fact, there is virtually no way to know if we have even thought of all the possible errors that may be present. However, one simple stratagem that can often detect unsuspected errors, and that always helps in estimating random errors, is the comparison of symmetry-related reflections. Even better is the use of quite independent experimental arrangements, such as different crystals, different wavelengths, etc. The only general rule is: there are no shortcuts to accurate data.

10.3.2   Resolution

Fully as important as accuracy is the experimental resolution. This is commonly measured by the maximum value of the reciprocal radius

$$H = d^* = |\mathbf{S}_1 - \mathbf{S}_0| = 2 \sin \theta / \lambda$$

for which reflections have been recorded (Stenkamp and Jensen 1984). In this notation, H is the length of the reciprocal-lattice vector $\mathbf{H}$ corresponding to the hkl Bragg reflection; $d^* = 1/d$ is the reciprocal interplanar spacing of the hkl lattice planes; $\mathbf{S}_0$ and $\mathbf{S}_1$ are vectors, each of length $1/\lambda$, along the incident and reflected rays, respectively; θ is the Bragg angle, which each of these vectors makes with the reflecting planes; and λ is the X-ray wavelength. Experience has shown that the most accurate electron-density maps are obtained from X-ray data extending to a reciprocal radius H_{max} not less than 2.2 to 2.5 Å^{-1} (the limit accessible with Mo Kα radiation being 2.8 Å^{-1}).

It was once supposed that high resolution is unimportant for electron-density studies, on the grounds that the deformation density is essentially confined to the valence orbitals and these are too diffuse to contribute to X-ray diffraction at large reciprocal radii. (The Fourier-transform relation (10.2) between electron density, in real space, and X-ray scattering, in reciprocal space, implies that the scattering from a sharp electron-density feature will extend over a broad range of reciprocal space while a diffuse density feature will scatter appreciably only at small reciprocal radii). But this argument is unsound for two reasons. Firstly, the demonstration that the valence contribution to the atomic X-ray scattering is concentrated at moderately small reciprocal radii (Stewart 1968, 1969) pertains explicitly to the *spherically averaged* free-atom densities. In bonded atoms, both experimental and theoretical deformation densities often show quite sharp non-spherical features, such as those associated with lone-pair orbitals or asymmetrically occupied d shells, that scatter much further out in reciprocal space (Coppens 1982). Secondly, it is important to include in the data a substantial number of high-order reflections that can be primarily attributed to scattering by the atomic cores, since it is only from these that we can unambiguously deduce the positions and vibrational amplitudes of the atomic nuclei.

The virtue of high resolution may be established by an alternative line of reasoning. For an atom in a general position, subject to no symmetry constraints, just determining its location and its anisotropic vibration (more correctly: *displacement*) amplitudes, in the harmonic approximation, requires 9 independent parameters. At least twice this many are needed for a reasonable description of the electron density in its near vicinity. A comfortable overdetermination of the adjustable parameters demands at least 10 to 20 observations per parameter. It follows that the number of independent reflection intensities measured should be several hundred times the number of atoms in the asymmetric unit. But the maximum number of independent reflections within a limiting reciprocal radius H_{max} is approximately

$$N_{ref} \cong (4\pi/3)(H_{max})^3 V/m \,,$$

where V/m is the volume of the asymmetric unit. Taking a typical value of 10 Å^3 for the volume per atom, we find the maximum ratio of reflections to atoms to be approximately $42(H_{max})^3$, with H_{max} in Å^{-1}. So for H_{max} equal to 2.2, 2.5, or 2.8 Å^{-1}, respectively, we have up to some 450, 650, or 920 reflections per atom accessible to measurement. Many of these, especially at the larger reciprocal radii, will likely be too weak for precise intensity measurement. But by cooling our crystal to a small fraction of its melting temperature T_m, we can greatly increase the proportion of higher-order reflections that will actually be measurable. At absolute temperature T, the intensity of a Bragg reflection at reciprocal radius H is roughly proportional (Lonsdale and El Sayed 1965) to the quantity $\exp[-\nu(T/T_m)H^2]$, where ν tends to lie between 1.5 and 3.0 Å^2 and the expression is approximately valid down to $T \cong 0.2\, T_m$. The exponential form of this temperature-dependence emphasizes the value of measuring at as low a temperature as can be achieved without loss of accuracy, especially when we are aiming for high-resolution data. For example, assuming $\nu = 2$ Å^2, we expect that lowering the temperature by one-fifth of T_m will enhance the average intensity by a factor of 12 at $H = 2.5$ Å^{-1}, a factor of 23 at 2.8 Å^{-1}.

The corresponding argument in real space asserts that to get the most out of an electron-density map, whether for precise location of the atoms or for a detailed description of the valence deformation, we don't want its sharpest features smeared either by excessive thermal vibration or by inadequate experimental resolution.

10.4 Deriving atomic parameters

Having measured our X-ray reflections, we wish next to evaluate the Fourier sum (10.2) for $\rho(\mathbf{r})$. Here we are confronted with what was once regarded as the central difficulty in X-ray crystallography, the phase problem. To evaluate $\rho(\mathbf{r})$ we require both the magnitudes and the phases of the structure factors $\mathbf{F_H}$, which are, in general, complex quantities. As shown in Chapter 6, Section 6.7, each structure factor can be written as

$$\mathbf{F_H} = F_{hkl}\exp\left(\mathrm{i}\alpha_{hkl}\right) = A_{hkl} + \mathrm{i}B_{hkl},$$

where the components A and B of $\mathbf{F}$ (omitting subscripts) are defined as

$$A = F\cos\alpha; \qquad B = F\sin\alpha.$$

We thus need the phase α of each structure factor in order to be able to resolve the quantity $\mathbf{F}$ into its real and imaginary components A and B. Only if the space group is centrosymmetric can we choose our origin so that the imaginary components B all vanish; the phase problem then reduces to the correct choice of sign for each A. Our experiment has provided us, we hope, with accurate values of the magnitudes $F_{\mathbf{H}}^{\mathrm{obs}}$, known as structure amplitudes, but no information at all about their phases $\alpha_{\mathbf{H}}$. Methods for obtaining this phase information are considered in Chapter 6, Section 6.10.

Let us, for the present, assume the phase problem has been solved and that we have been able to assign appropriate phases to all the measured structure factors. Via eqn (10.2) we now derive a map of the experimental electron density $\rho^{\mathrm{obs}}(\mathbf{r})$. This map will depict the true electron density $\rho(\mathbf{r})$ but for three sorts of defect: finite truncation of the Fourier series, eqn (10.2), experimental inaccuracy of the structure amplitudes, and errors in the estimated phases. Each of these defects is, furthermore, largely under our control and amenable to statistical evaluation. At this stage we have essentially solved our structure. The peaks of the electron density will mark the positions of the atoms, the peak heights will help us identify the several elements, and their shapes will tell us about their vibrational motion (or possible disorder). But how accurate is the information deduced in this way?

If the atoms were strictly at rest, our task would be much simpler. On the scale of interatomic distances, the nuclei have negligible size and may be treated as point charges. So with stationary atoms the electron density $\rho^{\mathrm{stat}}(\mathbf{r})$ would have — at infinite resolution — a sharp cusp at each nuclear site (Bingel 1963), establishing a well-defined position $\mathbf{R}_a$ for every atom. But vibrational smearing severely lowers and flattens these cusp densities. As a result the slope of the vibrationally averaged, *dynamic density* $\rho^{\mathrm{dyn}}(\mathbf{r})$ vanishes at each atomic peak, making the precise determination of its position much more problematic. Suppose we try to identify the *mean* position of an atom with the point of maximum $\rho^{\mathrm{dyn}}(\mathbf{r})$. This position is extremely sensitive to perturbations of various kinds. If the static density $\rho^{\mathrm{stat}}(\mathbf{r})$ is asymmetric around the cusp, the position of the maximum after vibrational smearing will probably not coincide with the mean position of the nucleus (Coulson and Thomas 1971). Moreover, any experimental inaccuracy in the function $\rho^{obs}(\mathbf{r}) \cong \rho^{dyn}(\mathbf{r})$ may severely shift the position of the density maximum. So simply measuring between positions of maximum $\rho^{obs}(\mathbf{r})$ will give us very inaccurate interatomic distances.

Evidently, a more reliable method would be to derive atomic positions from a least-squares fit to the electron density $\rho^{obs}(\mathbf{r})$ over an extended region encompassing each peak. Such a fit would make better use of all the available information and be far less susceptible to experimental error. However, this requires that we know the proper shape of the atomic peak to be fitted to the observed $\rho^{obs}(\mathbf{r})$. So, can we find a better model of the atomic densities than the spherical free atoms that comprise the promolecule?

10.5   The deformation density

In electron-density studies, the promolecule density $\rho^{pro}(\mathbf{r})$ defined by eqn (10.1) serves not only as the zero-order approximation; it is also a very convenient reference density. In practice, we rarely map the total electron density, whether derived experimentally or theoretically. Rather, we concentrate on the deformation density $\delta\rho(\mathbf{r})$, obtained by subtraction of $\rho^{pro}(\mathbf{r})$ from the total $\rho(\mathbf{r})$. This subtraction serves two purposes. Firstly, most chemically interesting density features are too small and too diffuse to be detectable in a map of the total density. We have a far better chance of seeing the subtle effects we are looking for if we remove the large atomic peaks and plot the remainder on a greatly expanded density scale. Secondly, the molecular deformation density presents a well-defined and conceptually illuminating picture of the electronic redistribution that must occur to transform a hypothetical assemblage of non-interacting atoms into a bound molecule. By subtracting that part of $\rho(\mathbf{r})$ that comes from the spherical atoms, we are effectively discarding the well-established atomic physics to concentrate on the chemically significant bonding interactions.

Sometimes the standard deformation density represents only the first of a series of successive decompositions of the total electron density into components that are more readily interpretable via simple chemical concepts. For example, Schwarz *et al.* (1985) and Kunze and Hall (1986) have analyzed the theoretical deformation densities of simple diatomic molecules by resolving them into components identified with such elementary concepts as atomic orientation, promotion, hybridization, covalency, and ionicity. Analogously, the deformation density of a composite system such as a hydrogen-bonded dimer, a crystalline hydrate, or an organometallic complex may be instructively resolved into components pertaining to its separate fragments plus an interaction term depicting their mutual perturbations (Hermansson and Lunell 1981; Martin *et al.* 1982; Hall 1986).

Since the deformation density $\delta\rho(\mathbf{r})$ is defined as a difference between two closely similar density functions, each of these must be determined with the highest accuracy in order that the resulting $\delta\rho(\mathbf{r})$ may be even qualitatively correct. For our experimental estimate of $\delta\rho(\mathbf{r})$ we often take the appropriate *residual density*, representing the difference, at the experimental resolution, between the observed density $\rho^{obs}(\mathbf{r})$ and the promolecule density $\rho^{pro}(\mathbf{r})$. The accuracy of $\rho^{obs}(\mathbf{r})$ depends on the accuracy of the measured X-ray intensities, on their reduction to observed structure amplitudes F_H^{obs}, and on the proper assignment of phases. If we have been meticulous in carrying out this part of our task, we can hope that $\rho^{obs}(\mathbf{r})$ will suffer mainly random errors whose magnitudes are statistically well characterized. More troublesome is the systematic bias that may afflict the promolecule density $\rho^{pro}(\mathbf{r})$ if it is calculated from systematically incorrect structural parameters. And these parameters are bound to be incorrect if they have been derived via an inadequate model of the electron density.

In particular, adoption of the promolecule model for fitting to the observed density $\rho^{obs}(\mathbf{r})$ amounts to the *a priori* assumption that the deformation density $\delta\rho(\mathbf{r})$ is, in fact, negligible. The fitting process will then optimize the atomic parameters in accordance with this assumption and so assure its apparent confirmation, i.e. it will produce structural parameters that tend to minimize the residual density (Coppens and Coulson 1967; Coppens *et al.* 1969; O'Connell 1969; Ruysink and Vos 1974). Suppose our crystal contains, for example, a pyramidal nitrogen atom having excess lone-pair density on the side of the nucleus opposite its ligands. If we try to fit to this asymmetric density distribution in $\rho^{obs}(\mathbf{r})$ a spherically symmetric model density, the center of the model peak will inevitably be displaced from the true nuclear position towards the lone-pair density (Dawson 1964). And when we subtract this displaced atomic density from $\rho^{obs}(\mathbf{r})$, the lone-pair peak will be artificially flattened, leaving a distorted residual density that has lost much of the asymmetry of the true deformation density (Fig. 10.1).

So we face a dual task. If we want to derive a reliable deformation density we need accurate, unbiased nuclear parameters. But we cannot derive an unbiased structure if we ignore the deformation density. As stated at the outset, the two kinds of information are linked together and must be pursued simultaneously.

10.6   Least-squares refinement

10.6.1   Reciprocal-space fitting

While it may be instructive to visualize the process of fitting a model density to the experimental $\rho^{obs}(\mathbf{r})$ in real space, in practice this fitting is carried out in reciprocal space, the space of the X-ray structure factors. For this

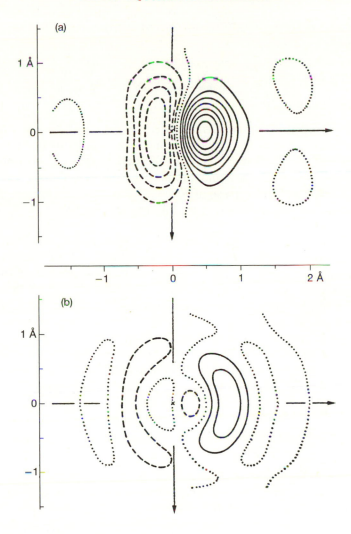

Fig. 10.1 Calculated effect of spherical-atom refinement on residual density in two-dimensional projection. Hypothetical 'prepared' nitrogen atom with lone-pair electrons in sp^3 hybrid orbital, isotropic mean-square displacement 0.025 $\mathring{A}^2$, as imaged by simulated two-dimensional X-ray data extending to $H_{max} = 1.3$ $\mathring{A}^{-1}$. Contour interval 0.1 e$\mathring{A}^{-2}$. From Dawson (1964). (a) Projected deformation density, from two-dimensional Fourier series, with correct atomic position and displacement parameters; (b) Projected residual density after unit-weight refinement on F with spherical-atom model, giving spurious atomic position (shifted 0.029 $\mathring{A}$) and anisotropic displacement parameters.

purpose we invert eqn (10.2) to obtain the corresponding inverse Fourier transform

$$\mathbf{F_H} = \int_V \rho(\mathbf{r}) \exp\left(2\pi i \mathbf{H} \cdot \mathbf{r}\right) dV, \tag{10.3}$$

where dV denotes a volume element d^3r in $\mathbf{r}$ space and the integration is over the unit-cell volume V. Using this, or a related equation based on a particular model of $\rho(\mathbf{r})$, we can calculate structure factors $\mathbf{F_H^{calc}}$ corresponding to an approximate structure and compare the calculated magnitudes with F_H^{obs}. Provided we find a reasonable level of agreement, we can progressively adjust our model structure for an optimum least-squares fit between F_H^{calc} and F_H^{obs}.

All this seems like a very roundabout approach to the derivation of the crystal structure from the measured structure amplitudes. But there are good reasons for performing the match on the structure amplitudes rather than on the electron density. One important consideration is that the quantities F_H^{obs} are more immediately related than is $\rho^{obs}(\mathbf{r})$ to the measured intensities and can be assigned statistical weights that are readily derived from the estimated random errors in these intensities. Moreover, we can usually assume that the errors in the various structure amplitudes are statistically independent of one another. This is never strictly true but it is a convenient approximation. It allows us to define the weighted *deviance*

$$\Delta = \sum_H w_H \left(F_H^{obs} - kF_H^{calc}\right)^2$$

as a simple sum over the N_{ref} independent observed reflections. This sum is a function of all the parameters governing the model density $\rho(\mathbf{r})$ that is substituted in eqn (10.3) for the evaluation of F_H^{calc}; hence it can be minimized by variation of these parameters. The weights w_H in the expression for Δ are taken as the reciprocals of the estimated variances of the corresponding F_H^{obs}. (Without the assumption of statistically uncorrelated errors in the structure amplitudes, the single summation for Δ would be replaced by a double sum and we should need a weight matrix (Hamilton 1964) derived by inversion of the full $N_{ref} \times N_{ref}$ covariance matrix of the measured F_H^{obs}.) The scale factor k must be included as an adjustable parameter since, in most experiments, only relative values of F_H^{obs} are determined. That is: the ratio between the measured intensity and F_H^2 depends on some factors that are known functions of $\mathbf{H}$ and others that are supposed constant and are combined into the unknown quantity k^2. The set of parameter values that minimizes Δ may be plausibly supposed to define the most probable values of these parameters as determined by the X-ray experiment.

This assumption will often rest on firmer ground if we choose a slightly different form for the deviance Δ, in which the quantities F_H^{obs} and F_H^{calc} are replaced by their squares. This has the virtue that the errors in $(F_H^{obs})^2$ are more likely to follow a Gaussian distribution, which is advantageous for the statistical analysis of the results. The weights w_H must then be chosen as the reciprocal variances of $(F_H^{obs})^2$.

A successful least-squares refinement gives us more than a set of structural parameters that accords best with the X-ray data. The minimum value of Δ provides a test of the correctness of the entire procedure. This value should be close to the quantity

$$N_{o-p} = N_{ref} - N_{par},$$

the number of excess observations over adjustable parameters. If the actual minimum of Δ slightly exceeds N_{o-p}, this may simply mean that we have somewhat underestimated the variances of the measured structure amplitudes F_H^{obs}. But if Δ is much larger than expected, the problem could be more serious. It could mean that our data are contaminated by severe systematic error, that the model we have used is grossly deficient, or, perhaps, that we have not located the global minimum of Δ.

Conversely, a *goodness-of-fit*

$$S = [\min(\Delta)/N_{o-p}]^{1/2}$$

close to 1 strengthens our confidence that our estimated variances are realistic, that our model provides an adequate approximation to the true electron density $\rho(\mathbf{r})$, and that the refinement has converged to the proper solution. We can then proceed to derive, from the output of our least-squares calculations, estimated variances as well as covariances of all the refined parameters. These are most important for estimating the accuracy of various derived quantities, such as bond lengths, bond angles, rigid-body molecular translation and libration amplitudes (if the molecule indeed behaves as a rigid body), deformation densities, dipole moments, and the like (Rees 1977).

It seems that by performing our least-squares fit on the structure amplitudes rather than on $\rho(\mathbf{r})$, we have somehow sidestepped the phase problem. In fact, the phase information is implicit in our structural model and appears explicitly in the calculated structure factors $\mathbf{F}_H^{calc}$, although it is only the amplitudes F_H^{calc} that we fit to the corresponding experimental quantities F_H^{obs}. Once the model parameters have been refined to their optimum values we can, if we wish, calculate a hybrid electron-density function, conventionally but somewhat imprecisely designated $\rho^{obs}(\mathbf{r})$, using the scaled experimental amplitudes F_H^{obs}/k together with the calculated phases α_H^{calc} in a Fourier summation via eqn (10.2).

10.6.2 The standard structure factor expression

For routine structural studies, the promolecule is generally adopted as an adequate model of the electron density. This allows us to replace $\rho(\mathbf{r})$ in eqn (10.3) by the promolecule density $\rho^{pro}(\mathbf{r})$ defined by eqn (10.1). We thus obtain, reversing the order of summation and integration,

$$\mathbf{F_H} = \sum_a \int \rho_a^{at}(|\mathbf{r} - \mathbf{R}_a|) \exp{(2\pi i \mathbf{H} \cdot \mathbf{r})} \, dv.$$

We can take an infinite integration volume for each atom and so restrict the summation to atoms having their centers in one unit cell. Defining $\mathbf{r}'_a = \mathbf{r} - \mathbf{R}_a$, we rewrite this as

$$\mathbf{F_H} = \sum_a \int \rho_a^{at}(r'_a) \exp{[2\pi i \mathbf{H} \cdot (\mathbf{R}_a + \mathbf{r}'_a)]} \, dv'_a.$$

We now introduce the atomic scattering factor

$$f_a(H) = \int \rho_a^{at}(r') \exp{(2\pi i \mathbf{H} \cdot \mathbf{r}')} \, dv',$$

which, like $\rho_a^{at}(r')$, is spherically symmetric, and so obtain the expression

$$\mathbf{F_H} = \sum_a f_a(H) \exp{(2\pi i \mathbf{H} \cdot \mathbf{R}_a)}.$$

However, this equation takes no account of the vibrational smearing of the electron density. The promolecule model implicitly treats each atom as tied rigidly to its nucleus (see Section 10.9.1). Hence we must convolute the static atomic density with the probability distribution of its nuclear position. Assume that each atom undergoes anisotropic harmonic vibration. In reciprocal space we then multiply each $f_a(H)$ by an anisotropic Debye–Waller temperature factor of the form

$$\tau_a(\mathbf{H}) = \exp{(-h_i \beta_a^{ij} h_j)},$$

in which the parameters β_a^{ij} are contravariant components of a symmetric second-rank tensor specifying the mean-square displacement amplitudes of atom a, and the tensor notation implies summation over the repeated indices i and j (see Chapter 8, Section 8.2). The standard expression for the calculated structure factor, for the vibrating promolecule model, thus becomes

$$\mathbf{F}_{\mathbf{H}}^{calc} = \sum_a f_a(H) \exp\left(-h_i\beta_a^{ij}h_j\right) \exp\left(2\pi i h_i x_a^i\right), \qquad (10.4)$$

where we have substituted

$$\mathbf{R}_a = x_a^i \mathbf{a}_i$$

so that

$$\mathbf{H} \cdot \mathbf{R}_a = h_i x_a^i = hx_a + ky_a + lz_a.$$

Least-squares minimization based on eqn (10.4) leads to refined, if somewhat biased, values of the structural parameters x_a^i and β_a^{ij} and their estimated variances and covariances.

10.7 Anomalous scattering

There is a further reason, apart from the phase problem and the weighting of the observational data, why we may be forced to derive structure factors $\mathbf{F}_{\mathbf{H}}^{calc}$ from a model density $\rho(\mathbf{r})$, for comparison with the observed structure amplitudes, rather than evaluating the electron density $\rho^{obs}(\mathbf{r})$ from the measured $F_{\mathbf{H}}^{obs}$. Equations (10.2) and (10.3) define the structure factor $\mathbf{F}_{\mathbf{H}}$ as the Fourier transform of the electron density $\rho(\mathbf{r})$. However, this is not the quantity that is actually measured in the diffraction experiment. Rather, the measured intensity depends on the *anomalous* structure factor, $\hat{\mathbf{F}}_{\mathbf{H}}(\lambda)$, which differs from the true, or *proper*, structure factor $\mathbf{F}_{\mathbf{H}}$ because of anomalous scattering. This is a specific resonant interaction of the X-ray photons with the lowest electronic energy levels of the crystal, the atomic core levels. The interaction produces a change in both the magnitude and the phase of the scattering from those atoms whose core energy levels are near the X-ray energy. Thus the atomic scattering factors $f_a(H)$ in eqn (10.4), which is the conventional approximation to eqn (10.3), must be replaced by the wavelength-dependent

$$\hat{f}_a(H, \lambda) = f_a(H) + \Delta f_a'(H, \lambda) + i\Delta f_a''(H, \lambda)$$

to yield the modified, anomalous structure factor

$$\hat{\mathbf{F}}_{\mathbf{H}}^{\text{calc}}(\lambda) = \sum_a \hat{f}_a(H, \lambda) \exp(-h_i\beta_a^{ij}h_j) \exp(2\pi i h_i x_a^i). \qquad (10.5)$$

This is the quantity whose magnitude should match the experimental structure amplitude of the *hkl* Bragg reflection. Accordingly, the deviance to be minimized in a least-squares refinement is

$$\Delta = \sum_{\mathbf{H}} w_{\mathbf{H}}\{[\hat{F}_{\mathbf{H}}^{\text{obs}}(\lambda)]^2 - [k\,\hat{F}_{\mathbf{H}}^{\text{calc}}(\lambda)]^2\}^2 \qquad (10.6)$$

and it is the minimum of this function that defines the goodness-of-fit

$$S = [\min(\Delta)/N_{\text{o-p}}]^{1/2}.$$

The proper structure factor $\mathbf{F}_{\mathbf{H}}$ of eqn (10.3) may be regarded as the short-wavelength limit of $\hat{\mathbf{F}}_{\mathbf{H}}(\lambda)$.

As eqns (10.4) and (10.5) illustrate, for a given model electron density $\rho(\mathbf{r})$ it is straightforward to calculate both the proper and the anomalous structure factors. The latter can be compared with the observed structure amplitudes and the discrepancies interpreted to yield an improved model for $\rho(\mathbf{r})$, usually by least-squares refinement. Without a model density, however, we have no way of converting the observed anomalous structure amplitudes $\hat{F}_{\mathbf{H}}^{\text{obs}}(\lambda)$ into the corresponding proper structure amplitudes $F_{\mathbf{H}}^{\text{obs}}$ needed for a Fourier synthesis of $\rho^{\text{obs}}(\mathbf{r})$. So, as with the phase problem, anomalous scattering requires us to proceed from an assumed $\rho(\mathbf{r})$ to the corresponding structure amplitudes $\hat{F}_{\mathbf{H}}^{\text{calc}}$ rather than from the observed amplitudes $\hat{F}_{\mathbf{H}}^{\text{obs}}$ to the experimental density $\rho^{\text{obs}}(\mathbf{r})$.

Only after the refinement has produced a final structural model, giving satisfactory agreement between calculated and observed magnitudes of the anomalous structure factors, can we reverse the process. Now we can use the difference between the calculated $F_{\mathbf{H}}$ and $\hat{F}_{\mathbf{H}}(\lambda)$ for each reflection to evaluate an adjustment

$$\delta_{\mathbf{H}}(\lambda) = \hat{F}_{\mathbf{H}}^{\text{calc}}(\lambda) - F_{\mathbf{H}}^{\text{calc}}$$

that we can apply to the corresponding observed $\hat{F}_{\mathbf{H}}^{\text{obs}}(\lambda)$ to obtain an estimate of the proper structure amplitude

$$F_{\mathbf{H}}^{\text{obs}} \cong \hat{F}_{\mathbf{H}}^{\text{obs}}(\lambda) - k\delta_{\mathbf{H}}(\lambda).$$

To this we attach the phase of $\mathbf{F}_{\mathbf{H}}^{calc}$, giving us the Fourier coefficient $\mathbf{F}_{\mathbf{H}}^{obs}$ to be inserted into eqn (10.2) for the evaluation of $\rho^{obs}(\mathbf{r})$ (Templeton 1955; Patterson 1963).

The effect of anomalous scattering may or may not be severe, depending on the elemental composition of the crystal, the wavelength λ, and the experimental accuracy we are aiming for. When a light-atom structure is studied with Mo Kα (or shorter-wavelength) radiation, the correction $\delta_{\mathbf{H}}(\lambda)$ is almost always negligible.

The difference between the proper structure factor $\mathbf{F}_{\mathbf{H}}$ and the anomalous structure factor $\hat{\mathbf{F}}_{\mathbf{H}}(\lambda)$ has a number of further practical consequences in X-ray crystallography. Friedel's law, which states that

$$\mathbf{F}_{-\mathbf{H}} = \mathbf{F}_{\mathbf{H}}^{*},$$

i.e. that a pair of structure factors related by inversion in the origin of the reciprocal lattice are complex conjugates of one another, follows from eqn (10.3) with the consideration that $\rho(\mathbf{r})$ is real. This law applies only to the proper structure factors $\mathbf{F}_{\mathbf{H}}$. It implies that every Friedel pair of reflections $\mathbf{H}$ and $-\mathbf{H}$ would, in the absence of anomalous scattering, be equal in intensity. This would make the diffraction pattern of every crystal centrosymmetric whether or not the crystal itself had a center of inversion. But because of the imaginary component $\Delta f_a''(H, \lambda)$ of the anomalous atomic scattering factor $\hat{f}_a(H, \lambda)$ appearing in eqn (10.5), a corresponding law does not hold, except approximately, for the anomalous structure factors $\hat{\mathbf{F}}_{\mathbf{H}}(\lambda)$. This breakdown of Friedel's law for the observed intensities permits an experimental discrimination between space groups with and without a center of inversion. And in chiral space groups it allows enantiomorphous structures to be distinguished (Bijvoet et al. 1951). More generally, it may lead to the assignment of the absolute sense of any polar axes (Coster et al. 1930). It is also useful as an aid in structure solution, permitting the location of anomalously scattering atoms from the Bijvoet differences between Friedel pairs of reflections (Rossmann 1961) and enhancing the power of probabilistic methods of phase determination (Hauptman 1982).

10.8 Overcoming parameter bias

Whether the ultimate goal of an X-ray crystallographic study is an approximate or a very precise nuclear structure, accompanied by detailed electron-density information, we will probably begin with a preliminary refinement using the promolecule model of eqn (10.5) — or eqn (10.4) if anomalous scattering is negligible at this stage. Once the least-squares refinement of this model has converged, we can begin to improve our description of $\rho(\mathbf{r})$ and so correct the bias arising from the assumption of spherical atomic densities.

One way to proceed would be to calculate a residual density map and examine it for direct evidence of non-spherical density features. Such a residual density can be evaluated most directly via eqn (10.2), with the structure factors $\mathbf{F_H}$ replaced by the structure factor differences

$$\Delta\mathbf{F_H} = \mathbf{F_H^{obs}}/k - \mathbf{F_H^{calc}}$$
$$\cong \hat{\mathbf{F}}_H^{obs}(\lambda)/k - \hat{\mathbf{F}}_H^{calc}(\lambda). \qquad (10.7)$$

The latter expression results from a treatment of the anomalous scattering correction $\delta_H(\lambda)$ that differs slightly from the one given above. The calculated structure factors appearing in these expressions, both proper and anomalous, refer to the promolecule model. Because of the bias imposed by the refinement process, we expect the residual density so derived to reveal only a greatly muted picture of the true deformation density (see Fig. 10.1). This may hint at the kinds of modification required in our model electron density but will not reveal their true magnitudes. We can, however, try to correct the model density by stages, gradually improving the promolecule model and monitoring our progress by examination of successive residual density maps.

10.8.1   High-order refinement

In fact, the earliest attempts to overcome this problem emphasized the opposite approach, concentrating on ways to avoid the bias in the structural parameters in the hope that a better structural model would lead to a better residual density. The first of these solutions was the high-order refinement method (Jeffrey and Cruickshank 1953). This relies on the 'frozen-core' approximation, according to which the atomic cores are practically inert to chemical influences (Bentley and Stewart 1974), together with the assumption (see Section 10.3.2 above) that the valence density contributes only to low-order X-ray scattering. Combining these two assumptions, we expect the high-order structure factors to be unaffected by the deformation density and thus to be calculated correctly via the promolecule model. All we have to do then is refine our structural parameters against the high-order X-ray data alone and hope that the resulting model, based essentially on the core scattering, will be free of bias (Stevens and Hope 1975).

When the structural parameters have been derived in this way from the high-order reflections, the resulting residual density is often designated the $X - X_{ho}$ density, in analogy to the corresponding $X - N$ density (see below). The name alludes to the source of the Fourier coefficients from which such a map is computed. These coefficients take the form (10.7) in which $\mathbf{F_H^{obs}}$, or $\hat{\mathbf{F}}_H^{obs}$, comes from the corresponding X-ray reflection intensity while $\mathbf{F_H^{calc}}$, or $\hat{\mathbf{F}}_H^{calc}$, is evaluated from parameters based on the high-order X-ray data.

A worrisome question in any application of this method is where to place the lower limit for the high-order reflections. The original authors (Jeffrey and Cruickshank 1953) suggested that the high-order refinement might include all data above $H_{min} = 1.0 \text{ Å}^{-1}$. Since that time, two developments have favored a progressive rise in the preferred value of the cut-off reciprocal radius H_{min}. On the one hand, computer-controlled diffractometers and especially the popularity of low-temperature intensity measurements have facilitated the measurement of high-resolution data. This has meant that a large H_{min} may still leave a sufficient number of high-order reflections above this threshold for a well-determined least-squares refinement. More importantly, accumulated evidence (Chandler and Spackman 1982; Coppens 1982; Hirshfeld 1984a; Swaminathan et al. 1984) that certain types of atoms may show quite sharp deformation-density features has persuaded experimenters to choose larger and larger H_{min} to try to avoid bias in the parameters of these atoms. Accordingly, recent studies have adopted values of H_{min} as high as 2.0 Å^{-1} (Coppens et al. 1984). It is evident that any value chosen must be essentially arbitrary and cannot correspond to any sharp division between valence and core scattering even for a particular atom, much less for an entire structure.

A special difficulty attaches to the location of hydrogen atoms. These lack core orbitals, making the high-order method, in principle, quite inapplicable to them. In fact, the atomic scattering factor of hydrogen has appreciable magnitudes only at quite low reciprocal radii. So, the very reflections we must discard to prevent bias in the parameters of the heavier atoms contain virtually all the available information about the hydrogen atoms. Accordingly, high-order refinements are usually performed with fixed hydrogen parameters, at values deduced from the low-order reflections or from neutron data or simply estimated from analogous structures. Nevertheless, with carefully measured low-temperature X-ray data, a value of H_{min} around 1.3 to 1.5 Å^{-1} has been shown to offer a delicate but workable compromise between totally unrefinable hydrogen parameters and appreciably biased parameters for some of the heavier atoms (Hope and Ottersen 1978; Seiler et al. 1984a).

10.8.2 The $X - N$ method

As an alternative to high-order X-ray refinement, Coppens (1967) introduced the use of neutron diffraction for the derivation of completely unbiased structural parameters. In many ways neutrons offer an ideal complement to X rays, being scattered essentially by the atomic nuclei rather than by the electrons (in non-magnetic materials). Thus neutron diffraction is blind to the chemical deformation of the atomic electron clouds and reveals directly the positions and displacement amplitudes of the nuclei alone. Consequently, when we use the structural parameters from a neutron-diffraction experiment to calculate F_H^{calc} for evaluation of the coefficients ΔF_H of an X − N Fourier synthesis, we circumvent entirely the problem of biased parameters. The resulting difference density should show a true picture of the deformation density $\delta\rho(\mathbf{r})$, apart from the effect of unavoidable experimental errors.

Neutron diffraction has the further merit, for certain purposes, of yielding comparable amplitudes of scattering for light and heavy elements. This permits, for example, the location of hydrogen atoms in the presence of much heavier elements (see Chapter 11). Conversely, neutron scattering cross sections may be quite different for isotopes of the same element.

Yet for all its clear advantages, the X − N method has not completely solved the problem of the derivation of unbiased deformation densities. One reason is the inconvenience of relying on access to a high-flux nuclear reactor as a radiation source. Even when such a source is readily available, counting rates are often low so that long experiments with large crystals (if these can be obtained) are needed to achieve reasonable accuracy in the intensity measurements. Competition for beam time may impose painful compromises, such as rapid data collection, with resultant poor precision, or the sacrifice of high resolution. Either option limits the accuracy of the final structural information. As a consequence, we may find ourselves having to choose between biased but precisely determined X-ray parameters and unbiased but far less precise neutron parameters.

A further difficulty arises in ensuring that the X-ray and neutron diffraction data are fully compatible. It is a common experience to find the two experiments yielding quite concordant atomic positions but systematically divergent displacement parameters β_a^{ij}, despite customary precautions against bias in the X-ray refinements (e.g. Bats *et al.* 1977; Scheringer *et al.* 1978; Epstein *et al.* 1982; Coppens *et al.* 1984). Sometimes the explanation may simply lie in a difference of temperature between the crystals in the two experiments. But often the reasons for the discrepancy are more subtle and harder to identify. Problems like uncorrected extinction error, or thermal diffuse scattering (see Chapter 7), which may be more severe in one experiment than in the other, are sometimes blamed for the difficulty, but such retrospective suspicions are rarely substantiated by direct demonstration of the supposed effects. In the absence of a fully convincing explanation, a common expedient is to scale the neutron displacement parameters, isotropically or anisotropically, to produce a satisfactory *average* match to the X-ray parameters. This may appear to yield acceptable results but it is clearly not a very sound base on which to construct an accurate deformation density.

An attractive alternative to complete reliance on either high-order X-ray data or neutron diffraction is to combine the two in a joint X + N refinement. In this way, atoms that contribute weakly to the X-ray scattering, especially hydrogen, will be predominantly located by the neutron data. Also, use of the neutron reflections lessens our dependence on the X-ray data, permitting the choice of a larger H_{min}, hence the inclusion of fewer X-ray reflections, so as to minimize the risk of bias from sharp deformation features. Reciprocally, the high-order X-ray data can compensate for a lower resolution, or lower precision, of the neutron data. But any such joint refinement is best undertaken after a preliminary refinement against each data set by itself. These separate refinements will reveal any incompatibility between the two experiments, verify the relative weights assigned to the two sets of data, and allow an estimate of the relative contributions of the two data sets to the determination of the various structural parameters (Coppens *et al.* 1981).

10.8.3 The phase problem again

The $X - X_{ho}$ and the $X - N$ method share a common weakness in dealing with non-centrosymmetric structures. Each Fourier coefficient ΔF_H is defined by eqn (10.7) as the difference between two complex quantities, which differ, in general, both in amplitude and in phase. Only for F_H^{calc} do we know both. For F_H^{obs} we derive the amplitude from the measured intensity (corrected, as necessary, for anomalous scattering) but the phase can only be estimated, for example by taking the phase of F_H^{calc}. In a centrosymmetric structure, where all phases are either 0 or π, this will almost always give the correct choice. But for non-centrosymmetric space groups it can only bias the residual density by systematically underestimating the magnitude of ΔF_H as well as falsifying its phase (Maslen 1968; Coppens 1974). Consequently, both these methods have been applied mainly to centrosymmetric structures. For a few non-centrosymmetric crystals, approximate measures have been adopted to provide improved phases for F_H^{obs} (Thomas *et al.* 1975; Ito and Shibuya 1977; Ottersen *et al.* 1980; Savariault and Lehmann 1980), but no fully satisfactory solution is known within the limitations of the promolecule model.

10.9 Deformation model refinement

Rather than treating the problem in stages, first trying to determine unbiased atomic parameters and then looking for electron deformation in the residual density, we can combine the structural and the electron-density information in a single flexible model. Such a model must describe the electron density parametrically, for example as an expansion in a set of local basis functions, of fixed or variable form, centered on the several atoms (Dawson 1967; Kurki-Suonio 1968). The relevant parameters are then subject to

adjustment, by least-squares minimization, along with the nuclear coordinates and displacement parameters. If the model possesses the requisite
flexibility, it will allow the refinement, against the full set of reflections, to
converge to a proper solution giving both unbiased structural parameters
and a faithful representation of the deformation density.

It is convenient to incorporate the promolecule as an explicit component
of the model density, expressing the total electron density as

$$\rho(\mathbf{r}) = \rho^{\text{pro}}(\mathbf{r}) + \delta\rho(\mathbf{r}) \tag{10.8}$$

and expanding the deformation density $\delta\rho(\mathbf{r})$ in the chosen basis, as

$$\delta\rho(\mathbf{r}) = \sum_{a,k} c_{ak}\rho_{ak}(\mathbf{r} - \mathbf{R}_a). \tag{10.9}$$

In this expansion, the index a labels the several atoms while k numbers the
distinct basis functions on each atom. The number of independent coefficients c_{ak} and, possibly, of additional parameters needed to define the basis
functions ρ_{ak} determines the degree of flexibility of the deformation model.

10.9.1   A convolution model

Usually we prefer to construct our electron-density model in two conceptually distinct stages. First, we model the *static* density, for example by
means of an expansion as outlined in eqns (10.8) and (10.9). This allows us
to incorporate information about the molecular symmetry and other theoretical or experimental constraints (see Section 10.9.2) that pertain specifically
to the static density. We then convolute this static density with a function
chosen to simulate the effects of vibrational smearing.

The required convolution is often based on the rigid-pseudoatom approximation (Stewart 1976), which assumes that the total electron density of a
molecule may be divided into atomic fragments each of which moves rigidly
with its nucleus in all vibrational modes. Suppose we let eqn (10.8) repre-
sent the static density $\rho^{\text{stat}}(\mathbf{r})$. Both terms on the right of this equation are
sums of atomic components, the promolecule density $\rho^{\text{pro}}(\mathbf{r})$ being defined
as such by eqn (10.1) while the static deformation density $\delta\rho^{\text{stat}}(\mathbf{r})$ takes
such a form on appropriate grouping of terms in eqn (10.9). Thus our model
has formally decomposed the total static density into atomic fragments. To
derive the dynamic density $\rho^{\text{dyn}}(\mathbf{r})$, we must convolute each such fragment,
assumed to constitute a rigid pseudoatom, with the (anisotropic) proba-
bility distribution of its nuclear position. Accordingly, to calculate structure factors in this approximation, we add the Fourier transform of each
atomic fragment of $\delta\rho^{\text{stat}}(\mathbf{r})$ to the corresponding free-atom scattering

factor $f_a(H)$, or $\hat{f}_a(H, \lambda)$, and multiply the sum by the appropriate temper-ature factor $\tau_a(\mathbf{H})$ as in eqn (10.4) or (10.5) (Hirshfeld 1971; Hansen and Coppens 1978). Other deformation models, in which some basis functions are centered off the nuclear positions, require a slightly different but essen-tially similar treatment (Coppens *et al.* 1971; Scheringer 1977).

But how good is the rigid-pseudoatom approximation? This question, which is basic to the validity of the convolution model, is best considered separately for rigid-body and for internal vibrations. In a molecular crystal much of the vibrational motion comes from rigid-body lattice modes in which each molecule moves as a rigid unit. For such motions it is entirely plausible that the molecular electron distribution should move rigidly with the nuclear framework. The only serious doubt is how well the smearing due to molecular *librations* can be simulated by translational vibrations of arbitrarily defined atomic fragments (Scheringer 1978). The approximation is probably adequate, and least arbitrary, where it matters most, near the atomic nuclei where the sharpest density gradients occur. Away from the nuclei, the approximation seems more doubtful but the density is smoother and so less sensitive to the details of the vibrational model. Of course, if libration amplitudes are appreciable, the curvilinear form (sometimes called 'bananicity') of the atomic motion must be properly treated by inclusion of higher-order terms in the temperature factors $\tau_a(\mathbf{H})$ (Pawley and Willis 1970; Scheringer 1979).

For internal vibrations, the assumption that each atomic fragment of electron density rigidly follows its nucleus is far less compelling. Since the promolecule density, by definition, strictly follows the nuclear motions, it remains to examine the validity of this approximation for the deforma-tion density. Calculations on several diatomic and triatomic molecules (Rozendaal and Ros 1982; Stephens and Becker 1983; Rozendaal and Baerends 1986) indicate that the deformation density is only slightly affected by bond-stretching and bending vibrations, mainly because these vibra-tions generally have small amplitudes. As the effect itself is small, it can be modeled quite satisfactorily by the pseudoatom approximation (Coulson and Thomas 1971; Epstein and Stewart 1979). Larger-amplitude vibrations that may occur in polyatomic molecules, such as torsional motions, probably do not differ greatly from rigid-body librations in their effects on the deformation density, but this assumption has yet to be tested directly.

10.9.2   Choosing a deformation model

Supposing we accept the approximations inherent in the convolution model, we must next decide on the number and form of the basis functions ρ_{ak} for the expansion (eqn (10.9)) of the static deformation density. Since the number cannot be increased indefinitely without endangering the convergence of the refinement, it is important to select a form that will properly model the expected deformation features with a reasonably compact expansion. General theoretical arguments, symmetry considerations, chemical intuition, and the like, as well as an examination of theoretical deformation densities for related small molecules, can provide important guidance in the choice of a suitable basis (Bentley and Stewart 1975, 1976; Chandler *et al.* 1980; Chandler and Spackman 1982). In particular cases, information from other kinds of experiments, such as electric field gradients or spectroscopic data, can be incorporated as constraints on the model deformation density (Schwarzenbach and Lewis 1982) or on the structural parameters (Eriksson *et al.* 1982). Theoretical or experimental constraints of this sort are especially valuable when they supply information that is not readily available from the diffraction experiment, such as the form of the deformation density close to the atomic nuclei. For example, the nuclear cusp condition (Bingel 1963) is often useful for preventing excessive correlations between deformation and displacement parameters (Eisenstein 1979). This is more effective, besides being more soundly based, than the frozen-core approximation that is built into some deformation models (Stewart 1976; Hellner 1977; Hansen and Coppens 1978), which make no explicit allowance for deformation of the atomic core densities. Similarly, the Hellmann–Feynman electrostatic theorem may be imposed as a constraint to permit the simultaneous refinement of core polarization parameters and atomic coordinates (Hirshfeld 1984*b*).

Several types of model have been in common use for the study of molecular crystals (e.g. Stewart 1969, 1973, 1976; Coppens *et al.* 1971; Hirshfeld 1971, 1977; Jones *et al.* 1972; Mullen and Hellner 1977; Hansen and Coppens 1978). Choice of an appropriate model for a particular structure, and of the degree of flexibility for a given type of model, involves such considerations as the chemical nature of the compound, the kind of information sought, and the quality of the X-ray data. Clearly, the more extensive and the more accurate our data — including neutron data if these are available for inclusion in a joint refinement — the more flexible a model we can use.

A fairly general rule is that too much flexibility is safer than too little. Using a model that is too restrictive, we run the risk of obtaining an apparently satisfactory solution that is, in fact, biased by the limitations of the model. An extreme example of this situation is the promolecule model. On the other hand, with more adjustable parameters than necessary, the only risk is that the refinement may fail to converge or, more likely, the refined parameters will have excessive variances. Either outcome will save us from unwarranted conclusions about the structure or the deformation density. And we retain the option of judiciously imposing constraints, thus reducing the number N_{par} of adjustable parameters, so as to lower the estimated variances, while checking that the goodness-of-fit S remains acceptable. Such constraints should be consistent with the results derived via the less restrictive models, within bounds determined by the estimated variances of their parameters.

10.9.3 Testing a refinement

Having refined our structure with whatever deformation model we have chosen, we need to know how reliable the results are. The first indication comes from the goodness-of-fit S or, more generally, from the statistics of the weighted discrepancies $w_H(\Delta F_H^2)^2$. In many cases we can expect these to have an approximately normal distribution, with a mean near N_{o-p}/N_{ref} (generally close to unity) and no systematic variation with H, F, or any other relevant variable. Secondly, we can examine the residual density for any signs of genuine deformation features that have been suppressed by the model. If the model used is adequate, we expect the residual density to show only random experimental noise. The only difficulty here—a nontrivial one—is to establish criteria for recognizing whether noise is essentially random or not. Further, the total electron density (eqn (10.8)) must be everywhere positive. This condition is not readily imposed as a constraint on the model but serves as a check on the validity of the refined deformation density. Another useful test is provided by the rigid-bond postulate (Rollett 1970; Hirshfeld 1976), which asserts that the mean-square displacement amplitudes of every pair of covalently bonded atoms should be nearly equal in the bond direction, especially when the two atoms have comparable masses. Failure of this test can sometimes point to a particular atom or bond where deficiencies of the deformation model may have been absorbed by the atomic displacement parameters β_a^{ij}. Having established that our refinement satisfactorily passes all these internal tests, we can then derive variances and covariances of the refined parameters in the standard way from the inverted least-squares matrix.

Too infrequently, several different deformation models are applied to the same set of diffraction data (Baert *et al.* 1982; Restori *et al.* 1987). Comparison among the results then allows a distinction between deformation features that are strongly model-dependent and those that are more likely to be inherent in the X-ray data. Such studies can also contribute to a better understanding of the weaknesses of particular models and so to efforts towards improving them.

Finally, the best test of an experimental deformation density is a comparison with a theoretically calculated $\delta\rho(\mathbf{r})$ for the same system (see, for example, Stevens 1980; Breitenstein *et al.* 1983; Swaminathan *et al.* 1984; Krijn and Feil 1988; Moeckli *et al.* 1988). Direct comparisons of this sort are not very common, however, largely because of the scarcity of systems that are amenable to accurate study both experimentally and theoretically. But the potential value of such a comparison is considerable. If the two $\delta\rho(\mathbf{r})$ maps agree in detail, we have strong support both for the experimental determination, including the X-ray measurements and the deformation model, and for the theoretical method with its inevitable approximations. Where they disagree, we can try to assess the relative contributions of experimental and theoretical errors and of genuine differences between, for example, the free molecule and the same molecule in a crystalline environment. The latter source of discrepancy may be particularly important in crystals with strong intermolecular interactions, e.g. hydrogen bonds. Several studies have attempted to account theoretically for the effects of such interactions on the molecular deformation density (Almlöf *et al.* 1973; Yamabe and Morokuma 1975; Kerns and Allen 1978; Johansen 1979; Stevens 1980; Hermansson and Lunell 1981, 1982; Breitenstein *et al.* 1983; Krijn and Feil 1988).

10.10   Alternative deformation maps

But how is the deformation density $\delta\rho(\mathbf{r})$ actually derived from the output of a deformation refinement? There are, in fact, several ways we can proceed, leading to alternative deformation maps for the same structure from the same X-ray data via the same least-squares refinement.

10.10.1   The difference density

Firstly, we can simply use the refined structural parameters to calculate structure factors for the promolecule model. Subtracting these from F_H^{obs}/k, as in eqn (10.7), will provide Fourier coefficients for evaluation of $\delta\rho^{obs}(\mathbf{r})$ in the same way as in the $X - X_{ho}$ and $X - N$ methods. However, this strategy offers important advantages over the $X - X_{ho}$ method. The use of a flexible deformation model has (or should have) eliminated the bias of the promolecule model without requiring an arbitrary selection of reflections attributable entirely to core scattering. Further, we can now use the calculated structure factors F_H^{calc} from the full deformation model to provide phases for F_H^{obs} that will, for a non-centrosymmetric crystal, be far more reliable than those calculated from the promolecule density (Thomas 1978). The procedure, then, is to calculate from the same structural parameters two sets of structure factors, $F_H^{calc} = F_H^{tot}$ corresponding to the total electron density (eqn (10.8)) and F_H^{pro} based on the promolecule density $\rho^{pro}(\mathbf{r})$ alone as in eqn (10.4). From the former we take only the phases, replacing the calculated amplitudes by F_H^{obs}/k, while the latter are subtracted from this result to yield the difference Fourier coefficients for insertion in eqn (10.2). The resulting $\delta\rho^{obs}(\mathbf{r})$ map will be as model-free as any we can deduce from the X-ray data, being dependent on the model only for the derivation of unbiased atomic positions and displacement parameters and, if the structure is non-centrosymmetric, for the assignment of phases to $\mathbf{F}_{\mathrm{H}}^{\mathrm{obs}}$. The price to be paid for this dubious advantage is that the difference map contains all the imperfection of the intensity data, in particular their experimental inaccuracies and their finite resolution.

10.10.2 *Model deformation densities, static or dynamic*

Alternatively, we can largely remedy these imperfections by calculating the deformation density directly from the model, simply inserting the refined expansion coefficients into eqn (10.9) for evaluation of $\delta\rho(\mathbf{r})$. In this way, we obtain a map from which the experimental error has been almost entirely filtered out by the smoothing process of the least-squares fit. Also, it depicts the model deformation density at infinite resolution — unless we have deliberately chosen to simulate the effect of series truncation by double Fourier inversion, i.e. by calculating $\mathbf{F}_{\mathrm{H}}^{\mathrm{def}}$ from the model deformation density (eqn (10.9)) via eqn (10.3) and then inserting these calculated coefficients, out to a limiting resolution H_{max}, into eqn (10.2) for evaluation of a filtered image of $\delta\rho(\mathbf{r})$. On the other hand, this model map suffers all the unknown deficiencies of our deformation model.

If we have adopted the convolution type of deformation model (see Section 10.9.1), we actually have a choice of two model deformation densities. Our refined parameters are then of two kinds. One set — the atomic coordinates and the deformation coefficients — represent the static deformation density via eqn (10.9), while the others — the atomic displacement parameters — describe the vibrational smearing in the convolution approximation. Hence one option is to use the first set alone, substituting in eqn (10.9) to model the static $\delta\rho^{\mathrm{stat}}(\mathbf{r})$. Alternatively, we can restore the vibrational smearing to produce a map of the dynamic deformation density $\delta\rho^{\mathrm{dyn}}(\mathbf{r})$, most readily by a process of double Fourier inversion (see above) in which the Debye–Waller temperature factors are included in the equation for $\mathbf{F}_{\mathrm{H}}^{\mathrm{def}}$ (Coppens and Stevens 1977).

With regard to the first of these options, much skepticism has been expressed about the possibility of deconvoluting the vibrational smearing from the experimentally observable dynamic density $\delta\rho^{\mathrm{dyn}}(\mathbf{r})$. It is undeniable that the form of the static deformation density $\delta\rho^{\mathrm{stat}}(\mathbf{r})$ so derived must depend crucially on the details of the deformation model, which is required to provide fine-scale information that is poorly represented, because of vibrational smearing and limited resolution, in the experimental data. Moreover, the vibrational convolution we have assumed is doubly questionable. The rigid-pseudoatom approximation is itself in doubt, and, further, we have no reason to suppose that eqns (10.8) and (10.9) define the most appropriate pseudoatom fragments for this approximation. Worse yet, even if this approximation should yield an acceptable fit to the observed dynamic density, and hence to the experimental structure amplitudes, this is no assurance that the implied separation into static density and vibrational smearing is valid. With all this uncertainty, it has been argued that we should content ourselves with mapping the dynamic deformation density $\delta\rho^{dyn}(\mathbf{r})$ and comparing this with a correspondingly smeared theoretical deformation density (Coppens and Stevens 1977; Stevens et $al.$ 1977).

Yet an experimental map of $\delta\rho^{stat}(\mathbf{r})$ is too valuable a prize to be lightly foregone. Such a map gives us a highly detailed and chemically informative picture of the electron distribution. In particular, this is the density that is most directly and critically comparable with a theoretically calculated deformation density. Because compact features in this map are appreciably sharper than those in the corresponding dynamic density $\delta\rho^{dyn}(\mathbf{r})$, it permits a much more crucial test of the agreement between theory and experiment, or between different experiments. So we have a strong incentive for trying to derive as reliable a static deformation density as we can. This will surely require unusually accurate, high-resolution X-ray data, together with a well proven deformation model. But if these conditions are met, and if our refinement satisfies all the tests we have devised for verifying its validity, we may begin to hope that the resulting static density $\delta\rho^{stat}(\mathbf{r})$ is as accurate as the least-squares covariance matrix implies. Until more experience has accumulated, however, only a detailed comparison with a theoretical deformation density is likely to establish how fully this hope is justified.

10.10.3   Extracting a molecular fragment

A major difference should be noted between any map derived by Fourier summation and one obtained from an expression such as eqn (10.9). The former preserves the full symmetry of the crystal, including its three-dimensional periodicity, while the latter is necessarily limited to a tiny fragment, depending on the number of terms included in the sum. If this sum comprises the deformation functions centered on the atoms of one molecule, we obtain what Coppens (1982) has called the 'pseudomolecule'. This is a largely imaginary fragment, defined by an arbitrary partitioning of the crystal according to our choice of basis functions ρ_{ak} in eqn (10.9). In particular, we have no experimental grounds for identifying the pseudo-molecule density with the electron distribution of an isolated molecule. In this respect a Fourier series, whether it depicts the difference density $\delta\rho^{obs}(\mathbf{r})$ or a static or dynamic model density, gives a more objective account of the situation in the crystal. However, when we attempt to esti-mate molecular dipole or quadrupole moments, etc., we cannot avoid the question of how to extract a single molecular entity from the continuous electron distribution in the crystal (Moss and Coppens 1980; Hirshfeld 1985). Unfortunately, this question does not seem to have a generally adequate answer.

10.11   Tetrafluoroterephthalonitrile

One of the best illustrations of the detailed agreement attainable between an experimental and a theoretical deformation density comes from recent work on tetrafluoroterephthalonitrile (or *p*-dicyanotetrafluorobenzene, **1**).

1

Several factors make this system particularly attractive for experimental electron-density mapping. It contains light atoms only, so that its core orbitals account for a small fraction of its electron population. Thus, the 'suitability factor' (Stevens and Coppens 1976) of crystalline tetrafluoroterephthalonitrile is

$$s = V / \sum (n_{core})^2 = 3.3 \text{ Å}^3,$$

a value high enough to permit quite precise mapping of the molecular deformation density. On the other hand hydrogen atoms are absent, sparing us the common difficulty of deducing hydrogen coordinates and displacement parameters from the X-ray data (Hirshfeld 1976). With no hydrogen bonds, or other strong local interactions between molecules, we expect minimal perturbation of the molecular charge distribution. Yet vibrational motion in the crystal (m.p. 472 K) is moderate and the molecules themselves are fairly rigid. The orthorhombic space group *Cmca* allows the measurement of general reflections in sets of eight symmetry equivalents. Also, the *mmm* (D_{2h}) molecular symmetry, being higher than the $2/m$ (C_{2h}) site symmetry, provides an opportunity for imposing symmetry constraints on a deformation model. Nevertheless, this high molecular symmetry does not preclude interesting chemical variety. While the asymmetric unit contains only two full atoms plus three half-atoms (in the crystallographic mirror plane), this simple molecule yet displays an extensive system of delocalized π orbitals, a benzene ring bearing two kinds of substituent, one linear and two trigonally hybridized carbon atoms, a C—F single bond, a C≡N triple bond, and three distinct carbon–carbon bonds, two of them aromatic, the other a conjugated single bond.

10.11.1 The experimental study

An unusually accurate set of X-ray data were measured at 98 K to a resolution $H_{max} = 2.3$ Å^{-1} (Dunitz *et al.* 1983) and processed with appropriate care (Seiler *et al.* 1984*b*). A high-order refinement against the 1406 measured reflections beyond $H_{min} = 1.7$ Å^{-1}, for which the goodness-of-fit S was 1.106, led to an $X - X_{ho}$ map showing most of the expected deformation-density features (see Chapter 7, Section 7.3). A deformation refinement (Hirshfeld 1984*a*) reduced S, for all 2387 independent reflections, from 4.764 to 1.140 and produced atomic coordinates and displacement parameters very close to those from the high-order refinement, satisfying the rigid-bond test within an r.m.s. deviation of 0.00013 Å^2. The resulting $\delta\rho^{stat}(\mathbf{r})$ map (Fig. 10.2) showed much the same peaks and valleys as the dynamic $X - X_{ho}$ map of Seiler *et al.* (1984*b*) (Fig. 7.14) plus much more sharply delineated deformation features around the fluorine nucleus. This map clearly reveals several compact density features associated with

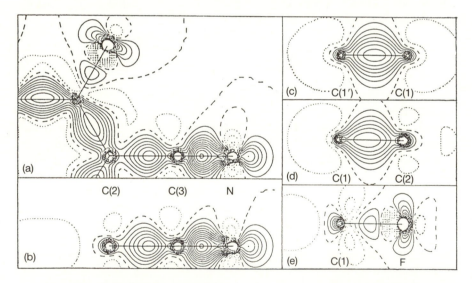

Fig. 10.2 Static deformation density $\delta\rho^{stat}(\mathbf{r})$ of tetrafluoroterephthalonitrile from deformation-model refinement against X-ray data of Seiler *et al.* (1984*b*). Contour interval 0.1 e Å^{-3}, terminated at 1.5 e Å^{-3}. From Hirshfeld (1984*a*). (a) Mean molecular plane; (b) Perpendicular section at $x = 0$ through C—C≡N group; (c) Perpendicular section through C(1')—C(1) bond; (d) Perpendicular section through C(1)—C(2) bond; (e) Perpendicular section through C(1)—F bond.

the orientation and hybridization of the fluorine valence orbitals as well as a very small peak in the $C-F$ bond. It also displays a subtle but consistent quinonoid pattern of variation in the extent of π bonding in the several bonds comprising the conjugated system.

10.11.2 Comparison with theory

For comparison, Delley (1986) calculated a theoretical deformation map of 1 by means of a local-density-functional approximation (Fig. 10.3). The degree of concordance between the experimental and theoretical deformation densities is quite remarkable. Detailed agreement is found both in the highly compact peaks and valleys very close to the fluorine nucleus and in the more diffuse regions of the outer π-electron distribution. The only significant discrepancy noted by Delley (1986) is in the lone-pair peaks on fluorine and nitrogen. These are so compact, in the theoretical map, that it is scarcely surprising that the deformation refinement has underestimated their peak densities and their sharpness. In fact, our deformation model

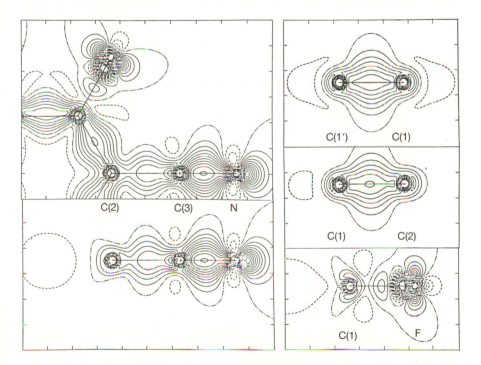

Fig. 10.3 Theoretical deformation density $\delta\rho(\mathbf{r})$ of tetrafluoroterephthalonitrile from local-density calculation. Same sections and contour interval as Fig. 10.2. Scale marks at intervals of 1 bohr = 0.5292 Å. From Delley (1986).

lacks sharp basis functions capable of describing such very compact features, but the experimental uncertainty in the nuclear regions is already so great, primarily because of large correlations between the structural and the deformation parameters, that a more flexible model would probably not produce much improvement. What is surprising is that the model has been as successful as it has in deconvoluting the vibrational smearing so as to reveal these compact features in the static deformation density. The peak densities in the bonds are slightly but systematically lower in the theoretical than in the experimental map. Also, the second moments of the partitioned atomic deformation densities (except the longitudinal moment on nitrogen) are generally larger in the theoretical calculation, most conspicuously in the out-of-plane direction (Delley 1986); this implies a greater contraction of the atomic charge clouds in the theoretical than in the experimental study. The present evidence does not allow us to identify the origin of these minor quantitative differences. However, the overall agreement between the X-ray and the local-density deformation densities is truly impressive and indicates that both must be of the highest accuracy.

The lesson for the X-ray crystallographer is that superior experimental data for a judiciously chosen crystal, combined with a properly flexible deformation model, *can* lead to a reliable, detailed map of the static deformation density. A few more such successful comparisons would help in establishing conditions so that an X-ray study might be trusted on its own to provide results of similar quality consistently and conclusively.

10.12 Acknowledgements

This chapter has benefited from the critical comments of several colleagues on successive earlier drafts. I am deeply grateful for the thoughtful suggestions of H.-B. Bürgi, J. D. Dunitz, M. Eisenstein, W. H. E. Schwarz, and K. N. Trueblood.

References

Almlöf, J., Kwick, Å., and Thomas, J. O. (1973). *J. Chem. Phys.*, **59**, 3901–6.

Baert, F., Coppens, P., Stevens, E. D., and Devos, L. (1982). *Acta Crystallogr.*, **A38**, 143–51.

Bats, J. W., Coppens, P., and Koetzle, T. F. (1977). *Acta Crystallogr.*, **B33**, 37–45.

Bentley, J. and Stewart, R. F. (1974). *Acta Crystallogr.*, **A30**, 60–7.

Bentley, J. and Stewart, R. F. (1975). *J. Chem. Phys.*, **63**, 3794–803.

Bentley, J. and Stewart, R. F. (1976). *Acta Crystallogr.*, **A32**, 910–4.

Bijvoet, J. M., Peerdeman, A. F., and van Bommel, A. J. (1951). *Nature*, **168**, 271–2.

Bingel, W. A. (1963). *Z. Naturforsch.*, **18a**, 1249–53.

Blessing, R. H. (1987). *Crystallogr. Rev.*, **1**, 3–58.

Breitenstein, M., Dannöhl, H., Meyer, H., Schweig, A., Seeger, R., Seeger, U., and Zittlau, W. (1983). *Int. Rev. Phys. Chem.*, **3**, 335–91.

Chandler, G. S. and Spackman, M. A. (1982). *Acta Crystallogr.*, **A38**, 225–39.

Chandler, G. S., Spackman, M. A., and Varghese, J. N. (1980). *Acta Crystallogr.*, **A36**, 657–69.

Coppens, P. (1967). *Science*, **158**, 1577–9.

Coppens, P. (1974). *Acta Crystallogr.*, **B30**, 255–61.

Coppens, P. (1982). Concepts of charge-density analysis: The experimental approach. In *Electron distributions and the chemical bond* (ed. P. Coppens and M. B. Hall), pp. 61–92. Plenum Press, New York.

Coppens, P. and Coulson, C. A. (1967). *Acta Crystallogr.*, **23**, 718–20.

Coppens, P. and Stevens, E. D. (1977). *Isr. J. Chem.*, **16**, 175–9.

Coppens, P., Sabine, T. M., Delaplane, R. G., and Ibers, J. A. (1969). *Acta Crystallogr.*, **B25**, 2451–8.

Coppens, P., Willoughby, T. V., and Csonka, L. N. (1971). *Acta Crystallogr.*, **A27**, 248–56.

Coppens, P., Boehme, R., Price, P. F., and Stevens, E. D. (1981). *Acta Crystallogr.*, **A37**, 857–63.

Coppens, P., Dam, J., Harkema, S., Feil, D., Feld, R., Lehmann, M. S., Goddard, R., Krüger, C., Hellner, E., Johansen, H., Larsen, F. K., Koetzle, T. F., McMullan, R. K., Maslen, E. N., and Stevens, E. D. (1984). *Acta Crystallogr.*, **A40**, 184–95.

Coster, D., Knol, K. S., and Prins, J. A. (1930). *Z. Phys.*, **63**, 345–69.

Coulson, C. A. and Thomas, M. W. (1971). *Acta Crystallogr.*, **B27**, 1354–9.

Dawson, B. (1964). *Acta Crystallogr.*, **17**, 990–6.

Dawson, B. (1967). *Proc. Roy. Soc.*, **A298**, 255–63 and 264–88.

Delley, B. (1986). *Chem. Phys.*, **110**, 329–38.

Dunitz, J. D., Schweizer, W. B., and Seiler, P. (1983). *Helv. Chim. Acta*, **66**, 123–33.

Eisenstein, M. (1979). *Acta Crystallogr.*, **B35**, 2614–25.

Epstein, J. and Stewart, R. F. (1979). *Acta Crystallogr.*, **A35**, 476–81.

Epstein, J., Ruble, J. R., and Craven, B. M. (1982). *Acta Crystallogr.*, **B38**, 140–9.

Eriksson, A., Hermansson, K., Lindgren, J., and Thomas, J. O. (1982). *Acta Crystallogr.*, **A38**, 138–42.

Hall, M. B. (1986). *Chem. Scripta*, **26**, 389–94.

Hamilton, W. C. (1964). *Statistics in physical science*, p. 128. Ronald Press, New York.

Hansen, N. K. and Coppens, P. (1978). *Acta Crystallogr.*, **A34**, 909–21.

Hauptman, H. (1982). *Acta Crystallogr.*, **A38**, 632–41.

Hellner, E. (1977). *Acta Crystallogr.*, **B33**, 3813–6.

Hermansson, K. and Lunell, S. (1981). *Chem. Phys. Lett.*, **80**, 64–8.

Hermansson, K. and Lunell, S. (1982). *Acta Crystallogr.*, **B38**, 2563–9.

Hirshfeld, F. L. (1971). *Acta Crystallogr.*, **B27**, 769–81.

Hirshfeld, F. L. (1976). *Acta Crystallogr.*, **A32**, 239–44.

Hirshfeld, F. L. (1977). *Isr. J. Chem.*, **16**, 226–9.

Hirshfeld, F. L. (1984*a*). *Acta Crystallogr.*, **B40**, 484–92.

Hirshfeld, F. L. (1984*b*). *Acta Crystallogr.*, **B40**, 613–5.

Hirshfeld, F. L. (1985). *J. Mol. Struct.*, **130**, 125–41.

Hope, H. and Ottersen, T. (1978). *Acta Crystallogr.*, **B34**, 3623–6.
Ito, T. and Shibuya, I. (1977). *Acta Crystallogr.*, **A33**, 71–4.
Jeffrey, G. A. and Cruickshank, D. W. J. (1953). *Quart. Rev.*, 7, 335–76.
Johansen, H. (1979). *Acta Crystallogr.*, **A35**, 319–25.
Jones, D. S., Pautler, D., and Coppens, P. (1972). *Acta Crystallogr.*, **A28**, 635–45.
Kerns, R. C. and Allen, L. C. (1978). *J. Am. Chem. Soc.*, **100**, 6587–94.
Krijn, M. P. C. M. and Feil, D. (1988). *J. Chem. Phys.*, **89**, 4199–4208 and 5787–93.
Kunze, K. L. and Hall, M. B. (1986). *J. Am. Chem. Soc.*, **108**, 5122–7.
Kurki-Suonio, K. (1968). *Acta Crystallogr.*, **A24**, 379–90.
Lonsdale, K. and El Sayed, K. (1965). *Acta Crystallogr.*, **19**, 487–8.
Martin, M., Rees, B., and Mitschler, A. (1982). *Acta Crystallogr.*, **B38**, 6–15.
Maslen, E. N. (1968). *Acta Crystallogr.*, **B24**, 1165–70.
Moeckli, P., Schwarzenbach, D., Bürgi, H.-B., Hauser, J., and Delley, B. (1988). *Acta Crystallogr.*, **B44**, 636–45.
Moss, G. and Coppens, P. (1980). *Chem. Phys. Lett.*, **75**, 298–302.
Mullen, D. and Hellner, E. (1977). *Acta Crystallogr.*, **B33**, 3816–22.
O'Connell, A. M. (1969). *Acta Crystallogr.*, **B25**, 1273–80.
Ottersen, T., Almlöf, J., and Hope, H. (1980). *Acta Crystallogr.*, **B36**, 1147–54.
Patterson, A. L. (1963). *Acta Crystallogr.*, **16**, 1255–6.
Pawley, G. S. and Willis, B. T. M. (1970). *Acta Crystallogr.*, **A26**, 260–2.
Rees, B. (1977). *Isr. J. Chem.*, **16**, 180–6.
Restori, R., Schwarzenbach, D., and Schneider, J. R. (1987). *Acta Crystallogr.*, **B43**, 251–7.
Rollett, J. S. (1970). Least-squares procedures in crystal structure analysis. In *Crystallographic computing* (ed. F. R. Ahmed), pp. 167–81. Munksgaard, Copenhagen.
Rossmann, M. G. (1961). *Acta Crystallogr.*, **14**, 383–8.
Rozendaal, A. and Baerends, E. J. (1986). *Acta Crystallogr.*, **B42**, 354–8.
Rozendaal, A. and Ros, P. (1982). *Acta Crystallogr.*, **A38**, 372–7.
Ruysink, A. F. J. and Vos, A. (1974). *Acta Crystallogr.*, **A30**, 503–6.
Savariault, J.-M. and Lehmann, M. S. (1980). *J. Am. Chem. Soc.*, **102**, 1298–303.
Scheringer, C. (1977). *Acta Crystallogr.*, **A33**, 426–9.
Scheringer, C. (1978). *Acta Crystallogr.*, **A34**, 905–8.
Scheringer, C. (1979). *Acta Crystallogr.*, **A35**, 838–44.
Scheringer, C., Kutoglu, A., and Mullen, D. (1978). *Acta Crystallogr.*, **A34**, 481–3.
Schwarz, W. H. E., Valtazanos, P., and Ruedenberg, K. (1985). *Theor. Chim. Acta*, **68**, 471–506.
Schwarzenbach, D. and Lewis, J. (1982). Refinement of charge-density models using constraints for electric field gradients at nuclear positions. In *Electron distributions and the chemical bond* (ed. P. Coppens and M. B. Hall), pp. 413–30. Plenum Press, New York.
Seiler, P., Martinoni, B., and Dunitz, J. D. (1984a). *Nature*, **309**, 435–8.
Seiler, P., Schweizer, W. B., and Dunitz, J. D. (1984b). *Acta Crystallogr.*, **B40**, 319–27.
Stenkamp, R. E. and Jensen, L. H. (1984). *Acta Crystallogr.*, **A40**, 251–4.
Stephens, M. E. and Becker, P. J. (1983). *Mol. Phys.*, **49**, 65–89.
Stevens, E. D. (1980). *Acta Crystallogr.*, **B36**, 1876–86.

Stevens, E. D. and Coppens, P. (1976). *Acta Crystallogr.*, **A32**, 915–7.

Stevens, E. D. and Hope, H. (1975). *Acta Crystallogr.*, **A31**, 494–8.

Stevens, E. D., Rys, J., and Coppens, P. (1977). *Acta Crystallogr.*, **A33**, 333–8.

Stewart, R. F. (1968). *J. Chem. Phys.*, **48**, 4882–9.

Stewart, R. F. (1969). *J. Chem. Phys.*, **51**, 4569–77.

Stewart, R. F. (1973). *J. Chem. Phys.*, **58**, 1668–76.

Stewart, R. F. (1976). *Acta Crystallogr.*, **A32**, 565–74.

Swaminathan, S., Craven, B. M., Spackman, M. A., and Stewart, R. F. (1984). *Acta Crystallogr.*, **B40**, 398–404.

Templeton, D. H. (1955). *Acta Crystallogr.*, **8**, 842.

Thomas, J. O. (1978). *Acta Crystallogr.*, **A34**, 819–23.

Thomas, J. O., Tellgren, R., and Almlöf, J. (1975). *Acta Crystallogr.*, **B31**, 1946–55.

Yamabe, S. and Morokuma, K. (1975). *J. Am. Chem. Soc.*, **97**, 4458–65.

11

Accurate crystal structure analysis by neutron diffraction

George A. Jeffrey

11.1 Introduction

Single-crystal neutron diffraction analysis is a long-established method of structure determination which is well documented in textbooks (Willis 1970; Brown and Forsyth 1973; Bacon 1975). Because a neutron-diffraction crystal structure analysis is a relatively expensive and time-consuming experiment, the method serves to complement the deficiencies of X-ray crystal structure analysis. The neutron scattering cross-sections of the atoms vary over a narrow range compared with the X-ray scattering factors, which increase with atomic numbers. Therefore, neutron diffraction is more effective for determining the atomic parameters of light atoms in molecules containing heavy atoms. This is particularly important in the study of hydrogen bonding, both in small molecules and in large biological macromolecules. It is also useful for distinguishing between atoms which are adjacent in the periodic table, where there is only a small proportional difference in the X-ray scattering factors. Since some atoms have magnetic neutron scattering factors, due to unpaired electrons, single-crystal neutron diffraction is also used extensively in the study of magnetic materials and phase transitions.

Because the nucleus is very small compared to the wavelength of the neutrons, the scattering power of the nucleus for neutrons does not fall off with scattering angle as do X-ray scattering factors. This makes single-crystal neutron diffraction a particularly powerful method for determining molecular structure with high precision, since the observation-to-parameter ratio can generally be well maintained.

11.2   Some comments on the molecular structure

The three-dimensional topology of a molecule, the so-called molecular structure, is completely described by a set of bond lengths, bond angles, and torsion angles. In a paper reporting the results of a crystal structure analysis, these quantities are given twice, as the molecular dimensions and in terms of fractional atomic coordinates with respect to the unit-cell dimensions. This duplication provides a useful check for typographical errors, which are relatively common in papers published prior to 1970. For this reason, the Cambridge Structural Database (see Chapter 15) is a more reliable source of crystal structural data for organic and organometallic molecules than the original papers, since the typographical errors have been detected, corrected when possible, and the discrepancies noted.

When the molecular structure is determined by single-crystal or powder neutron diffraction, the topology refers to the atomic coordinates of the nuclei. When the analysis is by means of X-ray diffraction, the structure is defined by the maxima in the electron-density distribution. With the exception of hydrogen atoms, it is only from very precise analyses that there are observable differences between these two molecular structures. Even then, the differences are only apparent for atoms such as oxygen where the lone-pair electron density is not symmetrical about the nucleus, or for atoms involved in multiple bonds. A systematic study of the differences in bond lengths from X-ray and neutron diffraction analyses of the same compound gave $r_X - r_N = -0.096(7)$ Å for C—H and $-0.155(10)$ Å for O—H bonds, whereas for the carboxyl C—OH bond the difference was $+0.0035(12)$ Å, and for C—O bonds in carbohydrates $+0.0054(8)$ Å (Allen 1986).

With hydrogen atoms, the difference is much greater. The hydrogen electron density is displaced away from the hydrogen nucleus in the direction of the covalently bonded atom. This results in a local dipole close to the nucleus. The more electronegative the bonded atom, the greater the displacement of the electron density. In this case, the discrepancy between the position of the hydrogen atoms from neutron and X-ray crystal structure analyses is very significant. For this reason it is common practice when deriving hydrogen bond lengths from X-ray analyses to normalize the X-ray

X—H distances to standard neutron-diffraction values, e.g. $r(O—H) =$ 0.97 Å, $r(N—H) = 1.00$ Å, $r(C—H) = 1.10$ Å (see Jeffrey and Lewis 1978; Taylor and Kennard 1983). The accuracy of locating the hydrogen atoms in X-ray analyses can be improved substantially by using this normalization procedure.

When the molecular dimensions are determined with high precision, i.e. with standard deviations less than 0.005 Å, it is important to recognize that the descriptor *molecular structure* has a different meaning for different investigators. For the theoretician calculating the molecular structure by *ab initio* MO methods, it is the 'nuclear' structure of the isolated molecule *at rest* (equilibrium structure, see Chapter 13), and so it is for the spectroscopist analyzing rotational and rovibrational spectra of very small molecules (see Chapter 4). For the gas-phase electron diffractionist, it is still the nuclear structure of the isolated molecule, but the internuclear distances are averaged over the vibrational motion (see Chapter 5). For the X-ray crystallographer, the structure is defined by the centroids of the electron-density distribution and refers to a molecule in an anisotropic crystal field, generally undergoing anisotropic thermal motion (see Chapters 6 and 8). The neutron crystallographer sees the same with respect to the nuclei. As regards the chemist, the molecule he is concerned with is most commonly a molecule in solution undergoing considerable isotropic motion in an isotropic solvent field.

At high levels of precision, *these results from different methods of investigation of molecular structure are not necessarily transferable*. Since X-ray crystal structure analysis has become a widespread analytical method in both chemistry and the biological sciences, the question of transferability of structural information from the crystal for use in the interpretation of chemical or biological problems is of central importance. There are circumstances where extreme caution has to be exercised in making chemical or biological inferences from the results of crystal structure analysis. In this chapter, we will give some examples where there is evidence that the intermolecular forces in the crystal have a profound effect on the molecular topology.

Of the three quantities which describe the structure of a molecule, the torsion angles are most deformable and the bond lengths least (see Chapter 19, Section 19.2). For instance, with cycloalkanes deforming a torsion angle by 3–4° from its equilibrium value requires about the same strain energy as changing a bond length by 0.01 Å or a valence angle by 1° (Dunitz and Waser 1972). However, as will be shown later, bond lengths are susceptible to shortening effects due to thermal motion, whereas bond angles and torsion angles are little changed.

In principle, the structure of a molecule in a crystal is *always* different from that of the molecule in solution. The question is, when are these differences large enough to be observable or chemically significant? This question is difficult to answer directly because the most important analytical method for determining molecular structure in solution, i.e. NMR spectroscopy, does not generally provide structural data of a precision comparable to that of the methods discussed in the present book. (This does not apply, of course, to NMR spectroscopy in a nematic mesophase (see Chapter 12).) It is only when there is a gross conformational difference that discrepancies between the results from proton coupling constants, for example, and X-ray crystallography are observed.

In this chapter, we will discuss how single-crystal neutron diffraction analysis can be applied to obtain highly accurate information on the geometry of small organic molecules. We will also see how this information compares with information obtained from other techniques of structure determination, and how it can be used to investigate the effect of crystal forces on molecular structures. The 'range' of the crystal diffraction methods is much greater than any other method of structure determination, going from small molecules of less than 100 daltons to very large molecules of several million daltons. High accuracy is difficult to achieve, however, at either end of the scale.

11.3 Neutron diffraction single-crystal structure analysis

11.3.1 Neutron vs. X-ray crystallography

Neutron single-crystal diffraction analysis has two major advantages over X-ray single-crystal analysis for molecular structure determination. One is that it determines the nuclear positions of the hydrogen atoms with a precision comparable to those of the non-hydrogen atoms. For small or medium-sized organic molecules, it therefore provides a more complete analysis. It also provides a complete description of the anisotropic thermal motion of the hydrogen atoms without which an accurate determination of bond length 'rest' values cannot be obtained. A second pragmatic advantage is that temperatures close to 10 K are easily attainable with the neutron diffraction facilities, whereas they are only rarely available on X-ray single-crystal diffractometers.

To offset these and other advantages shown in Table 11.1, there are a number of disadvantages, chief of which is having to schedule long experiments at a nuclear reactor facility and requiring large crystals, i.e. with dimensions of millimeters rather than hundredths of a millimeter. Because neutron diffraction analyses are more expensive and time-consuming than X-ray analyses, it is judicious to do as much preliminary X-ray structure analysis as possible, if these data are not already available in the literature. If the facilities are available for differential thermal calorimetry down to 10 K, this is useful to identify phase transitions that could destroy the

Table 11.1 Comparison of X-ray and neutron diffraction single-crystal analysis

X-ray diffraction	Neutron diffraction
— X rays available on demand from laboratory instruments	— Neutrons available only from national or international centers
— Data collection time, a few days (for routine work)	— Data collection time, a few weeks
— Temperatures below 120 K not generally available	— Temperatures down to 10 K conveniently available
— Hydrogen atoms poorly located, especially O—H. Accuracy ~ 0.1 Å	— Hydrogen positional parameters comparable in accuracy to C, N, and O, ~ 0.001 Å
— Cannot analyze anisotropic thermal motion or disorder of H atoms	— Analysis comparable to that of C, N, O
— Small crystals can be used (~ 0.01 mm^3, ~ 0.01 mg)	— Large crystals required (~ 4 mm^3, ~ 5 mg)
— Number of variable parameters $9N_X + 4N_H + 1$, where N_X is the number of non-hydrogen atoms, N_H the number of hydrogen atoms	— For comparable observation-to-parameter ratio, needs more observations. Number of variables $9N_T + 7$, where N_T is the total number of atoms and six anisotropic extinction parameters are used
— Not necessary to make deuterium substitution	— If the structure contains many hydrogens, deuterium substitution may be essential to reduce incoherent background
— Careful absorption corrections necessary	— Absorption negligible, except for crystals containing B, Cd, Sm, Li (corrections for H advisable for molecules with large H content)
— Extinction generally not serious for organic compounds	— Extinction serious and pervasive. Careful corrections necessary
— Radiation damage can occur and must be monitored by repeating selected measurements	— No radiation damage

crystals in the cooling process on the neutron diffractometer.

Selecting a single crystal which is suitable for a precise neutron diffraction analysis is not entirely predictable. Previous X-ray examination provides no guarantee. It is wise therefore to approach the nuclear reactor facility with a large number of potentially suitable crystals of several different compounds, to ensure that a valuable time-slot is not wasted. As Oliver Goldsmith (1766) remarked concerning wives, they should not be chosen 'for a fine glossy surface, but for such qualities as would wear (diffract) well.'

The neutron diffractometer facilities are fully equipped with the necessary software for data reduction, crystal structure determination and refinement. Accurate absorption corrections are advisable for molecules containing hydrogen, and are even more important if the major absorbing atoms shown in Table 11.1 are present in the molecule. Extinction corrections are much more important than for X-ray analyses. Extinction (see Chapter 7, Section 7.2.3) generally affects nearly all the diffraction intensities. Software based on the Becker and Coppens (1974, 1975) method is effective and most generally used. If the isotropic extinction model is adequate, it is preferred to the anisotropic model, which increases the number of variable parameters. For high-precision work, an observation-to-parameter ratio of better than 10:1 is desirable. Therefore monochromators that provide the shorter wavelengths are preferable, e.g. 1.0499 Å from a Be (002) monochromator. Some representative values of crystal sizes, effects of extinction, observation-to-parameter ratios, weighted agreement factors, and standard deviations of bond lengths are given in Table 11.2.

11.3.2   Thermal-motion corrections

For precision molecular structure determination, an analysis of the thermal motion of the individual atoms is essential so that appropriate corrections can be made to the bond lengths. Thermal-motion corrections to valence angles and torsion angles are an order of magnitude smaller and can generally be ignored. The theory of thermal-motion analysis is well developed (see Chapters 8 and 9; see also Johnson 1970a, b; Johnson and Levy 1974; Willis and Pryor 1975), and the necessary software is available (Johnson 1970b).

If the assumption of harmonic motion can be made, the analysis is often straightforward (Schomaker and Trueblood 1968). Without the assumption of harmonic motion, it becomes more complex. For this reason, the ability to carry out the crystal structure analysis at 10–20 K, where the thermal motion is closer to the zero-point value, is a significant advantage. At that temperature, the assumption that the atoms are undergoing harmonic motion is good, except for the hydrogen atoms. This is particularly important for small organic molecules which are liquids at room temperature. It is only in the more demanding area of experimental charge-density and

Table 11.2 Extinction and refinement data for some accurate crystal structure analyses by neutron diffraction[a]

| Compound, formula | Crystal volume (mm³) | Extinction | | Observation-to-parameter ratio | R_{wF}[c] | Standard deviations for representative bond lengths (Å) | | References |
		Reflections having $F^{obs}/F^{calc} \leq 0.95$	Degree of anisotropy[b]			$\sigma(C-N)$	$\sigma(N-H)$	
α-Cyanoacetohydrazide, C₃H₅N₃O	6.0	30%	3	15.8	0.042	0.0006	0.0010	Nanni et al. (1986)
Carbonohydrazide, CH₆N₄O	7.1	33%	7	14.8	0.046	0.0008	0.0010	Jeffrey et al. (1985a)
1,2,4-Triazole, C₂H₃N₃	3.5	2%	–[d]	16.8	0.050	0.0007	0.0014	Jeffrey et al. (1983)
Formamide oxime, CH₄N₂O	6.6	5%	2	15.6	0.023	0.0004	0.0008	Jeffrey et al. (1981a)
Monofluoroacetamide, C₂H₄FNO	4.8	22%	1	15.6	0.025	0.0004	0.0008	Jeffrey et al. (1981b)
Thioacetamide, C₂H₅NS	2.0	14%	43	12.6	0.047	0.0008	0.0014	Jeffrey et al. (1984)
1,2-Diformohydrazide, C₂H₄N₂O₂	3.3	1%	–[d]	15.1	0.024	0.0004	0.0007	Jeffrey et al. (1982)

[a] All studies have been carried out at 15–20 K.
[b] Highest ratio between principal axes of the anisotropic extinction ellipsoid.
[c] Defined as $R_{wF} = [\sum_H w_H (F_H^{obs} - F_H^{calc})^2 / \sum_H w_H (F_H^{obs})^2]^{1/2}$.
[d] One isotropic extinction parameter refined.

electrostatic potential analysis that the more sophisticated methods are necessary for analyzing the thermal motion. These methods, which employ the more complex Gram Charlier or Edgeworth expansions (see Johnson 1970b; Johnson and Levy 1974), are useful for describing the nuclear scattering density when it is curvilinear as, for example, in hindered-rotor CH_3 and NH_3^+ groups. These refinements are made at a heavy cost in the observation-to-parameter ratio. The third order c^{ijk} terms of the Gram Charlier expansion add ten more parameters per atom to the six U^{ij} terms. This type of refinement should only be used when there is an exceptionally high observation-to-parameter ratio and exceptionally precise intensity measurements, cf. the crystal structure of hexagonal ice I_h (Kuhs and Lehmann 1987).

11.3.2.1 The magnitude of the thermal-motion corrections The results of recent single-crystal neutron diffraction analyses of perdeuterated benzene (Jeffrey *et al.* 1987) and benzamide (Gao *et al.* 1991) at 15 and 123 K provide a measure of the magnitude of the thermal-motion corrections for the bonds not involving hydrogens (Table 11.3). At 15 K, the corrections are of the same order as the standard deviations, whereas at 123 K they can be five times greater. The corrected 'rest' values should be the same from both temperatures. The results in Table 11.3 suggest that the higher-temperature values are slightly overcorrected.

If the molecule is a single ring or a condensed ring system, the assumption that the thermal motion of the ring atoms is that of a rigid body is often valid and rigid-motion software can be used in the analysis. If the molecule is acyclic or consists of rings linked by acyclic bonds, a segmented-body analysis (Johnson 1970b) has to be used. A 'rule of thumb' test for rigid-body behavior is to compare the components of the thermal-motion tensor for pairs of atoms along the line joining them, or those of individual atoms normal to the line to the center of mass of the molecule. The former should be equal for each pair of atoms (see Chapter 8, Section 8.3), while the latter should increase proportionally to the distance from the center of mass of the molecule.

11.3.2.2 Thermal-motion corrections for hydrogen atoms The thermal-motion correction for bonds involving hydrogen atoms presents a different problem for two reasons. One is that the amplitude of motion of the hydrogen atoms is approximately 1.5 times greater than that of the atoms to which they are bonded, i.e. their probability ellipsoids are approximately twice as large, as shown in Fig. 11.1. The second problem is that the motion in the direction of the $X-H$ bond is anharmonic. The effects on the $X-H$ bond length of the wagging and anharmonic stretch motions are opposite in direction. The bond length correction, Δr, commonly used for the wagging or riding motion is (Busing and Levy 1964):

Table 11.3 Magnitude of thermal-motion corrections for C—C, C—N, and C=O bond lengths from neutron diffraction crystal structure analysis[a]

Bond	15 K			123 K		
	Standard deviation	Thermal-motion correction	Corrected bond length	Standard deviation	Thermal-motion correction	Corrected bond length
Deuterated benzene[b]						
C(1)—C(2)	0.0007	0.0008	1.3977	0.0011	0.0045	1.3988
C(2)—C(3)	0.0008	0.0007	1.3977	0.0011	0.0041	1.3995
C(1)—C(3')	0.0007	0.0009	1.3985	0.0011	0.0051	1.3973
		mean value	1.3980		mean value	1.3985
Benzamide, C_6H_5—$CONH_2$[c]						
C(1)—C(2)	0.0009	0.0012	1.4010	0.0013	0.0038	1.4021
C(2)—C(3)	0.0009	0.0005[d]	1.3953	0.0014	0.0014[d]	1.3947
C(3)—C(4)	0.0009	0.0011	1.3972	0.0013	0.0034	1.3970
C(4)—C(5)	0.0009	0.0011	1.3974	0.0014	0.0038	1.3990
C(5)—C(6)	0.0009	0.0005[d]	1.3927	0.0013	0.0014[d]	1.3920
C(6)—C(1)	0.0009	0.0012	1.4007	0.0012	0.0037	1.4008
		mean value	1.3974		mean value	1.3976
C(1)—C(7)	0.0009	0.0005[d]	1.4987	0.0012	0.0016[d]	1.4982
C(7)=O	0.0010	0.0010	1.2453	0.0014	0.0034	1.2462
C(7)—N	0.0008	0.0012	1.3400	0.0011	0.0035	1.3410

[a] All values are given in Å.
[b] Jeffrey et al. (1987).
[c] Gao et al. (1991).
[d] Note the corrections are small for the exocyclic bond C(1)—C(7) and ring bonds parallel to C(1)—C(7).

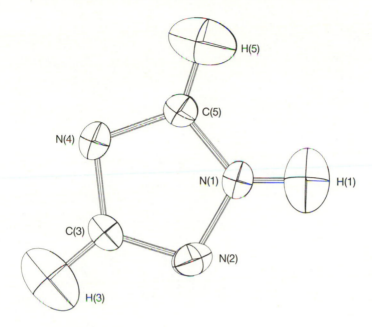

Fig. 11.1 Thermal ellipsoids at 99% probability (Johnson 1976) for 1,2,4-triazole from single-crystal neutron diffraction analysis at 15 K. From Jeffrey *et al.* (1983).

$$\Delta r = + (\langle u_H^2 \rangle_\perp - \langle u_X^2 \rangle_\perp)/2r(X-H).$$

For the anharmonic stretching motion, it is (see Kuchitsu and Bartell 1961):

$$\Delta r = - (3a/2)(\langle u_H^2 \rangle_\parallel - \langle u_X^2 \rangle_\parallel).$$

Here $\langle u_H^2 \rangle_\perp$ and $\langle u_X^2 \rangle_\perp$, $\langle u_H^2 \rangle_\parallel$ and $\langle u_X^2 \rangle_\parallel$, are the mean-square displacement amplitudes of the H and X atoms perpendicular to and parallel to the X—H bond, respectively. These quantities can be calculated from the experimental anisotropic displacement parameters. For C—H bonds, a value of $a = 1.98 \text{ Å}^{-1}$ taken from spectroscopic work (Kuchitsu and Morino 1965) has been used (see Craven and Swaminathan 1984; Jeffrey and Ruble 1984). For deuterated benzene, this gave a 'rest' value for the C—D bond lengths that agreed with the *ab initio* MO value calculated at the MP2/6–31G* level, within the experimental standard deviation, as shown in Table 11.4.

The compensatory effects of the riding and anharmonicity corrections are shown by the other examples given in Table 11.4. Part of the residual discrepancy with the theoretical values is due to the tendency for *ab initio* MO calculations at the HF/3–21G level to underestimate X—H bond lengths. In thioacetamide, where the methyl group is undergoing libration about the

Table 11.4 Riding (or segmented-body) motion and anharmonicity corrections for C—H and N—H bond lengths in some neutron diffraction analyses at 15 K[a]

Bond	Uncorrected X—H bond length	Corrections		Corrected bond length	Ab initio MO value[b]
		Riding or segmented-body motion	Anharmonicity		
Deuterated benzene[c]					
C(1)—D(1)	1.0879(9)	+0.0130	−0.0098	1.0911	–
C(2)—D(2)	1.0869(9)	+0.0130	−0.0119	1.0880	–
C(3)—D(3)	1.0843(8)	+0.0127	−0.0113	1.0857	–
			mean value	1.0883	1.087
Thioacetamide[d,e]					
N(1)—H(1)	1.0256(14)	+0.013	−0.020	1.019	0.999
	1.0263(14)	+0.016	−0.017	1.025	
N(1)—H(2)	1.0246(15)	+0.015	−0.019	1.021	1.000
	1.0231(14)	+0.016	−0.017	1.022	
C(2)—H(3)	1.0903(17)	+0.024	−0.017	1.097	1.083
	1.0880(17)	+0.030	−0.018	1.100	1.084
C(2)—H(4)	1.0903(16)	+0.026	−0.015	1.101	1.082
	1.0855(16)	+0.031	−0.020	1.097	1.084
C(2)—H(5)	1.0887(16)	+0.025	−0.017	1.097	1.082
	1.0832(16)	+0.031	−0.016	1.098	1.079
1,2,4-Triazole[f]					
N(1)—H(1)	1.0478(14)	+0.003	−0.017	1.033	0.993
C(3)—H(3)	1.0855(16)	+0.002	−0.016	1.072	1.061
C(5)—H(5)	1.0858(16)	+0.002	−0.015	1.073	1.063

[a] All values are given in Å.
[b] MP2/6–31G* for deuterated benzene; HF/3–21G for thioacetamide and 1,2,4-triazole.
[c] Jeffrey et al. (1987).
[d] Jeffrey et al. (1984).
[e] First line, molecule A; second line, molecule B.
[f] Jeffrey et al. (1983).

C—C bond, the C—H libration corrections are larger than for the hydrogen-bonded N—H bonds. However, in 1,2,4-triazole, where both C—H and N—H bonds are directly attached to the ring, there is no distinction between the N—H and C—H bond length corrections.

If possible, the neutron diffraction data should be collected at two temperatures, $\leq 15\,K$ and $\sim 120\,K$. There are two reasons for this. One is that it provides a check on the reliability of the thermal-motion corrections discussed above. Abnormal discrepancies in the corrected values are a signal that the thermal-motion parameters may also include some disorder. The second is that the 120 K data provide nuclear atomic parameters for any subsequent deformation density or electrostatic potential experiments where the data are generally collected using X-ray diffraction at liquid nitrogen temperature.

As shown in Table 11.3, the data on both deuterated benzene and undeuterated benzamide gave differences between the two sets of corrected values of the same order as the standard deviations for the 123 K data. The differences for the corrected C—D bond lengths ranged from 0.0004 to 0.0036 Å, as compared with the 123 K standard deviations of 0.0013 Å (Jeffrey *et al.* 1987).

11.4 Accurate structural studies of small organic molecules; the role of neutron crystallography

11.4.1 *What happens to molecules when they go from the solution to the crystal?*

A most interesting case is where the symmetry of the molecule in an isotropic solvent environment, or isolated in the gas phase, is higher than the point-group symmetry of the site it occupies in the crystal. In this circumstance, *the molecule is always distorted* in the crystal. The question is, are these distortions large enough to be experimentally observable?

Benzene is the classical example. The symmetry of the isolated molecule is D_{6h}. In the orthorhombic crystal, space group *Pbca* with $Z=4$, the crystallographic molecular symmetry is C_i. The molecules have non-crystallographic D_{3d} symmetry which is that of a chair conformation. In the latest neutron diffraction study at 15 K (Jeffrey *et al.* 1987), the distortion of the carbon ring to a chair conformation is barely significant at 5σ. But as shown in Table 11.5, the distortion of the hexagon of deuterium atoms to a chair conformation is highly significant at 20σ. Clearly the orthorhombic crystal field exerts a shear on the molecule, which significantly distorts the more easily deformable C—D bonds.

With molecules which do not possess high intrinsic symmetry, comparison between the molecular structure in the crystal and in solution is rarely

Table 11.5 Deviations from planarity in the crystal molecule of deuterated benzene at 15 K.[a] After Jeffrey *et al.* (1987)

Plane 1[b]		Plane 2[c]	
C(1), C(1′)	+, −0.0013(7)	C(1), C(1′)	0.0000
C(2), C(2′)	−, +0.0012(6)	C(2), C(2′)	0.0000
C(3), C(3′)	+, −0.0013(7)	C(3), C(3′)	+, −0.0039(8)
Plane 3[d]		Plane 4[e]	
D(1), D(1′)	+, −0.0080(10)	D(1), D(1′)	0.0000
D(2), D(2′)	−, +0.0072(9)	D(2), D(2′)	0.0000
D(3), D(3′)	+, −0.0064(9)	D(3), D(3′)	+, −0.0215(11)

[a] All values are given in Å. Primed and unprimed atoms are related by a crystallographic center of inversion.
[b] Through the six C atoms.
[c] Through C(1), C(1′), C(2), C(2′).
[d] Through the six D atoms.
[e] Through D(1), D(1′), D(2), D(2′).

possible. What is possible is comparison between the structure in the crystal and that calculated for the isolated molecule at rest by *ab initio* MO methods. When used at a sufficiently high level of approximation, these methods (to be described in Chapter 13) provide a standard for comparison with experimental measurements. They also provide a starting point for dynamical calculations aimed at exploring other energetically favorable conformations or the effect of solvation on the structure of molecules in solution.

From the experimentalist's point of view, a *sufficiently high level of approximation* is one where the difference in molecular dimensions from the preceding level of theoretical approximation is less than the standard deviation of the experimental measurement with which it is being compared. For high-precision single-crystal low-temperature neutron diffractometry, this is less than 0.001 Å in bond lengths and 0.05° in angles. As shown by the results in Table 11.6, this criterion requires calculations at the MP3/6–31G* level and higher. Even with modern supercomputers, this is not possible for molecules containing more than a few first-row non-hydrogen atoms.

This problem can be overcome to a certain extent by use of the 'offset' method (see Chapter 13, Section 13.4). This method makes the assumption that errors due to basis-set deficiency and electron correlation are atom-pair properties that are transferable between molecules containing the same atom pairs. The bond lengths obtained by a complete geometry optimization at the HF/3–21G level can then be corrected for basis-set deficiency and electron correlation by using results from the theoretical calculations at higher levels on simpler molecules. Comparisons between the theory and neutron diffraction analyses have justified this method, at least as an interim

Table 11.6 Effect of basis set and electron correlation on some calculated equilibrium bond lengths. After Hehre et al. (1986)

Bond	Calculated bond lengths[a]						Spread of values
	STO-3G	HF/3-21G	HF/6-31G*	MP2/6-31G*	MP3/6-31G*	CID/6-31G*	
$C-C$ in C_2H_6	1.538	1.542	1.527	1.527	–	–	0.015
$C=C$ in C_2H_4	1.306	1.315	1.317	1.336	1.334	1.328	0.030
$C\equiv C$ in C_2H_2	1.168	1.188	1.185	1.218	1.206	1.202	0.050
$C-N$ in CH_3NH_2	1.486	1.471	1.453	1.465	1.466	1.460	0.033
$C=N$ in CH_2NH	1.273	1.256	1.250	1.282	1.275	1.268	0.032
$C-O$ in CH_3OH	1.433	1.441	1.400	1.424	1.421	1.415	0.041
$C=O$ in H_2CO	1.217	1.207	1.184	1.221	1.210	1.205	0.037
$C-H$ in C_2H_6	1.086	1.084	1.086	1.094	–	–	0.010
in C_2H_4	1.082	1.074	1.076	1.085	1.086	1.084	0.012
in C_2H_2	1.065	1.051	1.057	1.066	1.066	1.065	0.015
in CH_3NH_2[b]	1.091	1.087	1.088	1.096	1.097	1.095	0.010
in CH_3OH[b]	1.094	1.082	1.084	1.094	1.095	1.092	0.013
$O-H$ in CH_3OH	0.991	0.966	0.946	0.970	0.967	0.963	0.045

[a] All values are given in Å.
[b] Mean values.

solution while we wait for more powerful computers.

Comparisons of results obtained by neutron crystallography and *ab initio* MO calculations have shown that for molecules containing strong hydrogen-bonding functional groups, the crystal-field effects can lead to major conformational differences between the isolated molecule and those in the crystal.

An example is α-cyanoacetohydrazide, $N\equiv C-CH_2-CO-NH-NH_2$ (Nanni *et al.* 1986). In the crystal, the observed conformation is the eighth in conformational stability, calculated at the HF/3–21G level. It is also the conformer with the largest calculated dipole moment. The calculated energy difference (30.3 kJ mol^{-1} at the HF/3–21G level, 36.6 kJ mol^{-1} at the HF/6–31G* level) is large enough that higher-level calculations are very unlikely to bring the conformer observed in the crystal to the lowest energy for the isolated molecule. Geometries have been calculated for the various conformers; the largest calculated change in bond lengths with conformation is for the C—CO and N—N bonds, of 0.019 Å. This is an order of magnitude greater than the experimental standard deviations or the thermal-motion corrections.

Another example of a conformational difference between the predicted isolated molecule and that in the crystal is provided by 1,2-diformohydrazide, OHC—NH—NH—CHO (Jeffrey *et al.* 1982). In the crystal, space group $P2_1/c$ with two centrosymmetric molecules in the unit cell, the molecules are planar with approximate C_{2h} symmetry (conformation **1**). *Ab initio* MO calculations at the HF/3–21G level predict that this conformer is 5.4 kJ mol^{-1} higher in energy than the non-planar conformer **2** with C_2 symmetry.

In hydrazine, H_2N-NH_2, the non-planarity is ascribed to the antibonding interaction of the π electrons in the planar conformation. In the formyl derivative, the formyl groups delocalize these electrons so that the planar and non-planar conformers are closer in energy. In addition, the planar conformer is stabilized by the stronger intramolecular $N-H\cdots O=C$ hydrogen bond: the $H\cdots O$ distance is 2.28 Å *vs.* 3.01 Å for the non-planar conformer. (In the crystal, this interaction is part of a cooperative network of hydrogen bonds which extends throughout the crystal structure.) As a consequence of both these effects, the energy difference between the non-planar and planar conformations is much lower in diformohydrazide than the 108 kJ mol^{-1} calculated for hydrazine at the HF/3–21G level.

In both diformohydrazide and α-cyanoacetohydrazide, the molecules in the crystal are linked by strong hydrogen bonds, coupled with strong polar interactions in the latter structure. These provide examples where hydrogen bonding or polar functionality might interfere with the transferability of structural information from the crystalline state or from *ab initio* MO calculations to the population of conformers in solution.

It is important to realize that under certain circumstances, the theoretical calculations will give results which have little relevance to the experimental measurements. Such an example is one of the simplest biologically interesting molecules, α-D-glucopyranose (**3**).

3

A theoretical calculation of the energy-optimized structure of α-D-glucopyranose will lead to a molecule with the maximum intramolecular hydrogen bonding (see, for example, Kroon-Batenburg and Kanters 1983). Each hydroxyl group will be oriented so as to form an intramolecular hydrogen bond to the oxygen atom on the adjacent carbon atom around the pyranose ring to form a finite chain of hydrogen bonds around the periphery of the sugar molecule. This is the hypothetical structure of α-D-glucopyranose in the gas phase, which does not exist due to decomposition above the melting point. It might also be the hypothetical structure of α-D-glucopyranose in a non-polar solvent in which it does not dissolve.

In order to make a theoretical calculation for comparison with the molecule in the crystal or in a polar solvent, it is necessary to prevent the intramolecular hydrogen bonding. This can be done by fixing the C—C—O—H torsion angles in an orientation so that the intramolecular hydrogen-bonding energy is a minimum. Otherwise, the intramolecular

Table 11.7 Effect of hydrogen bonding on covalent bond lengths

(a) Carboxylic acids

Compound	Technique[a]	Differences in bond lengths[b] (Å)			References
		$\Delta(C{=}O)$	$\Delta(C{-}OH)$	$\Delta(O{-}H)$	
Isolated dimer *vs.* monomer					
Formic acid	ED *vs.* ED	+0.003	−0.038	+0.052	Almenningen *et al.* (1969)
Acetic acid	ED *vs.* ED	+0.017	−0.030	–	Derissen (1971*a*)
Propionic acid	ED *vs.* ED	+0.021	−0.038	–	Derissen (1971*b*)
Crystal *vs.* monomer					
Acetic acid	ND[c,d] *vs.* MO[e]	+0.013	−0.029	+0.045	⎰ ND: Jönsson (1971) ⎱ MO: Jeffrey *et al.* (1985*b*)
Succinic acid	ND[d,f] *vs.* MO[e]	+0.031	−0.034	+0.036	ND: Leviel *et al.* (1981)

(b) Amides

Compound	Technique[a]	Differences in bond lengths[b] (Å)			References
		$\Delta(C{=}O)$	$\Delta(C{-}N)$	$\Delta(N{-}H)$	
Isolated dimer vs. monomer					
Formamide	MO[g] vs. MO[g]	+0.018	−0.023	+0.018	Jeffrey et al. (1981b)
Crystal vs. monomer					
Formamide	XD[h] vs. ED[i]	+0.027	−0.042	—	XD : Stevens (1978)
Formamide	XD[h] vs. MW[j]	+0.020	−0.026	—	ED : Kitano and Kuchitsu (1974)
Formamide	XD[h] vs. MO[g]	+0.027	−0.027	—	MW: Hirota et al. (1974)
					MO: Jeffrey et al. (1981b)
Acetamide	ND[d,k] vs. MO[g]	+0.034	−0.021	+0.038	Jeffrey et al. (1980)
Monofluoroacetamide	ND[d,l] vs. MO[g]	+0.030	−0.015	+0.040	Jeffrey et al. (1981b)

[a] ED, electron diffraction; MW, microwave spectroscopy; MO, ab initio MO calculations; XD, X-ray crystallography; ND, neutron crystallography.
[b] Dimer (or crystal) – monomer.
[c] At 133 K.
[d] Bond lengths are corrected for thermal-motion effects.
[e] The calculations are at HF/3-21G level and refer to acetic acid monomer.
[f] At 77 K.
[g] At HF/3-21G level.
[h] At 90 K (high-order refinement; bond lengths are not corrected for thermal-motion effects).
[i] r_g bond lengths.
[j] r_s bond lengths.
[k] At 23 K.
[l] At 20 K.

hydrogen bonding will modify the ring conformation, which in turn affects the bond lengths and valence angles giving a calculated geometry which is quite inappropriate for the molecule in the crystal or in solution.

11.4.2 Effect of intermolecular hydrogen bonding on covalent bond lengths

For very strong hydrogen bonds (Emsley 1980), such as occur in *ionic* complexes and crystal structures (e.g. $F-H\cdots F^-$, the pseudo hydrates, and the carboxylate hydrogen ions), hydrogen-bond formation profoundly affects the structure of the molecular and ionic species involved. For *molecular* structures, the effects are much smaller. The classical example of the effect of hydrogen bonding on covalent bond lengths is the lengthening of the $C=O$ bond and shortening of the $C-OH$ bond which is observed on formation of the carboxylic dimer in the gas phase (Almenningen *et al.* 1969; Derissen 1971*a*, *b*; see Table 11.7(a)). This is also observed in the crystal structures of carboxylic acids (see again Table 11.7(a)).

Another well-established example is the lengthening of the $C=O$ bond and shortening of the $C-N$ bond which occurs with $N-H\cdots O=C$ hydrogen bonding in the crystal structures of formamide, acetamide, and monofluoroacetamide. The dimers of these molecules are not sufficiently stable to be studied in the gas phase, but dimerization occurs as part of a hydrogen-bond network in the crystal. These results can then be compared with the electron diffraction and microwave spectroscopy data for the formamide monomer and theoretical calculations for the acetamide and monofluoroacetamide monomers in Table 11.7(b). The effect of the hydrogen bonding on the lengths of the covalent bonds is small, less than 0.05 Å, but well established, notwithstanding the different physical meaning of the quantities that are compared.

Yet another interesting case is that of α-cyanoacetohydrazide (Nanni *et al.* 1986), a molecule that we have already considered in Section 11.4.1. Experimental and calculated (HF/6-31G* level) bond lengths for the conformer that is found in the crystal are shown in Table 11.8. The differences between the experimental and calculated lengths for the $C(1)-N(1)$ bond, -0.021 Å, and $C(1)=O$ bond, $+0.049$ Å, are—at least in part —a consequence of the intermolecular $N-H\cdots O=C$ hydrogen bonding in the crystal. Calculations at HF/3-21G level for the formamide monomer and hydrogen-bonded dimer predicted comparable changes due to $N-H\cdots O=C$ hydrogen bonding (see Table 11.7(b)). There is also a significant lengthening of the $C\equiv N$ triple bond, $+0.020$ Å, which can be associated with its function as a hydrogen-bond acceptor in the crystal (there is evidence of $C-H\cdots N\equiv C$ hydrogen bonding, with $H\cdots N$ distances of about 2.50 Å). The experimental, thermally corrected $N-H$ and $C-H$

Table 11.8 Experimental and calculated bond lengths (Å) for α-cyanoacetohydrazide at 15 K. After Nanni *et al.* (1986)

Bond[a]	Experimental bond length		Calculated value[b]
	Uncorrected	Corrected to 'rest' values	
C(1)=O	1.2318(7)	1.235	1.186
C(1)—N(1)	1.3372(6)	1.339	1.360
C(1)—C(2)	1.5260(7)	1.528	1.533
N(1)—N(2)	1.4097(5)	1.411	1.381
C(2)—C(3)	1.4531(7)	1.455	1.466
C(3)≡N(3)	1.1531(6)	1.154	1.134
N(1)—H(1)	1.028(1)	1.023	0.996
N(2)—H(2)	1.017(1)	1.025	0.999
N(2)—H(3)	1.015(1)	1.022	0.999
C(2)—H(4)	1.095(1)	1.103	1.085
C(2)—H(5)	1.093(1)	1.097	1.085

[a] Atoms in the molecule are numbered as follows:
[b] *Ab initio* MO calculations at HF/6–31G* level for the conformation observed in the crystal.

bond lengths are ∼0.025 Å and ∼0.015 Å longer than the respective calculated values. This is qualitatively consistent with their function as hydrogen-bond donors in the crystal.

The shortening of the C—N bond in hydrogen-bonded amides is particularly important in relation to the effect of hydrogen bonding on the rigidity of the α-helices and β-sheets in proteins. Not only are these structures made more rigid by reason of the hydrogen bonding *per se*, but the hydrogen-bond formation indirectly increases the rigidity of the peptide link by reason of the additional π-bond character in the C—N bond. For other types of hydrogen bond the effects are predicted to be smaller. The rather strong N—H· · ·N hydrogen bond that occurs in crystalline imidazole has, however, an appreciable effect on the geometry of the molecule (Bencivenni *et al.* 1990).

11.4.3   Intrinsic structural features vs. crystal-field effects

Accurate experimental observations and *ab initio* MO calculations can be reinforcing for subtle features of molecular geometry. The study of π-bond anisotropy is an example where subtle features of molecular geometry observed by neutron crystallography are substantiated by the fact that the same features are predicted for the isolated molecule at rest. Deviations from π-bond planarity (so-called pyramidalization) will always occur when the mirror symmetry of the molecule is not a crystallographic mirror symmetry. Therefore, one crystal structure result alone does not distinguish between an intrinsic property of the molecule and a distortion due to the crystal field. If the deviations observed in the crystal are reproduced in magnitude and direction by the theoretical calculations for the isolated molecule, this provides good support for the hypothesis that the observation is a molecular property and not a crystal-field effect. This distinction could be made in a study of the pyramidalization of carbonyl carbons in asymmetrical structures of carboxylates, amides, and amino acids (Jeffrey *et al*. 1985*b*). In the case of the amino acid crystal structures, a statistical survey of a number of neutron diffraction studies showed a definite bias in the direction predicted by the theoretical calculations.

In the crystal structures of thioacetamide (Jeffrey *et al*. 1984), acetamide (Jeffrey *et al*. 1980), and deuterated nitromethane (Jeffrey *et al*. 1985*c*), all determined by low-temperature neutron crystallography, the methyl group is oriented with one C—H bond normal or nearly normal to the plane of the (thio)carbonyl or of the nitrogroup, as shown in Fig. 11.2. This leads to a pyramidalization of the sp^2-hybridized carbon or nitrogen atoms. In

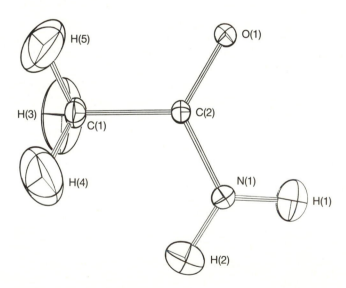

Fig. 11.2 Molecular structure of acetamide. Thermal ellipsoids at 75% probability (Johnson 1976) from single-crystal neutron diffraction analysis at 23 K. From Jeffrey *et al*. (1980).

monofluoroacetamide where the orientation of the fluoromethyl group is such that the molecule has nearly C_s symmetry, no pyramidalization is observed or predicted (Jeffrey *et al.* 1981*b*). Similarly, in one of the two different conformers in the crystal structure of thioacetamide (which has two crystallographically independent molecules in the asymmetric unit), the symmetry is close to C_s and no pyramidalization is observed or predicted (Jeffrey *et al.* 1984). These results are presented in Table 11.9.

Table 11.9 Experimental and calculated pyramidalization of the sp^2-hybridized central atom in rotamers of thioacetamide, acetamide, monofluoroacetamide and deuterated nitromethane

	τ^a (degrees)	$\langle\theta\rangle^b$ (degrees)
Thioacetamide[c]		
Experimental { Molecule A	177.7(1)	+0.03
Experimental { Molecule B	165.0(1)	+0.60
	180[d]	0
Calculated (HF/3–21G* level)	165[d]	+1.44
	150[d,e]	+1.76
Acetamide[f]		
Experimental	148.8[e]	+1.50
Calculated (HF/3–21G level)	{ 150[d,e]	+1.57
	{ 180[d]	0
Monofluoroacetamide[g]		
Experimental	179.0(4)	−0.25
Calculated (HF/3–21G level)	180[d]	0
Deuterated nitromethane[h]		
Experimental	−89.4(1)[e]	+1.48(9)
Calculated (HF/6–31G* level)	{ −89.1[e]	+1.66
	{ 0[d]	0

[a] τ is the torsion angle S—C—C—H(3) in thioacetamide, O—C—C—H(4) in acetamide, O—C—C—F in monofluoroacetamide, O(1)—N—C—D(2) in deuterated nitromethane.

[b] Average pyramidalization of the bonds $\mathbf{r}_i$, $\mathbf{r}_j$, and $\mathbf{r}_k$ from the sp^2-hybridized atom. The angle θ is defined as:

$$\theta = \cos^{-1}\left[\mathbf{r}_i \cdot (\mathbf{r}_j \times \mathbf{r}_k)/(r_i|\mathbf{r}_j \times \mathbf{r}_k|)\right] - \pi/2.$$

[c] Jeffrey *et al.* (1984)

[d] Assumed.

[e] This rotamer has a C—H or C—D bond normal (or very nearly normal) to the plane of the sp^2-hybridized atom.

[f] Jeffrey *et al.* (1980).

[g] Jeffrey *et al.* (1981*b*).

[h] Jeffrey *et al.* (1985*c*).

11.5   Is powder neutron diffractometry capable of providing accurate structural data?

Within the last ten years, pulsed neutron beams from spallation sources have become available and there is every indication that increasingly powerful sources can be built. With their potential for increasingly greater neutron flux, it seems likely that they will replace the socially less popular steady-state reactors as instruments for structural research. Hitherto, these instruments have been most effective for structural studies on inorganic compounds (Carrondo and Jeffrey 1988). However, the most suitable molecules for comparing the results of crystal structure analysis with those obtained by gas-phase techniques or *ab initio* MO calculations are small first-row element molecules without hydrogen-bonding functionality. For these the gas-phase techniques can give accurate results, the MO calculations can be extended to a high level of approximation, and in the absence of hydrogen bonding the crystal-field effects will be minimal. Such compounds are, of course, liquids at room temperature. It is not easy to grow neutron-diffraction quality crystals of low melting-point compounds or to manipulate them onto the cryostat of the neutron diffractometer. An alternative is to use powder neutron diffraction combined with Rietveld (1969) profile-fitting analysis. This reduces the extinction correction problem (and crystal twinning) but replaces it by a sample preparation problem. Simply freezing the liquid in the vanadium sample holder generally results in large crystals and preferred orientation. It is interesting to note that this was not a problem in the powder neutron diffraction work on the high-pressure ices. Reasonable agreement was obtained between the constant wavelength and time-of-flight powder diffraction analyses for D_2O ice VIII (Kuhs *et al*. 1984; Jorgensen *et al*. 1984), as shown in Table 11.10. The results are not strictly comparable, since one is at 10 K and the other at 269 K. These results are an order of magnitude less precise than those obtained by single-crystal analysis. They are, however, comparable to those provided by *ab initio* MO calculations with present-day computers.

A low-temperature powder sample preparation method which is effective, but uncomfortable, is to produce *snow* from a scent-spray onto a cold surface in a cold dry box and to scoop the snow, loosely packed, into the vanadium container. Where it has been used, it appears to avoid the problem of crystal size or preferred orientation. Attempts to use either constant-wavelength or time-of-flight powder neutron diffraction for the structure refinement of small organic molecules have hitherto been undecisive.

A powder neutron diffraction analysis of deuterated acetic acid using the profile analysis program SCRAP led to an anomalous geometry in which the C—OH single-bond length of 1.21(4) Å was smaller than the C=O double-bond length of 1.29(4) Å (Cooper *et al*. 1981; bond lengths

Table 11.10 Comparison of refinements of the structure of D_2O ice VIII by powder neutron diffractometry using constant-wavelength and time-of-flight methods

	Constant wavelength[a]		Time-of-flight[b]
Lattice parameters[c] ($\mathring{A}$)	At 10 K	At 263 K	At 269 K
a	4.656(1)	4.669(3)	4.6779(5)
c	6.775(1)	6.810(4)	6.8029(10)
Structural parameters ($\mathring{A}$, degrees)	At 10 K		At 269 K
$r(O-D)$	0.969(7)		0.973(11)
$r(D\cdots O)$	1.911(10)[d]		1.920(10)
$\angle O-D\cdots O$	178.3(7)		177(1)
$\angle D-O-D$	105.6(1.1)		104(2)
$r(O\cdots O)$ (hydrogen-bonded)	2.879(1)		2.8919(3)
$r(O\cdots O)$ (non-bonded)	2.743(9)		2.740(11)

[a] Kuhs *et al.* (1984).
[b] Jorgensen *et al.* (1984).
[c] Space group $I4_1/amd$.
[d] Calculated from the atomic and unit-cell parameters given in the original paper.

have been calculated from the published atomic parameters). The analysis was based on a set of constant-wavelength data collected at 12.5 K by Albinati *et al.* (1978). In contrast, an earlier analysis of the same data, using the Rietveld (1969) profile method, gave the expected values of $r(C-OH) = 1.32(2)$ $\mathring{A}$ and $r(C=O) = 1.23(2)$ $\mathring{A}$ (Albinati *et al.* 1978). Clearly, consistency with known stereochemistry to avoid false minima is an important factor in these powder diffraction analyses.

Both conventional (Trevino *et al.* 1980) and time-of-flight (Matias 1986) powder neutron diffraction analyses were attempted on deuterated nitromethane and the results could later be compared with a single-crystal neutron analysis (Jeffrey *et al.* 1985c). Table 11.11 shows the comparison between the experimental single-crystal values, the experimental constant-wavelength and time-of-flight powder diffraction values, and theoretical values calculated at the HF/6–31G* level. The refinement of time-of-flight powder data collected at 5.75 K gave good agreement for the bond lengths, but failed to converge on acceptable values for the N—C—D bond angles and the deuterium anisotropic temperature factors. This is believed to be due to the curvilinear nature of the deuterium thermal motion, i.e. the oscillatory motion of the CD_3 group. Although this did not present a problem with the single-crystal data with an observation-to-parameter ratio of 10:1, it did with the more limited data available from the powder method.

Table 11.11 Comparison between single-crystal, constant-wavelength and time-of-flight powder neutron diffraction results, and *ab initio* MO calculations for bond lengths of deuterated nitromethane

Bond	Single-crystal neutron diffractometry (15 K)[a]		Powder neutron diffractometry		*Ab initio* MO calculations (HF/6–31G* level)[a]
	Uncorrected	Corrected for thermal-motion effects	Constant wavelength (4.2 K)[b]	Time-of-flight (5.75 K)[c]	
C—N	1.4855(9)	1.488	1.481(3)	1.496(3)	1.478
N—O(1)	1.2270(9)	1.231	1.209(4)	1.223(3)	1.192
N—O(2)	1.2225(9)	1.225	1.223(4)	1.230(3)	1.192
C—D(1)	1.0751(13)	1.093	} 1.098(1)[d]	1.089(4)	1.080
C—D(2)	1.0736(14)	1.092		1.084(4)	1.076
C—D(3)	1.0739(13)	1.091		1.076(3)	1.076

[a] Jeffrey *et al.* (1985c).

[b] Trevino *et al.* (1980).

[c] Matias (1986). Based on intensity data up to $2\theta = 90°$, with individual isotropic temperature factors for C, N and O, and anisotropic temperature factors for D, for a total of 45 parameters.

[d] The methyl group was constrained to have threefold symmetry; the C—D bond length is corrected for libration.

Added in proof: A repetition of the time-of-flight powder diffraction experiment using the ISIS pulsed neutron facility at the Rutherford-Appleton Laboratory gave better results (David, Ibberson, and Jeffrey, unpublished work (1990)). The maximum difference in bond lengths from the single-crystal results was 0.0035 Å. The bond angles differed by less than 0.3°. The agreement in the anisotropic displacement parameters was similar to that obtained for deuterated benzene, shown in Table 11.12.

More promising results have been obtained using the ISIS pulsed neutron source at the Rutherford–Appleton Laboratory, England (David and Ibberson 1989). A time-of-flight powder diffraction analysis of deuterated benzene at 4.2 K gave results which were in excellent agreement with the single-crystal neutron analysis of Jeffrey *et al.* (1987). The mean and greatest difference in atomic coordinates from the two experiments was 0.003 Å and 0.006 Å. There was also good agreement between the anisotropic displacement parameters, as shown in Table 11.12; which also reports the values

Table 11.12 Anisotropic displacement parameters ($\text{Å}^2 \times 10^4$) for benzene at low temperatures

Atom	U^{11}	U^{22}	U^{33}	U^{12}	U^{13}	U^{23}
C(1)	90	66	89	0	7	3
	77	58	77	7	0	3
	79(2)	67(2)	88(2)	6(2)	7(2)	4(1)
	77(7)	42(6)	87(7)	−3(4)	5(5)	1(5)
C(2)	84	87	82	6	17	−2
	71	79	70	7	17	−2
	74(2)	81(2)	79(2)	9(2)	17(2)	0(2)
	71(7)	58(7)	68(6)	12(4)	26(5)	9(4)
C(3)	86	79	82	6	11	10
	73	72	70	−5	11	10
	81(2)	75(2)	82(2)	−3(2)	14(2)	10(1)
	83(7)	57(7)	92(7)	−1(4)	18(5)	0(5)
D(1)	216	173	212	38	39	11
	202	165	199	38	39	11
	224(3)	114(2)	239(3)	46(2)	25(2)	12(2)
	218(8)	121(7)	267(9)	31(5)	22(6)	19(5)
D(2)	184	226	215	56	56	5
	170	217	202	56	55	5
	183(2)	204(3)	208(3)	35(2)	88(2)	−8(2)
	170(8)	212(8)	225(9)	33(5)	120(6)	−2(6)
D(3)	228	208	171	−2	37	60
	215	199	158	−1	36	59
	214(3)	171(2)	199(3)	−18(2)	61(2)	58(2)
	241(9)	155(8)	214(8)	−20(5)	66(7)	75(6)

For each atom the values in the first and second line are from harmonic lattice-dynamical calculations on benzene at 15 K and 0 K, respectively (Filippini and Gramaccioli 1989). Values in the third line are from single-crystal neutron diffraction analysis of deuterated benzene at 15 K (Jeffrey *et al.* 1987). Values in the fourth line are from a time-of-flight powder neutron diffraction study of deuterated benzene at 4.2 K (David and Ibberson 1989).

calculated by Filippini and Gramaccioli (1989) using a harmonic lattice-dynamical model. Since the time-of-flight powder experiments are made in hours rather than days for the single-crystal experiment, this method may become that of choice for the neutron crystal structure analysis of small molecules which cannot be easily obtained as single crystals.

References

Albinati, A., Rouse, K. D., and Thomas, M. W. (1978). *Acta Crystallogr.*, **B34**, 2184-7.

Allen, F. H. (1986). *Acta Crystallogr.*, **B42**, 515-22.

Almenningen, A., Bastiansen, O., and Motzfeldt, T. (1969). *Acta Chem. Scand.*, **23**, 2848-64.

Bacon, G. E. (1975). *Neutron diffraction* (3rd edn). Clarendon Press, Oxford.

Becker, P. J. and Coppens, P. (1974). *Acta Crystallogr.*, **A30**, 129-47.

Becker, P. J. and Coppens, P. (1975). *Acta Crystallogr.*, **A31**, 417-25.

Bencivenni, L., Domenicano, A., Portalone, G., and Ramondo, F. (1990). Presentation at the *20th Congress of the Italian Crystallographic Association*, Genova, Italy. Abstracts, pp. 144-5.

Brown, P. J. and Forsyth, J. B. (1973). *The crystal structure of solids*, Chapters 2-4. Arnold, London.

Busing, W. R. and Levy, H. A. (1964). *Acta Crystallogr.*, **17**, 142-6.

Carrondo, M. A. and Jeffrey, G. A. (eds.) (1988). *Chemical crystallography with pulsed neutrons and synchrotron X-rays*, NATO ASI Series C, Mathematical and Physical Sciences, Vol. 221. Reidel, Dordrecht.

Cooper, M. J., Rouse, K. D., and Sakata, M. (1981). *Z. Kristallogr.*, **157**, 101-17.

Craven, B. M. and Swaminathan, S. (1984). *Trans. Am. Crystallogr. Assoc.*, **20**, 133-5.

David, W. I. F. and Ibberson, R. M. (1989). *ISIS Annual report*, pp. 43-6. Swindon Press, Swindon.

Derissen, J. L. (1971*a*). *J. Mol. Struct.*, **7**, 67-80.

Derissen, J. L. (1971*b*). *J. Mol. Struct.*, **7**, 81-8.

Dunitz, J. D. and Waser, J. (1972). *J. Am. Chem. Soc.*, **94**, 5645-50.

Emsley, J. (1980). *Chem. Soc. Rev.*, **9**, 91-124.

Filippini, G. and Gramaccioli, C. M. (1989). *Acta Crystallogr.*, **A45**, 261-3.

Gao, Q., Jeffrey, G. A., Ruble, J. R., and McMullan, R. K. (1991). *Acta Crystallogr.*, **B47**, 742-5.

Goldsmith, O. (1766). *The vicar of Wakefield*. Collins, Salisbury.

Hehre, W. J., Radom, L., Schleyer, P.v.R., and Pople, J. A. (1986). *Ab initio molecular orbital theory*. Wiley-Interscience, New York.

Hirota, E., Sugisaki, R., Nielsen, C. J., and Sørensen, G. O. (1974). *J. Mol. Spectrosc.*, **49**, 251-67.

Jeffrey, G. A. and Lewis, L. (1978). *Carbohydr. Res.*, **60**, 179-82.

Jeffrey, G. A. and Ruble, J. R. (1984). *Trans. Am. Crystallogr. Assoc.*, **20**, 129-32.

Jeffrey, G. A., Ruble, J. R., McMullan, R. K., DeFrees, D. J., Binkley, J. S., and Pople, J. A. (1980). *Acta Crystallogr.*, **B36**, 2292-9.

Jeffrey, G. A., Ruble, J. R., McMullan, R. K., DeFrees, D. J., and Pople, J. A. (1981*a*). *Acta Crystallogr.*, **B37**, 1381-7.

Jeffrey, G. A., Ruble, J. R., McMullan, R. K., DeFrees, D. J., and Pople, J. A. (1981*b*). *Acta Crystallogr.*, **B37**, 1885-90.

Jeffrey, G. A., Ruble, J. R., McMullan, R. K., DeFrees, D. J., and Pople, J. A. (1982). *Acta Crystallogr.*, **B38**, 1508-13.

Jeffrey, G. A., Ruble, J. R., and Yates, J. H. (1983). *Acta Crystallogr.*, **B39**, 388-94.

Jeffrey, G. A., Ruble, J. R., and Yates, J. H. (1984). *J. Am. Chem. Soc.*, **106**, 1571-6.

Jeffrey, G. A., Ruble, J. R., Nanni, R. G., Turano, A. M., and Yates, J. H. (1985*a*). *Acta Crystallogr.*, **B41**, 354-61.

Jeffrey, G. A., Houk, K. N., Paddon-Row, M. N., Rondan, N. G., and Mitra, J. (1985*b*). *J. Am. Chem. Soc.*, **107**, 321-6.

Jeffrey, G. A., Ruble, J. R., Wingert, L. M., Yates, J. H., and McMullan, R. K. (1985*c*). *J. Am. Chem. Soc.*, **107**, 6227-30.

Jeffrey, G. A., Ruble, J. R., McMullan, R. K., and Pople, J. A. (1987). *Proc. Roy. Soc.*, **A414**, 47-57.

Johnson, C. K. (1970*a*). An introduction to thermal-motion analysis. In *Crystallographic computing* (ed. F. R. Ahmed), pp. 207-19. Munksgaard, Copenhagen.

Johnson, C. K. (1970*b*). Generalized treatments for thermal motion. In *Thermal neutron diffraction* (ed. B. T. M. Willis), pp. 132-60. Oxford University Press.

Johnson, C. K. (1976). *ORTEP II*. Report ORNL-5138, Oak Ridge National Laboratory, Oak Ridge, Tennessee, USA.

Johnson, C. K. and Levy, H. A. (1974). Thermal-motion analysis using Bragg diffraction data. In *International tables for X-ray crystallography*, Vol. 4, *Revised and supplementary tables* (ed. J. A. Ibers and W. C. Hamilton), pp. 311-36. Kynoch Press, Birmingham.

Jönsson, P.-G. (1971). *Acta Crystallogr.*, **B27**, 893-8.

Jorgensen, J. D., Beyerlein, R. A., Watanabe, N., and Worlton, T. G. (1984). *J. Chem. Phys.*, **81**, 3211-4.

Kitano, M. and Kuchitsu, K. (1974). *Bull. Chem. Soc. Jpn.*, **47**, 67-72.

Kroon-Batenburg, L. M. J. and Kanters, J. A. (1983). *Acta Crystallogr.*, **B39**, 749-54.

Kuchitsu, K. and Bartell, L. S. (1961). *J. Chem. Phys.*, **35**, 1945-9.

Kuchitsu, K. and Morino, Y. (1965). *Bull. Chem. Soc. Jpn.*, **38**, 805-13.

Kuhs, W. F. and Lehmann, M. S. (1987). The structure of ice I_h. In *Water science reviews*, Vol. 1 (ed. F. Franks), pp. 93-170. Cambridge University Press.

Kuhs, W. F., Finney, J. L., Vettier, C., and Bliss, D. V. (1984). *J. Chem. Phys.*, **81**, 3612-23.

Leviel, J.-L., Auvert, G., and Savariault, J.-M. (1981). *Acta Crystallogr.*, **B37**, 2185-9.

Matias, P. M. H. M. (1986). *Neutron and X-ray diffraction studies of small organic molecules*. Thesis, University of Pittsburgh, USA.

Nanni, R. G., Ruble, J. R., Jeffrey, G. A., and McMullan, R. K. (1986). *J. Mol. Struct.*, **147**, 369-80.

Rietveld, H. M. (1969). *J. Appl. Crystallogr.*, **2**, 65-71.

Schomaker, V. and Trueblood, K. N. (1968). *Acta Crystallogr.*, **B24**, 63-76.

Stevens, E. D. (1978). *Acta Crystallogr.*, **B34**, 544–51.

Taylor, R. and Kennard, O. (1983). *Acta Crystallogr.*, **B39**, 133–8.

Trevino, S. F., Prince, E., and Hubbard, C. R. (1980). *J. Chem. Phys.*, **73**, 2996–3000.

Willis, B. T. M. (ed.) (1970). *Thermal neutron diffraction*. Oxford University Press.

Willis, B. T. M. and Pryor, A. W. (1975). *Thermal vibrations in crystallography*. Cambridge University Press.

12

Nuclear magnetic resonance spectroscopy and accurate molecular geometry

Peter Diehl

12.1   Introduction

The phenomenon of nuclear magnetic resonance, NMR (see, for example, Harris 1983), is based on the intrinsic spin angular momentum **P** of atomic nuclei such as, for example, ^{1}H, ^{2}H, ^{13}C, ^{14}N, ^{15}N, etc.:

$$P = |\mathbf{P}| = \hbar[I(I + 1)]^{1/2} \tag{12.1}$$

(where I is the spin quantum number of the nucleus, and $\hbar = h/2\pi$), which is quantized with respect to the direction of an applied magnetic field $\mathbf{B}_z$,

$$P_z = \hbar\, m_1 \tag{12.2}$$

(where $m_1 = I, I - 1, \ldots, -I$ is the magnetic quantum number). As a consequence of angular momentum and nuclear charge the nuclei have magnetic dipole moments:

$$\mu = \gamma h \left[I(I + 1)\right]^{1/2} \tag{12.3}$$

(where $\gamma = \mu/P$ is called the *magnetogyric ratio*), and their energy in the magnetic field,

$$U = -\mu_z B_z = \gamma \hbar m_I B_z, \tag{12.4}$$

is also quantized. Transitions between the energy levels with

$$\Delta U = h\nu = \gamma \hbar B_z \Delta m_I \tag{12.5}$$

(where $\nu = (\gamma/2\pi) B_z \Delta m_I$ and $\Delta m_I = \pm 1$) may be induced by a circularly polarized magnetic radiofrequency field at the Larmor frequency ν_i:

$$\nu_i = (|\gamma|/2\pi) B_z. \tag{12.6}$$

On the basis of eqns (12.1) to (12.6) NMR spectroscopy was originally (i.e. between 1946 and 1950) used prevalently to study nuclear properties. In such a study, the Larmor frequency for a nucleus i in a liquid was found to be dependent on its chemical environment, i.e. the *chemical shift* parameter (σ_i) was observed. Chemical shifts arise because of the shielding of the nuclei from the external magnetic field by the electrons. Within a year a further parameter, the *indirect spin–spin coupling* J_{ij}, leading to a splitting of resonance lines in liquids, was identified. Indirect coupling is transmitted via the molecular electrons. The calculation of both parameters depends upon a detailed knowledge of the molecular electron distribution and is consequently rather complex. Therefore chemical shift and indirect spin–spin coupling parameters were collected and the experimental values used empirically for the identification of molecular structure. A third, much simpler parameter, the *direct coupling* D_{ij}, i.e. the interaction between the nuclear magnetic moments through space, had been observed in solids for the first time in 1946 but, because of the broad transitions caused by the many interacting nuclei, the measurements had very low precision. However, modern multipulse, two-dimensional experiments, double-resonance methods with dilute spin pairs, and nutation spectroscopy have made possible the observation of direct spin–spin coupling with high resolution in single crystals (Menger *et al.* 1984) and amorphous materials (Horne *et al.* 1983). Usually, enriched samples (^{13}C, ^{15}N) have been used; work at natural abundance seems feasible but time consuming. Predominantly one-bond distances have been determined with a precision of the order of 1 per cent.

In the liquid, the direct coupling:

$$D_{ij} = -(\mu_0 \hbar \gamma_i \gamma_j / 8\pi^2) [(1/2) \langle (3\cos^2 \theta_{ijz} - 1)/r_{ij}^3 \rangle] \tag{12.7}$$

(where μ_0 is the permeability *in vacuo*, γ_i the magnetogyric ratio of nucleus

i, r_{ij} the distance between nuclei i and j, and θ_{ijz} the angle between the magnetic field $\mathbf{B}_z$ and the internuclear vector $\mathbf{r}_{ij}$) is averaged to zero, because $\langle 3\cos^2\theta_{ijz} - 1\rangle$ is zero for isotropic tumbling, i.e. for a spherical symmetry of the axes distribution. This is fortunate on one hand, because the much smaller chemical shifts and indirect couplings are thus revealed. On the other hand, important information on internuclear distances is lost. The isotropic tumbling also affects σ and $\mathbf{J}$. Their tensor properties are averaged out and only the traces are observed.

The obvious loss of information on precise molecular geometry, due to the averaging to zero of the direct coupling in liquids, induced NMR spectroscopists 28 years ago to investigate possibilities of orienting molecules by various means so that the disadvantage of the solid disappears, i.e. the intermolecular direct coupling is averaged out, but the intramolecular direct coupling averages to a non-zero value. The only successful method turned out to be the use of liquid crystals as orienting solvents (Saupe and Englert 1963; Englert and Saupe 1964).

12.2 Molecular orientation and direct dipole–dipole coupling

The nematic or smectic A phases of thermotropic (azo-, azoxy-, azine-type molecules, biphenyls, phenylcyclohexanes or bicyclohexanes) or lyotropic (for example, mixtures of alkyl sulphates, alcohols, sodium sulphate and water) liquid crystals can easily be oriented in magnetic fields of 10^{-1} Tesla. The molecular aggregates, as a consequence of their diamagnetic anisotropy, tend to have their axes of largest susceptibility perpendicular to the field; a uniaxial solvent is obtained (Fig. 12.1).

In the nematic phase there is order in the direction of molecular axes but not in the centers of gravity of the molecules. In the smectic A phase there is order in the direction of molecular axes and a layer structure, but no long-range ordering within the layer for the centers of gravity of the molecules (Fig. 12.1).

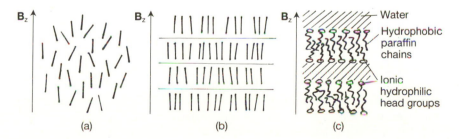

Fig. 12.1 Schematic representation of liquid-crystal phases in a magnetic field: (a) Thermotropic nematic phase; (b) Thermotropic smectic phase; (c) Lamellar lyotropic phase.

If a substance — gas, liquid, or solid — is dissolved in the oriented liquid crystal, the solute molecules interact with the highly ordered solvent, and although the molecules still rotate and diffuse, they show deviation from a spherical distribution of their axes with respect to the solvent axis and the applied magnetic field. This deviation is measured by an *order parameter* S_{ij}:

$$S_{ij} = (1/2) \langle 3\cos^2 \theta_{ijz} - 1 \rangle. \tag{12.8}$$

Depending upon the symmetry of the molecule, up to five different order parameters S_{ij} may be necessary to fully define its orientation. As a consequence of orientation the direct coupling D_{ij} is now observable (Emsley and Lindon 1975; Diehl 1985):

$$D_{ij} = -(\mu_0 \, \hbar \gamma_i \, \gamma_j / 8\pi^2) \, \langle 1/r_{ij}^3 \rangle \, S_{ij} \tag{12.9}$$

(assuming that r_{ij} and θ_{ijz} are independent of each other). The order parameter, which is typically of the order of 10^{-1}, and with it the direct coupling, depend on the temperature of the sample (roughly 1 per cent per K) and the concentration of the solution. Therefore concentration gradients must be avoided and the temperature must be kept constant.

12.3   Direct coupling and molecular geometry

As shown by eqn (12.9), the observable direct couplings are always products of the inverse cube of the internuclear distance and the degree of order of the internuclear axis, i.e. there are two unknowns for each measured parameter. As a consequence, it is not possible to deduce absolute distances but only ratios of distances, i.e. shapes of molecules and absolute molecular angles are determined. If internuclear distances are presented in the literature, they are always referred to an assumed basis derived by a different method.

Direct couplings of the order of 10^3 Hz between protons can easily be measured with an error of 10^{-1} Hz. The resulting precision of 10^{-4} in the ratio of dipolar couplings leads to a precision of roughly 10^{-5} in distance ratios, $\Delta r/r = (1/3) \Delta D/D$. It appears, therefore, easy to detect by this method differences of 10^{-4} Å with respect to a fixed basis. As discussed in Section 12.8, there are, unfortunately, at least three contributions which affect the observed couplings. Even if they are considered and the couplings corrected for, the accuracy is reduced to perhaps 10^{-3} Å in bond lengths and 10^{-1} degree in bond angles.

12.4 The NMR experiment

The NMR spectrometer (see Fig. 12.2) consists of a magnet, usually super-conducting (cryogenic) which should produce a strong magnetic field because the sensitivity (signal-to-noise ratio) increases with the field strength. The field must be highly homogeneous in order to provide the necessary resolution of liquid-crystal NMR of the order of 0.1 Hz in, for example, 250 MHz for protons. The field stability in time can be achieved by use of an electronic field-frequency lock.

The sample which is put into the probe is usually contained in a glass tube with a 5 mm diameter. It is rotated rapidly by means of an air spinner in order to average out residual field inhomogeneities.

The electronic components of the Fourier-transform NMR spectrometer are the transmitter, the receiver and the computer which handles the pulse programming, the signal acquisition and accumulation as well as the Fourier transformation.

The nuclear spin system is irradiated with pulses of monochromatic radio-frequency corresponding to the magnetic field (eqn (12.6)). The pulse length and amplitude are determined by the spectral width and the magnetogyric

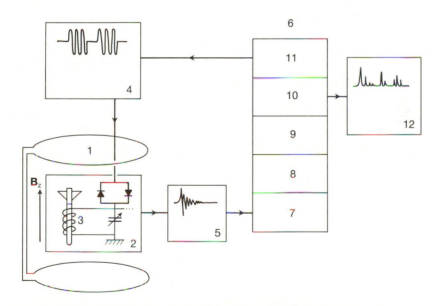

Fig. 12.2 Block diagram of a high-resolution Fourier-transform NMR spectrometer. 1: Super-conducting magnet; 2: Probe; 3: Sample; 4: Transmitter; 5: Receiver; 6: Spectrometer computer; 7: Analogue-to-digital conversion; 8: Signal acquisition and accumulation; 9: Fourier transformation; 10: Digital-to-analogue conversion; 11: Pulse programmer; 12: Recorder, oscilloscope.

ratio of the nucleus. The irradiation which causes spin flips induces a free induction decay (FID) of the nuclear magnetization in the time domain. The FID is amplified in the receiver and after an analogue-to-digital conversion fed into the computer. The computer must have a sampling rate (number of measured points per second) which is also determined by the spectral width, and the resolution depends upon the acquisition time for each measurement following the pulse.

When a sufficient number of FIDs is accumulated, the resulting signal is Fourier-transformed into the frequency domain and after a digital-to-analogue conversion the actual spectrum is obtained.

For ^{1}H spectra of oriented molecules the number of FIDs necessary for a good signal-to-noise ratio is between 1 and 10^2. If ^{13}C satellites at natural abundance are studied, the number of FIDs must be increased to 10^2–10^4. The corresponding times of measurement are one to a few minutes and a few hours, respectively. For ^{13}C spectra at natural abundance the low relative receptivity of the ^{13}C nucleus with respect to ^{1}H, 2×10^{-4}, causes a drastic increase of the measuring time.

By application of a second radiofrequency field, nuclei can be decoupled from each other so that the spectra are simplified.

12.5   Procedure of structure determination

To illustrate the procedure of structure determination, we have chosen a simple planar system of four protons, ethylene (Wasser and Diehl 1987; see Fig. 12.3 for the numbering of nuclei).

In isotropic solutions the system shows only one NMR transition, because the nuclei experience the same local molecular magnetic field (chemical shift). In the oriented system, there are three different direct couplings affecting the spectrum which now shows 12 transitions (Fig. 12.4). The spectrum of the liquid-crystal solvent, due to the very many interacting spins, usually forms an unresolved background.

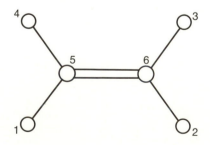

Fig. 12.3 Numbering of nuclei in ethylene.

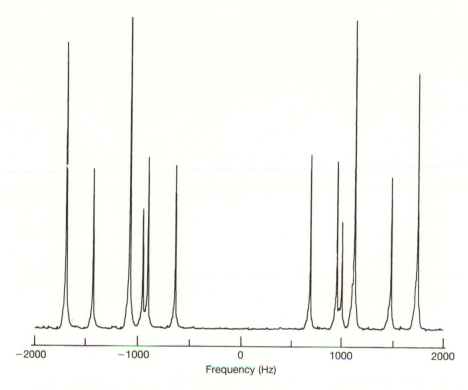

Fig. 12.4 ^{1}H NMR spectrum of ethylene, partially oriented in the liquid-crystal solvent ZLI 1167 (eutectic mixture of three *trans,trans*-4-cyano-4'-alkylbicyclohexyls). One FID, pressure 1 atm (101 kPa), Larmor frequency 250 MHz.

Sample preparation is easy. A few mole per cent (*ca.* 10) of the solute are dissolved in the heated isotropic phase of the liquid crystal which is then allowed to cool into the nematic phase. The smectic phase must be cooled in the magnetic field.

The analysis of the 4-spin proton spectrum is relatively straightforward and is performed iteratively with a computer program (Diehl *et al.* 1971*a*). The Hamiltonian which describes the spin system is:

$$\mathcal{H} = -(2\pi)^{-1}\sum_i \gamma_i\, \mathcal{I}_{zi}(1 - \sigma_i^{\text{iso}} - \sigma_i^{\text{aniso}})B_z$$

$$+ \sum_{i<j} \{ T_{ij}^{\text{iso}}[\mathcal{I}_{zi}\,\mathcal{I}_{zj} + (1/2)\,(\mathcal{I}_i^+\,\mathcal{I}_j^- + \mathcal{I}_i^-\,\mathcal{I}_j^+)]$$

$$+ T_{ij}^{\text{aniso}}[\mathcal{I}_{zi}\,\mathcal{I}_{zj} - (1/4)\,(\mathcal{I}_i^+\,\mathcal{I}_j^- + \mathcal{I}_i^-\,\mathcal{I}_j^+)]\}, \qquad (12.10)$$

with the following meaning of the symbols: σ_i^{iso}, chemical shift; σ_i^{aniso}, anisotropic contribution to the chemical shift; $T_{ij}^{iso} = J_{ij}$, indirect spin–spin coupling; $T_{ij}^{aniso} = J_{ij}^{aniso} + 2D_{ij}$, where D_{ij} is the direct spin–spin coupling; $\mathfrak{I}_z$, spin operator; $\mathfrak{I}^\pm = \mathfrak{I}_x \pm i\mathfrak{I}_y$, raising/lowering operator.

Unfortunately the measured quantity is $T_{ij}^{aniso} = J_{ij}^{aniso} + 2D_{ij}$, i.e. it contains contributions from the anisotropy of the indirect coupling. This, however, is generally small (see Section 12.8.2). Of the resulting parameters — chemical shifts, indirect couplings, and direct couplings — only the last ones, which carry the precise structural information, are of interest.

Before starting the analysis of the structure, it is necessary to check that a sufficient number of direct couplings is available to define the structure fully. In a 'rectangular' molecule like ethylene there exist two unknown degrees of order (S_{xx}, S_{zz}), and, with respect to a fixed basis, one unknown nuclear coordinate, i.e. there are three unknowns. Consequently, the three measured couplings exactly determine the problem.

In the next step a computer program (Diehl *et al.* 1971*b*) is used to reproduce the experimental couplings iteratively by variation of the order parameters and of the nuclear coordinates. The result of interest is the shape of the rectangle, together of course with its precision, which is determined by the errors of the direct couplings derived in the first part of the analysis. For ethylene in the liquid-crystal solvent EBBA (*N*-(*p*-ethoxybenzylidene)-*p*-butylaniline) we find, for example, $r_{14}/r_{12} = 0.7589(2)$.

Of course, most molecules of interest contain not only protons but also carbons. Unfortunately, ^{12}C has no magnetic moment. But in principle ^{13}C, which has spin $1/2$, could be enriched and its resonance observed. Luckily, with the use of Fourier-transform spectroscopy, NMR has become sensitive enough to detect ^{13}C in natural abundance. From the point of view of sensitivity an even better method is the observation of the ^{13}C satellites in proton spectroscopy. The proton spectra of molecules in which one of the carbon nuclei is ^{13}C are detected. In these, for ethylene, there are two additional direct coupling constants and one more unknown ^{13}C coordinate. The problem is now overdetermined and the entire structure can be measured, for example, $r_{56}/r_{12} = 0.5389(2)$.

In a small molecule like ethylene, the satellite spectra can easily be identified, although they are approximately fifty times weaker than the normal proton spectrum. In more complex molecules, difference spectroscopy (Canet *et al.* 1974) can be used to remove the strong spectrum. In this method, a spectrum in which the ^{13}C nuclei are decoupled from the protons is subtracted from the undecoupled one. The same procedure can also be applied to, for example, ^{15}N satellites.

As the spectrum complexity quickly increases with the number of nuclei (the number of intense transitions for n nuclei is approximately $n \cdot 2^{n-1}$), it is sometimes necessary to simplify the problem by deuterating a dominant portion of the molecule, decouple the deuterons from the protons and record the spectrum of the protons (Meiboom *et al.* 1972).

Also, with many transitions and many unknown direct couplings, the analysis of the spectrum may become very difficult. There are two alternative ways out of this particular problem. The first is to use automatic computer programs which have become available lately and for which the assignment of transitions is not necessary (Diehl *et al.* 1975; Diehl and Vogt 1976; Stephenson and Binsch 1980). The second is to use a computer program which keeps an approximate guessed molecular structure constant and varies only the degrees of order (Diehl and Lohner, unpublished results). Such a program finds an approximate solution more quickly and, because the number of unknown degrees of order is often three or less, the number of lines which must be assigned is much smaller than in the normal programs, where we have $n(n-1)/2$ direct couplings for n nuclei without symmetry. The approximate solution must finally be refined by a normal analysis with assignment of as many transitions as possible.

Nuclei with $I \geq 1$ (for example ^{14}N, ^{17}O, and the halogens) do not usually contribute to the spectra of nuclei with spin $1/2$, because they are effectively decoupled from, for example, the ^{1}H and ^{13}C nuclei by their quadrupole relaxation. Therefore their positions cannot be determined by NMR of oriented molecules. Exceptions are ^{2}H which has particularly small quadrupole coupling constants, and other nuclei in symmetrical environments, i.e. with small electric field gradient at the nuclear position.

12.6 Types of molecules studied so far

NMR spectroscopy of oriented molecules has so far been used to determine the structures of some 500 different molecules. Of these approximately 30 per cent have four magnetic nuclei, 20 per cent have five, 30 per cent six, 10 per cent seven, and 10 per cent eight or more. The resulting structural data are collected in books and reviews (Schumann 1980; Khetrapal and Kunwar 1982). Most of the observed spectra are of ^{1}H with and without satellites, but there are also some studies on ^{2}H, ^{19}F, and ^{31}P resonances. Even less used are ^{11}B, ^{13}C, ^{14}N, ^{15}N, ^{17}O, ^{111}Cd, ^{113}Cd, ^{119}Sn, and ^{199}Hg.

Unfortunately, of the many molecular structures determined so far only a small number have been derived from data corrected for harmonic vibration (see Section 12.8.3). For an even smaller number, of the order of ten molecules, the deformation corrections have also been performed (see Section 12.8.4). Only these solvent-independent data can be compared with results of different methods of structure determination. Some examples will be presented in Section 12.8.4.

12.7   Mixtures of conformers

NMR spectra are generally sensitive to all types of intramolecular motion as well as chemical exchange, if the rates are comparable to the observed splittings in the spectrum. In isotropic solvents these splittings, due to the chemical shift and indirect coupling, are typically between 10 and 10^3 Hz for protons, so that line broadening can be observed in the rate range of 1 to 10^4 Hz. In oriented systems the larger direct couplings expand the sensitivity by a factor of 10^3. The spectra of individual conformations may be observed if the rates are below the critical range. Above the critical range the observed average can be interpreted in terms of the structures of the conformers, their degrees of order and their concentrations (Luz 1985; Veracini and Longeri 1985).

For a molecule having only one conformation the direct coupling between nuclei i and j can be expressed in a general form through eqn (12.9), which can be written as:

$$D_{ij} = -K S_{ij} \langle 1/r_{ij}^3 \rangle. \tag{12.11}$$

If more than one conformer is present, eqn (12.11) becomes

$$D_{ij} = -K \sum_{\nu} P_{\nu} S_{ij}^{\nu} \langle 1/r_{ij\nu}^3 \rangle, \tag{12.12}$$

where P_{ν} is the concentration of conformer ν. In this relation a discrete number of conformers is assumed to exist at the steep potential minima. This approximation is clearly insufficient for flat potentials. Even with the simplifying assumption, the observed average only allows determination of the products $P_{\nu} S_{ij}^{\nu}$. Consequently the application of this method should be restricted to cases of only one conformer, several equivalent conformers, methyl rotation, ring inversion, bond shift, and bond rearrangement.

Many cases have been studied involving several conformers by use of further simplifying assumptions. If the external energy of molecular orientation is independent of the conformation, the observed average direct coupling is simply

$$D_{ij} = -K \langle S_{ij} \rangle \sum_{\nu} P_{\nu} \langle 1/r_{ij\nu}^3 \rangle, \tag{12.13}$$

where $\langle S_{ij} \rangle$ is an average degree of order. Although results derived on this basis agree roughly with data from other techniques, there is evidence contradicting the validity of eqn (12.13) (see Section 12.8.4).

12.8   The factors which limit the precision and accuracy of structure determination

12.8.1   Signal-to-noise ratio and line width

As pointed out in Section 12.3, there are limiting factors to the precision of structure determination by liquid-crystal NMR spectroscopy.

The first, of course, is the accuracy with which the direct couplings can be determined (see Section 12.3). Here, an improvement is possible with increased signal-to-noise ratio, for example, by the use of a strong magnetic field, high solute concentration and large sample volume. Usually the coupling between two protons is more precisely determined than between ^{13}C and proton in the satellite spectra without enrichment. Also a reduced line width of the transitions, i.e. low viscosity of the solvent, temperature stability and homogeneity (see Section 12.2), or sample spinning which homogenizes the magnetic field, will help — if the spinning does not destroy the liquid-crystal orientation.

12.8.2   Anisotropy of the indirect coupling constant

The interaction of nuclear spins is composed of the direct dipolar coupling and the indirect coupling via the electrons (see Section 12.1). Both are anisotropic, i.e. they depend upon the direction of the applied magnetic field with respect to the molecular axes. Consequently, in oriented molecules, the total measured anisotropic coupling contains also contributions from the indirect coupling (Lounila and Jokisaari 1982). These are rather complex. There is an orbital term stemming from the coupling of the nuclear magnetic moment with the electron orbital motion, a spin–dipolar term describing the dipole interaction between the magnetic moment of the nucleus and the electron spin angular momentum and, finally, the contact term due to the interaction between nuclear spin and electron spin at the nucleus. Fortunately, the contributions of the anisotropy of the indirect coupling to the observed 'direct' coupling are small and may therefore often be neglected. The order of magnitude of the relative contributions is 10^{-4} for a pair of protons, 10^{-4} to 10^{-3} for carbon–proton, but 10^{-2} for a pair of carbon atoms. In principle, it should be possible to measure the anisotropy of the indirect coupling and to make the necessary correction. In practice, however, it turns out that the measurement is very complex and imprecise. It depends upon the combination of data from various solvents. Obviously, here there is a first serious limitation of the accuracy. In general an accuracy of the order of 10^{-3} Å can rarely be exceeded.

12.8.3 Molecular vibration

It is well known that nuclei in molecules vibrate with considerable amplitudes. This motion affects the direct coupling in two ways. First, the internuclear distance r_{ij} is time-dependent and, second, the internuclear axes vary their directions and consequently their degree of order. Obviously the observed coupling D_{ij} must be corrected for this motion. Assuming that during a vibration the molecule does not rotate appreciably, D_{ij} can be expanded for small amplitudes of vibration as

$$D_{ij} = D_{ij}^{e} + d_{ij}^{a} + d_{ij}^{h}. \tag{12.14}$$

The terms D_{ij}^{e}, d_{ij}^{a}, and d_{ij}^{h} of eqn (12.14) correspond to equilibrium geometry, anharmonic correction, and harmonic correction, respectively.

As a simple example for a pair of nuclei on the symmetry axis of a system with rotational symmetry ($S_{xx} = S_{yy} = -\frac{1}{2}S_{zz}$) the following contributions are derived:

$$d^{h} = 6(C_{zz} - C_{xx})D^{e}/r^{2} \tag{12.15}$$

(where C_{zz} and C_{xx} are the quadratic average displacements, $\langle \Delta z^{2} \rangle$ and $\langle \Delta x^{2} \rangle$ respectively), and

$$d^{a} = -3\langle \Delta z \rangle D^{e}/r. \tag{12.16}$$

Vibration corrections can be calculated for all molecules for which a force field is known. Usually only the harmonic corrections are performed, because information on anharmonic potentials is scarce. The correction is calculated by a computer program (Sýkora *et al.* 1979) which derives normal coordinates and vibrational frequencies, evaluates the covariance matrix **C** of quadratic average displacements, and then calculates d_{ij}^{h}. The corrected direct couplings

$$D_{ij}' = D_{ij} - d_{ij}^{h} = D_{ij}^{e} + d_{ij}^{a} \tag{12.17}$$

lead to a molecular geometry which corresponds to an r_{α} structure (see Chapter 2, Section 2.4.1.3):

$$r_{\alpha} = r_{e} + \langle \Delta z \rangle_{T}, \tag{12.18}$$

i.e. the average projection of bond lengths onto the line joining the equilibrium positions of the nuclei is measured.

The relative harmonic vibration corrections to the direct couplings increase with r^{-2}; they are, therefore, particularly important for direct bonds, for example $C-H$. As an example, uncorrected and corrected direct couplings in ethylene are compared in Table 12.1. The geometry derived from the corrected couplings is compared in Table 12.2 with the r_z structure from high-resolution IR spectroscopy (Duncan 1974b). The relative correction may also become large in molecules with only two planes of symmetry, if a direct coupling happens to be small due to a fortuitous orientation of the internuclear axis. In every molecule, due to the tracelessness of the order tensor, there is always a conical set of directions (generally elliptical), such that D^e is zero for pairs of nuclei with internuclear vector parallel to a 'null' direction. The harmonic correction, however, is non-zero even for such pairs.

Table 12.1 Proton–proton, proton–carbon, and carbon–carbon direct coupling constants of ethylene in the liquid-crystal solvent EBBA[a] (Wasser and Diehl 1987)

Coupling constant[b]	Measured value (Hz)	Value corrected for harmonic vibration (Hz)	Relative correction (%)
D_{12}	−393.16(8)	−390.03	+0.8
D_{13}	−80.85(9)	−81.50	−0.8
D_{14}	563.13(7)	572.01	+1.6
D_{15}	211.96(9)	221.76	+4.6
D_{16}	−105.74(9)	−106.58	−0.8
D_{56}	−153.98(22)	−153.59	+0.3

[a] At a pressure of 3 atm (303 kPa). EBBA is N-(p-ethoxybenzylidene)-p-butylaniline.
[b] See Fig. 12.3 for the numbering of nuclei.

In general, the vibration corrections will not be extremely precise. They depend upon the force field used and the approximations which are made. With the assumption that the inaccuracy of the correction is 5 per cent, it can be seen that for a $C-H$ bond the relative uncertainty in the length is 10^{-3}, so that here also there is a limit on the accuracy of the order of 10^{-3} Å.

Equilibrium structures r_e can be derived, if cubic force constants are available (Lounila *et al.* 1987). Actually, a considerable part of the anharmonicity stems from the use of rectilinear coordinates for a motion which is purely harmonic when expressed in curvilinear coordinates. This 'transformation part' can be calculated from the harmonic force constants. It tends

Table 12.2 r_α structure of ethylene in the liquid-crystal solvent EBBA (Wasser and Diehl 1987); also given is the r_z structure from high-resolution IR spectroscopy (Duncan 1974*b*)

Parameter[a]	r_α structure in EBBA[b] (corrected for vibration)[c]	r_z structure[c]
r_{12}	2.4684[d]	
r_{13}	3.0916(2)	
r_{14}	1.8614(4)	
$r_{15} \equiv r(C-H)$	1.0874(3)	1.0870 ± 0.0020
r_{16}	2.1212(1)	
$r_{56} \equiv r(C=C)$	1.3438(4)	1.3384 ± 0.0010
$\angle 154 \equiv \angle H-C-H$	117.72(2)	117.37 ± 0.25

[a] See Fig. 12.3 for the numbering of nuclei.
[b] At a pressure of 3 atm (303 kPa).
[c] Distances are given in Å, angles in degrees.
[d] Assumed from the IR study.

to shorten the bonds because the distance between average positions of the nuclei moving on arcs of circles is shorter than the equilibrium value. On the other hand, the intrinsic anharmonic part of the stretching motion, because of the Morse-type potential, causes the average distance of nuclei to be larger than the equilibrium distance. These opposite effects tend to cancel out. As an example in benzene (Lounila *et al.* 1987) the transformation part shortens the C—H and C—C bonds by 1.4 and 0.2×10^{-2} Å respectively. The intrinsic part lengthens the bonds by 2.0 and 1.0×10^{-2} Å. The harmonic motions increase the apparent bond lengths by 1.7 and 0.3×10^{-2} Å, so that the observed bonds are longer than the equilibrium values by 2.3 and 1.1×10^{-2} Å.

12.8.4   Deformation of the molecule by the liquid-crystal solvent

In eqn (12.7), which defines the direct coupling, the measured entity is $\langle(3 \cos^2 \theta_{ijz} - 1)/r_{ij}^3\rangle$, which in eqns (12.8) and (12.9) is separated into $\langle 3 \cos^2 \theta_{ijz} - 1 \rangle$ and $\langle 1/r_{ij}^3 \rangle$. It is clear that this separation is only correct if r_{ij} does not depend upon θ_{ijz}, i.e. if the molecule is not deformed by the liquid-crystal solvent. However, since the molecules are oriented by the liquid crystal, there must be torques operating on the molecular bonds, torques which can also deform the molecule. The molecular shape consequently depends upon the angles between the molecular axes and the liquid-crystal axis. Actually there are definite proofs that this deformation really exists (Snijders *et al.* 1982, 1983). Molecules which have tetrahedral symmetry, as for instance methane, should normally not be able to orient; the direct couplings in this molecule should be zero. In fact proton–proton as well as proton–carbon coupling is observed, which can be attributed to a deformation of molecular shape, although the average structure is still tetrahedral.

A crude model for a pair of nuclei, which assumes exclusively a variation of the bond length, r_o, by an amount αr_o:

$$r = r_o \left[1 + \alpha \left(\cos 2\theta_z - 1\right)\right], \tag{12.19}$$

shows that the D^c which accounts for the deformation contains new terms depending upon α:

$$D^c = -K \, r_o^{-3} \left[1 + (20\alpha/7)S - 4\alpha/5 + \text{higher terms}\right]. \tag{12.20}$$

The relative correction

$$D^c/D^e = 1 + 20\alpha/7 - 4\alpha/5S \tag{12.21}$$

indicates that like the vibration correction, the deformation correction becomes particularly important for small degrees of order. For instance, for a bond length variation of 10^{-3} with a degree of order of 10^{-2}, the relative correction is 10^{-1}, i.e. the neglect of deformation induces an error of 10 per cent in the D value or 3 per cent in the distance.

Like the vibration correction, the deformation correction can now be performed by computer for all molecules for which a force field is known. The theory is described by Lounila and Diehl (1984*a, b*). In the general case, the orienting energy of the molecule, U_{ext}, depends upon the molecular orientation as well as on the deformation, which is expressed in normal coordinates Q_k. The interaction between the solute molecule and the liquid crystal is described by a tensor **A** which also varies with the normal vibrations. Consequently the resulting correction terms d_{ij}^d depend upon the derivatives of the direct coupling as well as the derivatives of the interaction tensors with respect to the normal coordinates. Also the normal frequencies of the vibrations ω_k enter the correction terms, because the addition of the linear interaction term $(\partial U_{\text{ext}}/\partial Q_k)Q_k$ to the quadratic harmonic vibration term $\omega_k^2 \, Q_k^2$ (anharmonic vibration is neglected) shifts the interaction potential minimum to a new equilibrium value $\bar{Q}_k$:

$$\bar{Q}_k = -(\partial U_{\text{ext}}/\partial Q_k)/\omega_k^2 \tag{12.22}$$

$$U_{\text{ext}} = -(1/2) \sum_{\alpha\beta} A_{\alpha\beta} \, (3 \cos \theta_\alpha \cos \theta_\beta - \delta_{\alpha\beta}) \qquad (12.23)$$

(where θ_α (θ_β) is the angle between the α (β) axis of the reference system ($\alpha, \beta = x, y, z$) and the director, and $\delta_{\alpha\beta}$ is the Kronecker delta);

$$d_{ij}^{d} = -K \sum_{k} (1/\omega_k^2) \left\{ (3/10) \sum_{\alpha\beta} (\partial \Phi_{\alpha\beta}^{ij}/\partial Q_k) \, (\partial A_{\alpha\beta}/\partial Q_k) + \right.$$

$$(12.24)$$

$$\left. (2/7) \sum_{\alpha\beta\mu} [3(\partial \Phi_{\alpha\mu}^{ij}/\partial Q_k) \, (\partial A_{\beta\mu}/\partial Q_k) - (\partial \Phi_{\mu\mu}^{ij}/\partial Q_k) \, (\partial A_{\alpha\beta}/\partial Q_k)S_{\alpha\beta}] \right\}$$

(where $\Phi_{\alpha\beta}^{ij} = \langle \cos \theta_{ij\alpha} \cos \theta_{ij\beta}/r_{ij}^3 \rangle$; $\theta_{ij\alpha}$ is the angle between the internuclear vector $\mathbf{r}_{ij}$ and the α axis of the reference system). The interaction tensor elements $A_{\alpha\beta}$, according to classical Boltzmann statistics, also determine the degrees of order of the molecule:

$$S_{\alpha\beta} = (1/Z_c) \int (3 \cos \theta_\alpha \cos \theta_\beta - \delta_{\alpha\beta}) \exp (-U_{\text{ext}}/k_B T) \, \mathrm{d}\Omega$$

$$(12.25)$$

$$Z_c = \int \exp (-U_{\text{ext}}/k_B T) \, \mathrm{d}\Omega, \qquad (12.26)$$

where k_B is Boltzmann's constant, and Ω denotes the orientation of the molecular axis system with respect to the laboratory fixed frame.

It is assumed that the **A** tensor can be expressed in terms of additive segmental interaction tensors corresponding to bonds or functional groups of the solute molecules and that bond stretching which contributes much less than bending can be neglected.

Finally, the fit of measured direct couplings leads to the simultaneous determination of the structure, the harmonic corrections and a set of interaction parameters, which, at the same time, define the degrees of order. Now there is a considerably increased number of unknowns and, usually, data from several liquid-crystal solvents must be combined.

Deformation corrections have already been applied to several molecules such as benzene (Lounila and Diehl 1984b), methyl iodide and methyl fluoride (Lounila et al. 1985), hydrogen cyanide (Dombi et al. 1984), tellurophene (Diehl et al. 1987a), cyclopropane (Kellerhals et al. 1987), ethylene and 1,1-difluoroethylene (Wasser and Diehl 1987), allene (Diehl et al. 1987b) and thiophene (Diehl et al. 1988).

Table 12.3 Distance ratios and angles in ethylene, showing the effect of different corrections (Wasser and Diehl 1987)[a]

	r_{14}/r_{12}	r_{56}/r_{12}	$\angle 154$ (degrees)
Solvent ZLI 1167 without vibration correction	0.7448(1)	0.5397(1)	116.57(2)
Solvent EBBA without vibration correction	0.7589(2)	0.5389(2)	117.43(2)
Solvent EBBA with vibration correction	0.7541(2)	0.5444(2)	117.72(2)
Result corrected for vibration and deformation[b]	0.7521(13)	0.5417(6)	117.30(3)
r_z values from high-resolution IR spectroscopy[c]	0.7525 ± 0.0022[d]	0.5422 ± 0.0011[d]	117.37 ± 0.25

[a] See Fig. 12.3 for the numbering of nuclei.
[b] Independent of solvent.
[c] Duncan (1974b).
[d] Calculated from the structural parameters given in the original paper.

Table 12.4 Comparison of structural results for allene and cyclopropane, as obtained by different techniques; the NMR results are corrected for harmonic vibration as well as deformation

Technique[a]	Type of structure	$r(C—H)/r(C—C)$	$\angle H—C—H$ (degrees)	References
(a) Allene, $H_2C=C=CH_2$				
NMR	r_α	0.8315(14)	118.12(5)	Diehl et al. (1987b)
IR	r_0	0.8309 ± 0.0010^{b}	118.17 ± 0.17	Maki and Toth (1965)
ED	r_a	$0.8247^{b,c}$	118.4^{c}	Almenningen et al. (1959)
(b) Cyclopropane, C_3H_6				
NMR	r_α	0.7154(16)	115.17(13)	Kellerhals et al. (1987)
RA	r_0	0.7163 ± 0.0024^{b}	114.0 ± 0.7	Butcher and Jones (1973)
ED + IR	r_z	0.7166 ± 0.0014^{b}	114.5 ± 0.9	Yamamoto et al. (1985)

[a] The following abbreviations are used: NMR, liquid-crystal NMR spectroscopy; IR, infrared spectroscopy; ED, electron diffraction; RA, Raman spectroscopy.
[b] Calculated from the r values given in the original paper.
[c] No error is given in the original paper.

It turns out that the deformation corrections are of similar importance to the vibration corrections, also becoming particularly large for small degrees of order. Results for ethylene are presented in Table 12.3.

In methyl iodide (Lounila *et al.* 1985), where the H—C—H angle, without deformation corrections, apparently varies with the solvent between 111.41° and 113.55°, the discrepancy is eliminated by the correction. The resulting solvent-independent angle is 111.58(1)°; the value from microwave spectroscopy is 111.67 ± 0.20° (Duncan 1974a). It is now also possible to determine the true maximum distortion of the molecule by the liquid crystal, which is a C—H bend of 0.1°. This true deformation is enlarged by the correlation effect to an apparent variation which is 20 times bigger, about 2°.

For all the molecules studied so far the dominant contribution to the deformation correction stems from C—H bending. Furthermore the anisotropy of C—H bond interaction tensors in all the molecules has been observed to be linearly dependent upon the anisotropy of the C—H bond interaction tensors of methane in the same solvents. This indicates that similar interactions, probably van der Waals', must be acting in all the molecules. It seems tempting to use an 'ideal' solvent for which the anisotropy of the C—H bond interaction tensors of methane disappears, in order to minimize the deformation effects (Jokisaari *et al.* 1984). However,

Table 12.5 Comparison of structural results for thiophene, as obtained by liquid-crystal NMR spectroscopy (corrected for harmonic vibration and deformation; Diehl *et al.* (1988)) and by microwave spectroscopy (Bak *et al.* 1961) (nuclei are numbered as shown in Fig. 12.5)

Distance ratios	NMR spectroscopy[a]	MW spectroscopy[b,c]
r_{12}/r_{23}	0.9964(9)	0.999 ± 0.005
r_{14}/r_{23}	1.7236(11)	1.719 ± 0.004
r_{15}/r_{23}	0.4087(6)	0.4082 ± 0.0008
r_{26}/r_{23}	0.4084(4)	0.4092 ± 0.0008
r_{56}/r_{23}	0.5197(5)	0.5187 ± 0.0009
r_{67}/r_{23}	0.5404(3)	0.5391 ± 0.0011
Angles (degrees)	NMR spectroscopy[a]	MW spectroscopy[b]
∠156	128.07(8)	128.68 ± 0.81
∠267	124.24(4)	124.27 ± 0.07
∠567	112.41(5)	112.45 ± 0.18

[a] r_α structure.   [b] r_s structure.
[c] Calculated from the structural parameters given in the original paper.

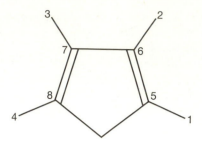

Fig. 12.5 Numbering of nuclei in thiophene.

such a solvent is not ideal, because deformation contributions due to C—H bends to various couplings do not disappear exactly with the methane couplings. Also different bond interactions, e.g. C=C, C≡N or C≡C, may contribute to the molecular distortion, even if the C—H contribution disappears. Finally, for small degrees of order, the relative deformation contribution to the direct couplings generally diverges, introducing drastic solvent effects even in the 'ideal' solvent.

So far, structures corrected for molecular deformation agree very well with those obtained by other methods of structure determination. Some examples are presented in Tables 12.4 and 12.5.

12.9   Combination of liquid-crystal NMR spectroscopy with other techniques of structure determination

Quite early in the development of liquid-crystal NMR spectroscopy Englert and Saupe (1969) suggested the use of a combination of NMR data with data from microwave spectroscopy in order to derive absolute distances and to improve the accuracy of structure determination. As an example they analyzed the molecular structure of acetonitrile. Lately electron diffraction data have also been included in such combinations (Cradock *et al.* 1988; Rankin 1988), and molecular structures have been derived by giving weights to all the fitted parameters which are inversely proportional to the squares of their uncertainties. In such combined fits, which have the advantage of using complementary information from the various methods (e.g. absolute molecular size from ED and MW, precise relative proton positions from NMR), the following criteria must be fulfilled for the NMR data:

1. Harmonic vibration as well as deformation corrections must be performed.

2. Measurement errors of direct couplings should generally be increased by 0.2 to 0.5 Hz because of the various approximations which are used in

the harmonic vibration correction and, particularly, in the deformation correction.

3. Direct couplings between directly bonded 'heavy' atoms such as C, N, and F should be excluded from the fit because of their anisotropic indirect coupling component (see Section 12.8.2).

4. Variances as well as covariances of the direct couplings should be considered.

12.10 Conclusions

NMR spectroscopy of oriented molecules is a fast and efficient method of structure determination for small molecules (10 magnetic nuclei or less) or small parts of larger spin systems. The accuracy which may easily be reached is of the order of 10^{-3} Å in distances and 10^{-1} degrees in angles. A higher precision of measurement is possible but not meaningful because of the neglect of J-coupling anisotropy, the imprecision of harmonic vibration corrections and the approximations introduced by the theory of correlated deformation.

It should be kept in mind that, unfortunately, the method is not dealing with free molecules but with molecules which must interact with the surrounding anisotropic medium in order to be oriented, and are deformed by this interaction.

References

Almenningen, A., Bastiansen, O., and Traetteberg, M. (1959). *Acta Chem. Scand.*, **13**, 1699–702.

Bak, B., Christensen, D., Hansen-Nygaard, L., and Rastrup-Andersen, J. (1961). *J. Mol. Spectrosc.*, **7**, 58–63.

Butcher, R. J. and Jones, W. J. (1973). *J. Mol. Spectrosc.*, **47**, 64–83.

Canet, D., Marchal, J.-P., and Sarteaux, J.-P. (1974). *Compt. Rend. Acad. Sci.*, **B279**, 71–3.

Cradock, S., Liescheski, P. B., Rankin, D. W. H., and Robertson, H. E. (1988). *J. Am. Chem. Soc.*, **110**, 2758–63.

Diehl, P. (1985). Molecular structure from dipolar coupling. In *Nuclear magnetic resonance of liquid crystals* (ed. J. W. Emsley), pp. 147–80. Reidel, Dordrecht.

Diehl, P. and Vogt, J. (1976). *Org. Magn. Reson.*, **8**, 638–42.

Diehl, P., Kellerhals, H. P., and Niederberger, W. (1971a). *J. Magn. Reson.*, **4**, 352–7.

Diehl, P., Henrichs, P. M., and Niederberger, W. (1971b). *Mol. Phys.*, **20**, 139–45.

Diehl, P., Sýkora, S., and Vogt, J. (1975). *J. Magn. Reson.*, **19**, 67–82.

Diehl, P., Kellerhals, M., Lounila, J., Wasser, R., Hiltunen, Y., Jokisaari, J., and Väänänen, T. (1987a). *Magn. Reson. Chem.*, **25**, 244–7.

Diehl, P., Baraldi, C., Kellerhals, M., and Wasser, R. (1987*b*). *J. Mol. Struct.*, **162**, 333-9.

Diehl, P., Ugolini, R., Kellerhals, M., and Wasser, R. (1988). *Mol. Cryst. Liq. Cryst.*, **159**, 267-76.

Dombi, G., Diehl, P., Lounila, J., and Wasser, R. (1984). *Org. Magn. Reson.*, **22**, 573-5.

Duncan, J. L. (1974*a*). *J. Mol. Struct.*, **22**, 225-35.

Duncan, J. L. (1974*b*). *Mol. Phys.*, **28**, 1177-91.

Emsley, J. W. and Lindon, J. C. (1975). *NMR spectroscopy using liquid crystal solvents*. Pergamon Press, Elmsford.

Englert, G. and Saupe, A. (1964). *Z. Naturforsch.*, **19a**, 172-7.

Englert, G. and Saupe, A. (1969). *Mol. Cryst. Liq. Cryst.*, **8**, 233-45.

Harris, R. K. (1983). *Nuclear magnetic resonance spectroscopy*. Pitman, London.

Horne, D., Kendrick, R. D., and Yannoni, C. S. (1983). *J. Magn. Reson.*, **52**, 299-304.

Jokisaari, J., Hiltunen, Y., and Väänänen, T. (1984). *Mol. Phys.*, **51**, 779-91.

Kellerhals, M., Diehl, P., Lounila, J., and Wasser, R. (1987). *J. Mol. Struct.*, **156**, 255-60.

Khetrapal, C. L. and Kunwar, A. C. (1982). Oriented molecules. In *Nuclear magnetic resonance*, Vol. 11 (ed. G. A. Webb), pp. 248-63. The Royal Society of Chemistry, London.

Lounila, J. and Diehl, P. (1984*a*). *J. Magn. Reson.*, **56**, 254-61.

Lounila, J. and Diehl, P. (1984*b*). *Mol. Phys.*, **52**, 827-45.

Lounila, J. and Jokisaari, J. (1982). *Progr. NMR Spectrosc.*, **15**, 249-90.

Lounila, J., Diehl, P., Hiltunen, Y., and Jokisaari, J. (1985). *J. Magn. Reson.*, **61**, 272-83.

Lounila, J., Wasser, R., and Diehl, P. (1987). *Mol. Phys.*, **62**, 19-31.

Luz, Z. (1985). Dynamics of molecular processes by NMR in liquid crystalline solvents. In *Nuclear magnetic resonance of liquid crystals* (ed. J. W. Emsley), pp. 315-42. Reidel, Dordrecht.

Maki, A. G. and Toth, R. A. (1965). *J. Mol. Spectrosc.*, **17**, 136-55.

Meiboom, S., Hewitt, R. C., and Snyder, L. C. (1972). *Pure Appl. Chem.*, **32**, 251-61.

Menger, E. M., Vega, S., and Griffin, R. G. (1984). *J. Magn. Reson.*, **56**, 338-42.

Rankin, D. W. H. (1988). Combined application of electron diffraction and liquid crystal NMR spectroscopy. In *Stereochemical applications of gas-phase electron diffraction* (ed. I. Hargittai and M. Hargittai), Part A, pp. 451-82. VCH, New York.

Saupe, A. and Englert, G. (1963). *Phys. Rev. Lett.*, **11**, 462-4.

Schumann, Ch. (1980). Nuclear magnetic resonance and electron spin magnetic resonance. In *Handbook of liquid crystals* (ed. H. Kelker and R. Hatz), pp. 426-509. Verlag Chemie, Weinheim.

Snijders, J. G., de Lange, C. A., and Burnell, E. E. (1982). *J. Chem. Phys.*, **77**, 5386-95.

Snijders, J. G., de Lange, C. A., and Burnell, E. E. (1983). *Isr. J. Chem.*, **23**, 269-81.

Stephenson, D. S. and Binsch, G. (1980). *Org. Magn. Reson.*, **14**, 226–33.

Sýkora, S., Vogt, J., Bösinger, H., and Diehl, P. (1979). *J. Magn. Reson.*, **36**, 53–60.

Veracini, C. A. and Longeri, M. (1985). Studies of solutes with internal rotors. In *Nuclear magnetic resonance of liquid crystals* (ed. J. W. Emsley), pp. 123–46. Reidel, Dordrecht.

Wasser, R. and Diehl, P. (1987). *Magn. Reson. Chem.*, **25**, 766–70.

Yamamoto, S., Nakata, M., Fukuyama, T., and Kuchitsu, K. (1985). *J. Phys. Chem.*, **89**, 3298–302.

13

Quantum mechanical determination of static and dynamic structure

James E. Boggs

13.1 Introduction

Recent years have witnessed an explosive proliferation in the use of all kinds of quantum chemical methods to evaluate molecular structures and vibrational dynamics. Semiempirical procedures have made major contributions in areas such as understanding the mechanisms of drug–substrate interactions, interpretation of organic mechanisms by study of reaction pathways and transition states, and in other applications where relatively rapid computation of molecular properties at a medium level of accuracy provides the needed information. However, this book is directed toward the experimental and theoretical methods that can provide structural information at the highest levels of accuracy and reliability so only *ab initio* theoretical methods or such methods with the computed parameters modified by minor empirical corrections are considered here. Accuracy in bond lengths to a few thousandths of an ångström and in bond angles to a few tenths of a degree may be taken as an appropriate goal.

Along with the requirement of seeking highly accurate results, emphasis is placed on approaches that can treat molecules of a size comparable to those studied by the experimental procedures discussed elsewhere in this book. As might be expected, the three factors of molecular size, required accuracy, and available computational capability are closely interrelated, so that much attention must be paid to achieving the optimum compromise among them for solving a given chemical problem.

There are a number of other limitations to computational structural chemistry in its present stage of development. Much current theoretical research is devoted to elimination of these barriers, but a thorough evaluation of all of these areas is beyond the scope of the present work. In this chapter, attention is limited to molecules containing only the lighter (pre-transitional element) atoms for which relativistic and spin considerations can safely be ignored. Also, attention is confined to molecules that are in or near one of their stable conformations. This specifically requires omission of the tremendously interesting field of study of reactive transition states, although these can be treated well by *ab initio* MO methods, at least for relatively small molecules (see Bernardi and Robb 1987). The subject is omitted here primarily because the experimental techniques around which this volume is centered cannot directly provide structural information on molecular configurations which are merely saddle points on the potential surface. Within these limitations imposed on the present discussion, there remain broad and important areas in which computational determination of molecular structures can be fully comparable to high-quality experimental determinations.

Many scientists reserve the words 'determination of structure' to refer only to results from an experiment, no matter how much theoretical interpretation may have gone into the analysis of that experiment, and speak of 'prediction' when referring to a theoretical result that does not rely directly on an experiment. Others use the word determination in a more general way to mean obtaining information, always necessarily of limited believability, whether it comes from experiment or from theory. The word is used in the latter sense here, since it is believed that an experimental result must be examined with as great skepticism as a computational result before it is accorded the tentative acceptance that we mean by scientific belief. Just as experimental methods vary in reliability, so do theoretical methods, and we wish to examine the latter question.

13.2 Quantum chemical background

There are numerous good introductory textbooks in quantum chemistry. A recent intermediate level text deserving mention is the book by Szabo and Ostlund (1982). More specialized treatments of direct relevance to the present topic are found in Volumes 3 and 4 of Schaefer's series on Modern Theoretical Chemistry (Schaefer 1977*a*, *b*) and Volumes 67 and 69 of Advances in Chemical Physics (Lawley 1987*a*, *b*). Other general references are given as appropriate below.

Time-independent molecular geometries and the expansion coefficients in the vibrational potential function could be calculated exactly from an exact solution of the Schrödinger equation and can be calculated approximately from an approximate solution. The entire problem comes down, then, to the question of whether useful accuracy in the calculation can be obtained from solutions that are practical with existing theoretical ingenuity and computational facilities. We shall first examine the approximations that are *commonly* made, although it must be emphasized that at least partial correction for some of these approximations can be made in special circumstances.

First, it is customary to begin with the Born–Oppenheimer approximation that the nuclear and electronic wavefunctions are separable. Next, all relativistic and spin coupling terms are omitted from the Hamiltonian. As mentioned above, these terms may be highly important for heavier atoms, but there is no evidence that they contribute significantly to calculations involving only atoms lighter than the transition elements. The total electronic wavefunction is then expanded in a set of real spin orbitals subject to the usual conditions of orthonormality. If the molecular system does not contain unpaired electrons, it is normally considered completely adequate to use a restricted Hartree–Fock (RHF) procedure in which a pair of electrons with opposite spin is associated with a single spatial orbital. Although some complexities are introduced, it is possible if desired to use the unrestricted (UHF) approach where this constraint is not imposed. The only really significant approximation introduced so far is the neglect of dynamic electron correlation. Unless correction is made for this, serious errors result in computed energies, molecular geometries, vibrational force constants, and other properties.

The second major error in the common approach follows in the next step which involves expansion of each of the individual molecular orbitals in some chosen orthonormal basis set (Szabo and Ostlund 1982). If the basis set could be infinite, the method would be exact, but in practice the basis set is not only finite but often rather small. In order to mimic the actual electron distribution using as few basis functions as possible, it is customary to choose atom-like functions, occasionally Slater but much more commonly Gaussian functions. Much effort has gone into evaluation of favorable basis sets of various sizes. A minimal basis set consists of one Slater-type function for each pair that is filled, or partially filled, in the atomic ground state or, more commonly, a set of Gaussian-type functions contracted to that number of functions. Thus a minimal basis set for carbon might be described as a $7s3p$ set contracted to $2s1p$. In general, the contraction coefficients and the Gaussian orbital exponents are optimized on atoms or simple hydrides and are carried over unchanged to molecules. If a basis set contains two sets of Slater functions or two sets of contracted Gaussian functions for each atomic shell, it is said to be double-zeta (e.g. a $9s5p$ set of Gaussian functions for carbon contracted to $4s2p$) with triple-zeta similarly defined. A more detailed discussion of such basis sets can be found in the review by Dunning and Hay (1977) or the one by Wilson (1987).

If a Gaussian basis set is to be expanded beyond double zeta, it is most useful (in terms of obtaining the greatest improvement in computed quantities with the fewest added terms) to add atomic functions with higher values of the angular momentum quantum number, i.e. d functions or even f functions. It should not be a matter of concern that these are added to atoms such as carbon, because their purpose is primarily to provide greater angular flexibility in the approximate description of the molecular electron distribution. They are particularly important in computing accurate bond angles around atoms with unshared electron pairs, such as nitrogen or oxygen, and in the accurate computation of dihedral angles involving such atoms.

One widely used variation of the basis set involves so-called split valence-shell bases which are double-zeta in the valence shell and minimal in the inner shells. In such bases, it is common to use identical orbital exponents for s and p functions. A description such as 3–3–21 for a sulfur atom basis, for example, indicates that 3 s-type Gaussian functions with predetermined, relatively large orbital exponents have been contracted with predetermined coefficients to make one minimal s-type orbital to describe the innermost shell of the sulfur atom. The second shell is described by a single s and single set of p_x, p_y, and p_z functions, each derived from 3 somewhat more diffuse Gaussian functions (functions with smaller negative exponents). The final 21 in the basis set description shows that the outermost shell of sulfur is described by two sets of s and p orbitals, one composed of two contracted Gaussians and the other a single uncontracted Gaussian. This basis set is thus minimal in the inner two shells but double-zeta in the valence shell. The addition of a final 'G' to the nomenclature for such a basis (3–3–21G) indicates nothing but a reminder that it is composed of Gaussian functions. Addition of polarization functions (d functions) is indicated by some series of *s, which unfortunately is not fully standardized. A careful author will specify precisely what basis set has been used in his calculations since the size and composition of the basis set is crucial in determining the accuracy of computed molecular properties.

While dynamic electron correlation is ignored in many computational determinations of structure, and this is often justifiable when relative values are sought, much modern research utilizes techniques in which at least partial correction for correlation effects is included. It is very important to understand the effect of neglect of electron correlation and the conditions under which the approximation of its neglect can be tolerated.

A great variety of methods exists for computing correlated wavefunctions and there is at the present time no consensus of opinion on which is most effective and efficient. One very widely used approach involves a perturbational expansion of the correlation energy (Møller and Plesset 1934). In this expansion, the Hartree–Fock energy is the sum of the zeroth and first-order terms, while the correlation correction first appears in the second-order energy. Second-order perturbation theory (often symbolized as MP2) is sometimes adequate, although there is evidence that it tends to over-correct for the correlation effect on some properties. Third- and fourth-order terms (MP3 and MP4) are sometimes computed, and MP4 generally gives very fine agreement with experiment.

Another common method for the inclusion of electron correlation is the configuration interaction (CI) technique (Foster and Boys 1960), often using single and double CI replacements (CISD). This method is variational and can be expanded with some difficulty to include triple replacements, but it has certain problems resulting from lack of size consistency. Numerous other approaches have been proposed and are in current use by different research groups.

13.3   Determination of molecular structure

The purpose of a calculation of molecular structure is to find the nuclear configuration, or one or more of the several nuclear configurations, which are local minima on the potential energy surface. At such a configuration, the forces on all the nuclei are zero and the matrix of second derivatives is positive definite. This is by definition the equilibrium (r_e) geometry. Comparison of such a structure with the variously averaged structures determined by experiment (see Chapter 2, Section 2.4) requires a careful consideration of the experiment and the manner in which it performs its particular kind of vibrational averaging.

Computational determination of the minimum-energy nuclear configuration proceeds in the following manner:

1. A nuclear geometry is assumed as an initial guess. The accuracy of this guess does not affect the accuracy of the final result, but it does affect the number of iterations which will be needed in the subsequent steps and it also determines which of several equilibrium geometries will be found in case the system has multiple possible conformations.

2. With the assumed nuclear geometry, the electronic Schrödinger equation is solved as described in the previous section. The accuracy with which this is done is the sole factor affecting the accuracy of the final result.

3. The approximate wavefunction obtained in the second step is used to calculate the expectation value of the first derivative of energy with respect to nuclear coordinates. This is the heart of the gradient technique, first developed by Pulay (1969), and described in detail in the reviews by Pulay (1977, 1987). It should be pointed out that it is essential to compute the full derivative and not just the Hellmann–Feynman contribution to it. The computation of energy derivatives is one of the major developments in quantum

chemistry in recent years since it has made application of quantitative MO techniques to larger molecules a practical matter.

4. A molecular force field is now assumed. Again, the accuracy of the guess makes no difference to the accuracy of the final result, but it does affect the amount of computing time needed to get there. Usually only the harmonic force constants are used and all coupling constants and anharmonic terms are taken as zero. The computed forces on the nuclei and the assumed force constants are now used to predict a new nuclear geometry. For details, see Pulay *et al.* (1979). Alternative approaches using analytical second and higher derivatives of the wavefunction are described by Pulay (1987).

5. Steps 2–4 are iterated until the forces become as close to zero as desired. The result is the fully relaxed equilibrium geometry of the molecule. Some computer programs are designed to carry out this iterative procedure automatically.

For practical use, several factors are important. Firstly, it is useful to optimize the geometry using a set of internal coordinates in which the vibrational force field is as nearly diagonal as possible so that the neglect of interaction terms in the assumed force field is reasonable. A suitable set is suggested in Pulay *et al.* (1979). Secondly, it is necessary to be sure that the zero-force geometry obtained is truly a minimum-energy geometry and not a saddle point of the energy surface. This is only a hazard if there are periodic motions such as ring inversion or internal rotation or if some molecular symmetry element can lead to a zero-force maximum in some coordinate. A test can be made in such a situation by making a small displacement along the suspicious coordinate. This will lead either to a restoring force or a force leading away from a saddle point. Alternatively, although with greater computational effort, the second derivatives of the energy can be calculated at the final geometry as a test for a true minimum.

Computational demands for the work can be quite varied, tied in as they are with the size of the molecular system under study and the imposed requirement for accuracy of the result. Calculations as described above are routinely run on computers of all sizes ranging from minicomputers to Cray vector processors. Ready access to the computer and available storage capacity are at least as important as computer speed. A great variety of computer programs are in use, each with their special merits. The most common is the GAUSSIAN system of programs that has been developed, maintained, and circulated through the dedicated efforts of the Pople group.†

† For information on the GAUSSIAN program system, contact Professor John A. Pople, Department of Physics, Carnegie–Mellon University, Pittsburgh, Pennsylvania 15213, USA.

The latest version, GAUSSIAN 90, is very convenient, powerful, and versatile.

13.4   Accuracy and the use of empirical corrections

From the viewpoint of the user of structural information, it is probably more useful to set a target of accuracy and inquire to what extent it can be achieved rather than simply to ask what is the highest accuracy that can be obtained. For molecules of a size sufficiently large to present complex structural and conformational questions, say up to about 20 atoms, the more sophisticated experimental techniques can give accuracies of a few thousandths of an ångström and a few tenths of a degree in favorable cases. This might then be considered a reasonable goal to strive for by computational methods. Higher accuracy would be nice, although it could not be checked by comparison with experiment but only by internal consistency (except for molecules containing no more than four or five atoms, like those considered in Chapter 4).

The stated accuracy goal can be achieved directly for small molecules. For example, recent calculations for HCN, HF, and NH_3 (Pulay *et al.* 1983*a*) gave at least the desired accuracy in bond lengths and angles using a large basis set (triple-zeta in the valence shell plus two sets of polarization functions) and a good treatment of electron correlation (all singles and doubles plus a coupled cluster correction). Unfortunately, calculations approaching this level are not possible for molecules in the larger size range considered here. The main problem is the treatment of electron correlation. Standard methods involving configuration interaction or MP4 perturbation theory have a time requirement varying with the number of electrons in the molecule as n^6. This means that a 1000-fold increase in the speed of computers would increase the size of the molecule that could be handled by a factor of only 3.2. Currently, the major thrust of theoretical work is the effort to solve the correlation problem in a way that is both computationally practical and sufficiently accurate. Larger computers alone won't do it.

If a molecule that contains over a dozen atoms is considered, the largest currently practical basis set is one that is approximately double-zeta with the addition of polarization functions. Such a basis gives bond angles that are accurate to approximately the stated goal. Even with the omission of the polarization functions, the angles are generally this accurate except for bond and dihedral angles around oxygen, nitrogen, and other atoms with unshared valence electron pairs. Unfortunately, the situation with bond lengths is not so satisfactory since these are computed with errors ranging up to a few hundredths of an ångström. No calculation is possible that can directly give the desired level of accuracy in bond lengths without use of both a large basis set and a good treatment of electron correlation and this is currently impossible for molecules of the size under consideration.

The situation is saved by the observation that errors in computed bond lengths are highly systematic, provided the basis set is of adequate size (generally double-zeta, at least in the valence shell). A C—H bond length, for example, is frequently calculated to be 0.005 Å smaller than the best experimental estimates (Pulay *et al.* 1979). Other bonds have similar characteristic errors, so that it is possible to add a pre-determined correction, or 'offset' value, to the computed length. There may be esthetic objections to such a 'contamination' of results from *ab initio* MO theory by the use of empirical data, but if the objective is to obtain meaningful information at this level of accuracy from theory, there is no other way. Only extensive comparison with high-quality experimental results can show how meaningful and how accurate the results are, and a large body of evidence has been accumulated with extremely encouraging results. A number of compilations of offset values exist (Pulay *et al.* 1979; Schäfer 1983).

The relationships can be visualized by sketching, as in Fig. 13.1, the error in computation of the length of a given type of bond against the size of basis set used. As shown, the infinite basis-set limit deviates from the true value if electron correlation is neglected. The shaded area represents the results obtained for calculations with the bond imbedded in a wide variety of

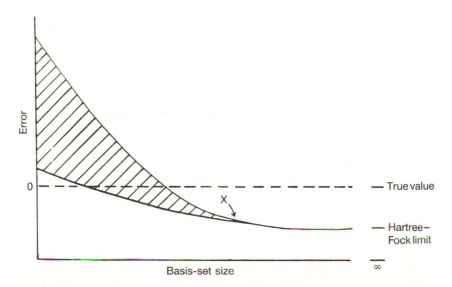

Fig. 13.1 Schematic representation of the error in calculating a single structural parameter (such as C—H distance) as a function of basis-set size in a variety of molecular environments. For wide families of compounds, the error falls within the shaded area. After Boggs and Cordell (1981).

molecular environments (as C—H in methane, benzene, imidazole, ethyl acetate, etc.). If a small basis set is used, the result might by chance be better in a given case than it is at the Hartree–Fock limit, but the error is unpredictable. Using the philosophy that a known, fixed error is to be preferred to a chance possibility of getting an exactly correct answer, we believe that calculations should be made at the point marked X, using the smallest basis set (for economy) that produces an adequately small variability in the error. In general, minimal basis sets or semiempirical calculations cannot give a variability small enough to satisfy the accuracy requirements stated above, although such calculations are extremely valuable for many other purposes.

Correction with offset values is necessary to obtain the best possible estimate available from theory for the *absolute* values of bond lengths. Most often, however, this is not the information that is required for purposes of chemical interpretation since the length of a chemical bond, by itself, is seldom of interest. What is of use is the *difference* in length of similar bonds in related systems; for example, how the relative C—C bond lengths in cyclopropane are affected by various substitutions on the ring. When the *ab initio* MO calculation is done at a level such that the error for a given type of bond is constant, these *relative* values are obtained directly without any empirical corrections. Questions of this type are, happily, both the ones that are of most chemical interest and most accurately answered directly by theory.

No systematic review exists of the structural results that have been obtained by *ab initio* MO methods. The results are probably now too numerous for such a review to be feasible. A very useful bibliographical source, however, is the compilation of references to *ab initio* MO publications given at the end of each year in a special volume of the Journal of Molecular Structure (Theochem).

13.5 Computational study of molecular vibrational dynamics

As seen in Chapter 2, Section 2.3, the vibrational potential energy of a molecule can be expanded as a function of internal displacement coordinates, q_i, in the form

$$V = V_{\text{ref}} + \sum_i g_i q_i + \frac{1}{2} \sum_i \sum_j F_{ij} q_i q_j + \frac{1}{6} \sum_i \sum_j \sum_k F_{ijk} q_i q_j q_k + \ldots$$

$$(13.1)$$

The g_i are forces acting on the nuclei, which are zero at the equilibrium geometry leaving the quadratic term as the first contributor to the change in energy on vibrational displacement. The expansion coefficient

$$F_{ij} = (\partial^2 V / \partial q_i \, \partial q_j)_e \qquad (13.2)$$

is an ordinary force constant if $i = j$ and a coupling constant if i and j are different. The coefficients F_{ijk} are the corresponding cubic anharmonic terms, etc. It is convenient to order the terms F_{ij} as a matrix, called the **F** matrix, where the force constants, F_{ii}, are the diagonal terms and the coupling constants, F_{ij}, are the off-diagonal terms.

Any complete, non-redundant set of coordinates can be used for the expansion of eqn (13.1). Different coordinate sets are convenient for different purposes, and the methods for converting between coordinate sets are well known (see again Chapter 2, Section 2.3). In many cases, it is useful to carry out calculations in coordinate sets which lead to an **F** matrix that is as nearly diagonal as possible. In other cases, it is simpler to use Cartesian coordinates. Symmetry coordinates are useful in simplifying the understanding of the vibrational spectrum, as long as the treatment is restricted to the harmonic oscillator approximation.

The F_{ij}, F_{ijk}, etc. terms can be computed by the same general methods described in Section 13.3 for computing the forces, g_i. Once an approximate wavefunction is obtained, the expectation values of the quantities shown in eqn (13.2) or the corresponding anharmonic constants can be calculated. Methods have been described for such calculations for a variety of types of wavefunctions—Hartree–Fock, multiconfiguration, CI, etc. Implementation of some of these in an efficient computer program is not trivial, but it has been done in most cases and the resulting demand on computer time is such that the computations are practical. Alternatively, the first derivative forces, g_i, can be calculated at the equilibrium and one or more geometries resulting from small displacement along each of the coordinates q_i, and the higher derivatives calculated numerically.

The question arises concerning how accurately vibrational force fields can be calculated with existing computational facilities and whether such results can be of use to supplement information obtainable by experiment. Again, it may be useful to set a target for desirable accuracy and see to what extent it can be reached. One might choose, at least as an interim goal, to accept the independent harmonic oscillator approximation as a limiting factor, which corresponds to being willing to accept an error of perhaps 2 per cent in the force constants, which is equivalent to 1 per cent in computed fundamental vibrational frequencies. The study of HF, HCN, and NH_3 referred to above (Pulay *et al.* 1983a) showed that direct calculation did not quite reach this level of accuracy for the diagonal harmonic constants even with the large basis sets and extensive CI treatment used there. On the other hand, it was found that the harmonic coupling constants and the cubic and quartic anharmonic constants were calculated directly at a level of accuracy within experimental uncertainty. Thus it appears that for these small molecules, direct computation may be a useful supplement to experiment by furnishing acceptable numbers for the small correction terms, leaving the main diagonal constants to be fitted to experimental data.

The level of quantum chemical treatment that can be done on larger molecules, however, gives very poor direct results for all force and interaction constants. Again, however, it is found that the errors tend to be systematic. Bond-stretching constants are always computed too large by approximately the same amount. Angle deformations are also computed with similar errors in related molecules. For some recent reviews, see Pulay (1981), Pulay *et al.* (1983*b*), and Fogarasi and Pulay (1984, 1985).

The most thoroughly developed scaling procedure derives a diagonal matrix of scale factors, C, to correct the computed force-constant matrix, F_{th}, according to $F = C^{1/2} F_{th} C^{1/2}$ (Pulay *et al.* 1983*b*). Coordinates that are similar in a chemical sense are scaled jointly, so that the number of independent parameters is smaller than the dimension of F. The coupling force constants are not usually scaled independently, but F_{ij} is scaled by $(C_i C_j)^{1/2}$. The small set of scale factors is then adjusted by least-squares refinement to give the best fit of the fundamental frequencies predicted by F to the observed fundamental frequencies. The force field, F, scaled by these optimized scale factors, has been called a 'scaled quantum mechanical' (SQM) force field (Pulay *et al.* 1983*b*).

The practical utility in deriving an SQM force field must lie in the degree to which the scale factors can be transferred from a molecule for which the spectrum is well known to a related molecule for which the spectrum is unknown. Since the scale factors represent the *errors* in the quantum mechanical calculation of the force constants, it would be expected that they would be much more reliably transferred than would the force constants themselves. It is also obvious that if the errors are very small (the factors approach 1), the accuracy of the method is vastly improved. Accordingly, it is advantageous to perform the calculations at the highest practical quantum mechanical level. Tests have shown that useful results can be obtained even from semiempirical calculations, but the scale factors are very far from unity and the accuracy of reproduction of spectra is considerably inferior to that which can be obtained from an *ab initio* MO calculation with a double-zeta basis set or better.

The use of SQM force fields to predict vibrational spectra has been tested quite extensively, and all of the results obtained so far indicate that the method can have important and widespread use. In one set of tests, the force fields of benzene (Pulay *et al.* 1981), pyridine (Pongor *et al.* 1984, 1985), naphthalene (Sellers *et al.* 1985) and aniline (Niu *et al.* 1985) were calculated at the double-zeta Hartree–Fock level. The force field of benzene was then fitted with a small set of scale factors to the known experimental spectra of the normal and deuterated forms of the molecule. It was then pretended that the spectra of pyridine, naphthalene, and aniline were unknown, and the scale factors derived from benzene were transferred without change to scale the computed force fields for these molecules. From the resulting SQM force fields, the spectra of normal and isotopically substituted forms were calculated, and comparison was made between the final computed results and experimental data. In all cases the agreement was well within the 1 per cent error in frequencies selected as a goal. In fact, the agreement was probably better than would be expected considering that the harmonic oscillator approximation was used, indicating that the scale factors transferred some correction for similarities in anharmonicity in the related molecules as well as transferring the error in the quantum mechanical calculation.

Several applications of SQM force fields may be suggested. Even without considering the transferability of the scale factors, the computed spectra can be of great assistance in the correct assignment of observed bands. It is commonplace in an infrared or Raman study of a molecule of the complexity under consideration to find that a number of fundamentals can be assigned with great certainty while others remain in doubt. Scale factors can be derived by fitting a computed spectrum to those transitions that are known with high certainty, giving accurate predictions for the others. Although the calculation of absorption intensities has not been mentioned, these can also be calculated and serve a very useful purpose in assisting in assignment of observed spectra.

In the context of the subject matter of this book, the accurate determination of molecular structure and the intercomparison of structures obtained by various methods, the importance of computational determination of molecular harmonic and anharmonic force fields is to obtain the parameters needed for correction of the various forms of vibrational averaging to a common basis. Relatively little use has been made of this possibility (a few examples are mentioned in Chapter 4), but it remains an interesting source of information for future exploitation.

13.6 The future

Very high-level quantum chemical *ab initio* calculations can obtain geometrical and vibrational molecular parameters with as high accuracy as is available from very good experiments, but such calculations are currently possible only for relatively small molecules. As discussed here, it is possible to extend the upper limits of molecular size for which reliable computational data can be obtained by making use of the systematic character of the computational errors and applying small transferable corrections which are empirically derived. While this approach is of great practical value in solving real chemical problems, it is not as satisfying as would be the direct production of the right answer by computation alone. Progress toward this goal for larger molecules must involve a major change in theoretical methods and there are a number of kinds of research being pursued that offer possibilities of revolutionary change.

For the largest existing computers, the limitations on *ab initio* MO calculations are really imposed by storage space for the nearly incomprehensible number of integrals that are calculated for large systems rather than by computer speed. Almlöf and Lüthi (1987) have demonstrated that their direct self-consistent field method can do Hartree–Fock level calculations on molecules as large as $C_{150}H_{30}$, using 1560 basis functions, by discarding the integrals and recalculating them every time they are used. This procedure makes use of the great speed of modern computers to compensate for their lack of the tremendous storage that would be needed to treat a molecule of this size by conventional methods. If the same approach can be developed for correlation calculations, there is reason to believe that rigorous calculations can be done for large molecules with no empirical corrections, provided a long computational time on a very large computer is available.

Among the concepts currently being investigated for speeding up the self-consistent field calculation by orders of magnitude, the pseudospectral method of Friesner (1987) is worthy of mention. It needs much further development before it can be of practical use, but it offers promise. The even more severe problem of accelerating the electron correlation part of the problem is under study in many laboratories. One promising approach, which is directed toward decreasing the n^6 dependence of time on molecular size rather than trying to develop methods that are faster for small molecules, is the local correlation approach (Saebø and Pulay 1985, 1987). It appears likely that use of this technique can make conventional high-quality methods (MP4, CISD, etc) practical for molecules containing a dozen or more atoms in the very near future.

13.7   Acknowledgements

The portions of the research described here that were carried out at the University of Texas were supported by a grant from the Robert A. Welch Foundation.

References

Almlöf, J. and Lüthi, H. P. (1987). *Theoretical methods and results for electronic structure calculations on very large systems: Carbon clusters.* University of Minnesota Supercomputer Institute, Report 87/20.

Bernardi, F. and Robb, M. A. (1987). Transition structure computations and their analysis. In: Lawley (1987*a*), pp. 155–248.

Boggs, J. E. and Cordell, F. R. (1981). *J. Mol. Struct. (Theochem)*, **76**, 329–47.

Dunning Jr., T. H. and Hay, P. J. (1977). Gaussian basis sets for molecular calculations. In: Schaefer (1977a), pp. 1–27.

Fogarasi, G. and Pulay, P. (1984). *Ab initio* vibrational force fields. In *Annual review of physical chemistry*, Vol. 35 (ed. B. S. Rabinovitch), pp. 191–213. Annual Reviews Inc., Palo Alto.

Fogarasi, G. and Pulay, P. (1985). *Ab initio* calculation of force fields and vibrational spectra. In *Vibrational spectra and structure*, Vol. 14 (ed. J. R. Durig), pp. 125–219. Elsevier, Amsterdam.

Foster, J. M. and Boys, S. F. (1960). *Rev. Modern Phys.*, **32**, 300–2.

Friesner, R. A. (1987). *J. Chem. Phys.*, **86**, 3522–31.

Lawley, K. P. (ed.) (1987a). *Ab initio methods in quantum chemistry*, Part I, Advances in Chemical Physics, Vol. 67. Wiley-Interscience, Chichester.

Lawley, K. P. (ed.) (1987b). *Ab initio methods in quantum chemistry*, Part II, Advances in Chemical Physics, Vol. 69. Wiley-Interscience, Chichester.

Møller, C. and Plesset, M. S. (1934). *Phys. Rev.*, **46**, 618–22.

Niu, Z., Dunn, K. M., and Boggs, J. E. (1985). *Mol. Phys.*, **55**, 421–32.

Pongor, G., Pulay, P., Fogarasi, G., and Boggs, J. E. (1984). *J. Am. Chem. Soc.*, **106**, 2765–9.

Pongor, G., Fogarasi, G., Boggs, J. E., and Pulay, P. (1985). *J. Mol. Spectrosc.*, **114**, 445–53.

Pulay, P. (1969). *Mol. Phys.*, **17**, 197–204.

Pulay, P. (1977). Direct use of the gradient for investigating molecular energy surfaces. In: Schaefer (1977b), pp. 153–85.

Pulay, P. (1981). Calculation of forces by non-Hellmann–Feynman methods. In *The force concept in chemistry* (ed. B. M. Deb), pp. 449–80. Van Nostrand-Reinhold, New York.

Pulay, P. (1987). Analytical derivative methods in quantum chemistry. In: Lawley (1987b), pp. 241–86.

Pulay, P., Fogarasi, G., Pang, F., and Boggs, J. E. (1979). *J. Am. Chem. Soc.*, **101**, 2550–60.

Pulay, P., Fogarasi, G., and Boggs, J. E. (1981). *J. Chem. Phys.*, **74**, 3999–4014.

Pulay, P., Lee, J.-G., and Boggs, J. E. (1983a). *J. Chem. Phys.*, **79**, 3382–91.

Pulay, P., Fogarasi, G., Pongor, G., Boggs, J. E., and Vargha, A. (1983b). *J. Am. Chem. Soc.*, **105**, 7037–47.

Saebø, S. and Pulay, P. (1985). *Chem. Phys. Lett.*, **113**, 13–8.

Saebø, S. and Pulay, P. (1987). *J. Chem. Phys.*, **86**, 914–22.

Schaefer III, H. F. (ed.) (1977a). *Methods of electronic structure theory*, Modern Theoretical Chemistry, Vol. 3. Plenum Press, New York.

Schaefer III, H. F. (ed.) (1977b). *Applications of electronic structure theory*, Modern Theoretical Chemistry, Vol. 4. Plenum Press, New York.

Schäfer, L. (1983). *J. Mol. Struct.*, **100**, 51–73.

Sellers, H., Pulay, P., and Boggs, J. E. (1985). *J. Am. Chem. Soc.*, **107**, 6487–94.

Szabo, A. and Ostlund, N. S. (1982). *Modern quantum chemistry: Introduction to advanced electronic structure theory*. MacMillan, New York.

Wilson, S. (1987). Basis sets. In: Lawley (1987a), pp. 439–500.

14

Molecular mechanics

Norman L. Allinger

14.1 History

The earliest reference known to the author on molecular mechanics is to be found in a paper by Andrews (1930), which has to do with vibrational spectroscopy. Andrews indicated that if we really understood the underlying fundamentals in this area, we should be able to calculate all of the kinds of things with which this discussion will be concerned. But this was in 1930, when people said that by solving the Schrödinger equation, we can find out essentially everything we want to know about chemistry and physics. The practical use of molecular mechanics (or of the Schrödinger equation) for determining molecular structures was still far in the future.

In 1946, three separate papers appeared, one by Hill, one by Ingold and coworkers, and a third one by Westheimer and Mayer, which taken together provide a very substantial basis for molecular mechanics (Dostrovsky *et al.* 1946; Hill 1946; Westheimer and Mayer 1946; see also De la Mare *et al.* 1955; Westheimer 1956). These papers showed that in the calculations commonly carried out at the time by vibrational spectroscopists, if one included in the force field van der Waals' interactions, one could calculate structures of molecules, and energy relationships between conformations and isomers and the like, in addition to vibrational information and thermodynamic information that follows from knowing the vibrational levels in molecules. In practice, only very simple or idealized kinds of cases could be studied at the time, because the calculations are beyond what can be managed without a computer. But the principles were firmly established.

Hendrickson (1961) published an application of these calculations to a study of the geometries of some medium-ring cycloalkanes, and he utilized a computer for the extensive calculations, and showed that indeed such calculations had become practical. He solved the problem for a special and somewhat exotic case, however, and it remained for Wiberg (1965) to publish the results and description of a general computer program that would begin with crude approximate starting coordinates for a molecule, and optimize these to find the actual molecular geometry. Wiberg's program, although primitive by present standards, was a real breakthrough, and may be viewed as the basis for all subsequent development in this area.

We will return to the workings and principles of such a computer program later. First, let us consider our objectives. In 1965 it seemed like the problems outlined below were assailable, although exceedingly difficult. Over the last twenty years or so, the assault has been very successful. For hydrocarbons, the solution to the problem seems to be in hand to a very good approximation, and the few remaining relatively small trouble spots are being rapidly eliminated. But for functionalized molecules, there are quite a few more problems, and these, while yielding to attack, certainly have not been solved as of yet.

The basic ideas of molecular mechanics are, in an overview, the following. A molecule can be described by a collection of atomic nuclei distributed on a potential energy surface. The Born–Oppenheimer approximation tells us that we can consider separately the motions of the electrons and the motions of the nuclei, because of the large difference in their masses. In an *ab initio* MO calculation (see Chapter 13) one positions the nuclei at some trial position, and calculates the electronic structure (wavefunction) of the molecule that gives the minimum energy for this nuclear arrangement. Then one calculates the forces on the nuclei, and moves the latter by some systematic iterative procedure to positions where the forces are zero, and the energy is a minimum. This nuclear arrangement corresponds to a stable geometry of the molecule. The problem with calculations of this kind is that the wavefunction is written as a combination of infinite series, and a practical computational solution to such a problem is faced with well-known difficulties.

Molecular mechanics considers the 'other part' of the Born–Oppenheimer surface. Explicit consideration of the eletrons is not necessary. It is assumed that the electrons provide a potential field, and we consider only the motions of the nuclei within this field. The potential surface, although unknown in general, can be described by a Taylor series expansion about the point (or points) of minimum energy (see Chapter 2, Section 2.3.1). This means that the energy of the molecule can be expressed in terms of a power series involving the atomic coordinates. Since the coordinates can be expressed as Cartesian coordinates, or equivalently, as internal coordinates, it means that we can write the energy of a molecule as a power series involving bond lengths, bond angles, and torsion angles, which are familiar from vibrational spectroscopy, plus interactions between atoms which are not bonded to each other or to common atoms (van der Waals' interactions).

All that has been said about molecular mechanics to this point is quite rigorous. But at this point some problems appear. A Taylor series expansion is accurate only when the series is infinitely long. Can we truncate the series to some reasonably small number of terms, and still have something which is a reasonable approximation of an actual molecule? Fortunately we can. Spectroscopists have long known that the harmonic approximation is pretty good. Adding one or two terms is usually sufficient to solve any real current problem.

But a Taylor series expansion contains parameters, which we must somehow determine. The parameters as we want to express them are the force constants for bending, stretching, and torsion. There are a number of structural parameters, the X_0 from which the ΔX are measured, which are bond lengths, angles, and torsion angles that correspond to 'natural' values (that is, values at which the energy is a minimum). Because vibrational spectra contain a limited number of lines, which is small relative to the number of atomic coordinates, we can always find a force field that will reproduce any observed spectrum. But to be of any use in molecular mechanics, the structural parameters must be transferable from one molecule to another. Otherwise, there will never be enough data to be able to draw any conclusions beyond the available experimental information. Are force constants transferable? Well, spectroscopists have long known that within a series of closely related compounds, they seem to be transferable to a high degree of precision. But when considering a broad general group of compounds, they are not transferable. In fact, it appears now that *force parameters* (not the *force constants*) in general are transferable to a very good approximation, not only between related compounds, but between all kinds of compounds, if one explicitly includes the van der Waals' interactions in the force field. Spectroscopic force fields traditionally include either no van der Waals' interactions, or at best, only those between atoms bound to a common atom (Urey–Bradley force field; see Lifson and Warshel (1968)). Because these interactions vary widely from one molecule to another in the general case, and because the spectroscopist does not include them as such, he instead varies the force constants to fit the experimental data. If the van der Waals' interactions are separately calculated and included in the force field, then the *force parameters* which are transferable between large groups of unrelated molecules are added to the van der Waals' parameters to give the correct *force constants*.

So while there are some strong similarities between force fields used in vibrational spectroscopy and those used in molecular mechanics, there are also some major differences as well. The force parameters needed for a molecular mechanics force field have numerical values similar to those force constants used by spectroscopists, but in general they differ somewhat, for the same reason that Urey–Bradley force constants differ from valence force constants. Namely, if the force fields (the equations used) are somewhat different, then the optimum values of the constants that go into those force fields will also be different.

In the force fields developed during the period roughly 1965–1975, different groups of workers were interested in different kinds of compounds and different kinds of problems, and each group optimized its force field to fit the kind of problem they were interested in. Lifson and Warshel (1968) said that one should really use a 'consistent force field'. That is, the force field should be optimized by fitting essentially all kinds of data, and all available data (Niketić and Rasmussen 1977). Certainly no one disagrees with this in principle. But in practice, Lifson and Warshel, as everyone else, were forced to look at only a relatively limited amount of data, for practical reasons (the total available data is far beyond the capabilities of current computers and manpower). The result was that several different force fields were developed (for example by Boyd (1968), Engler *et al.* (1973), Ermer and Lifson (1973), Allinger (1977), Bartell (1977), White and Bovill (1977)), which although roughly similar in terms of the functions and parameters used, differed significantly, and in some cases the predicted results differed very markedly. Different force fields each had their own area in which they did well, and other areas in which they did less well, and which force field was better in a particular area was strictly a matter of which area it was that one chose to talk about.

Is there a 'best' force field? I believe that most workers in the field think there is, but one cannot prove this. As time has gone on from about 1975 to the present, force-field modifications have occurred, and force fields have become more general, and have done better across broader areas than was possible in earlier years. Coincidentally, the different force fields have become more similar to one another. A convergence of force fields has been occurring, and the spread between them will probably continue to narrow as a function of time.

Why do force fields continue to differ from one another? Their developers continue to increase the size of their data sets, and to include in the sets more exotic compounds, which test potential functions further and further from their energy minima. Force fields are better at interpolating between existing data than they are at extrapolating from existing data. As the amount of data far from the energy minimum is accumulated, one can interpolate between these data to a greater extent, and extrapolate to a lesser extent, and hence the results become more reliable. It also is turning out that the force fields are becoming more similar to one another. Part of what is happening is surely that the parameters are highly correlated for most relatively strainless molecules, so that apparently very different force fields give the same predictions regarding observables. But these correlations tend to be broken down as one goes to more strained and exotic molecules. Consequently these originally different force fields tend to converge as the developers include more of the data on exotic compounds.

14.2   Objectives

Why do we want to carry out such calculations? Can we not determine geometries and other properties of molecules experimentally, or by *ab initio* MO calculations? The answers to these questions are clear enough, although somewhat involved. It turns out that each of the experimental methods has its strengths and weaknesses (see Chapters 3–12) and one or another may well be suited to a particular problem, or maybe not, depending upon the problem. *Ab initio* MO calculations now add another dimension to structure determination (see Chapter 13). But they can deal very well with some molecules, and less well or not at all with others. And they also give a different kind of structure, the equilibrium structure, whereas most experimental geometries correspond to some kind of thermal average over the vibrational motion.

Molecular mechanics offers still another alternative way to attack structural problems. There are certain kinds of situations in which this method can be counted upon to give very good results, and other kinds of situations which cannot be tackled at all. So it is one more tool of the structural chemist; one which can be advantageously used in different cases, depending upon the situation. The method does have one clear and very powerful advantage. If it can be used at all, that is, if all of the necessary parameters are available for the study of a specific compound, then this method is very fast and easy to apply, compared to other calculational or experimental methods. But if some of the parameters needed for the calculation have unknown values, then one may be forced to rather elaborate efforts in order to determine the necessary parameters. Often one can make parameter estimates which will be quite satisfactory and offer a quick and easy way to get a structure of adequate accuracy. But this gets to be more of an art than it is science, to know just what you can get away with.

14.3   Force fields

A number of reviews on molecular mechanics are available (Altona and Faber 1974; Dunitz and Bürgi 1975; Allinger 1976; Ermer 1976, 1981;

Warshel 1977; White 1978; Burkert and Allinger 1982; Ōsawa and Musso
1982). These, especially Burkert and Allinger (1982), may be consulted for
amplification of the discussion given here, and for additional references.

Building on the work of the vibrational spectroscopist, we know that the
force constants for stretching bonds are generally an order of magnitude
larger than the force constants for bending bond angles (see, for example,
Chapter 19, Section 19.2). Torsional force constants range from extremely
small to very large. We may consider dimethylacetylene as an example of
the first kind, where the torsional barrier is only a few hundredths of a kJ
mol^{-1} in height, and ethylene as the second kind, where rotation about the
double bond requires some 270 kJ mol^{-1}. There are intermediate cases, but
on the whole, barriers are usually either quite large (double bonds), or they
are of the order of 10–20 kJ mol^{-1} (single bonds). The latter deformations
are quite soft relative to bending. Hence when a molecule finds itself under
a great deal of tension because of intramolecular forces, it usually will tend
to relax by large torsional motions, modest angular distortions, and only
small bond stretchings.

14.3.1   Bond stretching

This can be written in terms of a Morse potential (see Chapter 2, eqn (2.9)),
or by a series expansion which is equivalent, which has a large quadratic
term, a smaller cubic term, a still smaller quartic term, etc. Modern force
fields always include quadratic terms to deal with ordinary molecules, and
those which have been developed to deal with highly congested molecules
will also include a cubic term. Thus the potential energy expansion will
be:

$$V_r = \frac{k_r}{2}(r - r_o)^2 - \frac{k_r'}{2}(r - r_o)^3. \tag{14.1}$$

Higher terms are probably not necessary for any molecule (ground state) so
far studied, except in the special case of hydrogen bonding. Hydrogen
bonds, when stretched very much, tend to come apart easily, so that if one
wants to include this kind of bond in molecular mechanics, a Morse-type
potential must be used.

14.3.2   Angle bending

The force constants here are smaller than for stretching, so that congested
molecules typically undergo relatively small bond stretchings, but larger
angular distortions. A quadratic potential function

$$V_\theta = \frac{k_\theta}{2}(\theta - \theta_o)^2 \qquad (14.2)$$

seems to be adequate up to at least 10° or 15° of bending, but it is less clear what happens at still larger bendings.

Most deformations larger than 15° in real molecules are toward smaller angles. Cyclobutane, for example, shows a deformation of the C—C—C angles of about 20° from tetrahedral. Such compounds can be dealt with perfectly well in force-field calculations, but usually one uses different force constants for small rings, and then a quadratic bending function still seems appropriate.

Angles which are opened by more than 15° from their natural values are rare in real compounds, and generally they occur in very complicated systems where it is difficult to know accurately everything else about the system, so that one can specifically look at the function required to represent angle bending. Work is in progress in this area.

14.3.3   Torsion

Organic molecules usually contain some combination of single and double bonds. The barriers about double bonds are usually mainly twofold (ethylene-type), and those about the single bonds are mainly threefold (ethane-type). But in general, one can write a Fourier series to describe the energy resulting from torsion. Spectroscopists frequently use a series containing up to six or so terms. If one wants to worry about energies on the order of a few hundredths of a kJ mol^{-1}, without explicit inclusion of the van der Waals' repulsions, about six terms may be required to fit experiment. In molecular mechanics it has been found that only one-, two-, and threefold terms are required for anything that we have so far tried to fit:

$$V_\phi = \frac{V_1}{2}(1 + \cos\phi) + \frac{V_2}{2}(1 + \cos 2\phi) + \frac{V_3}{2}(1 + \cos 3\phi). \quad (14.3)$$

This is partly because we do not worry about energies as small as a few hundredths of a kJ mol^{-1}, and partly because we do include van der Waals' interactions.

Equation (14.3) gives the terms that one would expect to be non-zero from elementary orbital considerations. While ethane and ethylene have very simple torsional functions because of their high symmetry, a molecule such as butane, CH$_3$—CH$_2$—CH$_2$—CH$_3$, wherein we consider rotation about the central C—C bond, can in principle be described by a Fourier series containing three terms. Early force fields assumed that only a threefold term was necessary. More recently onefold and twofold terms have been added. Although these are small in the case of a hydrocarbon, when they are included one can definitely fit the available experimental data much better than when they are omitted. For functionalized molecules any or all of the three terms may be large, depending on the particular functions involved.

One can rationalize the existence of these three terms in a torsional potential function in a straightforward manner (Radom *et al.* 1973). In ethane there is a threefold term which is a result of antibonding interactions between the hydrogens on opposite ends of the molecule. Any molecule which is alkane-like should show a qualitatively similar threefold torsional term. If we replace a hydrogen with a methyl, or a flourine, or something else, we might change the magnitude of this term, but we should not change its fundamental nature. Thus there should always be a threefold term in the torsional potential for rotation about a saturated bond. In ethylene there is a twofold term, which is due to the overlap of the p orbitals that make up the π system. In a molecule like butane, there should also be such a term, because the hydrogen attached to the second carbon is not exactly the same as the methyl, so there should be a twofold component to the torsional barrier. In butane this barrier component was not detected experimentally (Stidham and Durig 1986) until after it had been postulated from molecular mechanics studies (Allinger 1977; Allinger *et al.* 1977; Bartell 1977). It is rather small, about 1.8 kJ mol^{-1}. In molecules like 1,2-difluoroethane, the corresponding interaction is quite large, and was easily found by early *ab initio* MO calculations (Radom *et al.* 1973). But in principle, it should be present in all saturated bonds as a result of what is usually referred to as 'hyperconjugation'.

In butane, the two methyl groups show a mutual van der Waals' interaction, and if they were more polar, such as fluorines, they would show a dipole–dipole interaction. These two kinds of interactions each correspond to a onefold torsional component to the barrier. Any inadequacies in those parts of the calculation are therefore taken into account here.

So for molecules which contain polar groups, there are in general readily discernible one-, two-, and threefold torsional terms. For less polar molecules like butane, they may not be experimentally apparent, but it is believed that they in principle exist (although it is possible that some of the coefficients have values near zero). As it turns out, finding the necessary Fourier coefficients for the torsional terms is usually the most difficult part of developing a force field for a new class of compounds. Bond lengths and angles are usually available from known structural information, and force constants can be at least reasonably approximated from spectroscopic studies. Torsional barriers are sometimes known from microwave studies, or from thermodynamic studies, and more are now becoming available from Raman and far-infrared spectroscopy. But for molecules which have large values for two or more of the coefficients, one usually cannot formulate the force field without more experimental or *ab initio* MO information than can be found in the present literature.

14.3.4 Van der Waals' interactions

The interactions between atoms which are not bound to each other or to a common atom are generally referred to as *van der Waals' interactions*. These are ubiquitous. They lead to the non-transferability of spectroscopic force constants, and they have major effects on structure in congested molecules. They have also proven to be difficult to evaluate accurately. Molecules in which the van der Waals' interactions are large are generally complicated, show many degrees of freedom in which the molecule can relax, and hence for the most part the van der Waals' parameters are highly correlated with other parameters. In early force fields such as MM1, the relatively large *gauche* butane energy was fit by using large hydrogen–hydrogen van der Waals' repulsions (Wertz and Allinger 1974). After the advantages of one- and twofold torsional terms for this purpose were recognized, it was clear that much softer hydrogens could be used and they would allow much better fits to other kinds of data. A good *ab initio* MO calculation of the hydrogen–hydrogen repulsion was also published about this time (Kochanski 1973). If one looks in the older literature, these van der Waals' curves vary widely from one research group to another, but they have converged to a relatively narrow range more recently (Burkert and Allinger 1982).

Theory says that the van der Waals' potential should be composed of two parts. It is an electrostatic interaction, and the first term of the attractive part (between neutral atoms) is proportional to r^{-6}. This is sometimes called the *induced dipole–induced dipole interaction*. The higher terms in the expansion have so far been neglected in molecular mechanics. Lower-order terms exist only when the atom carries a charge.

The other term is just the repulsion which results when one tries to put two atoms in the same place at the same time. It is well represented over the range of interest by either an exponential, or by r^{-n}. Just what value should be used for n (or the corresponding quantity in the exponential) depends on just what it is one wants to fit. For the rare gases, and for closed-shell molecules where the whole molecule is considered, a value of $n = 12$ has been widely used and seems to work well over experimentally accessible ranges of r. For molecular mechanics, that value is definitely larger than optimum, and a value of about 9 is clearly better.

The van der Waals' function used in the MM2 program (Allinger 1977) is more or less typical, and is:

$$V_{vdw} = \epsilon [2.9 \times 10^5 \exp(-12.5r/r*) - 2.25(r*/r)^6]. \quad (14.4)$$

Here $r*$ is the sum of the van der Waals' radii of the interacting atoms, and ϵ is an energy parameter specific for each pair of atom types. Since this particular function was developed, still more congested molecules have been studied experimentally, and it is now clear that this function gives energies which are somewhat too high at very short distances. We now have a function which fits the newer data also, and only differs from the MM2 function in a major way at distances closer than 2 Å for hydrogens (MM3 force field: Allinger et al. (1989); Lii and Allinger (1989)). Unfortunately, when one changes the van der Waals' potential, one has to reoptimize the entire force field, and this is a nontrivial task.

The van der Waals' interactions between carbon atoms are also important and have been studied. In diamond, the bond-stretching forces are tending to contract the structure, while the van der Waals' forces are holding it apart. In graphite, the distance between the graphite planes is a result of only van der Waals' forces. Fitting these quantities pretty well defines the van der Waals' parameters for carbon, and then hydrocarbon structures pretty well define these parameters for hydrogen. But because of the parameter correlation previously mentioned, ranges, rather than points, are defined.

Another complication must be mentioned here. Experimentally or computationally, there are a lot of problems involved in deciding just where an atom is located. Much of the problem stems from the fact that atoms vibrate, and the amplitude of vibration is usually of the order of 0.05–0.10 Å. To define the position of the atom to 0.001 Å, one has to average over the vibrational motion. Depending on the calculation or experiment, this averaging is done in different ways, and hence the atomic positions obtained depend on the method used to obtain them. This fact must be recognized. If a bond length is determined with a precision of 0.0001 Å by two different experimental methods, it may have values that differ by perhaps 0.01 Å, and sometimes even more, due simply to the difference in the definitions used in specifying the atomic positions. One particular case is especially important here. This is the location of a hydrogen atom, or the equivalent, the determination of a C—H bond length. Microwave spectroscopy, electron diffraction, and neutron crystallography all determine the position of the atomic nucleus (and these different methods may well disagree in the location of this position, due to the vibrational averaging problem mentioned). But X-ray crystallography locates an ellipsoid of electron density (or a sphere, in the usual isotropic approximation for hydrogen) that best describes the hydrogen 'atom'. The center of this ellipsoid is found to be about 0.1 Å closer to carbon than the position of the hydrogen nucleus as measured by the other experimental methods. The reason is obvious.

Hydrogen has only one valence electron, and it is used in bonding. Hence it is pulled in towards the internuclear region, and the mean electron position will be offset from the proton towards the carbon, and as it turns out, by the relatively large amount of about 0.1 Å. Additional information on this effect can be found in Chapter 11, Section 11.2.

How should we deal with all of this in molecular mechanics? Again, different workers in the field have different objectives, and have looked at this problem in different ways. Our rationale and procedure have been the following. Since there are vastly more structures of molecules known from X-ray crystallography than from any other method (or, indeed, from all other methods put together), one clearly must for the most part compare the results of molecular mechanics calculations with results from X-ray work. But X-ray diffraction geometries suffer from all kinds of problems. One does not measure an average distance between atoms, one measures a distance between average atomic positions, and these quantities are not the same. They depend not only on the atoms in question, but also on their local environment. Fortunately, these discrepancies are generally small, and need not be worried about very much here.

To eliminate the effects of crystal forces, which vary from one crystal to another, one would like to study isolated molecules. (Once we understand the isolated molecules, we can pack them together, and study crystal forces as a separate problem.) Thus, one really wants to know what average internuclear distances are in the gas phase, and in electron diffraction work the r_g parameter means exactly this (see Chapter 5, Section 5.7). The quantities that we fit to are these r_g bond lengths. These quantities are precisely defined, and they can be accurately measured for small molecules under suitable conditions, so they provide very good standards for precise parameterization. But, very importantly, the bond lengths and angles obtained in this way are usually close to those obtained by X-ray diffraction experiments, with only a few exceptions (see below). However, *they are in principle different*, and for very accurate work these differences need to be remembered.

The major difference between bond lengths found by X-ray crystallography and by gas-phase techniques (or neutron crystallography) is in the C—H bond length, as discussed above. Here, one must decide precisely what one wants to talk about, and be consistent. The MM2 program uses the C—H bond length as defined by the nuclear positions. The X-ray diffraction bond lengths will be about 0.1 Å shorter. But in the force-field calculations, we have to put the 'atom' somewhere in order to calculate bond lengths, angles, etc. We put the 'atom' where the nucleus is, and use that position for all calculations, except for the van der Waals' calculation. At most distances (not too close) most of the interaction in the van der Waals' equation comes from the interaction of the electrons of one atom with the other atom.

This means that for the van der Waals' calculation we are not interested so much in the position of the nucleus as in the position of the electrons. The electron cloud is taken to be a sphere, located along the C—H bond, but moved away from the hydrogen nucleus towards carbon by about 0.1 Å. In MM2, this offset is 8.5 per cent of the C—H bond length, in the direction towards the attached carbon. Throughout the calculations, all van der Waals' interactions are based on the position of the electrons, as defined by this relationship, and all other quantities are calculated on the basis of the positions of the nuclei. Bartell also offsets the hydrogens in this fashion in his force field (Fitzwater and Bartell 1976). Williams has discussed this offset in crystal packing calculations (see Williams 1974, and references therein). Other force fields appear not to consider this offset, but place both the center of electron density and the nucleus at the same position. Thus when we see graphs in the literature illustrating comparisons of hydrogen–hydrogen repulsions as calculated by different force fields, they can be very misleading. In some force fields (MM2 (Allinger 1977), MUB-2 (Fitzwater and Bartell 1976)) the repulsion depends upon not only the distance from the nucleus, but also the direction of the approach, because of this effect. In other force fields, there is no directional dependence. Thus curves that look very different may in fact give numbers that are very similar when the calculation is actually carried out with the directional dependence included.

There is another difficulty in comparing X-ray diffraction structures with gas-phase structures, and this also stems from the fact that X-ray crystallography yields the position of the centroid of elecron density, and not the position of the nucleus. With an atom like carbon the difference between these two quantities is very small (see, for example, Chapter 18, Section 18.3.3.1), and is often ignored. With an atom like the nitrogen of an amine, this difference may amount to 0.01–0.02 Å, because the lone pair of electrons offsets the electron density from the nucleus. Accordingly, X-ray diffraction structures can differ somewhat from gas-phase structures, because of this effect. Note that neutron diffraction structures are determined by nuclear scattering, and hence they are, in this respect, comparable with the gas-phase structures rather than with the X-ray diffraction structures.

Finally, most X-ray diffraction structures given in the literature, retrieved in the Cambridge Structural Database (see Chapter 15), were determined at room temperature. The original authors usually provide the data necessary to correct these structures for the thermal librational motion, and thereby determine more accurate 'r_α-type' bond lengths (see Chapter 8, Section 8.5, and Chapter 18, Section 18.3.6). These thermal corrections to bond lengths are typically in the range of 0.005–0.015 Å, so again, if one is looking for accurate structures, the thermal motions must be taken into account. Better yet, one likes to have the structures determined at liquid-nitrogen temperatures, so that these corrections can be reduced to smaller values. Unfortunately, few structures in the current literature have been determined at low temperatures, and few have had their bond lengths corrected for thermal motion.

14.3.5 Additional terms

The four items mentioned in Sections 14.3.1–14.3.4 are included in all molecular mechanics force fields. But from this point on, there are a number of options available. In the most simple force fields, one stops here. But such a force field will fail to give results of experimental quality in many cases, and hence a number of second-order terms can be added. Different groups of workers have chosen different terms, and it is far from clear if there is any 'best' way to proceed beyond this point. In general, either the Urey–Bradley formulation has been used, or interaction constants have been added. The latter may be of the stretch-bend type, for example:

$$V_{r\theta} = \frac{k_{r\theta}}{2}\left[(r - r_{o}) + (r' - r'_{o})\right](\theta - \theta_{o}), \qquad (14.5)$$

and this particular interaction seems important in small and large rings, where severe angular deformation is enforced by the ring. Schachtschneider and Snyder (1963) showed that using a similar number of parameters, a Urey–Bradley force field and a valence force field with cross terms give approximately equivalent results. The same is probably true in molecular mechanics. The MM2 program happens to use a valence force field with interaction constants, with the stretch-bend interaction being the only one actually used. The interaction here is limited to a bending and a stretching which both involve a common atom. The sign of this force constant is such that as an angle is pinched together, the bonds stretch. Thus the effect of this term on the molecular structure is qualitatively the same as with a Urey–Bradley interaction constant.

14.3.6 Electrostatic terms

For saturated hydrocarbons it appears that one can neglect electrostatic terms. If they are included, the numerical values of the remaining parameters become quite different but the calculated observables seem to differ little if at all from those calculated without the inclusion of electrostatics. Hence most force fields do not consider electrostatics in saturated hydrocarbons. Good quality *ab initio* MO calculations indicate that hydrogens in alkanes are positive relative to carbons (Mulliken charges). But a Mulliken charge is the representation of a complicated quantity by a single number, and it is not obvious that the same number is appropriate for molecular mechanics. *Ab initio* MO calculations further show that the hydrogen attached to an sp^2 carbon is more positive than that in an alkane, while a hydrogen attached to an sp carbon is still more positive. We made the simplest approximation consistent with the data in the MM2 program; we took the C—H bond to be non-polar in alkanes, alkenes, and alkynes. To reproduce the desired electrostatic description of the molecule, one can either deal with point charges at nuclei, or with point dipoles in bonds. The MM2 program will do either, although we ordinarily use the point dipole approximation. The bond dipole moments were given magnitudes such that they correctly reproduce experimental dipole moments in simple cases. Only permanent moments are considered, not induced moments. This limits the accuracy of the electrostatic calculation. Additional studies have been carried out, aimed at finding better ways to include the induced moments in the calculation (Došen-Mićović *et al.* 1983). Much has been learned here, but much is still uncertain.

14.4 Computational methods

If we return to our view of the Born–Oppenheimer energy surface for our molecule, in the simplest case there is only one energy minimum. The molecular structure will correspond to the geometry for the nuclei at that energy minimum. Take the propane molecule as an example. How do we locate the energy minimum on the potential surface?

In principle, this is easy enough to do. The potential minimum is a place where the derivatives of the potential energy with respect to all of the coordinates are simultaneously equal to zero. So, again in principle, all we have to do is write down a trial structure for the molecule, which we can generate, for example, as in a mechanical molecular model, by using tetrahedral bond angles, standard bond lengths, and staggered torsional angles. Since we have equations which describe the change of the energy with the changes in internal coordinates, we can take the derivatives of the energy with respect to these coordinates, find which way the slopes lie, and then move down the slopes for each coordinate until we locate the energy minimum.

There are different ways in which this energy minimization can be carried out in practice. The earliest programs calculated the derivatives numerically, and used a steepest descent method for finding the energy minimum. This is a good method in the sense that it gets you to the energy minimum with few complications, but it is a poor method in the sense that it is very slow, and requires a great deal of computer time, compared with other methods to be described below. But it is sometimes used to partially optimize a very crude starting structure so as to get near enough to the energy minimum so that some of the more sophisticated methods will work well. The sophisticated methods always work well if one is near the energy minimum, but they may fail for various reasons if one is too far from the energy minimum.

The first improvement that was made in the above described method was to use analytically calculated derivatives instead of numerically calculated derivatives. This reduced the running time of the program by about a factor of ten or more, and was a very worthwhile improvement. Of course, a considerable amount of programming was required, because one needed to know not only what the energy functions were, but also the derivatives of these functions with respect to the coordinates. However, the one-time expenditure of programming effort paid a large benefit in terms of computer time eventually saved.

Pattern search methods can speed up the computation by perhaps a factor of two or three on the average. These are methods wherein if one is moving down a gentle slope, it may take a very large number of iterations to reach bottom. If the direction of the motion from one iteration is saved, and added to the direction of the motion for the next iteration, convergence can be accelerated.

A more sophisticated approach is to use not only the first derivative of the energy with respect to the coordinate, but also the second derivative. The first derivative measures the slope, and the second derivative the curvature, or the rate of change of the slope. Using both the first and the second derivatives, the improvement per iteration is very much greater than simply using the first derivative. But it is at the expense of calculating the second derivatives, which requires additional computing time, and a very large additional amount of programming time. The *Newton–Raphson method* (see Burkert and Allinger 1982) is the standard way of utilizing first and second derivative information, and it leads to the geometry of minimum energy with a net computing time improvement of a factor of two or three relative to first derivative methods. This method is powerful and leads to rapid convergence when one is near to an energy minimum, but it is less efficient, and may even fail to converge at all, if one is very distant from the energy minimum. The MM2 program uses a block-diagonal Newton–Raphson method, wherein instead of solving all of the equations involving the derivatives of the coordinates simultaneously, one atom is moved at a time. The geometric improvement per iteration is substantially less than in the full-matrix method, but the time required per iteration is very much less. In practice, it usually turns out that the block-diagonal method is faster by a factor of two or three than the full-matrix method. It is also less susceptible to hang ups of various kinds. But on the other side of the coin, it has not proven useful for locating transition states, which the full-matrix Newton–Raphson method does very well.

In summary, energy minima exist for each molecule. To locate those minima from trial structures is ordinarily quite straightforward and rapid with modern computer programs which are publicly available. With current ordinary computers of the VAX class, to minimize the energy and find the geometry of a small molecule containing say 10 or 15 atoms will take less than a minute of computer time. To find the structure of a larger but rather rigid molecule, such as a steroid, which contains perhaps 50 or 60 atoms, a few minutes will be required. A larger open chain or floppy molecule such as a large ring containing say 80 or 100 atoms may take an hour, and a 700 atom protein may require 24 hours or more. But obviously, these times are very short relative to the time required for even a minimal basis set *ab initio* MO calculation. The time factors between MM2 and STO-3G calculations for a very small molecule involving a few first-row atoms will be perhaps a factor of 1000. The *ab initio* MO calculation time will increase by something between the third and fourth powers of the number of atoms (actually the number of orbitals). The molecular mechanics calculation time increases as something between the square and the cube of the number of atoms, so the time advantage of molecular mechanics also increases rapidly with molecular size. It should be pointed out that one does not in principle obtain the same structures by the two methods. The molecular mechanics calculation gives a vibrationally averaged structure, while the quantum chemical calculation gives the equilibrium structure, at the bottom of the potential energy well. In principle one can calculate either one of these structures from the other (see Chapter 2, Section 2.4.1), but in practice this calculation (which is not very accurate at present, except for very small molecules) is rarely carried out.

We have run molecules containing up to 700 atoms without any special problems. Molecular mechanics calculations, as long as no manipulations of large matrices are required, work as well with large molecules as with small ones; they just take longer.

14.5 Molecular structures

The structures, and the rotational barriers, for small molecules such as ethane, propane, butane, isobutane, and neopentane are very well calculated by all modern force fields, since those compounds were used in determining the parameters that go into the force fields.

After hydrocarbons were dealt with adequately, including three- and four-membered rings, the calculations were extended to alkenes and alkynes, and subsequently to conjugated hydrocarbons. Next functional groups of all sorts were added, such as in alcohols, ethers, carbonyl compounds, carboxylic acids and derivatives, mercaptanes, sulfides, disulfides, amines, etc. Finally, aromatic heterocycles (pyridine, pyrrole, furan and thiophene) were added. A few less common functional groups have been studied to varying degrees, so that by now most molecules which contain the above features, as long as the compound is monofunctional, can be dealt with with reasonable accuracy in terms of structure. With polyfunctional compounds, there is a big problem, because so many potential parameters are required. There are in MM2 about 50 atom types (because a carbon which is sp^3 hybridized is one type, while sp^2 is a different type, etc. Hence, only the common elements require 50 types of atoms). To calculate how many possibilities there are for parameters when one has 50 atom types, if one looks at the torsion parameters, one can see that one can have type 1–1–1–1, and any one of these 1 can be replaced by any number up to 50 (excluding a few monovalent atoms). This means there are about 50^4 torsional parameters required to describe all of these combinations of atoms. Unfortunately, this number amounts to something more than six million! Obviously, most of these are not yet known, and are not likely to be known in the immediate future. Hence molecules which have relatively simple functionalization, that is functionalization where atoms from one function are not contained in the same four-atom segment as atoms from the other function, can mostly be dealt with now. But molecules which contain two functional groups within the same four atoms require all these additional parameters, and they mostly have not yet been studied.

14.6 Heat of formation

Chemists have for many years used bond energy schemes to predict heats of formation for molecules. A bond energy scheme works well for most molecules which are not highly strained. If the molecule is strained, for example by having a five-membered ring in it, then one must add another parameter for the five-membered ring, and with additional parameters like this, again heats of formation can be well calculated for a great many things. However, in general, strained molecules cannot be simply described by incremental additions, because they are more or less unrelated to other molecules in this respect. Obviously, molecular mechanics is an ideal way to calculate heats of formation. One uses a bond energy scheme, but one includes in the scheme the energy calculated in the molecular mechanics optimization. This latter quantity contains explicitly the strain energy of the molecule. When this is done, one can in fact calculate heats of formation with an accuracy that is competitive with good-quality experiments. The scheme works extremely well for hydrocarbons and monofunctionalized molecules. Again, if the functional groups are very close together, it is anticipated that they may show interactions which will change the energy in ways not allowed for by molecular mechanics. Hence, the calculation of heats of formation by molecular mechanics is currently limited to hydrocarbons, monofunctional molecules, and polyfunctional molecules in which the functions are not very close together along the chain. Presumably, being close together in space would be all right, as in that case the interactions would be adequately described by the long-range (van der Waals' and electrostatic) interactions.

14.7 Concluding remarks

Molecular mechanics is a currently available powerful tool for the determination of the structures of organic molecules. Like other methods, it has its limitations. However, if the method is applicable to the determination of the structure of a particular molecule, then it may well be the method of choice, since it is a method without competition in terms of speed. If parameters must be evaluated before a certain molecule can be studied, the investigator needs to weigh the difficulty in evaluating those parameters relative to the difficulty of determining the structure by some other method.

A potentially very powerful method, little used so far, but which one may expect to see increasingly used in the future, is to couple *ab initio* MO calculations with molecular mechanics calculations in the following way. Large molecules, which are not amenable to *ab initio* MO studies, may have their force fields determined by *ab initio* MO calculations on their small analogous components, and the composite force field from these components may then be put together and used in the molecular mechanics fashion.

References

Allinger, N. L. (1976). Calculation of molecular structure and energy by force-field methods. In *Advances in physical organic chemistry*, Vol. 13 (ed. V. Gold and D. Bethell), pp. 1–82. Academic Press, London.

Allinger, N. L. (1977). *J. Am. Chem. Soc.*, **99**, 8127–34.

Allinger, N. L., Hindman, D., and Hönig, H. (1977). *J. Am. Chem. Soc.*, **99**, 3282–4.

Allinger, N. L., Yuh, Y. H., and Lii, J.-H. (1989). *J. Am. Chem. Soc.*, **111**, 8551–66.

Altona, C. and Faber, D. H. (1974). Empirical force field calculations: A tool in structural organic chemistry. In *Dynamic chemistry*, Topics in Current Chemistry, Vol. 45, pp. 1–38. Springer, Berlin.

Andrews, D. H. (1930). *Phys. Rev.*, **36**, 544–54.

Bartell, L. S. (1977). *J. Am. Chem. Soc.*, **99**, 3279–82.

Boyd, R. H. (1968). *J. Chem. Phys.*, **49**, 2574–83.

Burkert, U. and Allinger, N. L. (1982). *Molecular mechanics*, ACS Monograph No. 177. American Chemical Society, Washington.

De la Mare, P. B. D., Fowden, L., Hughes, E. D., Ingold, C. K., and Mackie, J. D. H. (1955). *J. Chem. Soc.*, 3200–36.

Došen-Mićović, L., Jeremić, D., and Allinger, N. L. (1983). *J. Am. Chem. Soc.*, **105**, 1723–33.

Dostrovsky, I., Hughes, E. D., and Ingold, C. K. (1946). *J. Chem. Soc.*, 173–94.

Dunitz, J. D. and Bürgi, H.-B. (1975). Non-bonded interactions in organic molecules. In *Chemical crystallography*, MTP International Review of Science, Physical Chemistry Ser. 2, Vol. 11 (ed. J. M. Robertson), pp. 81–120. Butterworths, London.

Engler, E. M., Andose, J. D., and Schleyer, P.v.R. (1973). *J. Am. Chem. Soc.*, **95**, 8005–25.

Ermer, O. (1976). Calculation of molecular properties using force-fields: Applications in organic chemistry. In *Bonding forces*, Structure and Bonding, Vol. 27, pp. 161–211. Springer, Berlin.

Ermer, O. (1981). *Aspekte von Kraftfeldrechnungen*. Baur, Munich.

Ermer, O. and Lifson, S. (1973). *J. Am. Chem. Soc.*, **95**, 4121–32.

Fitzwater, S. and Bartell, L. S. (1976). *J. Am. Chem. Soc.*, **98**, 5107–15.

Hendrickson, J. B. (1961). *J. Am. Chem. Soc.*, **83**, 4537–47.

Hill, T. L. (1946). *J. Chem. Phys.*, **14**, 465.

Kochanski, E. (1973). *J. Chem. Phys.*, **58**, 5823–31.

Lifson, S. and Warshel, A. (1968). *J. Chem. Phys.*, **49**, 5116–29.

Lii, J.-H. and Allinger, N. L. (1989). *J. Am. Chem. Soc.*, **111**, 8566–75 and 8576–82.

Niketić, S. R. and Rasmussen, K. (1977). *The consistent force field: A documentation*. Springer, Berlin.

Ōsawa, E. and Musso, H. (1982). Application of molecular mechanics calculations to organic chemistry. In *Topics in stereochemistry*, Vol. 13 (ed. N. L. Allinger, E. L. Eliel, and S. H. Wilen), pp. 117–93. Wiley-Interscience, New York.

Radom, L., Lathan, W. A., Hehre, W. J., and Pople, J. A. (1973). *J. Am. Chem. Soc.*, **95**, 693–8.

Schachtschneider, J. H. and Snyder, R. G. (1963). *Spectrochim. Acta*, **19**, 117–68.

Stidham, H. D. and Durig, J. R. (1986). *Spectrochim. Acta*, **42A**, 105–11.

Warshel, A. (1977). The consistent force field and its quantum mechanical extension. In *Semiempirical methods of electronic structure calculation*, Part A, *Techniques*, Modern Theoretical Chemistry, Vol. 7 (ed. G. A. Segal), pp. 133–72. Plenum Press, New York.

Wertz, D. H. and Allinger, N. L. (1974). *Tetrahedron*, **30**, 1579–86.

Westheimer, F. H. (1956). Calculation of the magnitude of steric effects. In *Steric effects in organic chemistry* (ed. M. S. Newman), pp. 523–55. Wiley, New York.

Westheimer, F. H. and Mayer, J. E. (1946). *J. Chem. Phys.*, **14**, 733–8.

White, D. N. J. (1978). Molecular mechanics calculations. In *Molecular structure by diffraction methods*, Vol. 6 (ed. L. E. Sutton and M. R. Truter), pp. 38–62. The Chemical Society, London.

White, D. N. J. and Bovill, M. J. (1977). *J. Chem. Soc. Perkin Trans. II*, 1610–23.

Wiberg, K. B. (1965). *J. Am. Chem. Soc.*, **87**, 1070–8.

Williams, D. E. (1974). *Acta Crystallogr.*, **A30**, 71–7.

15

Crystallographic databases: retrieval and analysis of precise structural information from the Cambridge Structural Database

Frank H. Allen

15.1 Introduction

X-ray crystallography is the most widely used technique for the study of molecular structure at atomic resolution. Developments in theory, instrumentation and computer technology, described in earlier chapters of this volume, have led to a dramatic increase in the number of diffraction analyses over the past twenty years. A single-crystal study now takes days rather than months with the result that X-ray analysis is now a method of choice rather than a method of last resort. This is borne out by the data embodied in Table 15.1 and Fig. 15.1: over 100 000 full three-dimensional analyses have been reported in the literature, over 50 per cent of these have been published since 1980. The improved precision of routine data collection methods has also led to significant improvements in the precision of the resultant structures. The mean R factor for a sample of 38 742 diffractometer studies is 0.057,

Table 15.1   Crystal structure database holdings (January 1989)

Database[a]	Coverage	Number of entries
CSD	Organo-carbon compounds	73 893
ICSD	Inorganics, minerals	28 406
MDF	Metals, alloys	11 000[b]
PDB	Proteins, macromolecules	427

[a] CSD : Cambridge Structural Database, University of Cambridge, UK. ICSD : Inorganic Crystal Structure Database, University of Bonn, FRG. MDF : Metals Data File, National Research Council of Canada, Ottawa, Canada. PDB: Protein Data Bank, Brookhaven National Laboratories, USA.
[b] Includes 6000 published structure determinations and 5000 entries assigned to known structure types.

whilst an R of 0.02–0.04 is now commonplace. These figures may be compared with mean R factors for early photographic (0.102) and densitometer (0.086) studies.

A knowledge of existing results and experimental data is fundamental to the scientific method. In many cases valuable new information can be distilled by a systematic analysis of existing data. For example chemists are still using van der Waals' radii derived by Pauling (1940) from the relatively few crystal structures available at that time. However, as the results of crystal structure analyses became commonplace in the literature, systematic work of this type became less popular. This is not altogether surprising: the labour involved in locating relevant references, and in extracting and processing large quantities of numerical data presented a formidable obstacle. Nowadays this barrier has been almost completely removed by the availability of computer-readable crystallographic databases, and by the ongoing development of software packages to retrieve and analyse the stored results.

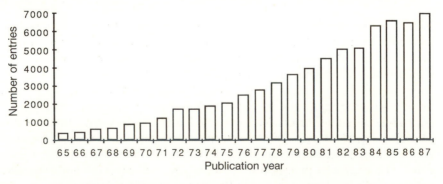

Fig. 15.1 Histogram showing the growth of Cambridge Structural Database (CSD), 1965–87, in terms of the number of entries published in each year.

15.2 Crystallographic databases: an overview

Crystallographers, perhaps mindful of the fundamental importance and wider implications of their results, have a long and continuous history of self-documentation. The 'Strukturbericht' (Ewald and Hermann 1929) began to summarize available results a mere 16 years after the publication of the first X-ray analysis. Printed compilations of this type are still of immense value and Watson (1977, 1987) has prepared a directory of such sources.

Computer-readable databases grew out of existing printed sources. Two of these: the Powder Diffraction Card Index (Jenkins and Smith 1987) and the NBS Crystal Data volumes (Mighell *et al.* 1987) transferred directly to computers. They are chemically comprehensive, but contain no atomic coordinates. They are used for identification purpose and for comparative studies based on lattice constants and symmetry.

The compilation of structural databases, designed to store crystal data *and* coordinates, together with bibliographic and chemical details of structures, began in the mid-1960s. They were designed not just for retrieval of literature references, but as a numeric resource to be used as the basis for further research. Four structural databases, CSD, ICSD, MDF and PDB, now cover the complete chemical spectrum. Their chemical coverage and current file holdings are summarized in Table 15.1. Full details of information content, software systems and availability of these databases are contained in a recent monograph prepared by the Data Commission of the International Union of Crystallography (Allen *et al.* 1987a). Apart from the chapters contained therein, other key references are: CSD: Allen *et al.* 1979; ICSD: Bergerhoff *et al.* 1983; MDF: Calvert and Rodgers 1984; PDB: Bernstein *et al.* 1977.

The remainder of this chapter is concerned with the principles of retrieval and analysis of information from crystallographic databases. The primary goal here is the derivation of precise structural information. For this purpose the Cambridge Structural Database (CSD) is taken as the basis, since much of the methodology has been developed in the area of organic and organometallic chemistry. Similar methods are, however, applicable in other areas. We begin with a brief overview of information content and available software tools.

15.3 Cambridge Structural Database (CSD)

15.3.1 Information content

CSD stores the primary results of full three-dimensional X-ray and neutron studies of organics, organometallics and metal complexes. The database is fully retrospective and is updated by some 600 new structures (entries) per month. The primary literature is abstracted together with any associated supplementary (deposited) data; CSD itself acts as a computerized depository for a number of journals. Over 500 primary sources are cited in CSD.

Each entry is identified by a reference code (Refcode) and the information content for any entry (Table 15.2) may be categorized as bibliography, chemical structure (as a connectivity table), and numeric structural data. A variety of chemical and numeric checks, the majority of which are computerized, ensure the accuracy of CSD input and of the published data. A major numerical check involves the recomputation of bond lengths from the primary data and their automated comparison with published values. More recently the 1:1 matching of chemical and crystallographic connectivity tables has enhanced the evaluation process. Similar check procedures are employed in the construction of all structural databases, although precise details depend upon the chemical nature (molecular, ionic, macromolecular) of the compounds stored in each system (see, for example, Altermatt and Brown 1985).

Table 15.2 Summary of information content of Cambridge Structural Database (CSD)

Bibliographic information
Compound name(s); molecular formulae; authors' names; journal reference; qualifying phrase(s) (e.g. neutron study, absolute configuration determined); chemical classification (1–86: for example 15 = benzene nitro compounds, 51 = steroids).

Chemical connectivity tables
Chemical structural diagram encoded in terms of:
  Atom properties
Atom sequence number (n); element symbol (el); number of connected non-H atoms (nca); number of terminal H atoms (nh); net atomic charge (ch); two-dimensional coordinates for graphical output (x, y).
  Bond properties
Node numbers of atoms i, j forming the bond (n_i, n_j); bond type for bond i–j (bt).

Numerical crystallographic data
Unit-cell parameters; space group and symmetry operators; covalent radii; atomic coordinates for 'crystal chemical unit' (symmetry-generated atoms bonded to asymmetric unit included); crystallographic connectivity; precision indicators; text comment (general remarks, details of disorder, summary of any errors located in the original paper).

15.3.2 Software systems for information retrieval

A variety of database architectures and software system designs have been applied to crystallographic databases. These fall into four main categories:

(1) simple serial search systems;

(2) screened serial search with possible indexed sequential (direct) access to main database entries;

(3) fully inverted hierarchical tree designs;

(4) implementation within a commercial database management system.

All four of these approaches have been applied to CSD and all have advantages and drawbacks when judged against criteria such as: in-house or dial-up use, flexibility of searching, portability over many computer systems.

Versions 1 and 2 of Cambridge software used design (1) with bibliography, chemical connectivity and numerical data held on three separate serial files (Allen *et al.* 1979). This permitted separate searches for text strings *or* for structural chemical fragments followed by retrieval of the numerical data for the 'hits'. An alternative strategy (3) is employed for CSD within the Chemical Databank System at the SERC Daresbury Laboratory (UK) (Machin 1985). Here searches are virtually instantaneous via examination of sorted lists of search items linked by pointers. Whilst (1) is slow, it is readily portable. By contrast (3) is very fast, but is machine dependent, can be difficult to update to cover new search terms, and uses disc storage considerably in excess of the original database size to hold the inverted lists and pointers.

Version 3 of CSD files and software (Allen and Davies 1988: Fig. 15.2)

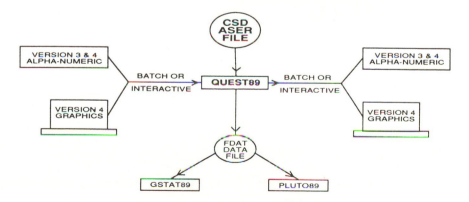

Fig. 15.2 Flowchart of version 4.1 of CSD software.

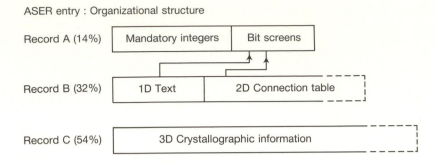

ASER entry : Organizational structure

Record A (14%) — Mandatory integers | Bit screens

Record B (32%) — 1D Text | 2D Connection table

Record C (54%) — 3D Crystallographic information

Fig. 15.3 Schematics of ASER file structure.

employs design (2) as a compromise between (1) and (3). Figure 15.3 shows that information is split between three logical records for each entry in the ASER file:

Record A   (a) *Key integers* or real numbers multiplied by 10^n (e.g. chemical class, year of publication, journal number, cell parameters, R factor, etc.)

(b) *Bit screens.* Bits are set on/off (1/0) to record the presence/absence of selected information items in record B (e.g. error status, neutron study, coordinate availability; text is screened by hash coding of contiguous letter pairs; chemical connectivity is screened by analysis of element and bond types, bonded pairs, bond-centred units, atom-centred units, ring systems, etc.)

Record B   *Searchable information.* This comprises all text and comment fields from Table 15.2, together with the chemical connectivity information.

Record C   *Numerical structural data.* Atomic coordinates, symmetry operators, atomic radii, etc.

Records of type A are always read sequentially and are the key to system efficiency. Each complete question (see Fig. 15.4) is analysed and a bit screen (plus other 'key integer' requirements) is generated to yield a record A_q which is compared with each 'target' record A_t on the master file. This comparison is rapid, and tells the search program (QUEST89: Fig. 15.2) whether a further (expensive) read of record B is required or not. All sub-features of the question, which correspond to permissible bit settings in the screen record, must be present in the file record A_t for this read operation to be executed. This does not automatically guarantee that the file entry is a hit, since the required features, although present, may not be arranged in the correct order. For example a search for C=C—S=O will generate bit settings for C=C and S=O and any entry containing these two features will satisfy the screening process. The entry only becomes a hit if a systematic atom-by-atom, bond-by-bond connectivity match shows that the two units are, in fact, linked as required by the question. The screening system is therefore designed to limit the demand for cpu-intensive character-matching or connectivity-matching functions. For most searches over 95 per cent of records B and C are skipped by this technique; many searches involving structural fragments achieve screenouts of 98 per cent and above. Record C is only read if a hit *is* recorded *and* the user wishes to create a subfile of numerical data (FDAT: Fig. 15.2) for later processing.

Figure 15.4 illustrates the general principles of the search system, and its ability to formulate and answer complex and detailed queries. A complete question is made up of a series of tests (here T1–T5) on individual information fields. The final QUEStion command consists of a Boolean combination of the tests using .AND. (or +), .OR. (or ,) and .NOT. (or −) operators. The speed of the system depends crucially on the efficiency of the screening process (described above) for searches involving text or chemical substructures. Screens are of two types: (a) those set by the user (e.g. those in the SCRE record of Fig. 15.4), and (b) those set automatically by the software. In the example (Fig. 15.4) 33 screens are deduced from an analysis of the connectivity specifications in T5: a few of these screened subfeatures are listed in Table 15.3.

The design of the system is such that the B and C records described above may form part of a simple sequential file, or they may be organised as separate indexed-sequential (direct access) files. The sequential file is very general, is suitable for any 32-bit computer, and may be stored on disk or magnetic tape. In the indexed-sequential mode disk storage is essential, and the A records form a separate sequential file. The 1989 database requires *ca.*150 Mb of disk storage.

Table 15.3 Examples of connectivity screens set for test T5 of Fig. 15.4 (33 set altogether)

Screen No.	Definition
264	.GE. 2 Oxygen atoms
274	C · O connection (any bond type)
314	Double bond
333	C—C single bond
387	O · C · O triplet (any bond type)
405	C–a–C–b–O (a = single bond, b = double or delocalised double)
448	C has .GE. 3 non-H connections, .GE. 1 are to C, .GE. 2 are to non-C

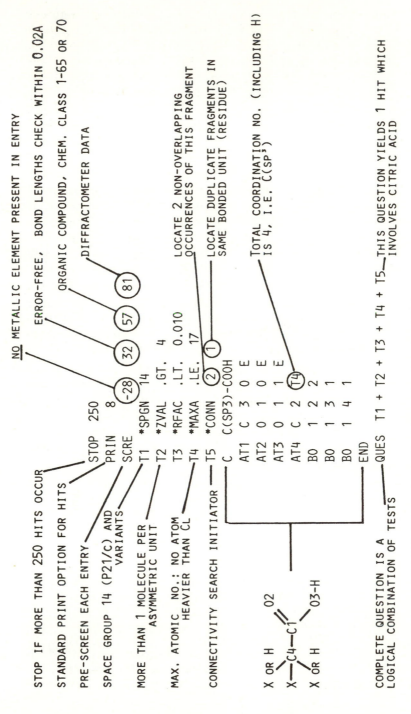

Fig. 15.4 Example of a search using program QUEST89.

Retrieval of information is performed by QUEST89 as it passes through the ASER file. The most important subfiles that may be generated are ASER and FDAT (Fig. 15.2). An ASER subfile can be reprocessed by QUEST89, permitting in-depth searches following an initial general search to create the subfile. For example a user may create a private subfile containing, say, all steroids; this file may then be the subject of any number of more specific (and rapid) searches. An FDAT file forms the input to the programs GSTAT89 and PLUTO89 (Fig. 15.2). The latter is well known to crystallographers (and others) for the generation of a wide variety of illustrations from three-dimensional coordinate data. GSTAT was originally developed (by W. D. S. Motherwell) for the calculation of intramolecular and intermolecular geometry for molecules and structural fragments (see Allen *et al.* 1979). It has since developed into a package within which that geometry may be analysed by statistical and numerical techniques (see Murray-Rust and Raftery 1985 and references therein). The package is complemented by a set of standalone statistical subroutines: the CAMAL library (Taylor 1986*a*). The systematic analysis of numerical data is fully described below, following a brief resumé of the query language used in QUEST89.

15.3.3 Search query construction

Input to the QUEST89 search program was initially (version 3) in the alphanumeric form illustrated in Fig. 15.4. Up to 200 tests (Tn) may be specified by the user and combined in the final QUEStion line by Boolean logic as described earlier. A number of other keywords (e.g. SAVE, PRINT) indicate the required output files or print options to be applied to the hits. There are five broad categories of information that can be specified on a Tn record via subkeyword mnemonics (*CLAS, *AUTH, etc.) which identify the required information item(s):

(1) Numeric searches: any of 38 individual numeric fields may be accessed and tested by use of the self explanatory operators .EQ., .NE., .LT., .LE., .GT., .GE., or by specification of a range: n_1-n_2.

(2) Text searches: character strings of the form 'string' (' = delimiter) may be located in any of 20 text fields.

(3) Chemical formula searches: a variety of options exist for the location of specified element(s) and/or element-count combinations. Any of 30 generalized group symbols based on the periodic table (7A = halogens, etc.) may be used here.

(4) Reduced-cell searches: input cell parameters (plus lattice type) are Niggli-reduced and matched against equivalent data stored in ASER.

(5) Chemical connectivity searches: the chemical connectivity tables may be searched for complete molecules or structural fragments. Atom and bond properties for the query are coded in a similar manner to those for ASER entries (Table 15.2, Fig. 15.4). Additional keywords permit the fragment and its environment to be restricted or relaxed in various ways via specification of (e.g.): special 'group' element symbols as at (3) above; cyclic/acyclic tagging of atoms or bonds; limiting coordination (hybridization) number; ranges of bond type, H count, etc.

A menu-driven graphics system (Version 4) was released in late 1989; it simplifies input in general, and input of structural fragments in particular. A variety of options is available in Version 3 for printed output (to hard copy or terminal screen); this has been augmented in Version 4 by graphical output of chemical diagrams for hit entries.

15.4   Systematic numerical analysis of crystallographic data

15.4.1   Calculation of molecular geometry

The program GSTAT has two very different modes of operation: a *basic* mode and a *fragment* mode. In its basic mode the program will calculate and print molecular geometry on an entry-by-entry basis. These lists can contain inter- or intramolecular distances, valence angles and torsion angles. In fragment mode the program will locate a specified chemical fragment within the crystallographic connection table stored in each FDAT entry. The fragment specified in GSTAT89 is usually the same as, or some subset of, the chemical substructure employed as a search term in QUEST89. However it must be remembered that the chemical connection tables employed by QUEST89 and the crystallographic connection tables used by GSTAT89 are different. The latter are established using distance criteria based upon covalent radii sums, hence by comparison with the chemical connectivity (Table 15.2) they lack bond type information, H-counts *may* be incomplete due to their non-location by the X-ray experiment, and charges cannot be assigned to atoms. The fragment specification in GSTAT (Fig. 15.5(a)) is a shortened version of that available in the chemical search process. The GSTAT definition may, however, be enhanced by use of TESTs based on distances, valence angles and torsion angles. Used singly, or together, these TESTs compensate for the lack of bond type information; they also go further in permitting the user to select out only those entries having specified geometrical characteristics of configuration or conformation.

Once a fragment has been coded, the user may then DEFine the geometrical parameters (Fig. 15.5(a)) which are to appear in a fragment

(a)

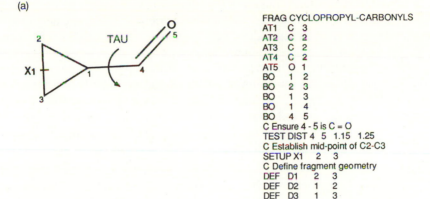

```
FRAG CYCLOPROPYL-CARBONYLS
AT1   C   3
AT2   C   2
AT3   C   2
AT4   C   2
AT5   O   1
BO    1   2
BO    2   3
BO    1   3
BO    1   4
BO    4   5
C Ensure 4 - 5 is C = O
TEST DIST 4  5  1.15  1.25
C Establish mid-point of C2-C3
SETUP X1   2    3
C Define fragment geometry
DEF   D1    2    3
DEF   D2    1    2
DEF   D3    1    3
DEF   #TOR      5    4    1    X1
C   Obtain absolute torsion angle
TRA    TAU  =  ABS  (#TOR)
```

(b)

Nfrag	Refcod	D1	D2	D3	TAU
1	ACMEPT	1.416	1.528	1.530	176.189
2	ACMEPT	1.416	1.522	1.529	176.206
...	......	.....	.....	.....	....
...	......	.....	.....	.....	....
265	SDPPCX	1.482	1.533	1.510	5.465
266	TOLIPO10	1.486	1.542	1.502	24.023
267	ZEHREJ	1.451	1.518	1.519	4.660
268	ZEHREJ	1.451	1.516	1.520	6.199
Mean		1.494	1.517	1.517	82.168
S.D.Sample		0.032	0.023	0.023	72.652
S.D.Mean		0.003	0.003	0.002	7.880
Minimum		1.416	1.460	1.440	0.422
Maximum		1.574	1.623	1.561	177.947
Nobs		268	268	268	268

Fig. 15.5 Example of (a) FRAGment input and (b) systematic tabulation output from GSTAT89.

geometry tabulation (Fig. 15.5(b)). The mnemonics employed by the user appear as column headings. A wide variety of parameters can be included in a tabulation: distances, angles and torsion angles (any of which may involve bonded and/or non-bonded atoms, midpoints and centroids), distances from mean planes, angles between vectors or planes, etc. Within GSTAT basic parameters may be transformed or combined by simple FORTRAN-like expressions; hence the sum of three valence angles (as a measure of planarity at N for example) may appear as a single parameter in the table. Most importantly some or all of the parameters thus generated may be selected for further statistical analyses by preceding their mnemonic by a % sign. The fragment approach, exemplified by Fig. 15.5, is not restricted to intramolecular units. It is possible to extend the crystallographic connection table, using non-bonded distance criteria, to include neighbouring molecules. This permits the study in geometrical terms of hydrogen-bonded systems and of non-bonded interactions.

Earlier in this chapter it was noted that an important database integrity check involved the 1:1 matching of chemical and crystallographic connection tables. Transfer of this facility from in-house to user software is an ongoing project. This will eliminate the need for fragment redefinition in GSTAT, and ensure that the fragments located by QUEST and by GSTAT are chemically identical.

15.4.2   Selection of entries: precision criteria

In most practical applications of any crystallographic database it is unlikely that all entries which satisfy search criteria will be candidates for geometrical analysis. Only the more precisely determined structures are normally included in this latter step. There are a number of ways in which this selection can be made. At a simple level one may exclude entries with residual checking errors (entries for which there remain some inconsistencies between crystal data, coordinates and published bond lengths), and one may also wish to exclude structures where disorder problems have been identified.

Within CSD there are two main indicators of precision: the crystallographic R factor and a numeric flag (AS: Table 15.4) which records the average estimated standard deviation (e.s.d.) of a $C-C$ bond length. For many years AS was derived manually from published e.s.d.s but, since 1985, individual atomic coordinate e.s.d.s have been stored in CSD and AS has been derived directly from these values. Table 15.4 presents a breakdown of AS over ranges of R factor for a large subset of CSD. Only 10 per cent of these entries fall into the minimum category of $AS = 1$ and $R \leq 0.040$. However a slight relaxation of these criteria to $AS = 1,2$ and $R \leq 0.060$ brings 42 per cent of entries into use. All of the criteria noted here can be

Table 15.4   Distribution of AS flag (see text) vs. R factor[a]

AS	Range of $\langle\sigma(C-C)\rangle$ (Å)	Number of entries in R-factor range				
		1–4%	4–5%	5–6%	6–8%	8–12%
0	not known	1147	1077	980	1342	880
1	0.001–0.005	3313	2347	1117	624	128
2	0.006–0.010	2792	2557	1999	1934	564
3	0.011–0.030	1392	1721	1750	2575	1425
4	>0.030	95	188	284	614	603

[a] For 33 448 CSD entries : diffractometer data, no disorder, no residual checking errors.

applied within QUEST89 or (later) within GSTAT89.

Two other factors emerge from Table 15.4. Firstly there are an unfortunate number of entries (16 per cent) in all R-factor ranges for which AS flags could not be assigned. Secondly the choice of AS range limits (in Å) is somewhat arbitrary, and the $AS = 3$ range may be rather too broad. A recent study (Allen and Doyle 1987) examined the relationship between R, $\langle \sigma(C-C) \rangle$ (the actual mean e.s.d. in Å) and Z_{max} (the maximum atomic number in the structure). The study used 4317 entries for which atomic coordinate e.s.d.s were available. Some results of this study are depicted graphically in Fig. 15.6 (a,b). They show the (expected) trend towards increased $\langle \sigma(C-C) \rangle$ as Z_{max} increases. If AS is used as a selection criterion, then the chemical nature of the problem being studied must be taken into account. Thus, for an R-factor limit of $ca.\,0.070$, an AS limit of 2 might be appropriate for 'organics' but should be increased to an AS limit of 3 for problems involving transition metal structures in which metal–metal or metal–ligand geometry is being studied. The data were also used to derive an empirical relationship of the form

$$\langle \sigma(C-C) \rangle = a + bR + cZ_{max}, \tag{15.1}$$

by which it is possible (a) to generate an empirical AS estimate for the current $AS = 0$ subset of Table 15.4, and (b) to provide an empirical selection criterion based upon $\langle \sigma(C-C) \rangle$ itself, rather than on arbitrary ranges. Equation (15.1) was derived by multiple linear regression using parameters available in CSD. For a few hundred entries the ratio N_p/N_r (N_p = number of refined parameters, N_r = number of reflections) was obtained from original papers. Inclusion of this ratio as an additional regression parameter yielded an equation which gave very much better agreement between actual and estimated values of $\langle \sigma(C-C) \rangle$.

The use of crystallographic e.s.d.s from individual structures as selection criteria or, particularly, as inter-structure comparison limits, has recently been studied by Taylor and Kennard (1986b). They used a small subset of 100 structures that had been determined independently by different research groups. Their main findings are pertinent to all users of crystallographic data: (a) e.s.d.s of non-H atom coordinates are typically underestimated by a factor of 1.4–1.45; (b) e.s.d.s for cell parameters are grossly underestimated; (c) e.s.d.s of heavy atom coordinates are less reliable than those of light atoms.

15.4.3 Averaging of fragment geometry

The AS, R criteria above may, of course, be used to select the most precise individual example(s) of a particular structure type. They are much more

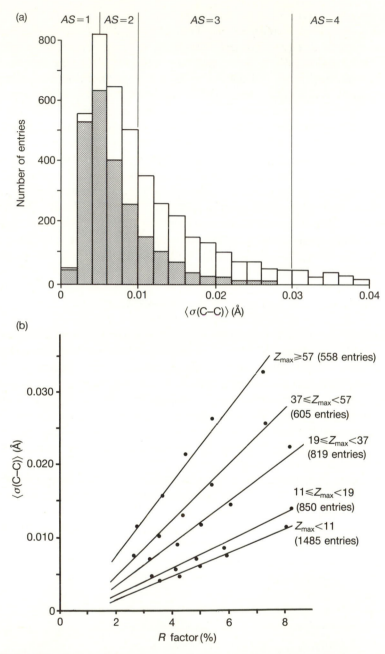

Fig. 15.6 (a) Distribution of the mean e.s.d. of a C—C bond, $\langle \sigma(C-C) \rangle$, for 4317 CSD entries having individual coordinate e.s.d.s. The shaded area represents 'organic' structures taken as having maximum atomic number (Z_{max}) less than 18. (b) Plot of $\langle \sigma(C-C) \rangle$ *vs.* R factor for ranges of Z_{max}. After Allen and Doyle (1987).

commonly used to ensure that only the more precise entries are represented in a fragment geometry table. There are some good statistical reasons for this approach arising out of the averaging process. For many applications it is preferable to use the mean geometry for a fragment than to use the geometry of that fragment from an isolated crystal structure (however good the precision of that structure may be). Often this is a necessity, since a structure of the required precision may not be available.

The derivation of mean molecular dimensions is an obvious use for crystallographic data. Compilations of 'standard' bond lengths (e.g. Sutton 1958, 1965; Allen *et al.* 1987*b*) provide valuable benchmarks against which new structural results may be compared. This approach has been extended to the derivation of mean geometries for complete chemical residues. Thus the pyranoses (Sheldrick and Akrigg 1980), nucleic acid bases (Taylor and Kennard 1982*a, b*), carboxylic acids, esters and amides (Borthwick 1980; Schweizer and Dunitz 1982; Chakrabarti and Dunitz 1982) and other groupings have been studied in this way. Given these mean values it is relatively simple (Taylor and Kennard 1982*b*) to derive orthogonal coordinates for 'standard' residues. Apart from their use in crystal structure analysis, as initial fragments in Patterson search methods (Egert and Sheldrick 1985), standard geometries are invaluable for the parameterization of empirical force fields and in model building for molecular graphics and drug design purposes.

The derivation of mean molecular dimensions has been thoroughly examined from a statistical standpoint (Taylor and Kennard 1983, 1985, 1986*a*). There are two obvious choices for averaging: the simple (unweighted) mean, $\langle x \rangle$, and the weighted mean, $\langle x(w) \rangle$, where for a sample of observations $i = 1 \rightarrow n$ we have:

$$\langle x \rangle = \sum_i x_i / n \tag{15.2}$$

$$\sigma(\text{sample}) = \left[\sum_i (\langle x \rangle - x_i)^2 / (n - 1) \right]^{1/2} \tag{15.3}$$

$$\sigma(\langle x \rangle) = \sigma(\text{sample}) / \sqrt{n} \tag{15.4}$$

$$\langle x(w) \rangle = \sum_i w_i x_i / \sum_i w_i \tag{15.5}$$

$$\sigma[\langle x(w) \rangle] = \left(1 / \sum_i w_i \right)^{1/2}. \tag{15.6}$$

The statistical analysis (Taylor and Kennard 1983) takes into account the relative effects of experimental errors and of environmental (e.g. crystal packing) effects. It is shown that the weighted mean (eqn (15.5)) is best where environmental effects are small, e.g. for 'hard' parameters such as bond lengths. Where environmental effects are large (e.g. torsion angles and, perhaps, valence angles, in flexible systems), then the unweighted mean (eqn (15.2)) is to be preferred. The second paper (Taylor and Kennard 1985), however, indicates that the unweighted mean may always be preferable if $\langle x \rangle$ is to be used in hypothesis tests.

The lack of e.s.d.s for the majority of CSD entries of course prevents the calculation of $\langle x(w) \rangle$ without recourse to the original literature. In the final paper (Taylor and Kennard 1986a) it is shown that the unweighted mean is quite acceptable, even for 'hard' parameters such as bond lengths, provided that the AS criterion (noted above) is used to restrict the range of variances of the original x_i. The authors suggest that restriction to $AS = 1,2$ is suitable for organic bond lengths, whilst the upper limit can be relaxed to include $AS = 3$ entries for 'softer' parameters. The simple summary statistics provided by GSTAT are perforce based upon eqns (15.2), (15.3), and (15.4).

15.4.4   Visual aids to data analysis

Mean geometry will always reflect the population from which it is derived. For precise work we attempt to define a unique chemical population, one in which changes in the chemical environment of the fragment are random rather than systematic. If this is so then each geometrical parameter in the fragment should follow a unimodal normal distribution. A simple histogram is an essential visual aid for examining the shape of a distribution and can be output on the line printer by GSTAT89. Some typical examples are illustrated in Fig. 15.7 and 15.8.

Figure 15.7 shows an otherwise unimodal distribution together with a single outlying observation. The origin of outliers (rogues) should always be examined, but sytematic errors in the structure are often the only likely cause. Occasionally a novel chemical effect can be invoked which may lead to further research. In normal circumstances the outlier is simply omitted and the averaging process repeated. The current version of GSTAT employs a two-pass averaging technique where individual contributors which are $>4\sigma$ (sample) from the mean in the first pass are omitted from the second pass. These automated attempts at outlier removal, although useful, are no substitute for visual inspection of the relevant histogram(s).

In Fig. 15.8 the apparently chemically unique fragment gives rise to a bimodal distribution for the C(aromatic)—N bond length. This is obviously due to the presence of $N(sp^3)$-pyramidal and $N(sp^2)$-planar contributors,

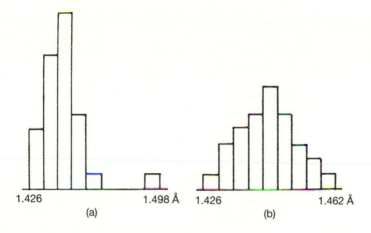

(a)   (b)

1.426   1.498 Å   1.426   1.462 Å

Fig. 15.7 (a,b) Effect of removal of outliers (contributors which are $>4\sigma$(sample) from the mean) for the C—C bond lengths in C(aromatic)—C≡N fragments. After Allen *et al.* (1987*b*).

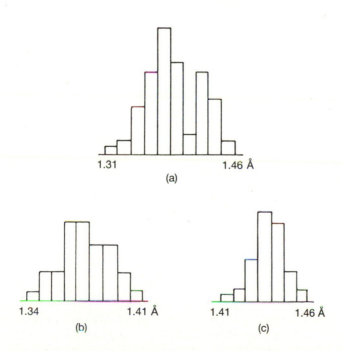

1.31   1.46 Å

(a)

1.34   1.41 Å   1.41   1.46 Å

(b)   (c)

Fig. 15.8 Resolution of bimodal distribution of C—N bond lengths in C(aromatic)—N—$[C(sp^3)]_2$ fragments using criterion of planarity at N (see text). (a) Complete distribution. (b) Distribution for planar N. (c) Distribution for pyramidal N. After Allen *et al.* (1987*b*).

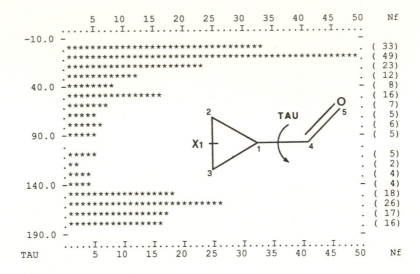

```
              5    10   15   20   25   30   35   40   45   50      Nf
         .....I....I....I....I....I....I....I....I....I.....
  -10.0 -                                                    -
         .*******************************                    . ( 33)
         .***************************************************. ( 49)
         .***********************                            . ( 23)
         .************                                       . ( 12)
   40.0 -********                                            - (  8)
         .****************                                   . ( 16)
         .*******                                            . (  7)
         .*****                                              . (  5)
         .******                                             . (  6)
   90.0 -*****                                               - (  5)
         .                                                   .
         .*****                                              . (  5)
         .**                                                 . (  2)
         .****                                               . (  4)
  140.0 -****                                                - (  4)
         .******************                                 . ( 18)
         .**************************                         . ( 26)
         .*****************                                  . ( 17)
         .****************                                   . ( 16)
  190.0 -                                                    -
         .....I....I....I....I....I....I....I....I....I.....
    TAU   5    10   15   20   25   30   35   40   45   50      Nf
```

Fig. 15.9 Distribution of the torsion angle O5—C4—C1—X1 (τ) for 202 occurences of the cyclopropylcarbonyl fragment.

together with some 'intermediates'. The resolution of this distribution is a little subjective but a planarity criterion based on the mean valence angles at N (108–115° and >117°) generates two normal distributions with significantly different mean values.

Figure 15.9 shows how a simple histogram can give a clear visual representation of a univariate conformational problem. From a consideration of the interactions between the orbitals of the carbonyl group and those of the ring (Hoffmann 1970) the C=O bond is expected to adopt a *cis*-bisected ($\tau \cong 0°$) or *trans*-bisected ($\tau \cong 180°$) conformation with respect to the ring. Figure 15.9 is graphic proof of this prediction over a large number of crystal structures.

Visual representations of bivariate distributions are given by simple two-dimensional scattergrams. These again can be simply generated in line-printer form via GSTAT.

15.4.5   Other statistical and numerical techniques

The statistical methodology for the analysis of crystallographic data is a rapidly developing area. Many of the methods being applied have their practical origins in fields where the analysis of large-volume numerical information is a frequent requirement: the social sciences, biology and economics are obvious examples. The wide variety of techniques is reflected in the contents (Table 15.5) of the CAMAL subroutine library (Taylor 1986a).

Table 15.5   Summary contents of the CAMAL subroutine library (Taylor 1986*a*)

Simple statistics
Analysis of variance; correlation and regression; hypothesis testing; descriptive statistics.

Multivariate statistics
Principal-component analysis; cluster analysis; multi-dimensional scaling; descriptive statistics.

Utility subroutines
Sorting and ranking; probability distributions; vector and matrix operations.

Regression and correlation analyses are essential in the quantification of relationships between parameters (see, for example, Allen 1984; Allen and Doyle 1987). Hypothesis testing is a fundamental statistical procedure which has been applied to the derivation of mean geometries (Taylor and Kennard 1985), studies of hydrogen bonding (see, for example, Taylor and Kennard 1984), and in a direct 1:1 comparison of geometries from X-ray and neutron studies (Allen 1986).

Statistical methods designed to analyse multivariate data sets are of particular importance in this area. Fragment geometry derived from CSD is often of this type, for example the sets of *n* torsion angles which define the conformation of an *n*-membered ring. Techniques such as principal-component analysis (see, for example, Murray-Rust and Motherwell 1978; Taylor 1986*b*) and a variety of clustering algorithms (Murray-Rust and Raftery 1985; Nørskov-Lauritsen and Bürgi 1985; Taylor 1986*b*) are designed to reduce the dimensionality of the problem and enable the user to locate and classify the conformational subgroups. A particular problem in this area is the effect of two-dimensional topological fragment symmetry. This gives rise to equivalent permutational sequences for derived three-dimensional geometrical parameters. The permutational symmetry must be taken into account before meaningful averages can be derived. Very recently (Allen *et al.* 1991) a set of symmetry-modified clustering algorithms have been prepared for inclusion in GSTAT.

15.5   Applications of CSD

Over 100 papers have been published which are based solely upon experimental data obtained from CSD. A number of reviews of research applications have appeared in the literature (Allen *et al.* 1983; Bürgi and Dunitz 1983; Taylor and Kennard 1984; Allen *et al.* 1987*a*—which includes a classified bibliography). This brief overview is designed to indicate the important contributions now being made by crystallographic results outside its own narrow confines.

Familiarity with the basic data and with computational methods meant that crystallographers were the first large-scale users of CSD. The user community has now expanded considerably to include scientists from a variety of disciplines. This is due to the extensive use of crystallographic results for molecular modelling purposes. Here data from CSD and from the Protein Data Bank are a useful, even essential, adjunct to computational and graphical techniques employed in the design of new molecules with specific biological activity. This process makes use of any study which provides systematic knowledge of molecular dimensions, molecular shape, or of the geometrical details of intermolecular interactions. The division between intramolecular and intermolecular results represents a convenient framework for this overview.

15.5.1 Intramolecular applications

Methods for the derivation of precise mean values for individual geometrical parameters or complete chemical residues have already been described. A recent large-scale application of the techniques (Allen *et al.* 1987*b*; Orpen *et al.* 1989) has resulted in an upgrading of earlier tabulations (Sutton 1958, 1965) of mean bond lengths. The organic section of this new table contains some 700 mean values for bonds involving B, C, N, O, F, Si, P, S, Cl, As, Se, Br, Te, I and H (neutron data only). The second section (Orpen *et al.* 1989), covers organometallics and metal complexes. The derivation of mean dimensions may be extended to studies of substituent effects via a comparison of precise mean values for closely related substructures. This approach has been used to study the substituent-induced distortion of the benzene ring (see Chapter 18) and a variety of bonding features of strained small-ring hydrocarbons (Allen 1987).

A topic of continuing interest, from both a statistical and a chemical viewpoint, is the classification of conformational preferences in flexible systems. Knowledge of these preferences is of particular importance in molecular modelling. For many systems a number of conformers are represented in the crystallographic data, each of which represents an energetically accessible form. Identification of major conformational components provides alternative starting points for model building, or for energy minimization via force-field calculations. For systems involving only one or two degrees of freedom classification is usually straightforward. However a distribution like that of Fig. 15.9 is of considerable interest in insecticide modelling, since the fragment is a common constituent of pyrethroids. For more complex fragments the multivariate statistical techniques cited in Section 15.4.5 are of prime importance.

15.5.2 Intermolecular applications

Crystallography is the only experimental technique which permits a systematic study of the geometrical aspects of intermolecular interactions. Results of these studies have broad chemical applicability and are of particular interest in molecular recognition, where possible interactions between a drug and a receptor site are being examined. Here there is a need for quantitative information on non-bonded atomic shapes, hydrogen-bond geometries, and on any directional preferences in the non-bonded approach of one group to another.

Crystal structure analyses have always been a primary source of information on the systematics of hydrogen bonding (see, for example, Hamilton and Ibers 1968). CSD has already been used extensively; specific areas of study include carbohydrates (Ceccarelli et al. 1981; Jeffrey and Mitra 1983), amino acids (Jeffrey and Maluszynska 1982), the $N-H\cdots O=C$ system (Taylor et al. 1984), proof of the (crystallographic) existence of $C-H\cdots X$ hydrogen bonds (Taylor and Kennard 1982c); the directionality of hydrogen bonding to sp^2 and sp^3 hybridized oxygen atoms (Murray-Rust and Glusker 1984); and studies of lone-pair directionality leading to an extended hydrogen-bond potential function for molecular mechanics applications (Vedani and Dunitz 1985). A recent review (Taylor and Kennard 1984) covers much of this work.

The geometry of non-bonded interactions is usually assessed in terms of non-bonded radii developed by Pauling (1940) or Bondi (1964). These values were based on limited experimental data and assume that (a) non-bonded atomic shapes are effectively spherical, and (b) the radii are additive and transferrable from structure to structure. A number of recent studies (e.g. Rosenfield et al. 1977; Britton and Dunitz 1980; Nyburg and Faerman 1985; Nyburg et al. 1987) indicate that these assumptions are inexact; there is evidence of asphericity in non-bonded atomic shapes, and the limiting non-bonded contact distance between two atoms depends upon their individual chemical environments and upon their direction of mutual approach. Directional preferences of approach are a key feature of studies of reaction pathways via the structure correlation principle (see Bürgi and Dunitz 1983). These studies represent one of the most important novel uses of crystallographic results and form the subject matter of Chapter 17.

15.6 Conclusion

It is now some twenty years since computerized files of crystallographic information began to be assembled in a systematic manner. At their inception all four of the structural databases noted in Table 15.1 were faced with a backlog of existing material, whilst simultaneously trying to assimilate current publications. An early priority was to develop systems and software for the acquisition, evaluation and management of the data. This was followed by systems for search, retrieval and analysis and for database distribution. It is not surprising, therefore, that research publications based upon the accumulated data only began to appear in the last decade.

It is encouraging that this volume contains other chapters concerned with databases and applications of crystallographic data. The subject matter presented in these chapters points quite clearly to the areas for development over the next few years. The database systems themselves will improve in terms of their information content, file organization, search systems and distribution media. The possibility of machine-readable submission of data directly to journals and databases must be fully investigated. It is only by this means that large-volume 'omissions', for example atomic displacement parameters in CSD, can be rectified for the future (if not for the backlog). There will be development also in the numerical, statistical and graphical methodologies used for data analysis. Techniques such as principal-component analysis and various clustering algorithms are already proving their worth in revealing the constituent patterns contained in multivariate datasets of structural data. These techniques will increase in importance as database sizes continue to increase. Finally, the range of research applications for crystallographic databases will continue to broaden. There is a symbiotic relationship between research applications and the developments in the database systems and associated analytical tools noted above. These developments make it easier to obtain basic data for the research; this in turn provokes new ideas, which often point up a need for more developments in the databases and software systems. This close relationship between research and development activities, already apparent over the past few years, is seen as the major force for progress over the next decade.

References

Allen, F. H. (1984). *Acta Crystallogr.*, **B40**, 64–72.

Allen, F. H. (1986). *Acta Crystallogr.*, **B42**, 515–22.

Allen, F. H. (1987). The Cambridge Structural Database as a research tool in chemistry. In *Modelling of structure and properties of molecules* (ed. Z. B. Maksić), pp. 51–66. Horwood-Wiley, Chichester.

Allen, F. H. and Davies, J. E. (1988). File structures and search strategies for the Cambridge Structural Database. In *Crystallographic computing 4: Techniques and new technologies* (ed. N. W. Isaacs and M. R. Taylor), pp. 271–89. International Union of Crystallography, Chester, and Oxford University Press.

Allen, F. H. and Doyle, M. J. (1987). *Acta Crystallogr.*, **A43** (suppl.), p. C291.

Allen, F. H., Bellard, S., Brice, M. D., Cartwright, B. A., Doubleday, A., Higgs, H., Hummelink, T., Hummelink-Peters, B. G., Kennard, O., Motherwell, W. D. S., Rodgers, J. R., and Watson, D. G. (1979). *Acta Crystallogr.*, **B35**, 2331–9.

Allen, F. H., Kennard, O., and Taylor, R. (1983). *Acc. Chem. Res.*, **16**, 146–53.

Allen, F. H., Bergerhoff, G., and Sievers, R. (eds.) (1987*a*). *Crystallographic databases*. International Union of Crystallography, Chester.

Allen, F. H., Kennard, O., Watson, D. G., Brammer, L., Orpen, A. G., and Taylor, R. (1987*b*). *J. Chem. Soc. Perkin Trans. II* (suppl.), pp. S1–S19.

Allen, F. H., Doyle, M. J., and Taylor, R. (1991). *Acta Crystallogr.*, **B47**, 29–40, 41–9, and 50–61.

Altermatt, D. and Brown, I. D. (1985). *Acta Crystallogr.*, **B41**, 240–4.

Bergerhoff, G., Hundt, R., Sievers, R., and Brown, I. D. (1983). *J. Chem. Inform. Comput. Sci.*, **23**, 66–9.

Bernstein, F. C., Koetzle, T. F., Williams, G. J. B., Meyer Jr., E. F., Brice, M. D., Rodgers, J. R., Kennard, O., Shimanouchi, T., and Tasumi, M. (1977). *J. Mol. Biol.*, **112**, 535–42.

Bondi, A. (1964). *J. Phys. Chem.*, **68**, 441–51.

Borthwick, P. W. (1980). *Acta Crystallogr.*, **B36**, 628–32.

Britton, D. and Dunitz, J. D. (1980). *Helv. Chim. Acta*, **63**, 1068–73.

Bürgi, H.-B. and Dunitz, J. D. (1983). *Acc. Chem. Res.*, **16**, 153–61.

Calvert, L. D. and Rodgers, J. R. (1984). *Comput. Phys. Commun.*, **33**, 93–8.

Ceccarelli, C., Jeffrey, G. A., and Taylor, R. (1981). *J. Mol. Struct.*, **70**, 255–71.

Chakrabarti, P. and Dunitz, J. D. (1982). *Helv. Chim. Acta*, **65**, 1555–62.

Egert, E. and Sheldrick, G. M. (1985). *Acta Crystallogr.*, **A41**, 262–8.

Ewald, P. P. and Hermann, C. (1929). *Strukturbericht 1913–28*. Akademische Verlagsgesellschaft, Leipzig.

Hamilton, W. C. and Ibers, J. A. (1968). *Hydrogen bonding in solids*. Benjamin, New York.

Hoffmann, R. (1970). *Tetrahedron Lett.*, 2907–9.

Jeffrey, G. A. and Maluszynska, H. (1982). *Int. J. Biol. Macromol.*, **4**, 173–85.

Jeffrey, G. A. and Mitra, J. (1983). *Acta Crystallogr.*, **B39**, 469–80.

Jenkins, R. and Smith, D. K. (1987). Powder diffraction file. In: Allen *et al.* (1987*a*), pp. 158–77.

Machin, P. (1985). Programming aspects of crystallographic data files: Interactive retrieval from the Cambridge Database. In *Crystallographic computing 3: Data collection, structure determination, proteins, and databases* (ed. G. M. Sheldrick, C. Krüger, and R. Goddard), pp. 106–18. Clarendon Press, Oxford.

Mighell, A. D., Stalick, J. K., and Himes, V. L. (1987). NBS Crystal Data: Database description and applications. In: Allen *et al.* (1987*a*), pp. 134–43.

Murray-Rust, P. and Glusker, J. P. (1984). *J. Am. Chem. Soc.*, **106**, 1018–25.

Murray-Rust, P. and Motherwell, W. D. S. (1978). *Acta Crystallogr.*, **B34**, 2518–26.

Murray-Rust, P. and Raftery, J. (1985). *J. Mol. Graphics*, **3**, 50–9 and 60–8.

Nørskov-Lauritsen, L. and Bürgi, H.-B. (1985). *J. Comput. Chem.*, **6**, 216–28.

Nyburg, S. C. and Faerman, C. H. (1985). *Acta Crystallogr.*, **B41**, 274–9.

Nyburg, S. C., Faerman, C. H., and Prasad, L. (1987). *Acta Crystallogr.*, **B43**, 106–10.

Orpen, A. G., Brammer, L., Allen, F. H., Kennard, O., Watson, D. G., and Taylor, R. (1989). *J. Chem. Soc. Dalton Trans.* (suppl.), pp. S1–S83.

Pauling, L. (1940). *The nature of the chemical bond* (2nd edn). Cornell University Press, Ithaca.

Rosenfield Jr., R. E., Parthasarathy, R., and Dunitz, J. D. (1977). *J. Am. Chem. Soc.*, **99**, 4860–2.

Schweizer, W. B. and Dunitz, J. D. (1982). *Helv. Chim. Acta*, **65**, 1547–54.

Sheldrick, B. and Akrigg, D. (1980). *Acta Crystallogr.*, **B36**, 1615–21.

Sutton, L. E. (ed.) (1958). *Tables of interatomic distances and configuration in molecules and ions*. Spec. Publ. No. 11. The Chemical Society, London.

Sutton, L. E. (ed.) (1965). *Tables of interatomic distances and configuration in molecules and ions* (Suppl.). Spec. Publ. No. 18. The Chemical Society, London.

Taylor, R. (1986a). *J. Appl. Crystallogr.*, **19**, 90–1.

Taylor, R. (1986b). *J. Mol. Graphics*, **4**, 123–31.

Taylor, R. and Kennard, O. (1982a). *J. Mol. Struct.*, **78**, 1–28.

Taylor, R. and Kennard, O. (1982b). *J. Am. Chem. Soc.*, **104**, 3209–12.

Taylor, R. and Kennard, O. (1982c). *J. Am. Chem. Soc.*, **104**, 5063–70.

Taylor, R. and Kennard, O. (1983). *Acta Crystallogr.*, **B39**, 517–25.

Taylor, R. and Kennard, O. (1984). *Acc. Chem. Res.*, **17**, 320–6.

Taylor, R. and Kennard, O. (1985). *Acta Crystallogr.*, **A41**, 85–9.

Taylor, R. and Kennard, O. (1986a). *J. Chem. Inform. Comput. Sci.*, **26**, 28–32.

Taylor, R. and Kennard, O. (1986b). *Acta Crystallogr.*, **B42**, 112–20.

Taylor, R., Kennard, O., and Versichel, W. (1984). *Acta Crystallogr.*, **B40**, 280–8.

Vedani, A. and Dunitz, J. D. (1985). *J. Am. Chem. Soc.*, **107**, 7653–8.

Watson, D. G. (1977). Indexed bibliography of crystallographic data publications. In *CODATA Directory of data sources for science and technology* (CODATA Bull., No. 24), pp. 15–37. CODATA, Paris.

Watson, D. G. (1987). Printed information sources in crystallography. In: Allen *et al.* (1987a), pp. 25–9.

16

The importance of accurate structure determination in organic chemistry

Georges Wipff and Stéphane Boudon

16.1   Introduction

Structural information is essential in chemistry not only for determining the geometrical arrangement of atoms within a molecule or a crystal, but also because every structure may tell us something about the electron distribution, the type and properties of bonds connecting the atoms in their potential energy minima, and, to some extent, about their reactivity (Dunitz 1979).

Chemists have not waited for structure determinations to develop a structural theory of the chemical bond and to write down chemical formulae which account for the reactivity and the presumed structure of molecules.

It has been a great success of chemical reasoning to think in terms of structure before it had been established by physical methods. In many cases, these formulae and speculated structures derived by deductive reasoning from chemical phenomena correspond closely to the observed atomic arrangements; a detailed example is given in Chapter 1. A great deal of progress has been made since the first structures of ionic or covalent crystals were determined by X-ray diffraction analysis (1912–13); nowadays the structure determination of organic molecules, including biologically important polymers, is readily achieved. The number of structures determined by X-ray crystallography is such that computer-assisted systematic searches in the Cambridge Structural Database, presently containing over 70 000 structures, permit a new approach to structure analysis (see Chapter 15).

In the present chapter, we are not concerned with structure determination as a proof of existence or conformation of molecular species, which is clearly of primary importance (Eliel *et al.* 1965). Instead we consider the significance of small changes in bond lengths as an expression of *electronic effects* (the most important one being electron delocalization), or reactivity patterns.

Nuclei and electrons are inseparable components of the molecule. The former are used to identify the molecule, but chemical concepts deal essentially with electrons, and electron displacement is at the heart of chemical transformations (Simonetta 1980; Salem 1982). This is why we will try, through the structure, to identify electronic effects described by the molecular orbital (MO) or valence bond (VB) language. We look mainly at these effects in terms of electron delocalization or relocalization; as in the resonance approach, departure from a reference structure is related to a mixing of appropriate localized electronic states. Using another dissection of electronic effects, we try to identify donor–acceptor interactions in molecules, since when two fragments interact, they are structurally perturbed, and electron density is transferred from one to the other. It is clear that other effects (steric, electrostatic, etc.) may contribute to the detailed structure and conformation. We will use donor–acceptor interactions as a guideline to read molecular structures in connection with subtle changes in local geometry and with conformational preferences.

Many examples have been chosen out of our personal interests. Because they involve electron reorganization, we will first comment briefly in Section 16.2 on electron delocalization and on the representations used to describe electron transfers. Without going into details, we will recall the role of theoretical calculations for structure analysis in Section 16.3. In Section 16.4, we will see examples where the structure of molecules in their ground state indicates incipient bond breaking, and crystal structures in which bond formation or breaking can be read. For the former no explicit reference will be made to conformation. In Sections 16.5–16.9, we will deal mainly with stereoelectronic effects related to the anomeric effect, which is conformation-dependent. Starting from experimentally characterized structures (Section 16.5) we proceed to structures of molecules in their ground state where stereoelectronic effects are present but less well quantified (Section 16.6); then to reactive species not amenable to experimental structural studies (Sections 16.7 and 16.8). In Section 16.9 we present examples of long-range stereoelectronic effects on reactivity, which suggest fine structural perturbations in the corresponding ground-state structures. The non-planarity of π systems provides an example where accurate structures coming after experiments or electronic theories explain the stereochemistry of additions to C=C and C=O bonds (Section 16.10). In most of these examples donor–acceptor interactions can be recognized or may serve at least as a guide to read the structures. For cyclopropane derivatives (Section 16.11) we see how this interpretation comes after accurate structures and chemical results are known.

These examples illustrate the constant interplay between the detailed arrangement of atoms within the molecules, the molecular behaviour (stability, reactivity) and the 'driving forces' (related to the electronic structure). These facets are rarely known simultaneously, but we hope to illustrate their close interrelation, and how small differences in molecular geometry reflect perturbations of the electronic structure and differences in reactivity.

16.2 Comments on electron delocalization and geometry

Since resonance has been an important concept in organic chemistry, we wish to mention briefly some recent views on this concept.

Let us consider first the case of aromaticity in cyclic systems. Based on the historic examples of benzene and cyclooctatetraene, it seems obvious to many chemists that electron delocalization is a driving force in chemistry, and that the rule of $4n + 2$ π-electrons is a quite general guide to identify an aromatic system. The cyclic structure of Li_6 obtained from *ab initio* MO calculations is delocalized with equal Li—Li bond lengths (D_{6h} symmetry); while H_6 prefers a localized structure of D_{3h} symmetry, made of three isolated dimers (Shaik and Hiberty 1985; Shaik *et al.* 1988). For N_6 a localized or delocalized structure is preferred, depending on the level of calculation (Ha *et al.* 1981; Huber 1982; Saxe and Schaefer 1983). And it has been argued (Epiotis 1983) that the π system of benzene is *destabilized*, rather than stabilized upon delocalization!

Another interesting comparison can be made for X—X—X systems with four electrons. Whereas the allyl anion, $C_3H_5^-$, has a C_{2v} structure with equal C—C bond lengths, the related H_3^- and I_3^- systems prefer respectively $C_{\infty v}$ and $D_{\infty h}$ arrangements (Shaik and Bar 1984).

Based on theoretical arguments and *ab initio* MO calculations (with inclusion of electron correlation by the configuration interaction technique), Shaik *et al.* (1987) came to the paradoxical conclusion that electron delocalization is seldom expected to be a significant driving force in chemistry. In particular, the π system of benzene, like that of cyclobutadiene, is reluctant to adopt geometries that lead to electron delocalization; in agreement with a previous analysis by Epiotis (1983), the symmetrical geometry of benzene is imposed by the σ framework, and the π system follows! Similarly, allylic resonance is forced upon the π system by the σ framework (Shaik *et al.* 1985).

Even if the relation between geometry and electronic structure may not be a simple matter, it remains that aromaticity of molecules in their ground state as well as in transition states is associated with some stabilization energy, and that the ease of electron shuffle from one fragment to the other depends on the molecular energy gap between the ground state and the excited state of the interacting fragments.

In MO language, the electrons delocalize from an occupied orbital A to a vacant orbital B. The larger the coupling between A and B (generally measured by a large A–B overlap) and the smaller the $E_A - E_B$ energy gap, the larger will be the electron delocalization from A to B. Frontier orbitals contribute thus mainly to electron transfer. Since a suitable proximity and orientation in space is required, there are stereoelectronic requirements for electron delocalization by resonance. In particular, the orbitals on A and B have to be aligned or parallel, rather than perpendicular or antiparallel. Stereoelectronic effects are trivial in aromatic systems, since planarity is required for conjugation. Intramolecular donor–acceptor interactions can be recognized in benzene (butadiene + ethylene π fragment), but not in antiaromatic molecules like cyclobutadiene. In terms of reactivity the 'allowedness' of $\pi_s^4 + \pi_s^2$ and 'forbidedness' of $\pi_s^2 + \pi_s^2$ pericyclic reactions is a stereoelectronic effect in that a given *supra-supra* orientation of the fragments is involved in the transition state; the $\pi_s^2 + \pi_a^2$ topology (i.e. *supra-antara* orientation) corresponds to an allowed reaction since it allows for electron transfer in the transition state (Woodward and Hoffmann 1970).

What this MO perturbational approach says is also expressed by the curved arrows used by chemists, which imply stereochemistry. Some of the arrows have obvious structural connotations (e.g. resonance, as in **1** and **2**); many of them, used to denote electron transfer and bond making or breaking, may also indicate trends in ground-state structures as in **3** and **4**. The arrows express trends of electrons to migrate, based on chemical knowledge; their structural implications have to be assessed experimentally. In MO language we may rationalize the structural effect of electron transfer by the bonding or antibonding character of the molecular orbitals involved

1   2   3   4

(e.g. on going from X_2 to X_2^+ the interatomic distance increases for N_2, but decreases for O_2; populating a σ_{C-X}^* orbital weakens the $C-X$ bond).

The less stable electrons (e.g. those at lone pairs or anionic centres) tend to relocalize into low-energy orbitals; thus stereoelectronic effects are important at heteroatoms, or high-energy bonds next to acceptor fragments. Transition states with electrons more delocalized than the corresponding ground states are typically unstable. Dewar (1984) argued that multibond reactions cannot normally be synchronous. Hence stereoelectronic effects on structures may be amplified in going from the ground state to the transition state, or to high-energy species.

Stereoelectronic effects on reactivity are difficult to quantify. To understand why different conformers behave differently under the same circumstances requires a high level of information (the reaction mechanism, the nature and energy of the intermediates and of the transition state, features concerning alternative paths on the energy surface), which is generally unavailable. In the following, we focus mainly on structural facts. An extensive review of stereoelectronic effects in organic reactions has been written by Deslongchamps (1983).

There has been much discussion on how to represent lone pairs. We will describe them by directional hybrids for the sake of simplicity, as a pure stereochemical guide. This is close to the MO representation for tetrahedral N and C^-, but not for the O and S lone pairs, described by energetically non-equivalent σ- and p-type orbitals (David et al. 1973). Both representations give the same total electron density, where individual lone pairs are quite indiscernible (Stevens and Coppens 1980) in agreement with the localization of four electrons of water in one loge rather than in two loges (Daudel et al. 1976).

16.3 On the use of experimental and theoretical methods for structure determination

Experimental and theoretical methods of structure determination complement each other. The former suffer from limitations inherent to each technique and to the physical state of the compound (see Chapters 3-12). They offer limited possibilities to determine the conformation and precise structure of transient species (carbocations, carbanions, radicals, transition states, etc.). On the other hand, theoretical calculations (mainly *ab initio* MO calculations for small molecules) offer the advantage that questions may be addressed where structural evidence is weak or unobtainable by experiment. The *ab initio* MO calculations can give meaningful results when electron correlation does not play an important role. Due to the increasing power of computers, full geometry optimizations using gradient techniques can be performed consistently and reveal structural trends in geometrical parameters as well as in vibrational spectra, and give insight into reaction mechanisms (see Chapter 13). The MO theoretical framework has also led to useful approaches to reactivity (frontier orbitals, donor–acceptor interactions, aromaticity, etc.) (Klopman 1974; Fleming 1976). Accurate experimental studies are thus required to validate the quantum mechanical calculations. These calculations have an important impact in organic chemistry: (i) as tools for structure elucidation and conformational analysis; (ii) to describe the electronic structure in connection with structural features and reactivity patterns; (iii) to analyse *electronic effects* in models, fragments or larger molecules; (iv) to give insight into intrinsic gas-phase properties of molecules, and indirectly, into environmental effects (solvent, packing forces, etc.).

Qualitative structural trends may also emerge from semiempirical calculations on approximate geometries (for example Walsh rules for $X-A-Y$ molecules, extended to larger systems; relations between bond order and bond lengths) (Albright *et al.* 1985).

Accurate experimental structures are also needed in order *to calibrate force fields* used in molecular mechanics calculations on ground-state molecules. The structures optimized by molecular mechanics result from a balance between effects previously recognized, parametrized and incorporated into the force field. They cannot thus reproduce structural features related to new electronic effects. When properly used, however, molecular mechanics can be a powerful tool for structure determination and conformational analysis (see Chapter 14).

16.4   Electron delocalization by electronegativity. Bond breaking in the ground state

Electrons move towards electronegative atoms or fragments, from a donor D to an acceptor A which possesses more stable empty orbitals. This effect is distance-dependent, but independent of orientation. The ionic/covalent character of the $D-A$ bond depends on the relative electronegativity of D and A; nevertheless its effect on the $D-A$ bond length is hardly predictable. There are some cases where a lengthening is observed concomitantly with charge separation.

16.4.1 The carbon–halogen bond

In alkyl halides RX, the C—X bond length increases from 1.781 ± 0.001 Å in CH_3Cl to 1.803 ± 0.002 Å in t-BuCl (Lide and Jen 1963). The bond length increases with increasing charge on the halogen. The difference in ground-state structures correlates with the differences in reactivity of Me and t-Bu halides: the C—X bond dissociation energies are 10–15 per cent lower with t-Bu than with Me. Thus the ground-state structure contains some 'bond breaking', related to the heterolytic $C^+ \cdots X^-$ bond cleavage: for a given X, the electron-releasing substituents on C which stabilize the carbocation formed after the cleavage cause a lengthening of the C—X bond.

16.4.2 The carbon–oxygen bond

A similar effect on the C—O bond length involving an sp^3 carbon atom has been revealed by the systematic analysis of ethers and esters in the Cambridge Structural Database (Allen and Kirby 1984). The C—O bond length in R—OR' increases with increasing electron-withdrawing ability of the OR' group. The values range from 1.418 Å to 1.475 Å depending on the nature of R and R'. For a given R these C—O bond lengths decrease linearly with the pK_a value of R'OH. The pK_a correlates generally with the logarithm of the rate constant for the reaction of C—O cleavage, and hence with the free energy of activation. The departure from a 'standard' structure is related to the way the molecule reacts; the longer the C—O bond, the more reactive the molecule.

These results may be pictured by resonance formulae (R—OR' ↔ R^+ $^-OR'$), but were hardly predictable since other effects (e.g. the change of hybridization at O from ethers to esters, which is expected to shorten the C—O bonds) might be considered. In fact, the number of bonds ($n = 2$) at oxygen seems to be preserved for a given R, i.e. the shorter O—R', the longer R—O. The chemical symbolism (curved arrows, resonance formulae) accounts for electronic features, geometric structure and reactivity. Taking the extreme case of **5** the very long C—O bond (1.496(5) Å) connects a

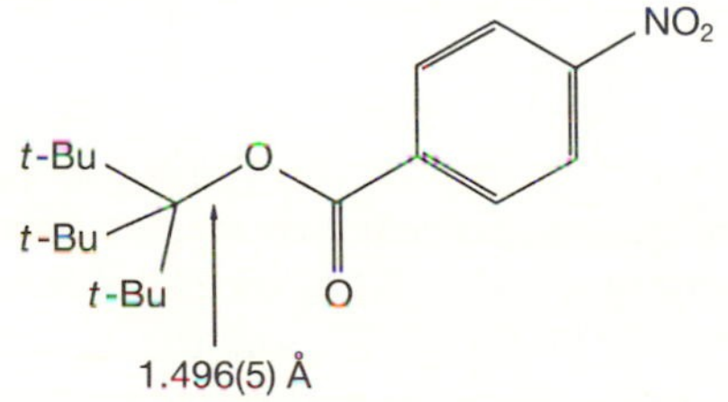

5

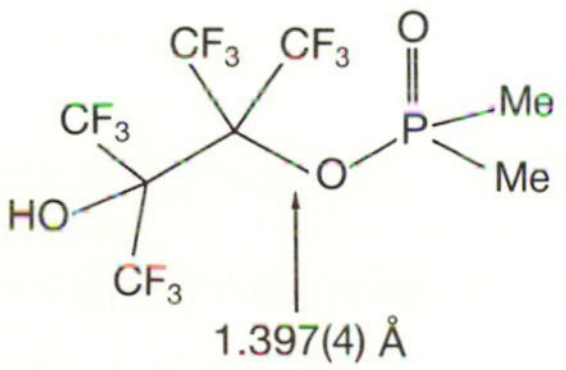

6

formally very good leaving group and a stabilized tertiary carbocation; while in **6**, where the C—O bond is unusually short (1.397(4) Å), the electron-attracting substituents at C inhibit the buildup of positive charge and the OP(O)Me$_2$ moiety is a poor leaving group (Allen and Kirby 1984).

From a mechanistic point of view, such a structure–reactivity relationship does not imply that the transition state is 'reactant-like'. To illustrate this comment, let us take one example dealing with ketone/enol tautomers, where one form is stabilized by geometric features. In the condensation of benzaldehyde with a series of 3-keto-5α-steroids **7**, leading to the corresponding 2-benzylideneketones **8**, the larger the ring flattening at C(2), the higher the rate of condensation (Guy *et al.* 1977; Allen *et al.* 1983). This is however a multistep reaction: enolization, followed by an addition-elimination, proton transfer, etc. The correlation found *confirms* that a planar carbon tends to be formed at C(2) in the transition state. The latter, however, is far from being reactant-like, since it involves elimination of OH$^-$ from a tetrahedral intermediate formed after addition of benzaldehyde.

16.4.3   Tetrahedral MX$_4$ and YMX$_3$ species (M = P, Al, S, Sn; X = O, S, N ...)

A study on the nature of the distortions around a variety of tetrahedral centres in crystals shows a relationship between Δr_2 (the variation of the M—Y bond length in YMX$_3$ as compared to MY$_4$), Δr_1 (the variation of the M—X bond length as compared to MX$_4$), and the flattening of the MX$_3$ site (Murray-Rust *et al.* 1975). The bond length variations Δr_2 and Δr_1 around different central atoms, plotted as a function of the Y—M—X angle, all lie on the same two curves (Fig. 16.1), which expresses the conservation of the number of bonds ($n = 4$). These curves can be interpreted as mapping the heterolytic M—Y cleavage, leading to a planar MX$_3$ species (SN$_1$ reaction); the more Y is electron-donating, the larger are Δr_1 and the Y—M—X angle, and the smaller is Δr_2, and conversely. These distortions parallel an electronic effect, the Y · · · MX$_3$ donor–acceptor interaction. Similar

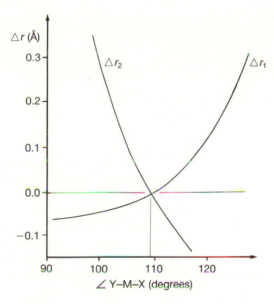

Fig. 16.1 Correlation of Δr_1 and Δr_2 with the Y—M—X angle in tetrahedral YMX$_3$ species (after Murray-Rust *et al.* 1975).

structural correlations have been found for linear triatomic fragments (for example N · · · I—I, I · · · I—I, S · · · S—S, O · · · H—O, H · · · H—H; Bürgi 1975) where the bond number ($n = 1$) is preserved, as well as for pentacoordinated cadmium complexes, X · · · CdL$_3$ · · · Y, where L are equatorial sulphur ligands, and the sum of the bond numbers equals 1 along the X · · · Cd · · · Y line (Bürgi 1973). In the latter example the ground-state structure contains geometric features of an SN$_2$ reaction, namely the stereochemical requirement for linear nucleophilic attack.

These deformations observed in crystals result from a balance between intrinsic properties (for example in R—Cl or R—OR′) and external forces (counterions, crystal packing, etc.). If one assumes that they lie on the minimum-energy paths for formation or cleavage of bonds they provide a valuable insight, independent of theory or calculations, of the coupling between structural and electronic reorganization away from the ground-state structures (Bürgi and Dunitz 1983).

16.5 Stereoelectronic effects in the X—C—Y fragment (X, Y = heteroatoms)

The C—X and C—Y bonds linked at a common carbon may differ markedly from those in the CH$_3$X or CH$_3$Y molecules, as a result of electro-negativity and conformational effects.

First the *electronegativity effect*: as the number of electronegative substituents on carbon increases, the bonds at carbon shorten. For example note the C—F bond distances in CH_3F (1.385 Å), CH_2F_2 (1.358 Å), CHF_3 (1.332 Å), CF_4 (1.317 Å), and the C—Cl bond distances in CH_3Cl (1.781 Å), CH_2Cl_2 (1.772 Å), $CHCl_3$ (1.761 Å), CCl_4 (1.755 Å) (all these values come from experimental studies of the isolated molecules; no error estimate is given in the original papers (Brockway 1937; Lide 1952)). In dimethoxymethane, $CH_3—O—CH_2—O—CH_3$, a gas-phase electron diffraction study (Astrup 1973) suggests that the CH_2—O bonds are 1.382(4) Å long, and the O—CH_3 bonds 1.432(4) Å. Recent *ab initio* MO calculations on polyfluorinated derivatives with central atoms A other than carbon (A = Be to O, and Mg to S) analyse these effects in terms of hyperconjugation and electrostatic charge-withdrawal effects (Reed and Schleyer 1987).

Secondly, bond lengths and stability depend on *conformational effects* related to the *anomeric effect*, as in the case of acetals and glucosides. An excellent account of the anomeric effect is given by Kirby (1983). There is a relation between the axial preference of the exocyclic C—OR bond in sugars and tetrahydropyran derivatives and the lengthening of the same bond (by about 0.02 Å) in going from equatorial (**9, 10**) to axial (**11, 12**) position. Note that the axial C—OR bond is antiperiplanar to a lone pair of the ring oxygen, while the equatorial C—OR bond is not.

This relation is confirmed by *ab initio* MO calculations performed on the $RO—CH_2—OR'$ species (R, R' = H, Me) taken in various conformations (Jeffrey *et al.* 1972, 1974, 1978). The (+ *sc*, + *sc*) conformation **13** (a model for α-D-pyranoside) and the (*ap*, + *sc*) conformation **14** (a model for β-D-pyranoside) mimic the O—C—O—R fragment with OR in axial and equatorial positions, respectively. Although the results depend on the basis set used and on the geometry constraints (e.g. tetrahedral O—C—O angle), the pattern of calculated bond lengths (see formulae **13, 14**, and **15**) reproduces the trends observed experimentally (Jeffrey *et al*. 1978; Kirby 1983; Fuchs *et al*. 1984; Jones and Kirby 1984; Bürgi and Dunitz 1987). The changes can be associated with the interaction of a lone pair of X, n_X, with the σ^*_{C-Y} molecular orbital (occurring when n_X is antiperiplanar to the C—Y bond), which strengthens C—X and weakens C—Y. The C—O bonds at the acetal centre are both shorter than the Me—O bonds, but the axial bond is longer than the equatorial.

The same order of stabilities, $(+sc, +sc) > (ap, +sc) > (ap, ap)$, is calculated with standard or optimized bond lengths. It parallels the number of $n-\sigma^*$ interactions in these structures (2, 1, 0 for **13**, **14**, **15** (or **16**, **17**, **18**), respectively), as well as resonance forms which can be written if the stereochemistry of the lone pairs is taken into account. Electrostatic dipole-dipole interactions also give the same order. Conformation-dependent variations in the O—C—O angles have been calculated (e.g. from 104.5° to 114.1° in HO—CH$_2$—OH; see Williams *et al*. (1981)), but are more difficult to relate to stereoelectronic effects, since steric interactions may influence the structure as well.

The knowledge of anomeric effects at acetal centres is now such that the results can be parametrized, and incorporated into molecular mechanics force fields, e.g. MM2 (Burkert and Allinger 1982). There are still small differences in C—O bond lengths depending on the standard used. However, the different techniques express the same effect: small changes in C—O bond length in the RO—CH$_2$—OR' system with the *same* substituents R and R' can be interpreted as if one O acts as a donor, and the other as an acceptor, depending on the relative orientation of the OR bonds and the lone pairs. The C—O$_{acceptor}$ bond is longer and more easily cleaved than the C—O$_{donor}$ bond; chemically, O$_{acceptor}$ is a better leaving group.

In the case of R ≠ R', the electrons indeed move towards the most electronegative group. The C—O bond lengths are linearly related to the pK_a of ROH (Briggs *et al.* 1984). They can be related to fractional bond orders, stretching force constants, and dissociation energies (Bürgi and Dunitz 1987; see also Chapter 17, Section 17.3.2). As shown below in the case of axial (19) and equatorial (20) tetrahydropyranyl acetals (R = aryl), increasing the electron-withdrawing ability of the OAr groups (and lowering the σ^*_{O-Ar} level) lengthens the C—OAr bond and shortens the endocyclic C—O bond in *both* the axial and equatorial compounds (Briggs *et al.* 1984).

In the axial compounds the ionic resonance contribution 19b increases with the $n_O-\sigma^*_{C-OAr}$ interaction, which is expected to be larger in tetra-hydropyranyl acetals than in β-glucopyranosides. This is because in the latter the oxygen substituents of the ring pull out electron density from the endocyclic oxygen lone pair which becomes less electron-donating; thus reducing the stereoelectronic effects on bond lengths.

In the equatorial compounds, 20, the C—OAr bond is not antiperiplanar to an oxygen lone pair, but to the C—O bond *p* which acts as an electron donor. When the acceptor capabilities of OAr increase, the C—O bonds *p* and *x* lengthen, and *q* shortens, as expressed by the VB ionic form 20b. This suggests a *conformation-dependent charge polarization*, which may be detected by ^{13}C NMR chemical shifts (Eliel *et al.* 1975).

The VB representations of the axial and equatorial compounds also express different *reactivities*. The ionic canonical form 19b represents the

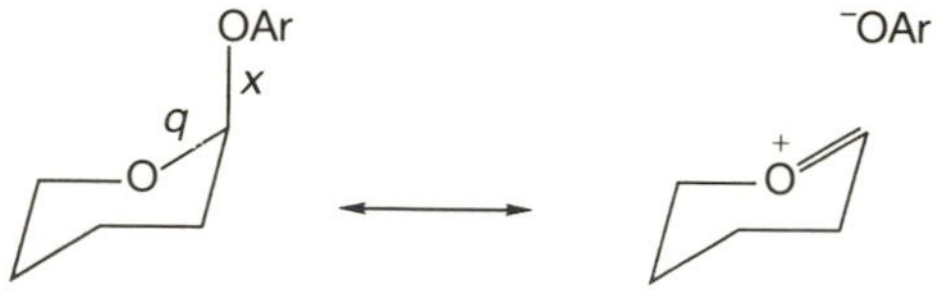

19a   19b

20a   20b

cleavage of the axial C—OR bond, rather than of the endocyclic C—O bond; while the equatorial compound **20** is progressing along the reaction coordinate towards a fragmentation reaction, with simultaneous cleavage of the bonds p and x.

In carbohydrate derivatives, **21**, examined through the Cambridge Structural Database, the exocyclic C—OR bond x is longer in the axial than in the equatorial position, but is *shorter* (especially in the equatorial compounds) than the endocyclic C—O bond q (Fuchs *et al.* 1984). The O—R bond is preferentially oriented in such a way that one oxygen lone pair is antiperiplanar to q; this is the *exo* anomeric effect, where the OR lone pair is the donor, and q the acceptor. In the acid hydrolysis of polysaccharides, however, the exocyclic C—OR bond is cleaved, rather then the endocyclic C—O bond q, presumably due to preferential protonation of OR. Although similar stereoelectronic effects are present in acetals and sugars, one cannot explain or predict structural effects or reactivity from one O—C—O fragment only.

21

Ab initio MO calculations on X—CH_2—YH systems (**22, 23**: X = Cl, F, OH; Y = S, O), confirm the close correspondence between the existence and magnitude of the anomeric effect in six-membered heterocyclic rings and the conformational behaviour of these acyclic model molecules (Wolfe *et al.* 1979). The conformer **23** is consistently more stable than **22** and possesses a longer C—X bond and a shorter C—Y bond. The anomeric effect is strongest for Cl—CH_2—OH, and weakest for HO—CH_2—SH, with the relative order: for X, Cl > F > OH, and for Y, O > S, in agreement with the ordering of the n_Y-σ^*_{C-X} interactions.

The calculated charge on X is larger in **23** than in **22**, which is consistent with the fact that charge transfer from the Y lone pair to the C—X bond occurs only in **23**. This agrees with ^{1}H and ^{17}O NMR (Augé and David

22   **23**

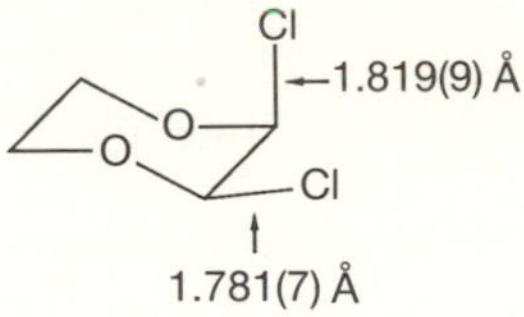

Cl
—1.819(9) Å

O—

O

Cl

1.781(7) Å

24

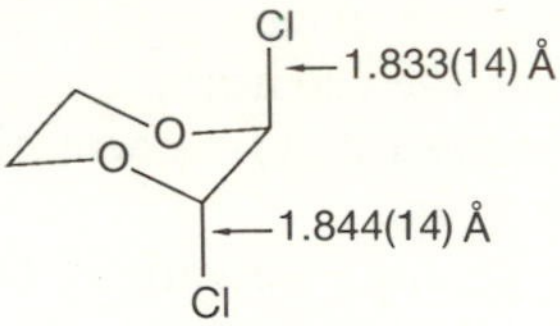

Cl
—1.833(14) Å

O—

O

—1.844(14) Å

Cl

25

1977; McKelvey *et al.* 1981), as well as Cl NQR spectroscopy data (Guibé *et al.* 1973). For instance, the axial chlorines of *cis* and *trans* 2,3-dichloro-1, 4-dioxanes (**24**, and **25**, respectively) resonate at lower NQR frequencies than the equatorial chlorine (Linscheid and Lucken 1970), and correspond to longer C—Cl bonds in the crystal (Altona and Romers 1963*a*, *b*).

16.6 Stereoelectronic effects on preferred conformation

There is a stereoelectronic preference for conformations in which the best donor lone pair is antiperiplanar to the best acceptor bond, but the extent to which the energy of this interaction parallels the total energy and over-comes other effects can hardly be predicted.

With sugars and analogues, the preference for axial, rather than equatorial electronegative substituents X at the anomeric carbon (Lemieux and Chü 1958; Zefirov and Shekhtman 1971) depends not only on X (e.g. OH, OMe, OAc), but also on the nature and orientation of other ring substituents, on the flexibility of the ring, and on solvent effects, thus making it difficult to assess the energy contribution of a particular stereoelectronic component (up to $8-12 \, kJ \, mol^{-1}$). In tetrahydropyran derivatives the axial preference of X decreases roughly in the order halides > PhC(O)O > AcO > AcS > MeO > RS > OH (Stoddart 1971). With X = N-derivatives, or protonated groups, the anomeric effect is reversed: effects other than the $n_O-\sigma^*$ inter-action have to be considered (e.g. dipole–dipole interactions or the donor properties of N).

Stereoelectronic effects are present in other ring systems where polar bonds are adjacent to ring heteroatoms (O, S, N) (Riddell 1980; Kirby 1983; Juaristy *et al.* 1986*a*, *b*), but they are difficult to separate from steric and other electronic effects. For instance, in 1,3-dithianes (**26**, Y = S) and dioxanes (**26**, Y = O) an axial substituent X in the 2-position suffers

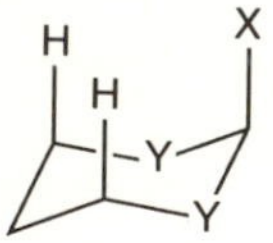

H   X

H

~Y~

Y

26

1,3-diaxial interactions which are respectively smaller and larger than in cyclohexane, due to differences in the puckering of the chair conformation and in the length of the C—O, C—S, and C—C bonds (Kirby 1983, pp. 21–25).

Cyclic esters of trivalent phosphorus acids (27) show a preference for the axial position of X, even for small alkyl groups (Bentrude *et al.* 1975), and this may be related to the n_P-σ^*_{C-O} interaction.

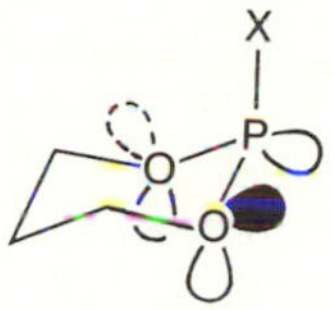

27

In 1,3-diaza- and 1,3,5-triazacyclohexanes a clear preference is found for *gauche* R—N—C—N fragments in which a nitrogen lone pair is antiperiplanar to a C—N bond, and dipole–dipole repulsions are minimized, as in **28** (Riddell 1980, pp. 90–92).

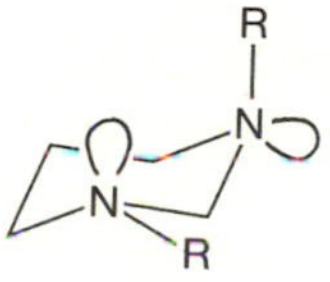

28

An anomeric effect centred at boron has been revealed by the X-ray diffraction study of a cyclic alkylchloroborane (Shiner *et al.* 1985). The B—Cl bond is in an axial rather than equatorial position; it is antiperiplanar to an oxygen lone pair and it is longer (1.890(6) Å) than any other B—Cl bond (these are usually in the range 1.72–1.88 Å).

Stereoelectronic effects may be present in NH_3: the N—H bonds are longer in the pyramidal than in the planar form (Lehn 1970), which may be attributed to n_N-σ^*_{N-H} interactions in the pyramidal form. Note that an alternative simple view on this change in bond length is the decrease of the covalent radius of nitrogen when the hybridization goes from sp^3 to sp^2.

Electrons can delocalize also from π bonds at unsaturated carbons into adjacent polar bonds. For example 2-halocyclohexanones prefer to be axial (29) rather than equatorial, with Br > Cl > F (Eliel *et al.* 1965, pp. 460–469). Methoxyl groups adjacent to C=C bonds are also oriented in the preferred conformation 30, which favours the interaction of the $\pi_{C=C}$ and σ^*_{C-O} molecular orbitals. The higher in energy $\pi_{C=C}$, the more stable the axial conformation as compared with the equatorial (Lessard *et al.* 1977).

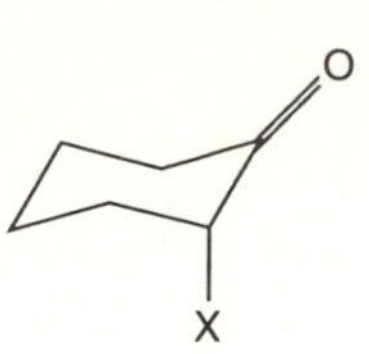

29 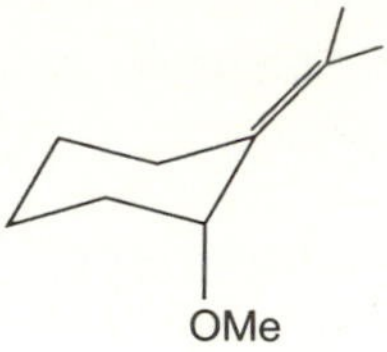   30

The preferred conformation of planar systems with sp^2 carbons (Simonetta and Carrà 1969; Schweizer and Dunitz 1982) can result partly from stereoelectronic effects. In the usual conformation of acids or esters, **31a**, the OR′ lone pair is *anti* to the C=O bond, and can thus interact with the $\sigma^*_{C=O}$ orbital. The *trans* conformer of formic acid (**31a**, R = R′ = H) is more stable than the *cis* conformer (**31b**, R = R′ = H). Microwave spectroscopy studies (Bjarnov and Hocking 1978; Davis *et al.* 1980; see also Chapter 3, Table 3.5) and *ab initio* MO calculations (Zirz and Ahlrichs 1981; Wiberg and Laidig 1987) consistently indicate that the C—OH bond in *trans* formic acid is about 0.01 Å shorter than in the *cis* conformer, while a difference of opposite sign is observed for the C=O bond. Imidate esters generally adopt the conformation **32a** (Fodor and Phillips 1975), but in some cases the conformation **32b** is preferred, which suggests that there is a driving force for the electronegative substituent on carbon, OR″, to be *anti* to the nitrogen lone pair. This preference is stronger with other electronegative substituents, like Cl (Kirby 1983, p. 73). For diazene, HN=NH, the *trans* conformer **33a** is calculated to be more stable than the *cis* conformer **33b**, presumably because of dipole–dipole interactions. Interestingly, the N—H bonds are longer in the *cis* conformer (where they are *anti* to the nitrogen

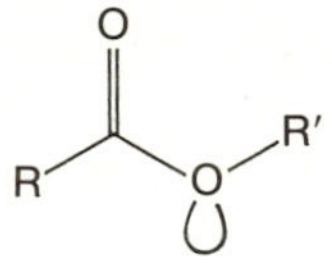

31a 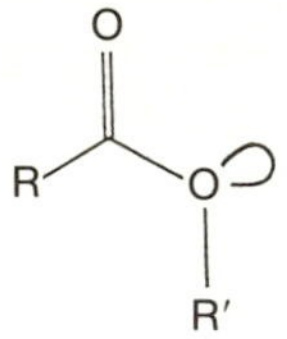   31b

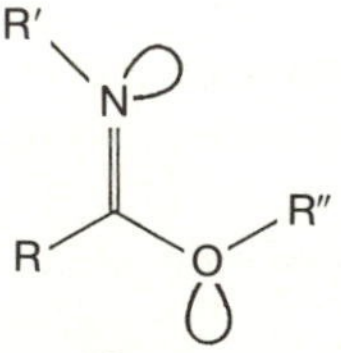

32a 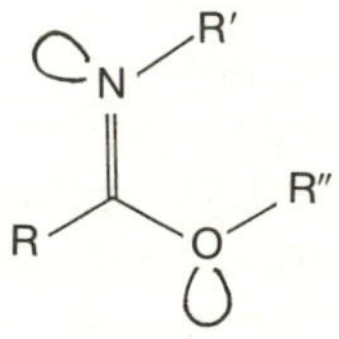   32b

33a   33b

lone pairs) than in the *trans* conformer (Ahlrichs and Staemmler 1976; Epiotis *et al.* 1977).

The observed *gauche* arrangement of compounds $R-X-Y-R'$ (Wolfe 1972), where X, Y are heteroatoms (e.g. H_2O_2, F_2O_2, H_2N-NH_2, H_2N-OH), is such that a lone pair of X is antiperiplanar to the $Y-R'$ bond. The enhanced nucleophilicity of $R-X-Y-R'$ nucleophiles (α effect) might also result from electron delocalization in the transition state.

The stereoelectronic effects on structure are much less documented in these systems than in acetal derivatives. Details of local geometry have been revealed recently by *ab initio* MO calculations with the gradient method. Conformational effects influence also $C=O$ and $C=N$ stretching frequencies, rotational barriers around $C-O$, NMR chemical shifts, ionization spectra, and reactivity (Deslongchamps 1983).

16.7   Amplification of stereoelectronic effects in transient $X-C-Y$ species

The stereoelectronic effects described above can in principle be observed experimentally from molecular geometries or equilibrium constants. This is not the case for transient species. *Ab initio* MO calculations on charged $X-C-Y$ models show that stereoelectronic effects are amplified by increasing either the electron demand at Y (e.g. upon protonation, metallation or formation of hydrogen bonds) or the electron donation at X (e.g. by making carbanions, oxyanions, etc.). A few examples will be discussed here.

16.7.1   The $O-C-O^+$ fragment in $HO-CH_2-OH_2^+$

Upon protonation of the oxygen atom, the contribution of the no-bond resonance form $HO-CH_2^+ \cdots OH_2$ increases, and the $C-O$ bond lengthens by 0.06–0.08 Å. This effect is conformation-dependent, as shown by the calculated $C-O$ bond distances in **34(a, b)** and **35(a, b)** (Wipff 1978). Oxygen protonation amplifies the primary anomeric effect on the leaving group and reduces the orientation effect of that group, as expected from the stabilization of the σ^*_{C-O} and n_O orbitals.

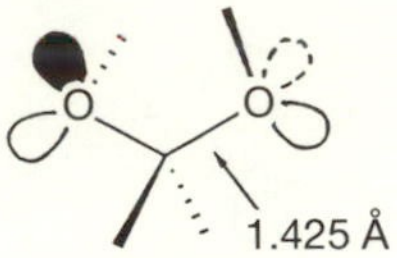

34a 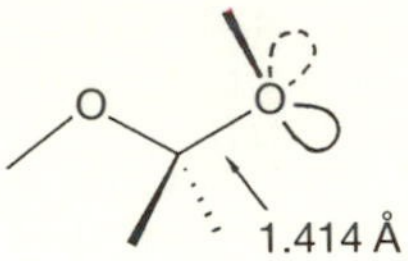   34b

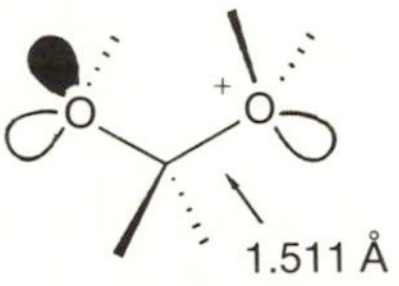

1.425 Å   1.414 Å

35a 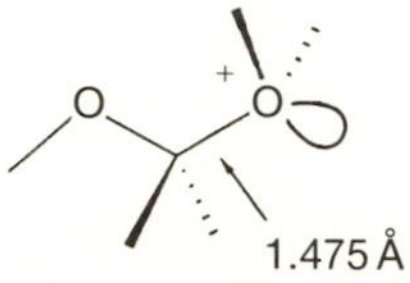   35b

1.511 Å   1.475 Å

16.7.2   The O—C—N⁺ fragment in HO—CH₂—NH₃⁺

Since NH_3 is a good leaving group, like H_2O in the previous example, the no-bond resonance form $HO-CH_2^+ \cdots NH_3$ contributes significantly to the ground-state structure, with stereoelectronic effects on the $C-N^+$ bond (**37a, b**) larger than on $C-N$ (**36a, b**) (Lehn and Wipff 1980*a*).

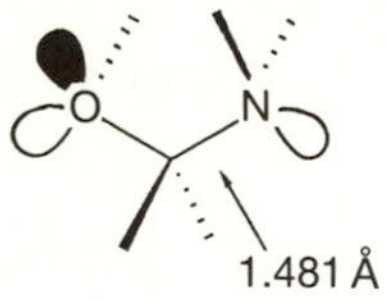

36a 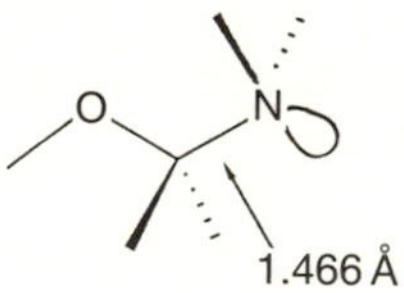   36b

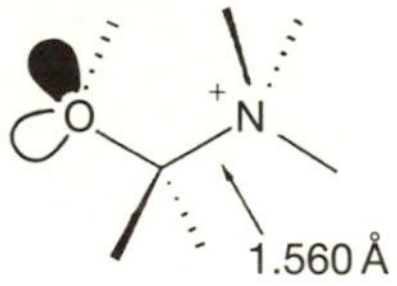

1.481 Å   1.466 Å

37a 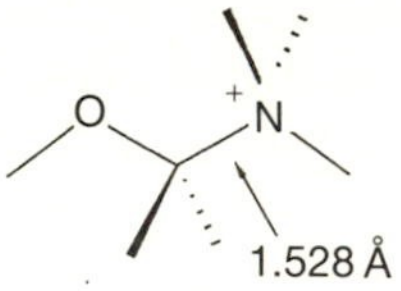   37b

1.560 Å   1.528 Å

16.7.3   Dependence of stereoelectronic effects on the environment

Any perturbation of the energies of n and σ^* orbitals modifies the donor–acceptor interaction and hence the bond lengths. For example, coordination of Mg^{2+} to N in $HO-CH_2-NH_2$ leads to stereoelectronic effects on the $C-N$ bond (**38a, b**) about twice as large as those found in the absence of cation (**36a, b**), and comparable to those observed upon N-protonation (**37a, b**).

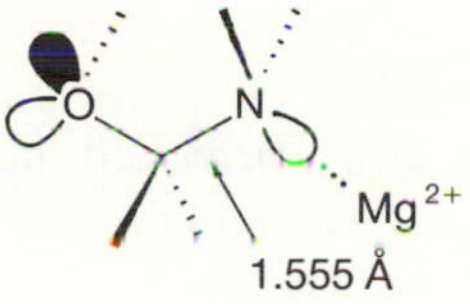

38a 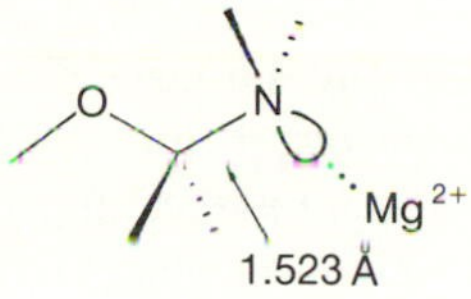   38b

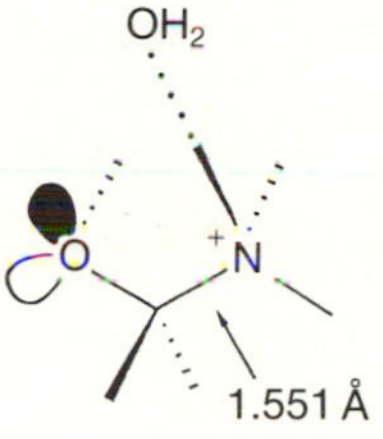

1.555 Å   1.523 Å

39 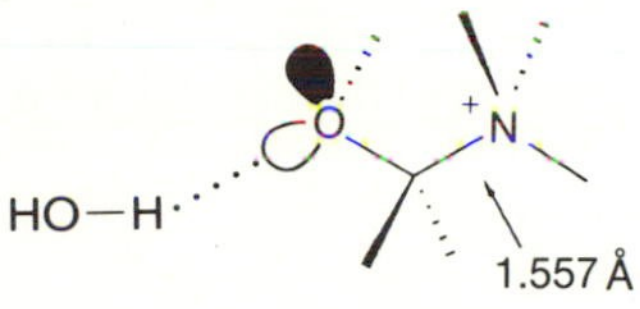   40

1.551 Å   1.557 Å

This is because the electron demand at the NH_2 moiety is increased by the presence of Mg^{2+}. When $HO-CH_2-NH_3^+$ is hydrogen-bonded to H_2O, the C—N bond length is slightly shorter than in free $HO-CH_2-NH_3^+$: coordination of H_2O to $-NH_3^+$ (**39**) reduces the electron demand at N, whereas coordination of H_2O to $-OH$ (as in **40**) makes O less electron-donating. Thus, solvation is expected to decrease the magnitude of stereo-electronic effects, but not their nature and conformation-dependence (Lehn and Wipff 1980*a*).

The enhancement of stereoelectronic effects on the C—N bond of $HO-CH_2-NH_2$ that occurs upon N-protonation in aqueous solution suggests that stereoelectronic effects should be amplified in the transition state of reactions where cleavage of a C—N bond is catalyzed by acids, for example amide hydrolysis.

16.7.4 *Stereoelectronic effects and reactivity at tetrahedral carbons*

There is no proof that stereoelectronic effects on reactants or intermediates *control* the formation or cleavage of tetrahedral species (Hosie *et al.* 1984; Perrin and Nuñez 1986); but the reaction is facilitated when the ground state correlates by a least-motion displacement with the transition state, and similar stereoelectronic effects operate in both. However, other effects (steric, solvent, catalysts, conformational flexibility) may obscure the kinetic data. Comparison of hydrolysis rates of axial *vs.* equatorial acetals (Kirby 1983, 1984) leads to the following conclusions: (i) if the transition state is late, and the C^+ intermediate closely similar for the cleavage of the two isomers, the relative reactivity is determined almost exclusively by the ground-state energies; thus the *less* stable equatorial isomer should react faster (Fig. 16.2(a)); (ii) if the transition state is neither very early nor very late and involves a large amount of C—O bond breaking, the axial isomer should react faster than the equatorial (Fig. 16.2(b)), provided they are not conformationally exchangeable (i.e. with an interconversion barrier lower than the activation energy); (iii) if the C—O cleavage is fast, the rate-limiting step of the reaction may well be the readdition of a nucleophile to C$^+$, rather than the C—O cleavage; (iv) in highly acidic conditions, the protonated axial isomer may become less stable than the equatorial (this is known as *reverse anomeric effect*).

The estimated relative rates for hydrolysis of **41** and **42** (1:1.3 × 10^{13}; Ar = 2, 4-dinitrophenyl) illustrate dramatically the effect of lone-pair assis-

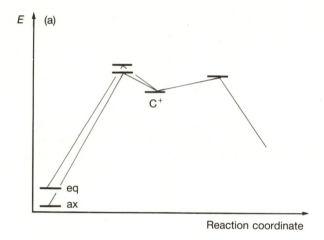

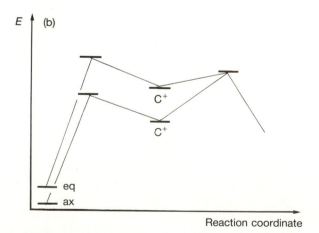

Fig. 16.2 (a,b) Energy profiles for equatorial/axial cleavage of acetals.

41   42

tance in the transition state for cleavage of **42**, and the stabilization of the corresponding C^+ intermediate by electron donation from the oxygen lone pair (Kirby 1983, 1984). An alternative explanation might simply be the non-planarity of the C^+ intermediate formed from **41**.

16.8 Charge delocalization in carbanions α to heteroatoms, $^-C{-}X{-}R$ (X = O, S, Se)

Carbanions gain stability when the negative charge is delocalized by electronegativity and polarization effects, related to the σ_{C-X}–σ^*_{C-X} excitation energy. The delocalization is conformation-dependent and is most effective when the C^- lone pair is antiperiplanar to the $X{-}R$ bond (Lehn and Wipff 1976; Lehn *et al.* 1977). The stabilization of **44** as compared to **43** is related to the n_{C^-}–σ^*_{X-R} interaction in **44**, and to the destabilizing interaction between antiperiplanar lone pairs (n_{C^-} and n_X) in **43**. The changes in bond lengths in going from **43** to **44**, as obtained by *ab initio* MO calculations (Lehn and Wipff 1976; Wolfe *et al.* 1983), reflect the increased acidity of the conjugated acids. The weight of the double-bond/no-bond resonance form **44b** increases when the size of X increases (Se > S > O) and the $X{-}R$ bond becomes more polarizable.

43   44a   44b

Ab initio MO calculations on the conjugated acids show that there are stereoelectronic effects on the $C{-}H$ bond lengths (for example in CH_3OH or CH_3NH_2, see Binkley *et al.* 1980); this is also supported by I R spectroscopy (Bellamy and Mayo 1976). Particularly interesting is the difference in kinetic acidity of diastereotopic hydrogens (Barbarella *et al.* 1978; Wolfe *et al.* 1984). The $C{-}H$ bond is longer and weaker when it is antiperiplanar to an O, S or N lone pair rather than to another bond; but H^+ is *less* easily abstracted, because the reaction is highly endothermic, and the transition state is more 'carbanion-like'. Note that there are subtle conformational effects on C—H bond lengths, even in the absence of heteroatoms: e.g. the C—H bonds of cyclohexane are calculated to be slightly longer in the axial than in the equatorial position (Schäfer *et al.* 1986). Stereoelectronic effects in hydrogen atom abstraction from a substituted cyclohexyl radical have been reported (Beckwith and Easton 1978).

When the polarity of the X—R bond is reversed (for example, X = CH_2, R = F, Cl, OR), the contribution of the double-bond/no-bond resonance form dominates and the X—R bond is cleaved. This gives formally the products of an E_{1cB} elimination reaction, for which a *trans* stereochemistry is required. *Ab initio* MO calculations on $CH_2F—CH_2^-$ indicate a direct elimination of F^- in the gas phase when the C—F bond is antiperiplanar to the C^- lone pair (Bach *et al.* 1979; Dixon *et al.* 1986).

Crystal structure studies of a number of α-sulphursubstituted organolithium compounds, containing the Li—C—S—R fragment (Amstutz *et al.* 1980, 1981*a*, 1984), confirm the effects on C—S bond lengths predicted by the theory. These structures, like others involving C—Li bonds, also provide an insight into the degree of aggregation in the crystal, to be compared with the aggregates characterized in solution by NMR spectroscopy or by other physical methods (see, for example, McGarrity and Ogle 1984; Jackman *et al.* 1987; Kaufman *et al.* 1987). It is also of interest to observe in the crystals of organolithium compounds how solvent molecules are coordinated to lithium in connection with the extent of delocalization of electron density from the carbanionic centre towards neighbouring groups.

Let us finally mention some X-ray diffraction studies on lithium, sodium, and potassium enolates of pinacolone (*t*-butylmethylketone), where, depending on the cation M^+, different types of aggregates and subtle changes in bond lengths are found (Amstutz *et al.* 1981*b*; Williard and Carpenter 1985, 1986). In all these species the enolate oxygen is coordinated to three alkali-metal ions (see **45**). One of the $M^+ \cdots O^-$ bonds is always antiperiplanar to the $C=CH_2$ double bond, and is approximately 0.1 Å shorter than the others. The $C=CH_2$ bond is shorter, and the O—C bond longer, with Li^+ than with Na^+ or K^+. The larger polarizing effect of Li^+, as compared to Na^+ or K^+, and the stronger $Li^+ \cdots O^-$ interaction stabilize the enolate with respect to its conjugated keto form. An alternative

45

view in terms of donor–acceptor interactions is to consider the electron demand at the cationic centre, which is larger for Li^+ than for Na^+ or K^+, and stabilizes the enolate structure $M^+ \ {}^-O-C(CMe_3)=CH_2$.

16.9   Long-range stereoelectronic effects on structure and reactivity

The W arrangement of bonds or lone pairs favours electronic transmission and coupling. In NMR spectroscopy, long-range $^4J_{HH}$ spin–spin couplings are observed when four σ bonds are in a W arrangement as in **46** or **47** (Barfield and Chakrabarti 1969). In terms of reactivity, anchimeric assistance by electron donors (lone pairs, double bonds) is observed in solvolysis (**48**), elimination (**49**), or fragmentation (**50**) reactions (Grob 1969; Lambert *et al.* 1979; Walling 1983). One would therefore expect to observe translations of these electronic effects into perturbations of the structures of these molecules in their ground state. These effects should be amplified upon increased electron donation at one end, or electron demand at the other.

46   47   48

49   50

Ab initio MO calculations on the $Cl-CH_2-CH_2-CH_2^-$ anion in various conformations (**51–54**) show that the C—Cl bond is longest and weakest when it is in a W relationship with the C^- lone pair (Lehn and Wipff 1980*b*). An interaction between the σ_{C-C^-} and σ^*_{C-Cl} orbitals might be considered as well, but it does not account for the orientation effect of the C^- lone pair. A similar, but smaller electronic effect is expected in the conjugated acids.

Overlap populations suggest that the C—Cl bond in **55** is weaker than in **56**, i.e. that the ground state of **55** may have some degree of C—Cl 'bond breaking' which is magnified in the transition state (Lehn and Wipff 1980*b*). Similarly, the C—Cl bond in **57** should be weaker than in **58**; indeed the solvolysis of **57** is about 10^9 times faster than that of **58** (Tabushi *et al.* 1975). Long-range stereoelectronic effects on structure are also expected in molecules undergoing a Grob fragmentation, **50** (Grob 1969; Grob *et al.* 1980).

16.10   Pyramidalization at the carbonyl group and in alkenes

There are electronic effects on the length and strength of the C=O bond. The I R carbonyl streching frequencies are sensitive to inductive and reson-ance effects, to the environment (hydrogen bonds, solvent effects, inter-action with Lewis acids), and to the strain at carbon (Bellamy 1980). π

donors stabilize the polar canonical form $C^+{-}O^-$ and lengthen the C—O bond. For instance in urea derivatives the C=O and C—N bond lengths vary through a range of 1.20–1.30 Å and 1.30–1.40 Å, respectively; the shorter C=O, the longer C—N, and vice versa (Blessing 1983). The C=O bond is longer in amides ($\sim$ 1.23 Å; Chakrabarti and Dunitz 1982) than in esters ($\sim$ 1.20 Å; Schweizer and Dunitz 1982); a weakening of this bond is also observed in *t*-butyl derivatives (Bellamy 1980). Concerning reactivity, the ease of nucleophilic additions on carbonyl groups can be related to the charge distribution on the C=O system, and hence to the length of the C=O bond.

More puzzling is the question of the planarity of the carbonyl group; it has long been taken for granted that carbonyl derivatives are planar in their ground state. However, X-ray crystallographic studies reveal that when a nucleophilic group (Nu = O, N) is close, but not directly bonded to the C atom (59), the C=O bond becomes longer and the system pyramidalizes (Bürgi *et al.* 1973; Bürgi and Dunitz 1983). The C · · · Nu distance may be unusually short, suggesting that the structure lies on the path for the addition of the nucleophile to the carbonyl group. Relations between the C · · · Nu distance, the pyramidalization at C, and the C=O bond length have been mapped and found to be similar to those obtained from *ab initio* MO calculations on the model system $H^- + H_2C{=}O$ or on related nucleophilic additions (Bürgi *et al.* 1974*a*, *b*). The lengthening of the C=O bond indicates that the ground state mixes into the (pyramidal) excited state obtained from an n–π^* electronic transition.

59

Experimental and theoretical results complement each other in indicating that the electron transfer from the nucleophile to the carbonyl group is stereoelectronically controlled. The real structures are too large to be optimized by *ab initio* MO calculations which, in addition, would not account for the effects of solvent on structure. On the other hand, the position of the nucleophile relative to the carbonyl group in the crystal might result from geometrical constraints.

Once a theoretical model has been validated, its extension to second-order effects (e.g. facial stereoselectivity) can be considered. In the transition state 60, electron donation from the nucleophile is favoured when the incipient

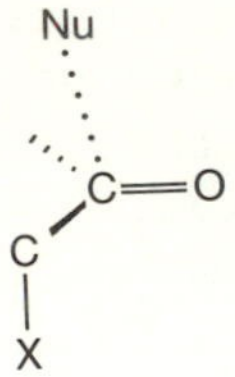

Nu

C=O

C

X

60

C · · · Nu bond is antiperiplanar to the best acceptor C—X bond, i.e. to the bond having the lowest-energy σ^*_{C-X} orbital (Anh and Eisenstein 1977; Anh 1980). As a result the C—X bond lengthens and the charge on X increases. This effect may be alternatively considered just as a torsional effect in the transition state, where staggered conformations must be as stable as possible. Other arguments have been advanced to account for the π-facial stereoselectivity in additions to cyclohexanones (Cieplak 1981) and methylenecyclohexanes (Johnson *et al.* 1987). The π–σ^* interaction is consistent with the axial preference of 2-halocyclohexanones **29** and with the small pyramidalization found by low-temperature neutron crystallography in the asymmetric rotamers of acetamide and thioacetamide (**61**, X = O, S) (Jeffrey *et al.* 1980, 1984; a detailed description is given in Chapter 11, Section 11.4.3). *Ab initio* MO optimizations and neutron diffraction studies of amino acids and dipeptides show a small pyramidalization (about 2°) in the expected direction, which is probably an intrinsic structural feature, rather than a crystal-field effect (Cieplak 1985; Jeffrey *et al.* 1985). Pyramidalization at the carbonyl group should increase with angle strain and with the electronegativity of the substituent (as observed for inversion barriers at nitrogen; see Lehn (1970)).

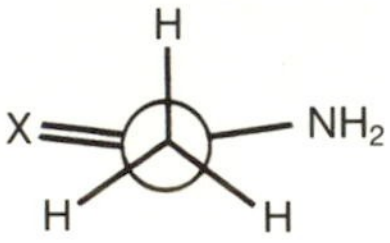

H

X   NH₂

H   H

61

Pyramidalization at C=C bonds can also be related to the π-facial stereo-selectivities observed in the addition reactions of cyclic alkenes (Huisgen *et al.* 1980; Huisgen 1981). Theory predicts that the π system of norbornene **62** has more developed *exo* than *endo* lobes (Inagaki *et al.* 1976), which may explain why many additions on the C=C bond are *exo*, rather than *endo*. Indeed, *ab initio* MO optimization leads to a non-planar π system with the C—H bonds slightly bent downwards (Wipff and Morokuma 1980; Carrupt

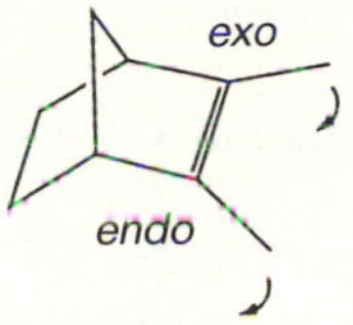

62

and Vogel 1985). Other MO calculations indicate that the olefinic C—H bonds of norbornadiene can be bent *endo* more easily than *exo* (Mazzocchi *et al.* 1980). Following this idea, systematic searches in the Cambridge Structural Database and *ab initio* MO geometry optimizations have confirmed the C=C pyramidalization in bicyclic alkenes, thus providing additional structural data which stimulate an analysis of the previously proposed electronic effects (Houk *et al.* 1983; Pinkerton *et al.* 1984).

16.11   Deformations of the cyclopropane ring

Considering cyclopropane as an electron donor (Walsh 1947, 1949; Hoffmann, 1964, 1965; Hoffmann and Davidson 1971) unifies structural effects and reactivity patterns, as shown by the following examples: (i) electron donation from the HOMO (3e′) is related to the stabilization of norcaradienes **64** compared to cycloheptatrienes **63** (Hoffmann 1970); (ii) the stability of the cyclopropyl carbocation depends on the conformation of the substituent (Skancke 1984); (iii) the C2—C3 bond length (see **65**) decreases and the C1—C2 and C1—C3 bond lengths increase in cyclopropane derivatives with π-acceptor substituents (C=O, C=C, C≡N) (Allen, 1980, 1981;

63   **64**

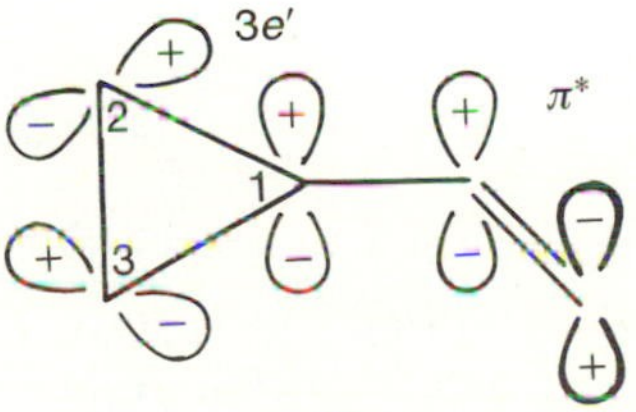

65

Allen *et al.* 1983). This relation is not surprising if one takes into account the electronic structure of the two fragments, and admits that electron density is transferred from the cyclopropane $3e'$ orbital into the π^* orbital of the substituent.

But what could be said *a priori* about this structural perturbation in the Cl, F, NH_2, Li, and Me derivatives? Parallel to theoretical speculations, it is essential that accurate molecular structures, determined by a cooperative effort of experimentalists and computational chemists, indicate deviations from D_{3h} symmetry which can be interpreted by more subtle electronic effects.

The exocyclic C—C bonds of cyclopropane derivatives are unusually short due to hybridization effects (the more the *s* character, the shorter the effective covalent radius and hence the bond) as well as conjugative effects. These have been assessed and compared with those occurring in alkyl and allyl analogs (Allen 1981; Allen *et al.* 1983). This analysis confirms the current views on the electronic structure of cyclopropane, particularly the *s/p* content of the hybrid orbitals at carbon associated with the endo- and exocyclic bonds (Coulson and Moffit 1949; Coulson and Goodwin 1962).

16.12   Conclusions

Accurate molecular structures are invaluable experimental facts. They are obtained *after* compounds are made, and some properties are known. They are like pictures of the molecules, which most often confirm and refine previous chemical knowledge. They serve also as a test and a stimulus for those who seek a deeper understanding of electronic effects in chemistry.

We have not dealt with the exciting field of weak intermolecular interactions (e.g. the geometry of hydrogen bonds), which have important implications like solvent or environmental effects on conformation and reactivity. Other important areas include the shape of biological systems, supermolecules, molecular assemblies, the design of receptors, and yet a number of other problems, to which accurate structures, obtained by experimental and theoretical approaches, bring a major contribution to our understanding. Chemistry was structural long before molecular geometries could be determined. Now, accurate structures help us to understand what is behind the geometry.

References

Ahlrichs, R. and Staemmler, V. (1976). *Chem. Phys. Lett.*, **37**, 77–81.
Albright, T. A., Burdett, J. K., and Whangbo, M.-H. (1985). *Orbital interactions in chemistry*. Wiley-Interscience, New York.

Allen, F. H. (1980). *Acta Crystallogr.*, **B36**, 81–96.

Allen, F. H. (1981). *Acta Crystallogr.*, **B37**, 890–900.

Allen, F. H. and Kirby, A. J. (1984). *J. Am. Chem. Soc.*, **106**, 6197–200.

Allen, F. H., Kennard, O., and Taylor, R. (1983). *Acc. Chem. Res.*, **16**, 146–53.

Altona, C. and Romers, C. (1963a). *Acta Crystallogr.*, **16**, 1225–32.

Altona, C. and Romers, C. (1963b). *Rec. Trav. Chim. Pays-Bas*, **82**, 1080–8.

Amstutz, R., Seebach, D., Seiler, P., Schweizer, B., and Dunitz, J. D. (1980). *Angew. Chem. Int. Ed. Engl.*, **19**, 53–4.

Amstutz, R., Dunitz, J. D., and Seebach, D. (1981a). *Angew. Chem. Int. Ed. Engl.*, **20**, 465–6.

Amstutz, R., Schweizer, W. B., Seebach, D., and Dunitz, J. D. (1981b). *Helv. Chim. Acta*, **64**, 2617–21.

Amstutz, R., Laube, T., Schweizer, W. B., Seebach, D., and Dunitz, J. D. (1984). *Helv. Chim. Acta*, **67**, 224–36.

Anh, N. T. (1980). Regio- and stereo-selectivities in some nucleophilic reactions. In *Organic chemistry, syntheses and reactivity*, Topics in Current Chemistry, Vol. 88, pp. 145–62. Springer, Berlin.

Anh, N. T. and Eisenstein, O. (1977). *Nouv. J. Chim.*, **1**, 61–70.

Astrup, E. E. (1973). *Acta Chem. Scand.*, **27**, 3271–6.

Augé, J. and David, S. (1977). *Nouv. J. Chim.*, **1**, 57–60.

Bach, R. D., Badger, R. C., and Lang, T. J. (1979). *J. Am. Chem. Soc.*, **101**, 2845–8.

Barbarella, G., Dembech, P., Garbesi, A., Bernardi, F., Bottoni, A., and Fava, A. (1978). *J. Am. Chem. Soc.*, **100**, 200–2.

Barfield, M. and Chakrabarti, B. (1969). *Chem. Rev.*, **69**, 757–78.

Beckwith, A. L. J. and Easton, C. (1978). *J. Am. Chem. Soc.*, **100**, 2913–4.

Bellamy, L. J. (1980). *The infrared spectra of complex molecules*, Vol. 2, *Advances in infrared group frequencies*, Chap. 5. Chapman and Hall, London.

Bellamy, L. J. and Mayo, D. W. (1976). *J. Phys. Chem.*, **80**, 1217–20.

Bentrude, W. G., Tan, H.-W., and Yee, K. C. (1975). *J. Am. Chem. Soc.*, **97**, 573–82.

Binkley, J. S., Pople, J. A., and Hehre, W. J. (1980). *J. Am. Chem. Soc.*, **102**, 939–47.

Bjarnov, E. and Hocking, W. H. (1978). *Z. Naturforsch.*, **33a**, 610–8.

Blessing, R. H. (1983). *J. Am. Chem. Soc.*, **105**, 2776–83.

Briggs, A. J., Glenn, R., Jones, P. G., Kirby, A. J., and Ramaswamy, P. (1984). *J. Am. Chem. Soc.*, **106**, 6200–6.

Brockway, L. O. (1937). *J. Phys. Chem.*, **41**, 185–95.

Bürgi, H.-B. (1973). *Inorg. Chem.*, **12**, 2321–5.

Bürgi, H.-B. (1975). *Angew. Chem. Int. Ed. Engl.*, **14**, 460–73.

Bürgi, H.-B. and Dunitz, J. D. (1983). *Acc. Chem. Res.*, **16**, 153–61.

Bürgi, H.-B. and Dunitz, J. D. (1987). *J. Am. Chem. Soc.*, **109**, 2924–6.

Bürgi, H.-B., Dunitz, J. D., and Shefter, E. (1973). *J. Am. Chem. Soc.*, **95**, 5065–7.

Bürgi, H.-B., Lehn, J.-M., and Wipff, G. (1974a). *J. Am. Chem. Soc.*, **96**, 1956–7.

Bürgi, H.-B., Dunitz, J. D., Lehn, J.-M., and Wipff, G. (1974b). *Tetrahedron*, **30**, 1563–72.

Burkert, U. and Allinger, N. L. (1982). *Molecular mechanics*, ACS Monograph No. 177. American Chemical Society, Washington.

Carrupt, P.-A. and Vogel, P. (1985). *J. Mol. Struct. (Theochem)*, **124**, 9–23.

Chakrabarti, P. and Dunitz, J. D. (1982). *Helv. Chim. Acta*, **65**, 1555–62.

Cieplak, A. S. (1981). *J. Am. Chem. Soc.*, **103**, 4540–52.

Cieplak, A. S. (1985). *J. Am. Chem. Soc.*, **107**, 271–3.

Coulson, C. A. and Goodwin, T. H. (1962). *J. Chem. Soc.*, 2851–4.

Coulson, C. A. and Moffit, W. E. (1949). *Phil. Mag.*, **40**, 1–15.

Daudel, R., Kapuy, E., Kozmutza, C., Goddard, J. D., and Csizmadia, I. G. (1976). *Chem. Phys. Lett.*, **44**, 197–203.

David, S., Eisenstein, O., Hehre, W. J., Salem, L., and Hoffmann, R. (1973). *J. Am. Chem. Soc.*, **95**, 3806–7.

Davis, R. W., Robiette, A. G., Gerry, M. C. L., Bjarnov, E., and Winnewisser, G. (1980). *J. Mol. Spectrosc.*, **81**, 93–109.

Deslongchamps, P. (1983). *Stereoelectronic effects in organic chemistry*. Pergamon Press, Oxford.

Dewar, M. J. S. (1984). *J. Am. Chem. Soc.*, **106**, 209–19.

Dixon, D. A., Fukunaga, T., and Smart, B. E. (1986). *J. Am. Chem. Soc.*, **108**, 4027–31.

Dunitz, J. D. (1979). *X-ray analysis and the structure of organic molecules*, Chap. 7. Cornell University Press, Ithaca.

Eliel, E. L., Allinger, N. L., Angyal, S. J., and Morrison, G. A. (1965). *Conformational analysis*. Wiley-Interscience, New York.

Eliel, E. L., Bailey, W. F., Kopp, L. D., Willer, R. L., Grant, D. M., Bertrand, R., Christensen, K. A., Dalling, D. K., Duch, M. W., Wenkert, E., Schell, F. M., and Cochran, D. W. (1975). *J. Am. Chem. Soc.*, **97**, 322–30.

Epiotis, N. D. (1983). *Pure Appl. Chem.*, **55**, 229–36.

Epiotis, N. D., Yates, R. L., Larson, J. R., Kirmaier, C. R., and Bernardi, F. (1977). *J. Am. Chem. Soc.*, **99**, 8379–88.

Fleming, I. (1976). *Frontier orbitals and organic chemical reactions*. Wiley-Interscience, Chichester.

Fodor, G. and Phillips, B. A. (1975). Constitution, configurational and conformational aspects, and chiroptical properties of imidic acid derivatives. In *The chemistry of amidines and imidates* (ed. S. Patai), pp. 85–155. Wiley-Interscience, London.

Fuchs, B., Schleifer, L., and Tartakovski, E. (1984). *Nouv. J. Chim.,* **8**, 275–8.

Grob, C. A. (1969). *Angew. Chem. Int. Ed. Engl.*, **8**, 535–46.

Grob, C. A., Bolleter, M., and Kunz, W. (1980). *Angew. Chem. Int. Ed. Engl.*, **19**, 708–9.

Guibé, L., Augé, J., David, S., and Eisenstein, O. (1973). *J. Chem. Phys.*, **58**, 5579–83.

Guy, J. J., Allen, F. H., Kennard, O., and Sheldrick, G. M. (1977). *Acta Crystallogr.*, **B33**, 1236–44.

Ha, T.-K., Cimiraglia, R., and Nguyen, M. T. (1981). *Chem. Phys. Lett.*, **83**, 317–9.

Hoffmann, R. (1964). *J. Chem. Phys.*, **40**, 2480–8.

Hoffmann, R. (1965). *Tetrahedron Lett.*, 3819–24.

Hoffmann, R. (1970). *Tetrahedron Lett.*, 2907–9.

Hoffmann, R. and Davidson, R. B. (1971). *J. Am. Chem. Soc.*, **93**, 5699–705.

Hosie, L., Marshall, P. J., and Sinnott, M. L. (1984). *J. Chem. Soc. Perkin Trans. II*, 1121–31.

Houk, K. N., Rondan, N. G., and Brown, F. K. (1983). *Isr. J. Chem.*, **23**, 3–9.

Huber, H. (1982). *Angew. Chem. Int. Ed. Engl.*, **21**, 64–5.

Huisgen, R. (1981). *Pure Appl. Chem.*, **53**, 171–87.

Huisgen, R., Ooms, P. H. J., Mingin, M., and Allinger, N. L. (1980). *J. Am. Chem. Soc.*, **102**, 3951–3.

Inagaki, S., Fujimoto, H., and Fukui, K. (1976). *J. Am. Chem. Soc.*, **98**, 4054–61.

Jackman, L. M., Scarmoutzos, L. M., and DeBrosse, C. W. (1987). *J. Am. Chem. Soc.*, **109**, 5355–61.

Jeffrey, G. A., Pople, J. A., and Radom, L. (1972). *Carbohydr. Res.*, **25**, 117–31.

Jeffrey, G. A., Pople, J. A., and Radom, L. (1974). *Carbohydr. Res.*, **38**, 81–95.

Jeffrey, G. A., Pople, J. A., Binkley, J. S., and Vishveshwara, S. (1978). *J. Am. Chem. Soc.*, **100**, 373–9.

Jeffrey, G. A., Ruble, J. R., McMullan, R. K., DeFrees, D. J., Binkley, J. S., and Pople, J. A. (1980). *Acta Crystallogr.*, **B36**, 2292–9.

Jeffrey, G. A., Ruble, J. R., and Yates, J. H. (1984). *J. Am. Chem. Soc.*, **106**, 1571–6.

Jeffrey, G. A., Houk, K. N., Paddon-Row, M. N., Rondan, N. G., and Mitra, J. (1985). *J. Am. Chem. Soc.*, **107**, 321–6.

Johnson, C. R., Tait, B. D., and Cieplak, A. S. (1987). *J. Am. Chem. Soc.*, **109**, 5875–6.

Jones, P. G. and Kirby, A. J. (1984). *J. Am. Chem. Soc.*, **106**, 6207–12.

Juaristi, E., Tapia, J., and Mendez, R. (1986*a*). *Tetrahedron*, **42**, 1253–64.

Juaristi, E., Valle, L., Valenzuela, B. A., and Aguilar, M. A. (1986*b*). *J. Am. Chem. Soc.*, **108**, 2000–5.

Kaufman, M. J., Gronert, S., Bors, D. A., and Streitwieser Jr., A. (1987). *J. Am. Chem. Soc.*, **109**, 602–3.

Kirby, A. J. (1983). *The anomeric effect and related stereoelectronic effects at oxygen*. Springer, Berlin.

Kirby, A. J. (1984). *Acc. Chem. Res.*, **17**, 305–11.

Klopman, G. (ed.) (1974). *Chemical reactivity and reaction paths*. Wiley, New York.

Lambert, J. B., Mark, H. W., Holcomb, A. G., and Magyar, E. S. (1979). *Acc. Chem. Res.*, **12**, 317–24.

Lehn, J.-M. (1970). Nitrogen inversion: Experiment and theory. In *Dynamic stereochemistry*, Fortschritte der chemischen Forschung, Vol. 15, pp. 311–77. Springer, Berlin.

Lehn, J.-M. and Wipff, G. (1976). *J. Am. Chem. Soc.*, **98**, 7498–505.

Lehn, J.-M. and Wipff, G. (1980*a*). *J. Am. Chem. Soc.*, **102**, 1347–54.

Lehn, J.-M. and Wipff, G. (1980*b*). *Tetrahedron Lett.*, **21**, 159–62.

Lehn, J.-M., Wipff, G., and Demuynck, J. (1977). *Helv. Chim. Acta*, **60**, 1239–46.

Lemieux, R. U. and Chü, N. J. (1958). Presentation at the *133th Meeting of the American Chemical Society*. Abstract 31N.

Lessard, J., Tan, P. V. M., Martino, R., and Saunders, J. K. (1977). *Can. J. Chem.*, **55**, 1015–23.

Lide Jr., D. R. (1952). *J. Am. Chem. Soc.*, **74**, 3548–52.

Lide Jr., D. R. and Jen, M. (1963). *J. Chem. Phys.*, **38**, 1504–7.

Linscheid, P. and Lucken, E. A. C. (1970). *J. Chem. Soc. Chem. Commun.*, 425–6.

McGarrity, J. F. and Ogle, C. A. (1984). *J. Am. Chem. Soc.*, **107**, 1805–10.

McKelvey, R. D., Kawada, Y., Sugawara, T., and Iwamura, H. (1981). *J. Org. Chem.*, **46**, 4948–52.

Mazzocchi, P. H., Stahly, B., Dodd, J., Rondan, N. G., Domelsmith, L. N., Rozeboom, M. D., Caramella, P., and Houk, K. N. (1980). *J. Am. Chem. Soc.*, **102**, 6482–90.

Murray-Rust, P., Bürgi, H.-B., and Dunitz, J. D. (1975). *J. Am. Chem. Soc.*, **97**, 921–2.

Perrin, C. L. and Nuñez, O. (1986). *J. Am. Chem. Soc.*, **108**, 5997–6003.

Pinkerton, A. A., Schwarzenbach, D., Birbaum, J.-L., Carrupt, P.-A., Schwager, L., and Vogel, P. (1984). *Helv. Chim. Acta*, **67**, 1136–53.

Reed, A. E. and Schleyer, P.v.R. (1987). *J. Am. Chem. Soc.*, **109**, 7362–73.

Riddell, F. G. (1980). *The conformational analysis of heterocyclic compounds*. Academic Press, London.

Salem, L. (1982). *Electrons in chemical reactions: First principles*. Wiley-Interscience, New York.

Saxe, P. and Schaefer III, H. F. (1983). *J. Am. Chem. Soc.*, **105**, 1760–4.

Schäfer, L., Ewbank, J. D., Klimkowski, V. J., Siam, K., and van Alsenoy, C. (1986). *J. Mol. Struct. (Theochem)*, **135**, 141–58.

Schweizer, W. B. and Dunitz, J. D. (1982). *Helv. Chim. Acta*, **65**, 1547–54.

Shaik, S. S. and Bar, R. (1984). *Nouv. J. Chim.*, **8**, 411–20.

Shaik, S. S. and Hiberty, P. C. (1985). *J. Am. Chem. Soc.*, **107**, 3089–95.

Shaik, S. S., Hiberty, P. C., Ohanessian, G., and Lefour, J.-M. (1985). *Nouv. J. Chim.*, **9**, 385–8.

Shaik, S. S., Hiberty, P. C., Lefour, J.-M., and Ohanessian, G. (1987). *J. Am. Chem. Soc.*, **109**, 363–74.

Shaik, S. S., Hiberty, P. C., Ohanessian, G., and Lefour, J.-M. (1988). *J. Phys. Chem.*, **92**, 5086–94.

Shiner, C. S., Garner, C. M., and Haltiwanger, R. C. (1985). *J. Am. Chem. Soc.*, **107**, 7167–72.

Simonetta, M. (1980). Bond formation and breaking: The heart of chemistry. In *Horizons in quantum chemistry* (ed. K. Fukui and B. Pullman), pp. 51–6. Reidel, Dordrecht.

Simonetta, M. and Carrà, S. (1969). General and theoretical aspects of the COOH and COOR groups. In *The chemistry of carboxylic acids and esters* (ed. S. Patai), pp. 1–52. Wiley-Interscience, London.

Skancke, P. N. (1984). *Int. J. Quantum Chem.*, **26**, 729–41.

Stevens, E. D. and Coppens, P. (1980). *Acta Crystallogr.*, **B36**, 1864–76.

Stoddart, J. F. (1971). *Stereochemistry of carbohydrates*, p. 72. Wiley-Interscience, New York.

Tabushi, I., Tamaru, Y., Yoshida, Z., and Sugimoto, T. (1975). *J. Am. Chem. Soc.*, **97**, 2886–91.

Walling, C. (1983). *Acc. Chem. Res.*, **16**, 448–54.

Walsh, A. D. (1947). *Nature*, **159**, 165 and 712–3.

Walsh, A. D. (1949). *Trans. Faraday Soc.*, **45**, 179–90.

Wiberg, K. B. and Laidig, K. E. (1987). *J. Am. Chem. Soc.*, **109**, 5935–43.

Williams, J. O., Scarsdale, J. N., Schäfer, L., and Geise, H. J. (1981). *J. Mol. Struct. (Theochem)*, **76**, 11–28.

Williard, P. G. and Carpenter, G. B. (1985). *J. Am. Chem. Soc.*, **107**, 3345–6.

Williard, P. G. and Carpenter, G. B. (1986). *J. Am. Chem. Soc.*, **108**, 462–8.

Wipff, G. (1978). *Tetrahedron Lett.*, 3269–70.

Wipff, G. and Morokuma, K. (1980). *Tetrahedron Lett.*, **21**, 4445–8.

Wolfe, S. (1972). *Acc. Chem. Res.*, **5**, 102–11.

Wolfe, S., Whangbo, M.-H., and Mitchell, D. J. (1979). *Carbohydr. Res.*, **69**, 1–26.

Wolfe, S., LaJohn, L. A., Bernardi, F., Mangini, A., and Tonachini, G. (1983). *Tetrahedron Lett.*, **24**, 3789–92.

Wolfe, S., Stolow, A., and LaJohn, L. A. (1984). *Can. J. Chem.*, **62**, 1470–5.

Woodward, R. B. and Hoffmann, R. (1970). *The conservation of orbital symmetry*. Verlag Chemie and Academic Press, Weinheim.

Zefirov, N. S. and Shekhtman, N. M. (1971). *Russ. Chem. Rev.*, **40**, 315–29.

Zirz, C. and Ahlrichs, R. (1981). *Theoret. Chim. Acta*, **60**, 355–61.

17

Structure correlations, reaction pathways, and energy surfaces for chemical reactions

Valeria Ferretti, Katharina C. Dubler-Steudle, and
Hans-Beat Bürgi

17.1 Introduction

The use of spectroscopy and diffraction experiments, of *ab initio* MO and molecular mechanics calculations to study molecular structure has been described in detail in Chapters 3–14. These techniques are based on the elementary physical laws describing the interactions of nuclei, electrons, neutrons and electromagnetic radiation and, at least in principle, on a theoretical definition of the concept of molecular structure, namely the Born–Oppenheimer energy surface (Chapter 2). Over the years a host of information on equilibrium structures and molecular vibrations, mainly in the solid and gaseous states, has accumulated and part of it is compiled in several data banks (Chapter 15). The information is of a physical nature in so far as it is being obtained by physical methods and to the extent that it pertains to individual compounds.

Chemical information may be extracted from these data through extensive comparisons between large numbers of related molecules. This raises several questions: What is a class of chemical compounds? In which respect and to what extent are different molecules chemically similar and therefore comparable? To the practicing chemist the answers may be obvious; to the theorist there is no answer. In their textbook on 'Elementare Quantenchemie' Primas and Müller-Herold (1984) state 'Der Begriff der Stoffklasse konnte bis heute nicht auf die Quantenmechanik reduziert werden' (the notion of a class of compounds could not – up to this day – be reduced to quantum mechanics). In this situation we resort to empirical criteria for classifying chemical structures as similar; specifically, the notions of chemical fragment and fragment environment will be introduced (Section 17.2). The precision of concepts necessary for accurate structure determination by physical methods is largely lost in chemical structure comparisons. The latter nevertheless yield information which is not accessible if we restrict ourselves to a physical point of view.

In this chapter we attempt to interpret structural differences between related molecules in terms of energy and chemical reactivity. We explain and illustrate the principle of structural correlation, the interpretation of molecular structure as an incipient stage of a chemical reaction, and the mapping of a reaction pathway from crystal structure data (Section 17.3.1; for reviews see Bürgi (1975); Dunitz (1979); Bürgi and Dunitz (1983)). In favourable cases structural data can be combined with vibrational force constants, activation energies and other kinetic information to delineate the main features of an energy surface describing the breaking and making of bonds (Section 17.3.2). The appearance of such surfaces is an expression of electronic structure. These aspects will not be considered in detail since they are discussed extensively in Chapter 16.

17.2 Concepts

17.2.1 Molecule and molecular fragment

Most chemists consider the molecules as the elementary building blocks of a chemical compound. They define them in terms of a few simple rules with

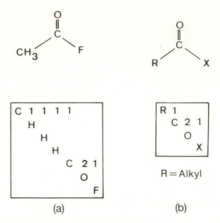

Fig. 17.1 Line formulae and incidence matrix representations for (a) acetyl fluoride, and (b) the family of alkylsubstituted carbonyl compounds with any substituent X.

respect to the number of nearest-neighbour interactions between atoms, the magnitude of corresponding interatomic distances (connectivities) and represent them in terms of the familiar line formulae or the more abstract incidence matrices. Such a matrix contains on the main diagonal the types of atomic nuclei in the molecule, whereas the off-diagonal elements reflect neighbourhood relationships usually coded in terms of some model of chemical bonding (single, double or fractional bond, non-bond, hydrogen bond, etc., Fig. 17.1 (a)).

For the purpose of structural comparisons the notion of a molecule is unnecessarily restrictive. It has been recognized for a long time that the geometry of certain fragments within molecules depends only slightly on environment. Examples are bonded and non-bonded distances between two atoms, bond angles spanned by three atoms or dihedral angles between four atoms. To obtain a suitable definition of a *molecular* or *structural fragment*, we rely on the widespread chemical practice which dissects molecules into a constant backbone, frame or fragment and variable functional groups, substituents or ligands. We describe a family of molecules with a common fragment in terms of a modified incidence matrix. It contains the atoms and bonds which are the same for all members of the family and a number of variable substituents (R, X, Fig. 17.1(b)). The constant part of the matrix defines the fragment. The scope and limitation of a structural comparison is fixed by giving the list of variable substituents to be considered.

17.2.2   Environment

Anything that is not part of the molecule or molecular fragment as defined above constitutes its *environment*. The environment of a molecule may be empty (as in an ideal gas); it may be a host phase (as in liquid and solid solutions), or it may be composed of other copies of the molecule under investigation (as in pure liquid, crystalline or amorphous phases). In the crystalline state different environments are available through polymorphs, solvates, inclusion compounds and other crystalline phases composed of more than one type of molecules. For a structural fragment the environment is comprised mainly of the variable substituents and, usually less important, of all neighbouring molecules.

Differences in environment induce larger or smaller differences in the structure of a molecule (Chapter 19) or molecular fragment. These differences are usually small enough not to wipe out the identity of the fragment, but sufficiently large to blur it to an extent which can be determined experimentally and interpreted in terms of the flexibility and chemical reactivity of the fragment. For the purpose of structural comparisons the environment is not included explicitly in the analysis. It merely serves as a source of perturbations of the structure of a molecule or molecular fragment.

17.2.3 Description of fragment structure, configuration space and structural similarity

Once a molecule or fragment is distinguished from its environment, a set of coordinates is chosen to describe its geometric structure. Chemical coordinates such as bond lengths, bond angles and torsion angles are convenient. In general $3N-6$ independent quantities are needed for an N-atomic molecule, but redundant ones are sometimes included. For certain problems some of the coordinates may be irrelevant, e.g. the distance from an atom of the fragment to an atom of a variable substituent; others may show insignificant variation; these are simply excluded from structural comparisons. The use of cartesian coordinates with respect to some kind of molecular axis system has also been discussed. Some problems and advantages of this alternative are described by Murray-Rust and Raftery (1985).

Each structural parameter defines a coordinate axis of configuration space and each molecule or fragment is represented by a single point in this space, with coordinates corresponding to the values of its bond distances, angles, etc. Different fragment structures are represented by different points and distances between these are taken as a measure of structural similarity; small distance corresponds to high similarity. Several technical problems arising from this measure of similarity and ways to circumvent them have been discussed elsewhere (Nørskov-Lauritsen and Bürgi 1985).

17.2.4   Cluster analysis, principal-component analysis, standard structure and reference structure

Usually, data points in configuration space are distributed neither randomly nor uniformly, but cluster in one or several locations. The clusters may be identified with the help of cluster analysis and their shapes analyzed by principal-component analysis (Chatfield and Collins 1980). The occurrence of clusters indicates that only specific combinations of distances and angles, i.e. specific fragment structures, are likely to occur. Often a large fraction of the total structural variance within a cluster can be expressed in terms of a small number of principal components, i.e. of highly correlated changes in distances and angles. It is not uncommon that such correlations take the form of a line or curve in configuration space and are therefore essentially one-dimensional.

The centre or some weighted mean of a cluster is often referred to as *standard structure*. The latter is important in X-ray crystallographic studies of proteins and nucleic acids where experimental data may be insufficient to resolve individual atoms. Ways to extract standard structures from a family of equal structural fragments have been described (Taylor and Kennard 1983, 1985; see Chapter 15, Section 15.4.3). For certain problems it is more convenient to specify a *reference structure*, corresponding to either the structure of a single molecule with a specific pattern of substituents or to a prominent point in configuration space, often one that corresponds to a fragment structure of high symmetry.

17.2.5   Structural correlation, incipient reaction and mapping of reaction pathway

Cluster and principal-component analysis are systematic methods to uncover correlations between many structural parameters from many different molecules (all containing the same fragment). They are methods to produce what has been called structural or structure–structure correlation. The details of the correlations are often reminiscent of the structural changes known or suspected to occur during a chemical reaction. In favourable cases structural correlations extend very far into configuration space and may map a reaction coordinate or pathway from reactants to products. If a cluster extends over a relatively small part of configuration space, it is sometimes said that its members represent incipient stages of a chemical reaction (for reviews see: Bürgi (1975); Dunitz (1979); Bürgi and Dunitz (1983)).

17.2.6   Structure and energy

The basic assumption behind the interpretation of structural correlations as mapping a reaction path is that similar structures are associated with similar potential energy surfaces. Given an arbitrary reference structure and its associated energy surface, changes in crystal or fragment environment are expressed in terms of perturbing forces deforming the reference fragment to a point where the perturbing and restoring forces are balanced. For a given magnitude of the force the deformation will be larger in directions of slowly increasing potential energy than in directions of steeply increasing energy. The former often coincide with reaction paths, the latter with displacements perpendicular to reaction paths. Thus, observed distributions of structural parameters found in different crystal environments provide qualitative information about the shapes of low-energy regions of potential energy surfaces (Bürgi and Dunitz 1983, 1988).

Absolute deformation energies cannot be derived from structural data alone. At best, ratios of energy-related quantities may be obtained, e.g. the ratio between a perturbing force and a restoring force (Bürgi and Dunitz 1988). In certain favourable cases, when reasonable assumptions can be made about the direction of perturbing forces (not their magnitudes), it is possible to estimate ratios of force constants (Bürgi and Shefter 1975; Nørskov-Lauritsen *et al.* 1985).

17.2.7   Structure and reactivity

For a few series of related molecules undergoing the same type of chemical reaction activation energies are available in addition to structural information. The resulting structure–reactivity correlations typically show large changes in activation energies associated with relatively small but systematic changes in reactant structure. Structural information about one molecule in the series may be combined with force constants, kinetic and thermodynamic data to parametrize an approximate energy surface for the reaction. By applying simple perturbations on such an energy surface, changes of structure and activation energy are calculated for related molecules and compared to the corresponding experimental quantities. If calculation and experiment agree in detail an explicit relationship between reactant structure and reactivity has been established which may be used to predict the behaviour of related molecules (Dubler-Steudle and Bürgi 1988*a*, *b*).

17.3   Case studies

Many of the results, possibilities and limitations which have emerged from the study of structural correlations, of reaction pathway mappings and of structure–reactivity correlations are conveniently illustrated by one important type of chemical transformation: nucleophilic addition/elimination at carbonyl carbon. A general scheme (Scheme 17.1) shows two alternative pathways: (i) elimination of a nucleophile X from 1 followed by addition

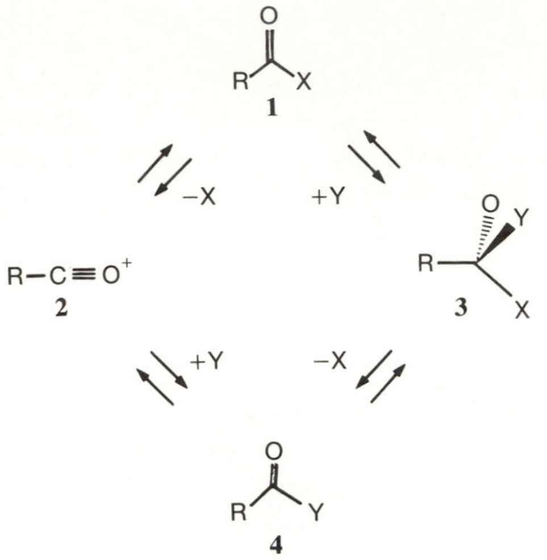

Scheme 17.1

of another nucleophile Y leading, first, to a linear oxocarbenium intermediate, **2**, and then to a modified carbonyl product, **4**; (ii) the reverse
sequence, namely addition of a nucleophile Y followed by elimination of X
leading to the product **4** by way of a tetrahedral intermediate **3**. The three
types of species with two-, three- and four-coordinate carbon have been
characterized structurally. If X equals Y, the reaction **1 ⇌ 2** is symmetric
to the reaction **4 ⇌ 2**, as are **1 ⇌ 3** and **4 ⇌ 3**. From the requirement of
microscopic reversibility the reaction path **1 → 2** or **1 → 3** is equal to the
reverse of the reaction path **2 → 1** or **3 → 1**. Since we are interested primarily
in general aspects of these paths, aspects that are independent of the nature
of X and Y, it is sufficient to analyse **1 → 2** and **1 → 3**.

17.3.1   Reaction pathways

17.3.1.1   Substituent effects on the structure of RCOX molecules   Structural data obtained by microwave spectroscopy, gas-phase electron diffraction and *ab initio* MO calculations have been retrieved from the literature
for molecules of the type RCOX (**1**; R = H, CH_3; X = any substituent
(Ferretti and Bürgi, to be published)). The constant fragment includes either
H, C and O or CH_3, C and O, the bonds between them and the bond link-
ing X to C. The structural parameters considered to be relevant in the present context include the angles $\alpha = \angle R-C-X$, $\beta = \angle R-C=O$,
$\gamma = \angle X-C=O$ and the distance $r(C=O)$. Results published until 1984
are collected in Tables 17.1 and 17.2 if the structure determination was performed without assumptions about the angles and distances at the carbonyl

Table 17.1 Selected structural parameters for HCOX molecules[a]

X	∠H–C–X (α)	∠H–C=O (β)	∠X–C=O (γ)	r(C=O)	r(C–X)	Technique[b]	Parameter type[c]	References
F	107.7	130	122.3	1.188	1.346	ED + MW	r_g, $\angle_z$	Huisman et al. (1969)
OCOH	{ 108.8	125.3	125.9	1.197	1.360 }	MW	r_s	Vaccani et al. (1977)
	112.5	126.7	120.8	1.183	1.388 }			
Cl	110.0	126.5	123.5	1.188	1.760	MW	r_s	Takeo and Matsumura (1976)
OH (mon.)	111.9	123.3	124.8	1.201	1.340	MW	r_e	Davis et al. (1980)
OH.H₂O	110.8	123.8	125.4	1.212	1.340	MO (4–21G*)	r_e	Schäfer et al. (1982)
OCH₃	109.3	124.9	125.8	1.200	1.334	MW	r_s	Curl (1959)
OC₂H₅	109.91	124.95	125.14	1.204	1.347	MO (4–21G)	r_e	Klimkowski et al. (1984a)
OC₃H₇	110.00	124.75	125.25	1.204	1.345	MO (4–21G)	r_e	Klimkowski et al. (1984a)
NH₂	112.7	122.5	124.7	1.220	1.350	MW	r_s	Hirota et al. (1974)
H	116.2	121.9	121.9	1.203	1.100	ED + MW	r_e	Duncan (1974)
C₂H₅	115.01	120.76	124.23	1.211	1.513	MO (4–21G)	r_e	Klimkowski et al. (1984b)
C₃H₇	115.09	120.70	124.21	1.211	1.512	MO (4–21G)	r_e	Klimkowski et al. (1984b)
CH₃	115.3	120.5	124.2	1.207	1.512	ED + MW	r_z	Iijima and Tsuchiya (1972)

[a] Bond distances are given in Å, angles in degrees. Experimental errors have been estimated in different ways by the various authors and can be found in the original papers.

[b] ED: gas-phase electron diffraction; MW: microwave spectroscopy; MO: *ab initio* MO calculations.

[c] For definition of parameter types see Chapter 2, Section 2.4.

Table 17.2 Selected structural parameters for CH_3COX molecules[a]

X	∠CH₃–C–X (α)	∠CH₃–C=O (β)	∠X–C=O (γ)	r(C=O)	r(C–X)	Technique[b]	Parameter type[c]	References
$OCOCH_3$	108.3	130.0	121.7	1.183	1.405	ED	r_a	Vledder et al. (1971)
F	110.5	128.8	120.7	1.185	1.362	ED + MW	r_g, ∠$_z$	Tsuchiya (1974)
Cl	111.6	127.2	121.2	1.188	1.798	ED + MW	r_g, ∠$_z$	Tsuchiya and Iijima (1972)
Br	111.0	126.7	122.3	1.184	1.977	ED + MW	r_g, ∠$_z$	Tsuchiya and Iijima (1972)
OH (mon.)	110.6	126.6	122.8	1.214	1.364	ED	r_g, ∠$_a$	Derissen (1971)
OH (dim.)	113.0	123.6	123.4	1.231	1.334	ED	r_g, ∠$_a$	Derissen (1971)
CN	114.2	124.6	121.2	1.205	1.474	ED + MW	r_z	Sugie and Kuchitsu (1974)
H	115.3	124.2	120.5	1.207	1.114	ED + MW	r_z	Iijima and Tsuchiya (1972)
$NHCH_3$	114.1	124.1	121.8	1.224	1.386	ED	r_g, ∠$_α$	Kitano et al. (1973)
NH_2	115.1	122.9	122.0	1.220	1.380	ED	r_g, ∠$_α$	Kitano and Kuchitsu (1973)
CH_3	116.0	122.0	122.0	1.210	1.517	ED + MW	r_z	Iijima (1972)
$OSiH_3$	115.9	122.4	121.7	1.214	1.358	ED	r_a	Barrow et al. (1981)

[a],[b],[c] See corresponding footnotes in Table 17.1.

carbon atom. Figures 17.2 and 17.3 show scatterplots of the angles α and β. The correlation between α and β is obvious in spite of the widely different origin and physical meaning of the geometrical parameters used. It covers a range of about 10° in both angles. A quantitative investigation using principal-component analysis yields two principal components; the bigger one accounts for 79 per cent (R = H) and 93 per cent (R = CH_3) of the total variance and shows eigenvectors (α, β, γ) of (-0.715, 0.699, 0.015) and (-0.699, 0.715, -0.016), respectively. They express the negative correlation between α and β with slopes of -0.98 and -1.02, respectively. The component in γ is very small, implying that this angle remains nearly constant for a given substituent R. If R = H, γ lies in the range between 121 and 126° with a mean of 124.2°; if R = CH_3, γ is between 121 and 123°, mean 121.8°. For the six entries in Tables 17.1 and 17.2 with the same X, the average difference $\langle \gamma_H - \gamma_{CH_3} \rangle$ is 2.0°, similar to the difference of *ipso*-carbon angles in benzene and toluene (1.9°, from Table 18.5; a detailed analysis and interpretation of structural substituent effects in benzene derivatives is given in Chapter 18).

There is another regularity in Figs. 17.1 and 17.2: poor σ-donating substituents X with high electronegativity are found at large values of β and small values of α; conversely, for good σ donors β is small and α is large. The

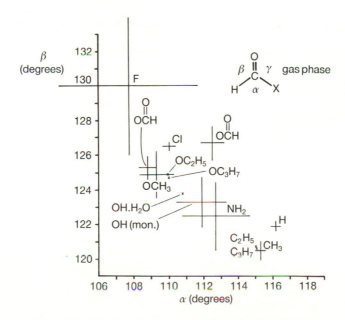

Fig. 17.2 Scatterplot of bond angles β vs. α for HCOX molecules. The nature of X is given next to each datapoint.

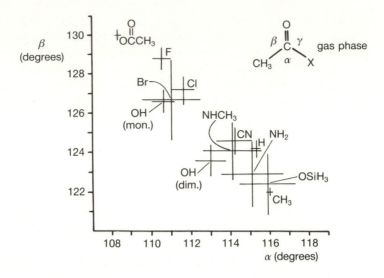

Fig. 17.3 Scatterplot of bond angles β *vs.* α for CH_3COX molecules. The nature of X is given next to each datapoint.

sequence of substituents is halides and anhydrides first, then esters, acids and amides and, finally, ketones and aldehydes. This trend may be interpreted in terms of electronic structure. The arguments resemble those presented in a discussion of bond angles at carbonyl carbon in lactones (Nørskov-Lauritsen *et al.* 1985). In ketones (aldehydes) the highest occupied molecular orbital is dominated by the *p*-type lone pair on O and destabilized by antibonding admixtures of C—C (C—H) bonding orbitals in a symmetric way (**5**). With increasing electronegativity of X the symmetry is lost. The *p*-type lone pair is now destabilized by only one C—C (C—H) bonding orbital and increasingly stabilized by a C—X antibonding orbital (**6**). The destabilizing C—C (C—H) contribution is decreased by the in-plane angle deformation of α and β. There is only a small decrease in γ (e.g. CH_3COCH_3, $\gamma = 122.0°$; CH_3COF, $\gamma = 120.7°$), presumably because the X · · · O interaction is antibonding.

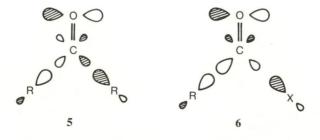

In an RCOX molecule with a small ring, e.g. a lactone or lactam, the R−C−X angle (α) is forced to be small. In such molecules it is found that β is always much larger than γ (138 vs. 126° in β-lactones, 122 vs. 118° in ϵ-lactones; Nørskov-Lauritsen et al. 1985), in agreement with the experimental data given here for RCOX compounds.

17.3.1.2 Incipient elimination 1 → 2

The correlation between bond angles described above is reminiscent of the angular changes that have to occur when RCOX (1) decomposes into RC≡O⁺ (2) and X⁻. $\beta = \angle\,$R−C=O will increase and eventually approach 180°, $\alpha = \angle\,$R−C−X will decrease whereas $\gamma = \angle\,$X−C=O may be expected to remain in the range between ~126 and ~100°. The upper limit of 126° corresponds to the maximum γ value observed for RCOX compounds, the value of 100° corresponds to the lower limit of angles observed for nucleophilic attack at a carbonyl carbon atom (for reviews see Bürgi (1975); Dunitz (1979); Bürgi and Dunitz (1983)). It is significant that the structures of RCOX are more product-like if X is a good leaving group − one that is susceptible to elimination under the action of a strong Lewis acid − than if X is a poor leaving group. Indeed, RC≡O⁺ is obtained from RCOF or RCOCl by reaction with SbF_5 and its structure has been determined (Chevrier et al. 1972). RCOF and RCOCl are further along the pathway leading to RC≡O⁺ than CH_3COCH_3, their structures may be viewed as incipient stages of elimination. To our knowledge Bent (1968) in a review of donor–acceptor interactions was the first to introduce this interpretation of structure.

It is concluded that differences in molecular structure carry at least qualitative information about differences in chemical reactivity.

17.3.1.3 Reaction path for elimination 1 → 2

Using the principle of structural correlation the results obtained so far may be used to delineate an approximate pathway for elimination. Such a pathway is specified by explicit relationships among the distance and angle changes leading from 1 to 2.

With respect to α and β one might be tempted to simply extrapolate the linear relationship found from principal-component analysis. In the limit of complete dissociation of X this would lead to $\beta = 180°$, $\alpha = 57°$ and $\gamma = 123°$. The extrapolation is seen to be inadequate for molecules 7 and 8 (Wallis and Dunitz 1984; Procter et al. 1981) which are considered as models for an almost broken C · · · X bond. The empirical relationship $\alpha \cdot \beta = [\,(360 - \gamma)/2]^2$, with $\gamma = 123°$, accounts much better for the observed bond angles at the digonal centres of 7 and 8 (calculated angles in brackets). The lesson to be learned from this simple exercise, namely that interpolation is preferable to extrapolation, is trivial but indispensable for the construction of realistic reaction paths from structural data.

Obtaining a relationship between distance changes and angular defor-

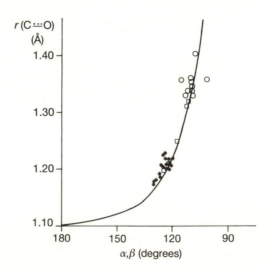

2.443(2) Å
[2.46]
γ
O ‒ ‒ ‒ N   1.099(2) Å
α   β   [1.10]
N

α=85.9(1)° [82.8]
β=169.6(1)°
γ=104.4(1)° [107.6]

7

2.594 Å
CH₃   [2.63]
γ
O ‒ ‒ ‒ C   1.138(5) Å
α   β   [1.10]

α=82.7(4)° [80.8]
β=173.8(4)°
γ=103.5(4)° [105.4]

8

mations presents some difficulties. There is a variety of C—X bonds; these are comparable only if appropriate reference distances are also available (Bürgi 1973). To avoid such complications we restrict the analysis to the cases where X = OR′. Figure 17.4 shows a scatterplot of α vs. $r(C-O)$ and β vs. $r(C=O)$ including the gas-phase data for CH_3COX, $HCOX$ and averages obtained by Borthwick (1980) from crystal structure determinations of $RCOO^-$, $RCOOH$ and $RCOOCH_3$ molecules. The data follow an empirical relationship of the form $r(C \dot{\cdot} \dot{\cdot} O) = 1.10 + 0.0031 \exp[0.0625(180 - \alpha, \beta)]$ Å. The constant 1.10 Å has been chosen to reproduce the observed distances of the $RC \equiv O^+$ fragment (Chevrier *et al.*

Fig. 17.4 Scatterplot of R—C ⋯ O angles vs. adjacent C ⋯ O distances. Stars: β (= ∠ R—C=O) vs. $r(C=O)$; circles: α (= ∠ R—C—OR′) vs. $r(C-O)$; squares: average α, β, and r values for $RCOOCH_3$, $RCOOH$, and $RCOO^-$ from Borthwick (1980).

1972). The validity of the equation in ranges where no data are available $(1.1 < r(C \cdots O) < 1.2 \text{ Å}, r(C \cdots O) > 1.4 \text{ Å})$ may be tested as before with the help of 7 and 8. The $C—C \equiv N$ and $C—N \equiv N$ fragments are bent only slightly ($\sim 170°$), the triple bonds are therefore not expected to be lengthened significantly (calc. ~ 0.006 Å). The $O \cdots C$ and $O \cdots N$ distances are predicted close to the observed values—surprisingly close considering that the empirical equation was obtained for $C—O$ bonds and had to be extrapolated from the observed range by ~ 1 Å. The two $C—O$ distances are also related via Pauling's relationship $r_i = 1.334 - 0.2164 \ln n_i$ (Pauling 1947) and the rule of constant bond order $n_1 + n_2 = 3$ (Johnston 1960), at least approximately.

Together the three relationships $f[\alpha, \beta]$, $f[\alpha, r(C—O)]$ and $f[\beta, r(C=O)]$ define a line in four-dimensional configuration space spanned by $r(C=O)$, $r(C—O)$, α and β. The observed structures are all within 1–2° or 0.01–0.02 Å from this line which maps the energetically favourable regions of configuration space. Following the principle of structural correlation, it is interpreted as an approximate reaction path for the transformation $1 \rightarrow 2$. To summarize we mention two points. Firstly, the analysis of correlations between structural parameters has provided a model for the reaction path which may be extended by interpolation into regions of configuration space where no structures have been observed. Secondly, since transition states for elimination have to lie on or close to the reaction path the range of their possible structures has been narrowed down to a limited region of configuration space. More accurate locations can only be determined with the help of some kind of energetic information (see Section 17.3.2).

17.3.1.4 Elimination reactions in related molecules α-Substituted enamines, **9**, have been studied for various substituents X ranging from CH_3 and CN to I, Cl and F (van Meerssche *et al.* 1979). The structure determinations show a number of features that parallel those discussed for the RCOX compounds. The $C—C—N$ angle increases from 119.4 for $X = CH_3$ to 129.6° for $X = F$, while the $C—C—X$ angle decreases from 124.2 to 117.8°. The $N—C—X$ angle varies only little, from 116.4 to 112.6°. The molecules with the electronegative substituents may be regarded as

9   10

representing incipient stages of the elimination reaction to the corresponding keteneiminium products, **10**. The latter are easily obtained in slightly polar solvents in the presence of weak Lewis acids (Ag^+, BF_3; Ghosez and Marchand-Brynaert 1976).

The torsion angles $C-C-N-R_4$ indicate that the lone pair on nitrogen is inclined by $\sim 34°$ to the CCXN plane for $R = CH_3$ but by only 17, 7 and 5° for $R = F$, Cl and I, respectively; for the latter it is almost in the CCXN plane, antiperiplanar to X. The $C-X$ distances are 0.02, 0.05, 0.05 and 0.11 Å longer in **9** than in the corresponding $CH_2=CHX$ molecules. The conformation about the $C-N$ bonds and the concomitant elongation of the $C-X$ bonds are evidence for a $n_N - \sigma^*_{C-X}$ interaction between N and the $C-X$ bond, i.e. a stereoelectronic assistance facilitating the cleavage of this bond.

Crystal structures of lithium ester enolates (**11**, $X = OR$: Seebach *et al.* 1985), lithium amide enolates (**11**, $X = NR_2$: Bauer *et al.* 1985; Laube *et al.* 1985) and lithium enolates of ketones (**11**, $X = R$: Amstutz *et al.* 1981) have also been determined. The $C-C-O(Li)$ angles are 128–125°, 125° and 124–122°. The $C-C-X$ angles are 115–118, 118–122 and 122–123°, respectively. The molecules with the best leaving groups, the ester enolates ($X = OR$), are furthest along the reaction coordinate to ketene formation as indicated by the largest $C-C-O(Li)$ angles. There is evidence, albeit not very strong, for the transient existence of ketenes, **12**, or ketene-like intermediates during the reaction of one ester enolate with butyllithium (Seebach *et al.* 1985).

17.3.1.5 Reaction path for addition **1** → **3** The findings described for the elimination **1** → **2** above are entirely analogous to our earlier results on the nucleophilic addition of amines to carbonyls and on the elimination of alcoholate ion from acetals or ketals. These reaction pathways also show characteristic relationships among bond angles and between angles and distances describing the geometry about the reacting carbon atom. There is no need to repeat these results here, since they have been reviewed extensively elsewhere (Bürgi 1975; Dunitz 1979; Bürgi and Dunitz 1983).

17.3.2 Potential energy surfaces

So far the discussion has concentrated on the structural aspects of reaction paths and on qualitative interpretations of structure in terms of relative reactivity. A more complete description of a reaction coordinate has to take into account energy as well. In most cases energetic information is not available from structural studies. However, there is a wealth of such information from other sources. Vibrational spectroscopy, for example, yields force constants which describe the change of energy in the immediate neighbourhood of the equilibrium structure. The relative energies of reactants and products are available from thermodynamic measurements. Transition-state energies have been measured by microwave spectroscopy for conformational interconversions and by kinetic methods for various types of bond breaking–bond making reactions.

Here we attempt to combine structural data and whatever is available in terms of energetic quantities. The aim is to obtain an approximate description of the reaction coordinate in structure–energy space and, if possible, of the energy surface in the immediate neighbourhood of this coordinate. The spontaneous cleavage of tetrahydropyranyl acetals, $13 \rightarrow 14$, serves as

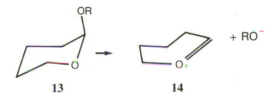

Table 17.3 Exocyclic (r_1) and endocyclic (r_2) C—O distances in tetrahydropyranyl acetals **15–22** (Jones and Kirby 1984). Distances q_E from average of compounds **21** and **22**. Estimated energies of activation $E^\ddagger$ for cleavage of the exocyclic C—O bond.

Compound	r_1 (Å)	r_2 (Å)	$q_E{}^a$ (Å)	$E^\ddagger$ (kJ mol^{-1})
15	1.476	1.379	0.080	72.7
16	1.466	1.377	0.073	81.9
17	1.458	1.383	0.063	89.9
18	1.448	1.385	0.054	103.2
19	1.433	1.405	0.029	123.3
20	1.427	1.398	0.030	119.1
21	1.411	1.428	0.000	162.6
22	1.406	1.416		

^{a} Defined as $[(r_1 - r_{10})^2 + (r_2 - r_{20})^2]^{1/2}$. The reference distances $r_{10} = 1.409$ Å and $r_{20} = 1.422$ Å are averages from compounds **21** and **22**.

an example. Craze and Kirby (1978) have measured the rate of this reaction for a series of leaving groups and found that the rate increases as the pK_a value of the leaving alcoholate, phenolate or benzoate groups decreases (free energy relationship). From accurate structure determinations it was found that the ketal C—O distance in the ring decreases and that to the leaving group increases with decreasing pK_a. Jones and Kirby (1984) concluded that the exocyclic bond distance becomes longer and the activation energy lower as the leaving group becomes more electron withdrawing (Table 17.3). Note that for quite small, but significant changes in ground-state structures free energies of activation change dramatically.

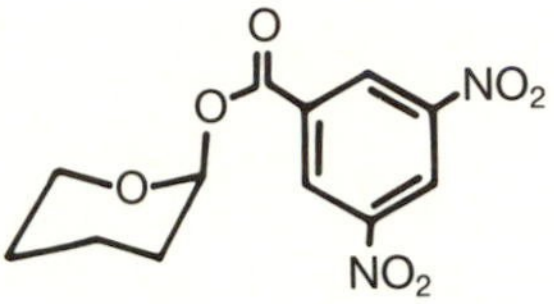

15

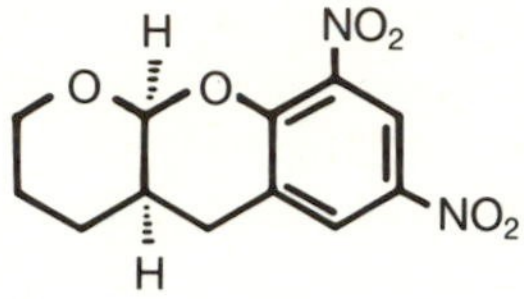

16

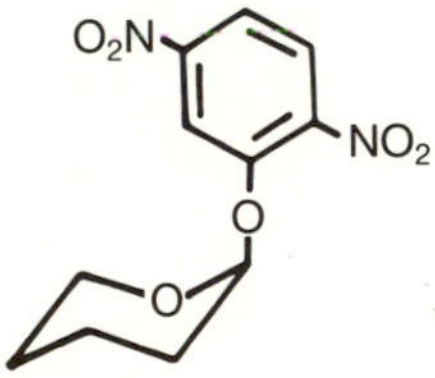

17

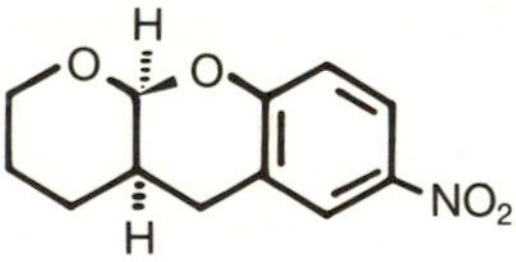

18

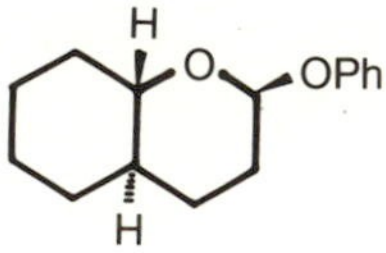

19

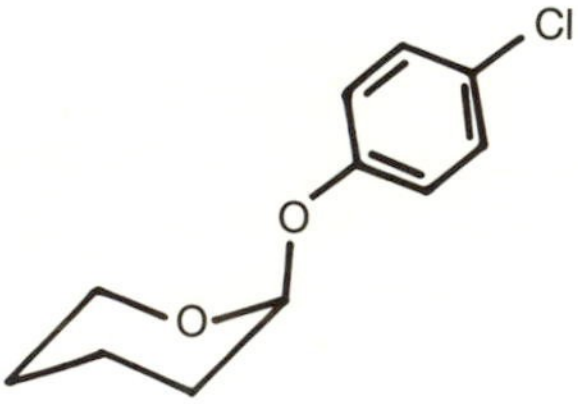

20

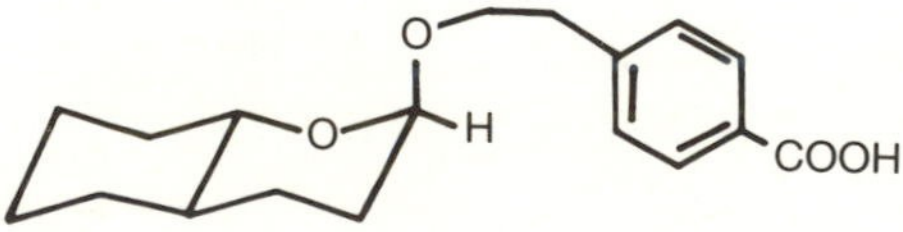

21

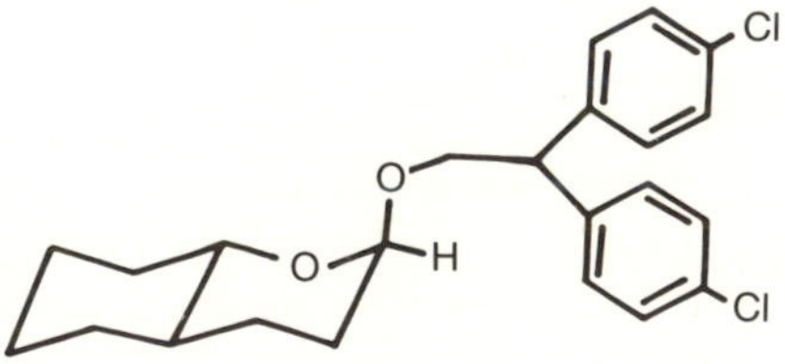

22

17.3.2.1   One-dimensional model of spontaneous cleavage of tetrahydro-
pyranyl acetals (Dubler-Steudle and Bürgi 1988b)   To start with, energy E
is expressed as a function of the reaction coordinate q

$$E(q) = k_{qq}q^2/2 + k_{qqq}q^3 - E_0^{\ddagger}.$$

This is the simplest polynomial showing the necessary extrema, a minimum
at $q = q_E = 0$ and a maximum at $q = q_{TS}$. The constant k_{qq} is the force
constant along the reaction path and $E_0^{\ddagger}$ is the activation energy for an
arbitrary reference molecule; k_{qqq} is an anharmonicity constant obtained
from the equilibrium condition $dE/dq = 0$ evaluated at the transition state
where $q = q_{TS}$ (Fig. 17.5). The effect of small changes of the leaving group
is represented by a linear perturbation

$$E(q) = k_{qq}q^2/2 + k_{qqq}q^3 - E_0^{\ddagger} + k_q q.$$

The constants are the same as for the unperturbed molecule, except for
the new quantity k_q which determines the degree of perturbation. A
sufficiently small value of k_q will not change the general appearance of
the reaction coordinate. The new equilibrium will be shifted to $q_E(k_q)$,
the transition state to $q_{TS}(k_q)$. The activation energy changes to

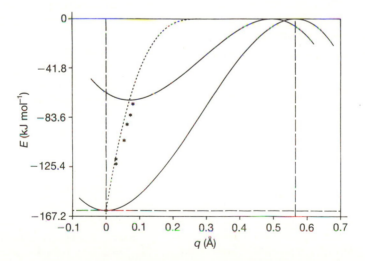

Fig. 17.5 Correlation between change in ground-state structure (q) and activation energy (E)
in tetrahydropyranyl acetals **15–22**. Stars: experimental values; dotted line: values calculated
from one-dimensional model; solid lines: energy profiles of spontaneous acetal cleavage for
two different ground-state structures. The variable q is $[(r_1 - r_{10})^2 + (r_2 - r_{20})^2]^{1/2}$,
where r_{10} and r_{20} are C—O reference distances. After Dubler-Steudle and Bürgi (1988b).

$E^\ddagger = E_0^\ddagger [1 - 2q_E(k_{qq}/6E_0^\ddagger)^{1/2}]^3$ (for $k_q < 0$, $q_E > 0$ and $E^\ddagger < E_0^\ddagger$). Numerical computations require the choice of a reference molecule to fix $E_0^\ddagger$, e.g. **21** or **22**, where $R = CH_2R'$ ($E_0^\ddagger = 162.6\,kJ\,mol^{-1}$, Table 17.3). k_{qq} is given a value that is typical for C—O bonds ($510\,N\,m^{-1}$). Equilibrium coordinates q_E and activation energies $E^\ddagger$ are then calculated as a function of the perturbation k_q. The resulting curve (Fig. 17.5, dotted line) is compared to experimental values of $E^\ddagger$ and q_E (Fig. 17.5, stars). The latter is obtained from $q_E = [(r_1 - r_{10})^2 + (r_2 - r_{20})^2]^{1/2}$ where r_{10} and r_{20} are the C—O distances in the reference molecule (1.409, 1.422 Å; Table 17.3). The agreement between calculated and experimental quantities shows that the model is able to predict the dependence of activation energy on small structural changes from data for a single molecule.

It is concluded that the model potential is a reasonable approximation to the real potential along the reaction path, at least in the region of the minimum and with respect to transition-state energy. The strong dependence of $E^\ddagger$ on q_E is seen to result from the fact that the perturbation k_q affects the reference potential everywhere along the reaction path, only slightly at q_E, but more strongly at q_{TS}. As a spinoff the model provides an estimate of the distance between q_E and q_{TS} (~0.55 Å). The model does not say, however, how much r_1 is shortened and how much r_2 is lengthened on going from the ground state to the transition state. Such information may be obtained from a more detailed treatment taking into account both distances.

17.3.2.2 Two-dimensional model of spontaneous cleavage of tetrahydropyranyl acetals (Dubler-Steudle and Bürgi 1988b) To study the change of both C—O distances along the reaction path the potential energy is expressed as a function of r_1 and r_2. As before the simplest energy surface with a minimum and a transition state is a third-order polynomial

$$E(\Delta r_1, \Delta r_2) = k_{11}(\Delta r_1^2 + \Delta r_2^2)/2 + k_{12}\Delta r_1 \Delta r_2$$
$$+ k_{111}(\Delta r_1^3 + \Delta r_2^3) + k_{112}(\Delta r_1^2 \Delta r_2 + \Delta r_1 \Delta r_2^2)$$
$$- E_0^\ddagger$$

with $\Delta r_1 = r_1 - 1.409$ Å, $\Delta r_2 = r_2 - 1.422$ Å. To simplify the problem, Δr_1 and Δr_2 have been constrained to be equivalent.

The quadratic force constants are taken from normal-coordinate analyses and from *ab initio* MO calculations for related molecules ($k_{11} = 530$, $k_{12} = 27\,N\,m^{-1}$). The cubic force constants ($k_{111} = 2.2 \times 10^{12}$, $k_{112} = 6.7 \times 10^{12}\,N\,m^{-2}$) are obtained from the following conditions: (1) together with k_{11} and k_{12} they must reproduce the experimental activation energy $E_0^\ddagger$; (2) the distances in the transition state have to obey the rule of constant bond order, $n_1^\ddagger + n_2^\ddagger = 2$, where $n_1^\ddagger$ and $n_2^\ddagger$ are the Pauling bond orders

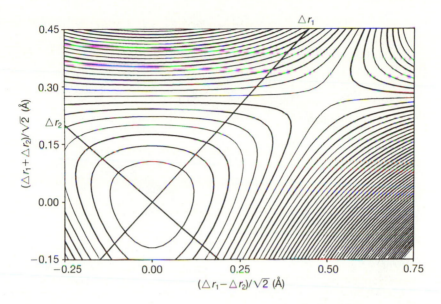

Fig. 17.6 Energy contour diagrams for the spontaneous cleavage of tetrahydropyranyl acetals as a function of the two acetal C—O bond distances. Minimum and transition state are shown (note that Δr_1 and Δr_2 are oriented diagonally). After Dubler-Steudle and Bürgi (1988*b*).

in the transition state; (3) the stretching force constant calculated with the above potential for $r_2^{\ddagger}$ has to follow Badger's rule ($k_{22} = c(r_2 - d)^{-3}$; Badger 1934) or Herschbach and Laurie's relationship ($r_2 = a - b\ln k_{22}$; Herschbach and Laurie 1961), where a, b, c, d are known constants. The last condition is redundant but serves as a useful test for the quality of the energy surface in areas far from the minimum. Part of the energy contour diagram calculated with the parameters given above is shown in Fig. 17.6. It displays the minimum and the transition state for one of the two acetal C—O bonds.

The perturbed potential is of the form

$$E(\Delta r_1, \Delta r_2) = k_{11}(\Delta r_1^2 + \Delta r_2^2)/2 + k_{12}\Delta r_1 \Delta r_2$$
$$+ k_{111}(\Delta r_1^3 + \Delta r_2^3) + k_{112}(\Delta r_1^2 \Delta r_2 + \Delta r_1 \Delta r_2^2)$$
$$- E_0^{\ddagger} + k_1(\Delta r_1 - \Delta r_2).$$

The negative sign in the perturbation reflects the fact that a change of the substituent R lengthens one C—O bond but shortens the other. As for the one-dimensional model Δr_{1E}, Δr_{2E} and $E^{\ddagger}$ are evaluated for increasing

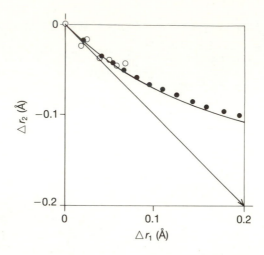

Fig. 17.7 Acetal C—O distances perturbed by variation of the leaving group. Empty circles: observed; filled circles: calculated from two-dimensional model. The solid line is a curve of constant bond order. The diagonal is at $\Delta r_1 = -\Delta r_2$ [$\Delta r_i = r_i - r_{i0}$ ($i = 1, 2$); the r_{i0} are C—O reference distances].

perturbations k_1. A plot of $(\Delta r_{1E}^2 + \Delta r_{2E}^2)^{1/2}$ ($= q_E$) *vs.* $E^{\ddagger}$ is practically indistinguishable from that given in Fig. 17.5 (dotted line). Calculated distances Δr_{E1}, Δr_{E2} agree with the experimental ones (Fig. 17.7, filled and empty circles, respectively).

The detailed agreement between calculated and experimental quantities, structural as well as energetic, leads us to conclude that the two-dimensional potential is a good approximation to the real energy surface around the minimum and at the transition state. It is especially pleasing that the anharmonicity built into the model reproduces the non-linear changes in Δr_{1E} and Δr_{2E} (Fig. 17.7, filled and empty circles). The two-dimensional model thus represents a significant improvement over the one-dimensional model. For the reference molecule the distances in the transition state are predicted at $r_1^{\ddagger} = 1.409 + 0.537 \cong 1.95$ Å and $r_2^{\ddagger} = 1.422 - 0.153 \cong 1.27$ Å; the corresponding value of q_{TS} is ~ 0.56 Å, similar to the result from the one-dimensional model (~ 0.55 Å). The transition-state structure obtained here compares well with those calculated by *ab initio* MO techniques for the (water-assisted) decomposition of $H_2CO_3 \cdot H_2O$ into $CO_2 \cdot 2H_2O$ (Nguyen *et al.* 1987), $CH_2{=}C(OH)_2 \cdot H_2O$ into $CH_2{=}C{=}O \cdot 2H_2O$ (Nguyen and Hegarty 1984) and $CH_2(OH)_2 \cdot H_2O$ into $CH_2{=}O \cdot 2H_2O$ (Williams *et al.* 1983).

17.3.3 Correlation between ground-state structure and activation energy. Determination of the transition-state structure

In the previous paragraphs a method was illustrated to combine structural data, force constants and activation energies with simple algebraic expressions to obtain models of energy surfaces describing chemical reactions. The models accomplish two things: (1) they rationalize a sometimes dramatic dependence of reactivity (activation energy) on small changes in ground-state bond distances. Typical ranges are 0.05–0.10 Å, much larger than the precision (and often accuracy) attainable in carefully done structure analyses; (2) geometrical structures of transition states may be determined. This is a type of information that traditionally has been reserved to quantum chemical calculations. The latter are feasible only for relatively small systems. Our approach is applicable to molecules of any complexity, provided the necessary structural, vibrational and kinetic data are available. It may be generalized to deal with reaction products as well. This may be achieved by including higher-order terms in the polynomial energy expression or by introducing other types of potential functions, e.g. Morse potentials. The necessary force constants are accessible through additional kinetic data, e.g. isotope effects, product ratios, free energy relationships and other types of mechanistic information.

There is no *a priori* reason for the perturbation to be linear; it could become steeper or shallower on the way from the ground to the transition state. Its behaviour along the reaction coordinate provides a basis for a rational discussion of concepts like ground-state and transition-state (de-)stabilization: a perturbation levelling off between the former and the latter would have to be associated with (enthalpic) ground-state (de-)stabilization and vice versa for the reverse situation.

The models presented above may be considered as an adaption of molecular mechanics techniques (see Chapter 14) to the description of chemical reactions. Such techniques depend, as does our model, on a set of appropriate potential constants. Since we were interested primarily in understanding an observed structure–reactivity correlation in tetrahydropyranyl acetals, we have tried to take these constants, as far as possible, from sources not related to these acetals. However, our analysis could be reversed and the data on the tetrahydropyranyl acetals be used to determine an optimized set of potential constants. The possibility of obtaining anharmonic constants (Bürgi 1973) is especially attractive since these are generally difficult to determine.

17.4   Discussion

17.4.1   Accuracy of structural data

Accuracy is not crucial for mapping reaction pathways, because the change of structural parameters along such a path is usually several times larger than their experimental uncertainty. For such mappings it is advantageous to choose a fuzzy fragment definition, i.e. a large range of chemically different substituents inducing many different perturbations and sampling an extended region of the energy surface.

The situation is different for structure–reactivity correlations where large differences in activation energy are often accompanied by small differences in structure which in some cases amount to no more than a few hundredths of an ångström. In such cases structural data of the highest attainable precision, $\sigma = 0.001$–$2 \, \text{Å}$, are clearly desirable.

17.4.2   Dynamic implications: possibilities and limitations

Dynamic implications of structural data have been illustrated by several examples in Section 17.3. Future research aimed at the determination of reaction pathways or transition-state structures and at the interpretation of structure–reactivity relationships would seem to depend mainly on the researchers' imagination in formulating a chemically meaningful question.

Our experience indicates that the scope and limitation of this type of work depends to a certain extent on fragment definition. If it is kept fuzzy by admitting a broad range of different substituents on the invariant frame the correlations tend to be fuzzy too, but there is usually plenty of data. As the fragment definition becomes more restrictive, the correlations tend to sharpen, but the database may become sparse. Fuzziness of correlations is not necessarily a disadvantage because the diversity of perturbations implicit in fuzzy correlations distorts a fragment structure not only along a reaction path but also in directions perpendicular to it. As a result more can be learned about the shape of low-energy regions of a potential energy surface.

A more serious problem arises for investigations dealing with structure and reactivity. Although there is a wealth of data of either kind, it frequently happens that the kinetic data are available for one molecule, the structure for a related one. Furthermore retrieving kinetic data and judging their reliability or even their mechanistic implications is much more difficult than retrieving structural data. The former are scattered throughout the literature, the latter are available in critically edited data banks (see Chapter 15).

In spite of such difficulties we are convinced that new and exciting inter-pretations of molecular structure are waiting to be unravelled by people willing to penetrate the interfaces between theory and experiment, between structure and property, between the static and the dynamic.

References

Amstutz, R., Schweizer, W. B., Seebach, D., and Dunitz, J. D. (1981). *Helv. Chim. Acta*, **64**, 2617–21.

Badger, R. M. (1934). *J. Chem. Phys.*, **2**, 128–31.

Barrow, M. J., Cradock, S., Ebsworth, E. A. V., and Rankin, D. W. H. (1981). *J. Chem. Soc. Dalton Trans.*, 1988–93.

Bauer, W., Laube, T., and Seebach, D. (1985). *Chem. Ber.*, **118**, 764–73.

Bent, H. A. (1968). *Chem. Rev.*, **68**, 587–648.

Borthwick, P. W. (1980). *Acta Crystallogr.*, **B36**, 628–32.

Bürgi, H.-B. (1973). *Inorg. Chem.*, **12**, 2321–5.

Bürgi, H.-B. (1975). *Angew. Chem. Int. Ed. Engl.*, **14**, 460–73.

Bürgi, H.-B. and Dunitz, J. D. (1983). *Acc. Chem. Res.*, **16**, 153–61.

Bürgi, H.-B. and Dunitz, J. D. (1988). *Acta Crystallogr.*, **B44**, 445–8.

Bürgi, H.-B. and Shefter, E. (1975). *Tetrahedron*, **31**, 2976–81.

Chatfield, C. and Collins, A. J. (1980). *Introduction to multivariate analysis*. Chapman and Hall, London.

Chevrier, B., Le Carpentier, J.-M., and Weiss, R. (1972). *J. Am. Chem. Soc.*, **94**, 5718–23.

Craze, G.-A. and Kirby, A. J. (1978). *J. Chem. Soc. Perkin Trans. II*, 354–6.

Curl Jr., R. F. (1959). *J. Chem. Phys.*, **30**, 1529–36.

Davis, R. W., Robiette, A. G., Gerry, M. C. L., Bjarnov, E., and Winnewisser, G. (1980). *J. Mol. Spectrosc.*, **81**, 93–109.

Derissen, J. L. (1971). *J. Mol. Struct.*, **7**, 67–80.

Dubler-Steudle, K. C. and Bürgi, H.-B. (1988a). *J. Am. Chem. Soc.*, **110**, 4953–7.

Dubler-Steudle, K. C. and Bürgi, H.-B. (1988b). *J. Am. Chem. Soc.*, **110**, 7291–9.

Duncan, J. L. (1974). *Mol. Phys.*, **28**, 1177–91.

Dunitz, J. D. (1979). *X-ray analysis and the structure of organic molecules*, Chapters 7–10. Cornell University Press, Ithaca.

Ghosez, L. and Marchand-Brynaert, J. (1976). α-Haloenamines and keteneiminium salts. In *Iminium salts in organic chemistry*, Advances in Organic Chemistry, Vol. 9, Part I (ed. H. Böhme and H. G. Viehe), pp. 421–532. Wiley, New York.

Herschbach, D. R. and Laurie, V. W. (1961). *J. Chem. Phys.*, **35**, 458–63.

Hirota, E., Sugisaki, R., Nielsen, C. J., and Sørensen, G. O. (1974). *J. Mol. Spectrosc.*, **49**, 251–67.

Huisman, P. A. G., Klebe, K. J., Mijlhoff, F. C., and Renes, G. H. (1969). *J. Mol. Struct.*, **57**, 71–82.

Iijima, T. (1972). *Bull. Chem. Soc. Jpn.*, **45**, 3526–30.

Iijima, T. and Tsuchiya, S. (1972). *J. Mol. Spectrosc.*, **44**, 88–107.

Johnston, H. S. (1960). Large tunnelling corrections in chemical reaction rates. In *Advances in chemical physics*, Vol. 3 (ed. I. Prigogine), pp. 131–70. Interscience, New York.

Jones, P. G. and Kirby, A. J. (1984). *J. Am. Chem. Soc.*, **106**, 6207–12.

Kitano, M. and Kuchitsu, K. (1973). *Bull. Chem. Soc. Jpn.*, **46**, 3048-51.

Kitano, M., Fukuyama, T., and Kuchitsu, K. (1973). *Bull. Chem. Soc. Jpn.*, **46**, 384-7.

Klimkowski, V. J., Schäfer, L., and Bohn, R. K. (1984*a*). *J. Comput. Chem.*, **5**, 175-81.

Klimkowski, V. J., van Nuffel, P., van den Enden, L., van Alsenoy, C., Geise, H. J., Scarsdale, J. N., and Schäfer, L. (1984*b*). *J. Comput. Chem.*, **5**, 122-8.

Laube, T., Dunitz, J. D., and Seebach, D. (1985). *Helv. Chim. Acta*, **68**, 1373-93.

Murray-Rust, P. and Raftery, J. (1985). *J. Mol. Graphics*, **3**, 50-9.

Nguyen, M. T. and Hegarty, A. F. (1984). *J. Am. Chem. Soc.*, **106**, 1552-7.

Nguyen, M. T., Hegarty, A. F., and Ha, T.-K. (1987). *J. Mol. Struct. (Theochem)*, **150**, 319-25.

Nørskov-Lauritsen, L. and Bürgi, H.-B. (1985). *J. Comput. Chem.*, **6**, 216-28.

Nørskov-Lauritsen, L., Bürgi, H.-B., Hofmann, P., and Schmidt, H. R. (1985). *Helv. Chim. Acta*, **68**, 76-82.

Pauling, L. (1947). *J. Am. Chem. Soc.*, **69**, 542-53.

Primas, H. and Müller-Herold, U. (1984). *Elementare Quantenchemie*, p. 311. Teubner Studienbücher, Stuttgart.

Procter, G., Britton, D., and Dunitz, J. D. (1981). *Helv. Chim. Acta*, **64**, 471-7.

Schäfer, L., van Alsenoy, C., Scarsdale, J. N., Sellers, H. L., and Pinegar, J. F. (1982). *J. Mol. Struct. (Theochem)*, **86**, 267-75.

Seebach, D., Amstutz, R., Laube, T., Schweizer, W. B., and Dunitz, J. D. (1985). *J. Am. Chem. Soc.*, **107**, 5403-9.

Sugie, M. and Kuchitsu, K. (1974). *J. Mol. Struct.*, **20**, 437-48.

Takeo, H. and Matsumura, C. (1976). *J. Chem. Phys.*, **64**, 4536-40.

Taylor, R. and Kennard, O. (1983). *Acta Crystallogr.*, **B39**, 517-25.

Taylor, R. and Kennard, O. (1985). *Acta Crystallogr.*, **A41**, 85-9.

Tsuchiya, S. (1974). *J. Mol. Struct.*, **22**, 77-95.

Tsuchiya, S. and Iijima, T. (1972). *J. Mol. Struct.*, **13**, 327-38.

Vaccani, S., Roos, U., Bauder, A., and Günthard, Hs. H. (1977). *Chem. Phys.*, **19**, 51-7.

Van Meerssche, M., Germain, G., Declercq, J. P., and Colens, A. (1979). *Acta Crystallogr.*, **B35**, 907-13.

Vledder, H. J., Mijlhoff, F. C., Leyte, J. C., and Romers, C. (1971). *J. Mol. Struct.*, **7**, 421-9.

Wallis, J. D. and Dunitz, J. D. (1984). *J. Chem. Soc. Chem. Commun.*, 671-2.

Williams, I. H., Spangler, D., Femec, D. A., Maggiora, G. M., and Schowen, R. L. (1983). *J. Am. Chem. Soc.*, **105**, 31-40.

18

Structural substituent effects in benzene derivatives

Aldo Domenicano

18.1 Introduction

The purpose of this chapter is to show how the availability of *accurate* geometrical information on a class of related molecules—the substituted derivatives of benzene—has fostered new lines of structural research. It would have been possible to select some other example, but this one has at least two merits. First, it refers to a fundamental class of organic molecules; second, it involves important contributions from most of the techniques of structure determination described in this book.[†]

Substituting a hydrogen atom with a functional group causes a perturbation in the valence electron distribution of a molecule. This affects a number of chemical and physical properties. Changes occur in, e.g. reaction rates,

---

[†] It may be added that the study of structural substituent effects in benzene derivatives has been a joint research interest of the editors of this book for many years.

equilibrium constants, dipole moment, spectral parameters, and, last but not least, molecular geometry. While reactivity and spectral parameters may vary by orders of magnitude, the distortions of the molecular framework are usually small. With benzene derivatives they amount to no more than a few hundredths of an ångström for bond distances and some degrees for bond angles. They can thus be determined only through accurate experiments, or derived by high-level *ab initio* MO calculations with full geometry optimization.

On a relative scale the changes in bond angles tend to be larger than those in bond distances. This is due to the fact that the energy required to enlarge a $C-C-C$ angle by, say, $1°$, is some 3–4 times smaller than that required to stretch a $C-C$ bond by 0.01 Å. Another advantage of bond angles is that they are generally better determined than bond distances.

The effects of substitution on molecular geometry have been extensively investigated with benzene derivatives. These small, fairly rigid molecules are ideally suited to accurate structural studies, and the high symmetry of the hexagonal framework is instrumental in allowing the determination and analysis of distortional effects. But other systems have of course been studied, among these cyclopropane derivatives (Allen 1980, 1981; see also Chapter 16, Section 16.11), cyclobutane derivatives (Allen 1984), and the $R-CO-X$ species (see Chapter 17, Section 17.3.1.1).

18.2 Historical note

The first example of a ring distortion in a benzene derivative is found in a paper by Keidel and Bauer (1956) on the molecular structure of phenylsilane, $C_6H_5-SiH_3$, as determined by gas-phase electron diffraction. These authors found that the agreement between experimental and theoretical radial distributions could be improved substantially by allowing the internal ring angle at the *ipso* position to become $117.4°$ rather than $120°$. Subsequent studies by X-ray crystallography have confirmed that an *ipso* angle of about $117°$ is typical of silicon substitution in benzene derivatives (see Domenicano *et al.* 1975*b*).

Curiously, this early observation did not stimulate further research. Detailed structural studies of monosubstituted benzene derivatives in the gaseous phase, based on the substitution method of microwave spectroscopy, only began to appear after a dozen years (Nygaard *et al.* 1968 (fluorobenzene); Casado *et al.* 1971 (cyanobenzene); Høg 1971 (nitrobenzene); Lister *et al.* 1974 (aniline); Cox *et al.* 1975 (phenylacetylene); Michel *et al.* 1976 (chlorobenzene); Larsen 1979 (phenol); Amir-Ebrahimi *et al.* 1981 (toluene)). In the meantime, the molecular structures of hundreds of substituted derivatives of benzene had been determined by X-ray crystallography accurately enough to show at least some of the effects of substitution on the ring geometry. Almost invariably, however, the determination of these effects was not the primary aim of the X-ray diffraction studies, and the small deviations from the reference geometry of the unsubstituted benzene ring were seldom noticed.

The first attempts to rationalize the ring deformations were published in the sixties (Carter *et al.* 1966; Nygaard *et al.* 1968). More general conclusions were derived after a decade through the analysis of many structural results, mostly obtained by X-ray crystallography (Domenicano *et al.* 1975*a*, *b*; Domenicano and Murray-Rust 1979; Domenicano and Vaciago 1979; Norrestam and Schepper 1981). A full statistical treatment of the structural information available for monosubstituted benzene rings followed after a few years (Domenicano *et al.* 1983).

Today the study of structural substituent effects in benzene derivatives is an active field of research. All techniques of accurate structure determination are being used to augment the structural information available for these molecules and to investigate the electronic effects of the substituents through their geometrical effects. A comprehensive review of the results obtained from gas-phase studies has appeared recently (Domenicano 1988).

18.3 Accuracy requirements

The *accurate* measurement of the benzene ring distortions caused by substitution is a difficult task. Each technique of structure determination has its own systematic errors, whose existence is not always made clear in scientific papers, and tends to escape the attention of the non-specialist. These errors are often of the same order of magnitude as the structural changes caused by substitution and should not be ignored. Moreover, the geometries produced by the various experimental techniques correspond to different types of vibrational averaging, as discussed in detail in Chapters 2–5 and 8–12. This is not *per se* a systematic error, but should always be considered when comparing results obtained by different techniques.

Here we will briefly review the shortcomings and some specific important merits of the various techniques of structure determination, when applied to the accurate measurement of structural substituent effects in benzene derivatives.

18.3.1 *Microwave spectroscopy studies*

The molecular structure obtained by the isotopic substitution method of microwave spectroscopy (as described in Chapter 3, Section 3.4) is generally considered to be a very good approximation of the equilibrium structure (Costain 1958, 1966; but see Chapter 4, Section 4.4.1, for a more critical view). However, the accuracy of the isotopic substitution method is severely

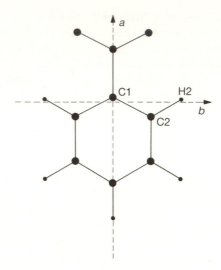

Fig. 18.1 The principal axes system of nitrobenzene. Note the small *a* coordinates of atoms C1, H2, and C2.

reduced when the atoms involved have small coordinates (i.e. less than about 0.4 Å in absolute value) in the inertial reference framework. This is often the case of atoms C1, H2, and/or C2 in a monosubstituted benzene ring of C_{2v} symmetry (an example is shown in Fig. 18.1).

Although the small coordinates are most often derived by alternative procedures (see again Chapter 3, Section 3.4), there is persuasive evidence that microwave spectroscopy may not be the method of choice for determining the geometry of the *ipso* region in monosubstituted benzene rings. For example, the values of the internal ring angle at the *ipso* position obtained by microwave spectroscopy for chlorobenzene (Michel *et al*. 1976) and nitrobenzene (Høg 1971) have been shown to be in error by 1–2°! (see Schultz *et al*. 1980; Domenicano *et al*. 1990). This shortcoming is particularly unfortunate, as the largest geometrical changes caused by substitution tend to occur in the *ipso* region. On the other hand, there is little doubt that the geometry of the *para* region of the ring is determined with great accuracy by the isotopic substitution method.

18.3.2   Gas-phase electron diffraction studies

The relatively low molecular symmetry makes the monosubstituted derivatives of benzene difficult subjects for electron diffraction studies. The carbon skeleton of a benzene ring of C_{2v} symmetry (Fig. 18.2) has, in general, three different C—C bond distances (*a, b, c*) and four different

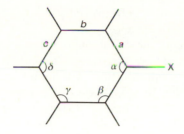

Fig. 18.2 Lettering of the C—C bonds and C—C—C angles in a monosubstituted benzene ring of C_{2v} symmetry.

angles ($\alpha, \beta, \gamma, \delta$). These seven internal coordinates are linked by two equations of geometrical constraint:

$$\alpha + 2\beta + 2\gamma + \delta = 4\pi, \tag{18.1}$$

$$a \sin(\alpha/2) + b \sin(\beta + \alpha/2 - \pi) = c \sin(\delta/2). \tag{18.2}$$

Five independent parameters are therefore required to describe the ring geometry.

The accurate determination of these five geometrical parameters *solely* by electron diffraction is a hopeless task. Particularly difficult is the measurement of the small differences in the lengths of the a, b, and c bonds, as they seldom exceed 0.02 Å. It is clear, therefore, that *in the electron diffraction study of a monosubstituted derivative of benzene full advantage must be taken of any reliable piece of structural information that is available from other techniques.*

If a substitution geometry has been determined by microwave spectroscopy, the internal angle at the *para* position of the ring, δ, and the differences between those C—C bond distances that do not involve atoms with small coordinates may be taken from the microwave study, and imposed as constraints in the least-squares analysis of the electron diffraction data (see, for example, Domenicano *et al.* 1990). But even if a substitution geometry is not available, the details of the ring deformation are expected to be more reliable when the refinement is based conjointly on the electron diffraction intensities and the rotational constants from spectroscopic studies (see, for example, Almenningen *et al.* 1985). A discussion of the problems involved in this type of conjoint refinement is found in Chapter 5, Section 5.8.

Combined analysis of electron diffraction intensities, rotational constants from spectroscopic studies, and dipolar coupling constants from liquid-crystal NMR spectra is a promising method, which has given encouraging results for a variety of molecules (Rankin 1988; see also Chapter 12, Section

12.9). It has been applied recently to the study of chlorobenzene (Rankin 1988; Cradock *et al.* 1990).

With some substituents the differences between the lengths of the ring C—C bonds are determined with reasonable accuracy by *ab initio* MO calculations with geometry optimization, although their absolute values may not be reliable (see Section 18.3.5). These differences can be imposed as constraints in the analysis of the electron diffraction data. Such procedure is termed MOCED (Molecular Orbital Constrained Electron Diffraction) and has been applied successfully to the study of a variety of organic molecules (Schäfer *et al.* 1988).

We will see in Section 18.5 that the statistical analysis of a large sample of monosubstituted benzene rings, mostly studied by X-ray crystallography, has revealed a linear relationship between the internal ring angles α and β. This entirely empirical relationship is firmly established, and may serve as a valuable constraint in the analysis of electron diffraction data (see, for example, Portalone *et al.* 1987; Domenicano *et al.* 1990).

The symmetrically *para*-disubstituted and *sym*-trisubstituted derivatives of benzene, $1,4\text{-}C_6H_4X_2$ and $1,3,5\text{-}C_6H_3X_3$, are better suited for electron diffraction studies than the corresponding monosubstituted derivatives, because many of the atom–atom interactions double (or treble) in the electron scattering from these molecules. Moreover, the higher symmetry of the carbon skeleton lowers the number of independent parameters that define its geometry. Note that the gas-phase structure of these molecules can only be determined by electron diffraction, since they lack a permanent dipole moment and are, therefore, inaccessible to microwave studies.

18.3.3   Crystallographic studies

There are two main sources of systematic error in benzene ring geometries obtained by X-ray crystallography: they are the asphericity shifts of the ring carbons and the effects of solid-state thermal motions. Only the second is relevant to geometries obtained by neutron crystallography.

18.3.3.1   Asphericity shifts   These shifts originate from the fact that X rays — unlike neutrons — are scattered by electrons rather than by nuclei. The non-spherical distribution of valence electrons causes the positions of the centroids of electron density of the ring carbons to differ systematically from the nuclear positions. The effect is especially important for bond distances since it causes the C—C bonds of the benzene ring to appear *shorter* by several thousandths of an ångström, as compared to true internuclear separations; it is somewhat less important for bond angles. As discussed in Chapter 10, Section 10.8.1, the asphericity shifts can be removed if enough high-order reflections have been measured. A refinement based solely on high-order reflections yields atomic positions much closer to nuclear positions, since the contribution of valence electrons to the scattering of X-rays falls more rapidly than that of core electrons as the scattering angle increases.

Of course, a delicate question arises: where to place the lower limit for the high-order reflections. Any value chosen for the minimum reciprocal radius, $H_{min} = (2\sin \theta / \lambda)_{min}$, is arbitrary, as there is no sharp division between the scattering of valence and core electrons. For atoms with lone pairs of electrons, such as N or O, and when low-temperature X-ray data are used, even a value of H_{min} as high as 1.5 $\mathring{A}^{-1}$ may prove unsatisfactory, as the rather sharp lone-pair density still diffracts appreciably at these high H values. This is not the case of the ring carbons, because their valence shell contains only the relatively diffuse bonding electrons. If room-temperature data are used, it appears that a value of H_{min} of 1.1–1.2 $\mathring{A}^{-1}$ suffices to remove most of the scattering of the bonding electrons, thus providing reasonably accurate positions for these atoms (see, for example, Brunvoll *et al.* 1984; Colapietro *et al.* 1984*a*, *b*). However, low-temperature X-ray diffraction studies are highly desirable, as they allow the measurement of many more high-order reflections with non-zero intensity than is the case in room-temperature studies.

18.3.3.2 Solid-state thermal motions   The effect of librational rigid-body motions on molecular geometry is described in detail elsewhere in this book (Chapter 8, Sections 8.4 and 8.5). In the case of a monosubstituted benzene ring we can visualize three different components of libration: they may be described as (i) yawing, (ii) pitching, and (iii) rolling (see Fig. 18.3 (Domenicano *et al.* 1983)).

The main effect of yawing and pitching is that atoms near the *para* position will have the highest thermal motion. Bond lengths will apparently decrease, the effect being more pronounced for the c bonds and less pronounced for the a bonds (see Fig. 18.2 for the lettering of the ring C—C bonds). The changes in the ring angles are likely to be much smaller.

By contrast, rolling leads to a shortening of only a and c and also results in a decrease of α and δ and an increase of β and γ. Again, the effect on bond angles is less pronounced than that on bond distances. This component of libration is important in *para*-disubstituted benzene rings (see, for example, Brunvoll *et al.* 1984; Colapietro *et al.* 1984*b*, 1987).

In most benzene derivatives studied at room temperature the motion of the ring and of at least some of the non-hydrogen atoms bonded to it is well approximated by the harmonic rigid-body model of Schomaker and Trueblood (1968). The libration corrections can therefore be applied with reasonable confidence. These corrections are by no means marginal for the ring C—C bonds; they may amount to more than 0.01 $\mathring{A}$, and are often quite

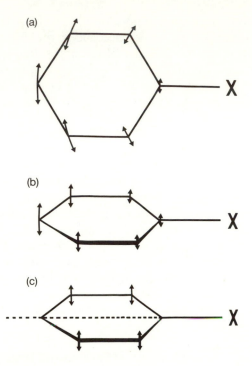

Fig. 18.3 Libration components of a phenyl group: (a) yawing; (b) pitching; (c) rolling. From Domenicano *et al.* (1983).

different for the *a*, *b*, and *c* bonds. *They must be taken into account if one wishes to investigate the effects of substitution on bond distances.* Low-temperature studies are again highly desirable, as they substantially reduce the extent of the correction. A detailed example concerning benzamide, studied at two different temperatures by neutron crystallography, is found in Chapter 11, Table 11.3.

18.3.4 NMR spectroscopy studies

Liquid-crystal NMR spectroscopy has been applied repeatedly to determine the molecular structure and ring distortions of substituted benzene derivatives. In some cases the results are consistent with those obtained by gas-phase and solid-state studies. This is, for example, the case of fluorobenzene (Jokisaari *et al.* 1981); a detailed presentation of the geometry of this molecule, as obtained by various techniques of structure determination, will be found in Section 18.3.6. In other cases, however, the results are con-spicuously at variance with those obtained by other techniques. An out-standing example is given in Table 18.1. It is clear that (until the origin of

Table 18.1 Internal ring angles γ and δ of nitrobenzene as determined by different techniques[a]

Technique[b]	Type of structure	γ	δ	References
MW	r_s	120.3^c	120.2^c	Høg (1971)
ED	r_a	120.5 ± 0.2	120.2 ± 0.4	Domenicano et al. (1990)
XD[d]	$\approx r_\alpha$	$120.3\,(2)$	$120.4\,(2)$	Domenicano and Murray-Rust (1979)
XD[d]	$\approx r_\alpha$	$120.3\,(1)$	$120.5\,(1)$	Domenicano et al. (1990)
MO (HF/6-31G)	r_e	120.1	120.4	Bock et al. (1985a)
MO (HF/6-31G*)	r_e	120.1	120.4	Domenicano et al. (1990)
NMR (ZLI 1167)	r_α	$117.8\,(4)$	$123.0\,(6)$	Catalano et al. (1984)
NMR (PCH)	r_α	$117.6\,(2)$	$123.3\,(2)$	Catalano et al. (1984)

[a] All values are in degrees.
[b] The following abbreviations are used: ED, electron diffraction; MW, microwave spectroscopy; XD, X-ray crystallograpy; NMR (. . .), liquid-crystal NMR spectroscopy (with the nematic mesophase specified in parentheses); MO (. . .), ab initio MO calculations with geometry optimization (with the level specified in parentheses).
[c] Least-squares standard deviations are less than 0.1°.
[d] By regression from the molecular structures of polysubstituted benzene derivatives.

such discrepancies is ascertained) liquid-crystal NMR spectroscopy can hardly qualify as a reliable method to determine the structural effects of substitution in benzene derivatives.

18.3.5　Studies by quantum mechanical calculations

Ab initio MO calculations with geometry optimization give the equilibrium structure of a molecule; i.e. the structure of the isolated molecule in the hypothetical motionless state, at the minimum position of the potential energy well. However, the actual calculated bond distances depend critically on the basis set used and on the treatment of electron correlation. A detailed example, concerning some simple molecules, is found in Chapter 11, Table 11.6. For a molecule of the size of a benzene derivative, where adequate treatment of electron correlation is presently out of reach, *absolute* values of bond distances can only be obtained by using empirical 'offset' corrections (see Chapter 13, Section 13.4). Most often, however, the interest is for *relative* bond distances: for instance, one may wish to know the differences between the lengths of the *a*, *b*, and *c* bonds in a monosubstituted benzene derivative with C_{2v} symmetry. These differences can be obtained without any empirical correction, provided the calculation is done at such a level that the error for a given type of bond is constant.

As regards the internal ring angles, calculations at the HF/6-31G level for a number of monosubstituted benzene derivatives with common functional groups give values that — in several cases — agree with the best experimental results within a few tenths of a degree (Bock *et al.* 1985*a*). In some cases, however, significant discrepancies occur, especially for the ring angle α. A difference of about $2°$ between the experimental and calculated values of α occurs with cyanobenzene, a molecule for which an accurate experimental geometry is available (Casado *et al.* 1971; Portalone *et al.* 1987). Calculations at the HF/6-31G* (5D) and 6-31G** levels (Bock *et al.* 1985*b*) fail to improve the agreement with the experiment; this strongly suggests that the discrepancy in the value of α is more likely to originate in the neglect of electron correlation than in basis-set truncation.

18.3.6　Comparison of structural results obtained by different techniques

Comparing structural results obtained for the same molecule by different techniques of structure determination is of special importance in accurate structural work. As seen in Table 18.1, this is a very powerful way of uncovering systematic errors. If these are identified and properly treated, *real* structural differences may eventually show up, e.g. variations in the geometry of a molecule caused by changes in the state of aggregation.

Apart from systematic errors, the differences in physical meaning of the geometrical parameters produced by various techniques should also be kept in mind. For gas-phase studies these differences are discussed in detail elsewhere in this book (Chapter 2, Section 2.4.1; Chapter 5, Section 5.7) and will not be reviewed here. For crystallographic studies — once the systematic effects mentioned earlier (Section 18.3.3) have been properly treated — the corrected interatomic distances correspond to separations between average nuclear positions, the average being taken over the internal motions of the molecule at the temperature of the diffraction experiment. Thus the resulting structure is of the r_α type. An r_α structure is also obtained by liquid-crystal NMR spectroscopy, see Chapter 12.

Ideally, when comparing results obtained by different techniques, all geometries should be reported to a common basis, e.g. the r_e or r_α^0 structures. This would require, however, a detailed knowledge of the force field of the molecule, and is seldom done for a system of the size of a substituted benzene derivative.

The quality of the agreement obtained when accurate structural results produced by different techniques are directly compared is exemplified in Table 18.2. This table reports the geometry of the heavy-atom skeleton of fluorobenzene, as obtained by different methods in different laboratories. Since 1968 fluorobenzene has been the subject of several high-quality experimental and theoretical studies, involving most of the techniques considered in this book.

Table 18.2 shows at once that *the agreement is much better for bond angles than for bond distances*. The values of the internal ring angles from five different experiments agree within a few tenths of a degree, with the exception only of the angles α and β from one of the two liquid-crystal NMR studies. Note that the X-ray diffraction results are in agreement with those obtained by gas-phase techniques. Thus there seem to be no strong intermolecular interactions in the solid state, in agreement with the relatively low melting point of the compound, 232 K.

The *ab initio* MO calculations yield values of γ and δ that reproduce the experimental results. They are less successful, however, in reproducing the large distortion of the *ipso* region of the ring caused by the F substituent. This is probably due to the neglect of electron correlation (see Bock *et al.* 1985*b*).

As regards bond distances, the poor quality of the agreement is seen in the values of $\langle r(C-C) \rangle$ and $r(C-F)$ in Table 18.2. *Differences* between bond distances are somewhat more reliable. They consistently indicate that the *a* bonds are shorter than the *b* bonds by about 0.01 Å, and that the *b* and *c* bonds have nearly equal lengths.

It is reasonable to conclude that, for a relatively rigid system such as a substituted benzene ring, $C-C-C$ angles can be reliably determined to

Table 18.2 Internal ring angles and distance parameters of fluorobenzene as determined by different techniques[a,b]

Technique[c]	Type of structure	α	β	γ	δ	$\langle r(C-C) \rangle$	$a-b$	$b-c$	$r(C-F)$	References
MW	r_s^d	123.4	117.9	120.5	119.8	1.392	−0.012	−0.002	1.354	Nygaard et al. (1968)
ED	$\angle_a, r_g$	123.4	118.0	120.2	120.2	1.396	−0.012[e]	−0.002[e]	1.356	Portalone et al. (1984)
XD[f]	$\approx r_\alpha$	123.4	118.0	120.5	119.6	-	-	-	-	Domenicano and Murray-Rust (1979)
NMR (ZLI 1167)[g]	r_α	124.3	117.5	120.4	119.8	1.389	−0.015	−0.006	1.371	Jokisaari et al. (1981)
NMR (ZLI 1132)[g]	r_α	123.1	118.1	120.3	120.0	1.395	−0.016	+0.004	1.355	Jokisaari et al. (1981)
MO (HF/4-21G)	r_e	122.4	118.6	120.3	119.8	1.381	−0.009	−0.002	1.369	Boggs et al. (1982)
MO (HF/6-31G)	r_e	123.0	118.2	120.3	120.0	1.385	−0.011	−0.001	1.376	Bock et al. (1985a)
MO (HF/6-31G**)	r_e	122.4	118.4	120.5	119.8	1.383	−0.007	−0.001	1.331	Bock et al. (1985b)

[a] Bond distances are given in Å, angles in degrees.

[b] Experimental errors are not shown. They have been estimated in different ways by the various authors and can be found in the original papers. The present table is not aimed at showing whether or not the geometrical parameters of the benzene ring obtained by different techniques are consistent within their estimated errors, but rather at ascertaining whether they can be determined reliably to within a few tenths of a degree (for bond angles) and a few thousandths of an ångström (for bond distances). For this reason bond angles are presented with one decimal figure and bond distances with three, although in some cases one more figure appears in the original papers.

[c] See footnote b to Table 18.1 concerning abbreviations.

[d] The a coordinates of F and C2 are not substitution coordinates.

[e] Assumed from the microwave study.

[f] By regression from the molecular structures of polysubstituted benzene derivatives.

[g] The c bond distance was assumed to be 1.3976 Å.

within 0.1–0.2°. *Differences in physical meaning are apparently of marginal importance at this level of accuracy.* However, this statement applies strictly to bond angles; a similar conclusion does not apply to bond distances.

18.4 Benzene ring deformations in monosubstituted derivatives

18.4.1 Analysis of the structural changes

In a C_6H_5X molecule the distortion of the carbon skeleton of the benzene ring conforms generally to C_{2v} symmetry. Further distortion to C_s symmetry is only appreciable with some strongly asymmetric functional groups, e.g. X = OMe (Di Rienzo *et al.* 1976; Konschin 1983*b*), O H (Larsen 1979; Konschin 1983*a*), CH=CH$_2$ (Bock *et al.* 1985*c*). When C_{2v} symmetry is retained the following geometrical changes are found to occur (Domenicano *et al.* 1975*a*, 1983):

(1) *Bond distances*: the *a* bonds (see Fig. 18.2) lengthen, or shorten, with respect to unsubstituted benzene. The *b* and *c* bonds seem to be scarcely, if at all, affected.

(2) *Bond angles*: the internal ring angles α and β (and, to a lesser extent, γ and δ) deviate from the ideal value of 120°, with $\Delta\beta \cong -\Delta\alpha/2$.

The parameter that is most affected by the presence of the substituent is the angle α. Its values span a range of 13°, from 112° to 125°, depending upon the electronic properties of X. We have seen in the preceding section that, unlike C—C bond distances, C—C—C angles can be measured with considerable accuracy, within ± 0.1° in favourable cases. All this makes α an interesting indicator of electronic substituent effects.

The angle α is particularly sensitive to the σ-inductive effect of the substituent. Within a row of the periodic table it increases linearly with the electronegativity of the substituent (Fig. 18.4). The regression lines in Fig. 18.4 may be used to evaluate the electronegativity of functional groups from the observed ring deformation (Domenicano *et al.* 1975*b*; Schultz *et al.* 1981). The increase in α is associated with a decrease in β, a small increase in δ, and a shortening of the *a* bonds.

When the substituent is a π-electron donor the angle α decreases with increasing conjugation. This effect — which also involves other geometrical changes — is not as large as the previous one, but is certainly significant and well documented (Domenicano and Vaciago 1979). A striking example is given by the molecular structure of diphenylaminotriphenylmethane, Ph_3C—NPh_2, as determined by low-temperature X-ray crystallography (Hoekstra and Vos (1975); but see Domenicano and Vaciago (1979) for the interpretation of the results). In this molecule the N atom has a planar configuration. Its two phenyl groups have different conformations in the

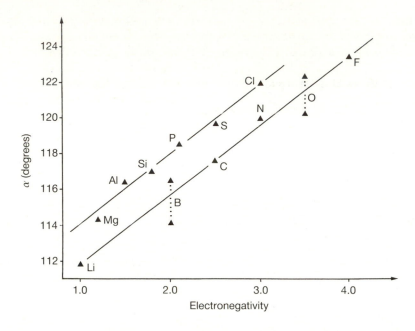

Fig. 18.4 Empirical correlation between the angle α of a monosubstituted benzene ring and the electronegativity of the substituent. After Domenicano *et al.* (1980).

Table 18.3   Effect of increasing π donation from a NR_2 substituent on the geometry of the benzene ring (Domenicano and Vaciago 1979)

Bond distances (Å)

	a	*b*	*c*	$r(C-N)$
$\tau = 75°$	1.398	1.395	1.393	1.438
$\tau = 12°$	1.411	1.392	1.393	1.405
Variation	+0.013	−0.003	0.000	−0.033

Bond angles (degrees)

	α	β	γ	δ
$\tau = 75°$	119.4	120.2	120.3	119.7
$\tau = 12°$	117.0	121.1	121.3	118.2
Variation	−2.4	+0.9	+1.0	−1.5

The effect is seen by comparing the geometries of the two phenyl groups bonded to N in Ph_3C-NPh_2; τ is the angle of twist between the plane of the ring and the coordination plane of N. The geometry of this molecule has been determined by low-temperature X-ray crystallography (Hoekstra and Vos 1975); least-squares standard deviations are about 0.001 Å for bond distances and 0.1° for bond angles.

crystal: they are twisted by 75° and 12°, respectively, out of the coordination plane of N. Their geometries are compared in Table 18.3. The different conformation strongly affects the extent of conjugation of the ring with the substituent, as indicated by the different lengths of the two C—N bonds. As regards the ring deformation, it is seen that π donation from the substituent causes the angles α and δ to decrease by 2.4° and 1.5°, respectively, and the angles β and γ to increase by 1°. These angular changes are associated with a lengthening of the a bonds by about 0.01 Å.

The changes of the ring geometry that occur when a typical π-acceptor substituent goes from the orthogonal to the planar conformation are apparently much smaller than those mentioned above. A detailed study has been carried out recently for nitrobenzene, by *ab initio* MO calculations at the HF/6-31G and 6-31G* levels as well as X-ray diffraction analysis for some

Table 18.4 Effect of increasing π abstraction by the NO_2 substituent on the geometry of the benzene ring (Domenicano *et al.* 1990)

Calculated bond distances (Å)

	a	b	c	$r(C-N)$
$\tau = 90°$	1.378	1.385	1.386	1.462
$\tau = 0°$	1.383	1.383	1.387	1.458
Variation	+0.005	−0.002	+0.001	−0.004[a]

Calculated bond angles (degrees)

	α	β	γ	δ
$\tau = 90°$	122.8	118.3	120.1	120.3
$\tau = 0°$	122.3	118.5	120.1	120.4
Variation	−0.5	+0.2	0.0	+0.1

Experimental bond angles (degrees)

	α	β	γ	δ
$\tau = 84°$	123.4	117.8	120.1	120.7
$\tau = 2-21°$	122.7	118.1	120.3	120.5
Variation	−0.7	+0.3	+0.2	−0.2

The effect is seen by comparing the geometries of the benzene ring in the planar ($\tau = 0°$) and orthogonal ($\tau = 90°$) conformations of nitrobenzene, as determined by *ab initio* MO calculations at the HF/6-31G* level. Bond angles for the two conformations have also been determined experimentally, by X-ray diffraction analysis of a number of properly selected derivatives (see the original paper for details). The uncertainties of the experimental angles are 0.1-0.2°.

[a] This difference is less reliably determined than the others, as it shows some dependence on the basis set used. The experimental value is −0.005(3) Å.

properly selected derivatives (Domenicano *et al.* 1990). The results produced by the two techniques consistently indicate that the angular changes are minute and virtually confined to the *ipso* region (see Table 18.4). The variation of the angle α which occurs in going from the orthogonal to the planar conformation is four to five times smaller than the corresponding variation in Ph_3C-NPh_2.

Also the variation in the length of the $C-N$ bond is much smaller, *ca.* -0.005 Å *vs.* -0.033 Å. This indicates that the actual extent of conjugation in the planar equilibrium conformation of nitrobenzene is small, in agreement with reactivity data: the resonance parameter σ_R^0 has a large negative value, -0.52, for a strong π donor like NMe_2, while the values for NO_2 and other typical π-acceptor substituents are all close to $+0.15$ (Ehrenson *et al.* 1973).

A statistical analysis of the geometrical variance of a large sample of monosubstituted benzene rings, mostly studied by X-ray crystallography (Domenicano *et al.* 1983), has substantially increased our knowledge of the ring deformations caused by substitution. The sample was selected so as to cover a wide spectrum of substituent effects. Since accurate structural data were seldom available for derivatives with simple functional groups, most molecules in the sample had large, relatively unusual molecular fragments as substituents. The results of the statistical analysis are outlined below.

The angular variance of monosubstituted benzene rings with first-row substituents is fully described by two orthogonal components of distortion, involving angular changes in different ratios. The distortion that accounts for most of the variance is a concerted change of the internal angles α, β, and, to a lesser extent, δ:

$$\Delta\alpha : \Delta\beta : \Delta\gamma : \Delta\delta = 1.00 : -0.69 : 0.06 : 0.26°. \qquad (18.3)$$

This distortion is related to the electronegativity of the substituent. When bond distance variation is included in the model, a change of the *a* bond distances is also involved:

$$\Delta\alpha : \Delta a = 1.00° : -0.0027 \text{ Å}. \qquad (18.4)$$

The second distortion involves mainly the internal angles γ and δ:

$$\Delta\alpha : \Delta\beta : \Delta\gamma : \Delta\delta = 0.08 : 0.19 : -0.73 : 1.00°, \qquad (18.5)$$

and appears to be controlled to a large extent by the π-donor/acceptor strength of the substituent.

Only the first distortion is of importance for benzene derivatives with second-row substituents.

Note how *the separation of the two distortions originates directly from the statistical treatment, without any previous chemical assumption.* Also note that increasing the extent of conjugation between the ring and the substituent, as in going from $\tau = 75°$ to $\tau = 12°$ in Table 18.3, corresponds to a variation of the ring geometry that consists of a blend of the two distortions. This is due to the fact that the shortening of the C—N bond caused by conjugation is associated with a decrease of the σ-electron-withdrawing properties of the substituent.

The existence of only two orthogonal components of angular distortion implies that the four different internal angles of a benzene ring of C_{2v} symmetry are related by *two* equations of constraint. One of these is the exact geometrical constraint expressed by eqn (18.1). The other is entirely empirical, and is found to be:

$$\Delta\beta = -0.591(7)\Delta\alpha - 0.301(15)° \qquad (18.6a)$$

or

$$\Delta\beta = -0.615(11)\Delta\alpha - 0.384(19)°, \qquad (18.6b)$$

for first- and second-row substituents, respectively. (The numbers in parentheses are the standard deviations of the coefficients, given as units in the last digit). These correlations are of high quality, as seen from Fig. 18.5.

The physical meaning of eqn (18.6) is that the angular distortion caused by a substituent at the *ipso* position of the ring is largely relaxed by a distortion of opposite sign at the *ortho* positions. The difference between the calculated coefficients for first- and second-row substituents is small and of marginal importance. Note that in both cases *unsubstituted benzene* ($\Delta\alpha = \Delta\beta = 0$) *definitely lies off the regression line.*

Unfortunately only limited information about the changes in C—C bond distances caused by substitution could be obtained from the statistical analysis. A distortion consisting of a simultaneous lengthening/shortening of all the C—C bonds of the ring is indicated as statistically significant by principal-component analysis. But most of it arises from uncorrected thermal-motion effects (which may differ considerably from one molecule to another) and other inhomogeneities of the sample.

It is reasonable to expect that the structural effects of substitution are not confined to the carbon skeleton of the ring, but extend to the C—H bonds. However, the accurate determination of the geometrical changes of these bonds is beyond the possibilities of some of the experimental techniques of structure determination, such as electron diffraction and X-ray crystallography. *Ab initio* MO calculations at the HF/6-31G level (Bock *et al.* 1985a) indicate that the two C—H bonds in closest proximity of the substituent are the most affected, but the changes of their length with respect

(a)

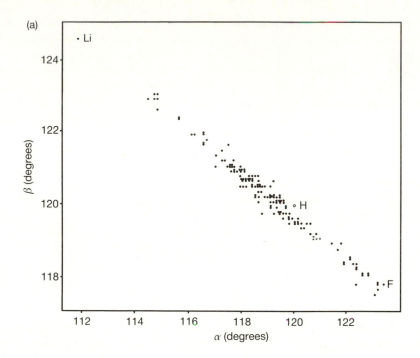

(b)

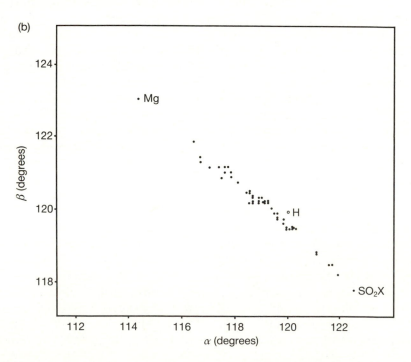

to unsubstituted benzene are small, a few thousandths of an ångström at most.

18.4.2 Interpretation of the structural changes

A qualitative interpretation of the distortions of the *ipso* region of the benzene ring caused by substitution was proposed in terms of hybridization effects (Carter *et al.* 1966; Nygaard *et al.* 1968). In *sp*-, *sp²*-, and *sp³*-hybridized carbons the *p* character tends to concentrate in those hybrid orbitals that point towards more electronegative substituents (Walsh 1947; Bent 1961). This follows from the fact that *p* electrons are held more loosely than *s* electrons, since they penetrate the inner core less than *s* electrons do, thus experiencing a smaller effective nuclear charge. In the case of a benzene derivative a shift of electron density from the *ipso* carbon to a σ-electron-withdrawing substituent is best accomplished through an increase in the *p* character of the hybrid orbital that points towards the substituent. This implies a decrease in the *p* character of the other two hybrid orbitals and leads, therefore, to a shortening of the *a* bonds and an increase of the angle α. The interpretation is easily extended to substituents which interact with the ring at the π level (Domenicano *et al.* 1975*a*).

A pictorial rationalization of the benzene ring deformations has been given (Domenicano *et al.* 1975*a*) in terms of the VSEPR model (Gillespie 1972; Gillespie and Hargittai 1991). According to this model a σ-electron-withdrawing substituent polarizes the σ_{C-X} molecular orbital, so that the bonding electron pair takes up a decreasing amount of space in the valence shell of the *ipso* carbon and interacts less strongly with the two neighbouring σ-bonding pairs. This causes the *a* bonds to become shorter, and the angle α to increase with respect to the value, α_0, expected for a substituent having the same electronegativity as the ring (Fig. 18.6(a, b)). Note that α_0 may differ from 120°, since the spatial requirements of a bonding electron pair are not necessarily the same for a $C-X$ bond and a $C-H$ bond. The reverse pattern of effects occurs with a σ-electron-releasing substituent (see Fig. 18.6(c)).

If the π system of the substituent interacts with the π system of the ring, there will be a tendency for the electron density of σ_{C-X} to increase, due to the shorter interatomic distance. This will, in turn, increase the repulsive interactions between the electron pairs of the $C-X$ and *a* bonds, so that the angle α becomes smaller, and the *a* bonds longer, with respect to the values that would be found with the same substituent in the perpendicular conformation.

Fig. 18.5 Scatterplot of the angle β against α for monosubstituted benzene rings with (a) first-row substituents and (b) second-row substituents. The correlation coefficients are -0.991 and -0.993 on 149 and 50 data points, respectively. The reference position of unsubstituted benzene is marked with H. From Domenicano *et al.* (1983).

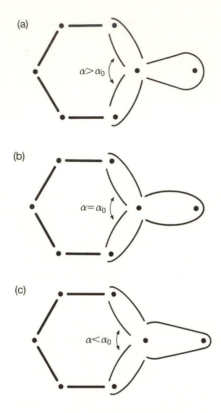

Fig. 18.6 Consequences of valence-shell electron-pair repulsions on the angle α of a monosubstituted benzene ring: (a) a σ-electron-withdrawing substituent; (b) a substituent having the same electronegativity as the ring; (c) a σ-electron-releasing substituent. From Domenicano *et al.* (1975*a*).

18.5   Benzene ring deformations in polysubstituted derivatives

A substantial body of experimental evidence indicates that, in most polysubstituted derivatives of benzene, the angular distortion of the ring may be interpreted, at least to a first approximation, as arising from the superposition of separate, independent contributions from each substituent (see, for example, Domenicano *et al.* 1975*a*; Domenicano and Murray-Rust 1979; Norrestam and Schepper 1981). Additional evidence—relative to bond distances as well as angles—has been obtained more recently by *ab initio* MO calculations at the HF/6–31G level (George *et al.* 1986; Bock *et al.* 1987).

   Within this approximation the contribution of a substituent X to the angular distortion of a polysubstituted benzene ring is expressed by a set of

Table 18.5 Angular substituent parameters for important functional groups

(a) From gas-phase studies[a]

Substituent	$\Delta\alpha$	$\Delta\beta$	$\Delta\gamma$	$\Delta\delta$
F	3.4	−2.0	0.3	0.0
NO$_2$	3.4	−2.3	0.5	0.2
CN	1.8	−1.2	0.3	0.0
NH$_2$	−0.6	0.1	0.7	−1.1

(b) From solid-state studies[b]

Substituent	$\Delta\alpha$	$\Delta\beta$	$\Delta\gamma$	$\Delta\delta$
F	3.4	−2.0	0.5	−0.4
NO$_2$	2.9	−1.9	0.3	0.4
Cl	1.9	−1.4	0.6	−0.2
NH$_3^+$	1.8	−1.2	0.4	−0.1
SO$_2$Me	1.6	−1.3	0.2	0.6
CN	1.1	−0.8	0.3	−0.1
OMe	0.2	−0.6	1.1	−1.1
OH	0.2	−0.4	0.6	−0.6
COOH	0.1	−0.2	0.1	0.2
NHAc	−0.1	−0.3	0.7	−0.6
N=NR	−0.1	−0.3	0.5	−0.4
COOR	−0.6	0.2	0.3	−0.3
COMe	−1.0	0.4	0.2	−0.3
NH$_2$	−1.2	0.2	1.0	−1.3
CH=CHR	−1.8	0.8	0.3	−0.4
Me	−1.9	1.0	0.4	−0.8
Ph	−2.3	1.0	0.6	−0.9
NMe$_2$	−2.4	0.6	1.4	−1.7

[a] Only those substituents have been included for which the four angular parameters are consid:red to be reliable to within 0.2–0.3° (estimated by the present author, based on the agreement between results obtained by different techniques). For F and CN, the values given are taken from Domenicano (1988); they are averages of results obtained by microwave spectroscopy and electron diffraction. The values for NH$_2$ are from a study by microwave spectroscopy (Lister *et al.* 1974); those for NO$_2$ from a recent study by electron diffraction (Domenicano *et al.* 1990).

[b] After Domenicano and Murray-Rust (1979). The angular parameters have been derived by a least-squares procedure from many structural results on polysubstituted benzene derivatives, studied almost exclusively by X-ray crystallography. For planar polyatomic substituents the values given refer to conformations where the dihedral angle between the plane of the ring and that of the substituent is never greater than 42°, and usually much less. C_{2v} symmetry has been assumed throughout. Standard deviations are 0.1–0.2°.

angular parameters, defined as the deviations from 120° caused by X in the ring angles of C_6H_5X. If the distortion in C_6H_5X conforms to C_{2v} symmetry the angular parameters will be $\Delta\alpha_x$, $\Delta\beta_x$, $\Delta\gamma_x$, and $\Delta\delta_x$. These four quantities are not independent, due to the exact geometrical constraint expressed by eqn (18.1), and, further, to the empirical constraint of eqn (18.6).

Ideally, the angular parameters of a functional group should be determined by direct measurement on the monosubstituted derivative in the gas phase. Unfortunately only very few of these molecules have so far been studied with the necessary accuracy (see Table 18.5(a)).

By contrast, the geometries of many polysubstituted derivatives of benzene with common functional groups have been determined by X-ray crystallography. While the accuracy is generally low for bond distances, it is often adequate for angles. By using multiple regression it has proved possible to derive from the observed ring angles the angular parameters of many functional groups (Domenicano and Murray-Rust 1979; Norrestam and Schepper 1981). Some of the values obtained are reported in Table 18.5(b). The agreement with the gas-phase results of Table 18.5(a) is reasonably good – but see Section 18.7 for a possible origin of the discrepancies.

Table 18.6 Breakdown of the rule of additivity of angular distortions in (a) *p*-nitroaniline and (b) *p*-diaminobenzene[a]

(a) *p*-Nitroaniline (Colapietro *et al.* 1982)

Parameter	Expected value			Experimental value (crystal)
	I[b]	II[c]	III[d]	
∠C—C(NO$_2$)—C	122.3	121.6	121.9	120.9(2)
∠C—C(NH$_2$)—C	119.6	119.2	119.7	118.4(2)
r(C—NO$_2$)	1.486	–	–	1.434(2)
r(C—NH$_2$)	1.402	–	–	1.355(2)

(b) *p*-Diaminobenzene (Colapietro *et al.* 1987)

Parameter	Expected value[e]	Experimental value (free molecule)
∠C—C(NH$_2$)—C	118.3	119.8 ± 0.2
r(C—NH$_2$)	1.402	1.424 ± 0.005

[a] Bond distances are given in Å, angles in degrees.
[b] Obtained by superimposing the gas-phase structures of aniline (Lister *et al.* 1974) and nitrobenzene (Domenicano *et al.* 1990).
[c] Based on the angular parameters of Table 18.5(b).
[d] Based on the angular parameters of Norrestam and Schepper (1981), as given in Table 5 of the original paper.
[e] Obtained by superimposing on itself the gas-phase structure of aniline (Lister *et al.* 1974).

The angular parameters of Table 18.5 can be used to predict the ring angles in polysubstituted derivatives. In many cases the predicted angles agree with the experimental results within a few tenths of a degree.

Systematic deviations from additivity, amounting to no more than $1-2°$, have been found to occur with some benzene derivatives where two or more substituents interact through the benzene ring (see, for example, Colapietro *et al.* 1982; Párkányi and Kálmán 1984; Anulewicz *et al.* 1987; Colapietro *et al.* 1987; Maurin and Krygowski 1988). A reference case is *p*-nitroaniline, where a strong π donor is *para* to a typical π acceptor. Here the differences between experimental and predicted ring angles (Table 18.6(a)) support the view that π donation from the amino group is enhanced by the nitro group; i.e. that the canonical form 1 contributes substantially to the structure of the molecule. The opposite pattern of deviations from additivity is shown by the gas-phase structure of *p*-diaminobenzene (Table 18.6(b)). This fits the idea that π donation from the amino group is reduced when a second amino group is placed in the *para* position. Note that, with both molecules, the angular distortions are consistent with the changes occurring in the lengths of the $C-N$ bonds. These changes are certainly large enough to overcome the differences in physical meaning, arising from the different techniques used for structure determination.

$$H_2N = \!\!\!\!=\!\!\!\!= NO_2^-$$

1

An analysis of bond distance changes in polysubstituted benzene rings has also been carried out (Norrestam and Schepper 1981), based again on the hypothesis that each substituent gives a separate contribution to the ring deformation. Here, however, the rather small value of the effects and the lack of adequate corrections for systematic errors in most of the experimental data used make the results of the analysis much less reliable than that with bond angles. It will certainly be appropriate to repeat the analysis when more accurate experimental information will become available.

A specific class of polysubstituted derivatives of benzene has been the subject of extensive gas-phase electron diffraction work during the last decade. These are the symmetrically *para*-disubstituted derivatives, $1,4-C_6H_4X_2$, that have been studied in Budapest and Rome; an analysis of the results has appeared (Domenicano 1988).

In Section 18.3.2 it has been shown that the symmetrically *para*-disubstituted derivatives of benzene are better suited for electron diffraction studies than the corresponding monosubstituted derivatives. At the same time in these molecules some of the structural effects of substitution may

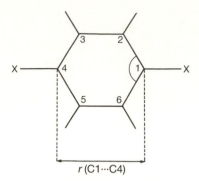

Fig. 18.7 Numbering of the ring carbons in a symmetrically *para*-disubstituted benzene derivative.

superimpose and become more pronounced. This is, for example, the case of any change occurring in the C1 · · · C4 non-bonded separation; also the changes of the ring angles at the *ipso* and *para* positions may reinforce each other, since they have equal sign for many substituents.

In these studies, the carbon skeleton of the ring has been assumed to have D_{2h} symmetry, which implies that only three parameters are required to describe its geometry. These may be chosen in several ways; a choice that is particularly convenient for interpreting the structural effects of substitution is the following (see Fig. 18.7): (i) the internal angle at the place of substitution, $\angle C2-C1-C6$; (ii) the 1–4 non-bonded separation, $r(C1 · · · C4)$; and (iii) the difference between the lengths of the two non-equivalent C–C bonds of the ring, $\Delta(C-C) = r(C1-C2) - r(C2-C3)$.

The first two parameters are determined accurately by electron diffraction, as their variation is associated with changes in several of the longest atom–atom separations. Unfortunately $\Delta(C-C)$ is poorly determined, since $r(C1-C2)$ and $r(C2-C3)$ have very close values.

The parameters $\angle C2-C1-C6$ and $r(C1 · · · C4)$ are linearly related, as shown by the scatterplot of Fig. 18.8. The gradual decrease of $\angle C2-C1-C6$ from 123.5° (X = F) to 115.7° (X = SiMe₃) is accompanied by a simultaneous increase of $r(C1 · · · C4)$, from 2.71 Å to 2.90 Å. Note that the point corresponding to unsubstituted benzene lies on the regression line.

The concerted change of $\angle C2-C1-C6$ and $r(C1 · · · C4)$ indicates that while some substituents (like F) are able to distort the benzene ring by *pushing* C1 and C4 towards the ring centre, other substituents (like SiMe₃) cause the opposite type of distortion, by *pulling* C1 and C4 away from the ring centre. The much smaller variation of $\Delta(C-C)$, ~0.03 Å, as compared to 0.19 Å for $r(C1 · · · C4)$, indicates that the relatively large distortion of the ring that takes place along the C1 · · · C4 axis is accompanied by only

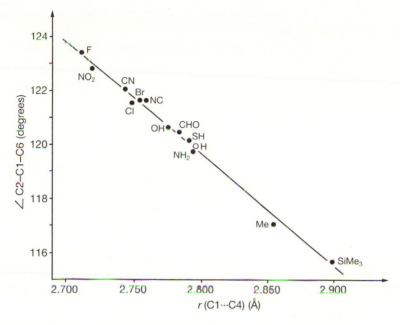

Fig. 18.8 Scatterplot of the angle C2−C1−C6 against the non-bonded distance r(C1 · · · C4) for symmetrically *para*-disubstituted benzene rings. The correlation coefficient is -0.997 on 12 data points, all from gas-phase studies. The reference position of unsubstituted benzene is marked with H. After Domenicano (1988).

minor changes in the lengths of the C−C bonds. This variation of the C−C bonds is interesting, but − in our opinion − the information at hand, including that from solid-state studies, is not accurate enough to warrant a detailed analysis. Nevertheless, an interpretation has been attempted (see Krygowski 1984).

18.6 Structural *vs.* electronic substituent effects

The interpretation of benzene ring deformations in terms of electronic interactions between the ring and the substituent makes it interesting to see how the various ring angles depend on inductive and resonance effects. These are usually quantified by means of empirical parameters derived from reactivity data, such as the *inductive parameter* σ_I (for which a single, universally accepted scale is in use) and the *resonance parameter* σ_R (for which a number of different scales exist − σ_R^0, σ_R^+, σ_R^-, $\sigma_{R(BA)}$ . . . − to be used with different types of reactions; see, for example, Ehrenson *et al.* (1973)).

By regressing the angular parameters of Table 18.5(b) against the corresponding σ_I and σ_R^0 values it has been possible to ascertain the role that

inductive and resonance effects have in determining the ring deformations. This procedure leads to the following conclusions (Domenicano and
Murray-Rust, unpublished observations (1981); Domenicano (1988)):

(1) the relatively large angular changes that occur at the *ipso* and *ortho*
    positions of the benzene ring ($\Delta\alpha$, $\Delta\beta$) are controlled primarily by the
    inductive effect of the substituent;

(2) the minute changes that occur at the *meta* position, $\Delta\gamma$, are controlled
    almost entirely by the resonance effect;

(3) the rather small changes at the *para* position, $\Delta\delta$, are controlled by a
    $\sim 1{:}1$ blend of the two effects.

The quality of the fits that are obtained is exemplified in Fig. 18.9, where
the values of $\Delta\alpha$ from Table 18.5(b) are plotted against the inductive
parameter σ_I.

These findings are consistent with the results of the statistical analysis of
monosubstituted benzene rings described in Section 18.4.1 (see eqns (18.3)
and (18.5)). Note that the two approaches are entirely different, and are
based on different sets of experimental data.

A similar analysis has been carried out by Krygowski *et al.* (1986), using
calculated ring angles (HF/STO-3G level), and adopting Mulliken σ and π
electron charge densities at the respective ring carbons as explanatory
variables. The results give further support to the present conclusions, which

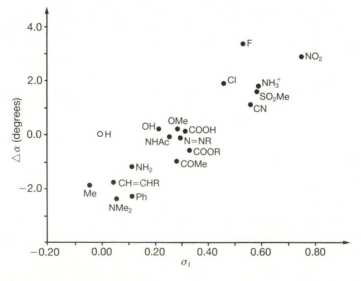

Fig. 18.9 Plot of the angular parameter $\Delta\alpha$ against the inductive parameter σ_I for monosubstituted benzene derivatives in the solid state. After Domenicano and Murray-Rust (1979).

are based entirely on experimental data.

Structural substituent effects have been used repeatedly to investigate inductive and resonance effects in benzene derivatives (see, for example, Anulewicz *et al.* 1987; Krygowski 1987; Domenicano *et al.* 1990; Krygowski and Turowska-Tyrk 1990). Geometrical parameters such as those of Table 18.5 are a particularly valuable addition to traditional reactivity parameters, since they:

(1) measure the effect of a substituent not on a remote reaction centre, but directly on the carbon skeleton of the benzene ring in the ground state;

(2) do not depend on the choice of appropriate reaction series;

(3) are not disturbed by solvent effects, insofar as they are measured in the free molecule;

(4) refer to a fixed, well-known intermolecular environment, if they are measured in the crystal;

(5) can be related to the conformation of the substituent, since rotation about the C—X bond may affect the extent of conjugation and hence the ring geometry.

18.7   Exploring the role of intermolecular interactions

The interaction of a substituent with the benzene ring should not be considered as a mere intramolecular effect, unless the molecule is in the gaseous phase. Intermolecular forces in a condensed phase may influence the ring–substituent interaction. This is more likely to occur with π-donor or π-acceptor functional groups, since π electrons are more easily polarized than σ electrons by interactions with adjacent molecules.

As we have seen in the previous sections, any change in the electronic interaction of a substituent with the ring is associated with a change of the ring geometry. The influence of intermolecular interactions on substituent effects may thus be investigated by comparing the geometry of the benzene ring in the free molecule with that in the crystal molecule (Domenicano 1988).

With respect to traditional solution studies, based on the measurement of suitable physical parameters in different solvents, this approach has the advantage of providing results that refer to a well-defined intermolecular environment. There is a major difficulty, however: the changes of the ring geometry that occur in going from the free molecule to the crystal molecule are minute and hard to measure accurately. An appropriate assessment of the various systematic effects inherent in each technique of structure determination should be made prior to the comparison. Of course, it is preferable to compare bond angles rather than bond distances.

The internal ring angles at the place of substitution, $\angle C-C(X)-C$, for a number of benzene derivatives that have been studied with reasonable accuracy in the gaseous and solid states, are presented in Table 18.7. All molecules in this table possess at least one strong π-donor or π-acceptor functional group. In the crystal they are held together by either strong dipole–dipole interactions or networks of intermolecular hydrogen bonds. While the ring angles of the free molecules all come from direct experiment, those of the crystal molecules are often based on angular parameters derived by regression techniques. This does not imply lower accuracy; it only implies that the crystal environment of the functional group should be taken as the 'average environment' experienced by that group in the crystals of the various species used in the regression.

Inspection of Table 18.7 shows that the angle $C-C(X)-C$ is always slightly larger in the free molecule as compared with the crystal molecule. Most of the individual differences are very small, but a pattern emerges when they are taken together.

In terms of Pauling's V B model, a natural explanation is that the contribution of polar canonical forms like, for example, **2, 3, 4,** and **5** is enhanced by the crystal field. Indeed, these forms make the dipole–dipole interactions stronger; also, they give rise to stronger intermolecular hydrogen bonds, due to the increased acidity of the NH_2 or OH protons (Colapietro *et al.* 1987; Portalone *et al.* 1987; Domenicano *et al.* 1988, 1990). Note that in molecules such as, for example, fluorobenzene (see Table 18.5), chlorobenzene (results reviewed by Domenicano (1988)), *p*-dichlorobenzene (Schultz *et al.* 1980), and *p*-dimethylbenzene (Domenicano *et al.* 1979) the effect is not observed: the $C-C(X)-C$ angles measured in the gas phase have essentially the same values as those based on the angular parameters of Table 18.5(b). In these species the functional groups have average crystal environments that involve weaker interactions, as compared with the functional groups of Table 18.7.

It is clear, however, that the quality of the structural information at hand is still rather poor for this type of study. Other, more accurate experimental studies, both in the gaseous phase and in the crystal, and also theoretical calculations on appropriate model systems, are needed for a better understanding of the role of intermolecular interactions in the structural chemistry of these systems.

Table 18.7 Effect of strong intermolecular interactions between functional groups on the benzene ring geometry

Compound	Internal ring angle $C-C(X)-C$ (degrees)		Relevant references
	Free molecule[a]	Crystal molecule	
Cyanobenzene	121.82(5)[b], 121.9 ± 0.3	121.1(2)[c]	Casado et al. (1971); Portalone et al. (1987)
p-Dicyanobenzene	122.1 ± 0.2	121.6(1)[d], 121.3(2)[e], 121.0(3)[c]	Guth et al. (1982); Colapietro et al. (1984b)
Nitrobenzene	123.4 ± 0.3	122.9(2)[c], 122.7(2)[f]	Domenicano et al. (1990)
Aniline	119.4 ± 0.2[b]	118.8(2)[c], 118.7(2)[f]	Lister et al. (1974)
p-Diaminobenzene	119.8 ± 0.2	117.9(1)[d], 117.5(3)[c], 117.9(3)[f]	Colapietro et al. (1987)
Phenol	120.9 ± 0.4[b], 121.4 ± 0.2	120.2(2)[c], 120.2(1)[f]	Larsen (1979); Domenicano et al. (1988)
p-Dihydroxybenzene	120.7 ± 0.2	119.7(1)[d], 119.6(3)[c], 119.8(2)[f]	Domenicano et al. (1988)
sym-Trihydroxybenzene	122.4 ± 0.2	121.4(3)[c], 121.5(2)[f]	Domenicano et al. (1988)

[a] By gas-phase electron diffraction (unless otherwise specified).
[b] By microwave spectroscopy.
[c] Based on the angular parameters of Table 18.5(b).
[d] Direct determination by X-ray crystallography.
[e] Direct determination by neutron crystallography.
[f] Based on the angular parametes of Norrestam and Schepper (1981), as given in Table 5 of the original paper.

18.8   Acknowledgements

I wish to express my gratitude to my colleagues István Hargittai and Gustavo Portalone for their helpful comments on the first version of this chapter, and to Mrs Clara Marciante for the preparation of the figures.

References

Allen, F. H. (1980). *Acta Crystallogr.*, **B36**, 81–96.
Allen, F. H. (1981). *Acta Crystallogr.*, **B37**, 890–900.
Allen, F. H. (1984). *Acta Crystallogr.*, **B40**, 64–72.
Almenningen, A., Brunvoll, J., Popik, M. V., Sokolkov, S. V., Vilkov, L. V., and Samdal, S. (1985). *J. Mol. Struct.*, **127**, 85–94.
Amir-Ebrahimi, V., Choplin, A., Demaison, J., and Roussy, G. (1981). *J. Mol. Spectrosc.*, **89**, 42–52.
Anulewicz, R., Häfelinger, G., Krygowski, T. M., Regelmann, C., and Ritter, G. (1987). *Z. Naturforsch.*, **42b**, 917–27.
Bent, H. A. (1961). *Chem. Rev.*, **61**, 275–311.
Bock, C. W., Trachtman, M., and George, P. (1985a). *J. Mol. Struct. (Theochem)*, **122**, 155–72.
Bock, C. W., Trachtman, M., and George, P. (1985b). *J. Comput. Chem.*, **6**, 592–7.
Bock, C. W., Trachtman, M., and George, P. (1985c). *Chem. Phys.*, **93**, 431–43.
Bock, C. W., Domenicano, A., George, P., Hargittai, I., Portalone, G., and Schultz, Gy. (1987). *J. Phys. Chem.*, **91**, 6120–7.
Boggs, J. E., Pang, F., and Pulay, P. (1982). *J. Comput. Chem.*, **3**, 344–53.
Brunvoll, J., Colapietro, M., Domenicano, A., Marciante, C., Portalone, G., and Hargittai, I. (1984). *Z. Naturforsch.*, **39b**, 607–9.
Carter, O. L., McPhail, A. T., and Sim, G. A. (1966). *J. Chem. Soc. (A)*, 822–38.
Casado, J., Nygaard, L., and Sørensen, G. O. (1971). *J. Mol. Struct.*, **8**, 211–24.
Catalano, D., Forte, C., and Veracini, C. A. (1984). *J. Magn. Reson.*, **60**, 190–8.
Colapietro, M., Domenicano, A., Marciante, C., and Portalone, G. (1982). *Z. Naturforsch.*, **37b**, 1309–11.
Colapietro, M., Domenicano, A., Marciante, C., and Portalone, G. (1984a). *Z. Naturforsch.*, **39b**, 1361–7.
Colapietro, M., Domenicano, A., Portalone, G., Schultz, Gy., and Hargittai, I. (1984b). *J. Mol. Struct.*, **112**, 141–57.
Colapietro, M., Domenicano, A., Portalone, G., Schultz, Gy., and Hargittai, I. (1987). *J. Phys. Chem.*, **91**, 1728–37.
Costain, C. C. (1958). *J. Chem. Phys.*, **29**, 864–74.
Costain, C. C. (1966). *Trans. Am. Crystallogr. Assoc.*, **2**, 157–64.
Cox, A. P., Ewart, I. C., and Stigliani, W. M. (1975). *J. Chem. Soc. Faraday Trans. II*, **71**, 504–14.
Cradock, S., Muir, J. M., and Rankin, D. W. H. (1990). *J. Mol. Struct.*, **220**, 205–15.
Di Rienzo, F., Domenicano, A., Portalone, G., and Vaciago, A. (1976). Presentation at the *2nd Yugoslav-Italian Crystallographic Conference*, Dubrovnik,

Yugoslavia. *Izvj. Jugoslav. Centr. Kristalogr.*, **11**, A102-4.

Domenicano, A. (1988). Substituted benzene derivatives. In *Stereochemical applications of gas-phase electron diffraction* (ed. I. Hargittai and M. Hargittai), Part B, pp. 281-324. VCH, New York.

Domenicano, A. and Murray-Rust, P. (1979). *Tetrahedron Lett.*, 2283-6.

Domenicano, A. and Vaciago, A. (1979). *Acta Crystallogr.*, **B35**, 1382-8.

Domenicano, A., Vaciago, A., and Coulson, C. A. (1975a). *Acta Crystallogr.*, **B31**, 221-34.

Domenicano, A., Vaciago, A., and Coulson, C. A. (1975b). *Acta Crystallogr.*, **B31**, 1630-41.

Domenicano, A., Schultz, Gy., Kolonits, M., and Hargittai, I. (1979). *J. Mol. Struct.*, **53**, 197-209.

Domenicano, A., Hargittai, I., and Schultz, Gy. (1980). Presentation at the *8th Austin Symposium on Molecular Structure*, Austin, Texas, USA. Abstracts, pp. 17-21.

Domenicano, A., Murray-Rust, P., and Vaciago, A. (1983). *Acta Crystallogr.*, **B39**, 457-68.

Domenicano, A., Hargittai, I., Portalone, G., and Schultz, Gy. (1988). *Acta Chem. Scand.*, **A42**, 460-2.

Domenicano, A., Schultz, Gy., Hargittai, I., Colapietro, M., Portalone, G., George, P., and Bock, C. W. (1990). *Struct. Chem.*, **1**, 107-22.

Ehrenson, S., Brownlee, R. T. C., and Taft, R. W. (1973). A generalized treatment of substituent effects in the benzene series: A statistical analysis by the dual substituent parameter equation. In *Progress in physical organic chemistry*, Vol. 10 (ed. A. Streitwieser Jr. and R. W. Taft), pp. 1-80. Wiley-Interscience, New York.

George, P., Bock, C. W., and Trachtman, M. (1986). *J. Mol. Struct. (Theochem)*, **137**, 387-90.

Gillespie, R. J. (1972). *Molecular geometry*. Van Nostrand-Reinhold, London.

Gillespie, R. J. and Hargittai, I. (1991). *The VSEPR model of molecular geometry*. Allyn and Bacon, Boston.

Guth, H., Heger, G., and Drück, U. (1982). *Z. Kristallogr.*, **159**, 185-90.

Hoekstra, A. and Vos, A. (1975). *Acta Crystallogr.*, **B31**, 1716-21 and 1722-9.

Høg, J. H. (1971). *A study of nitrobenzene*. Thesis, University of Copenhagen, Danemark.

Jokisaari, J., Kuonanoja, J., Pulkkinen, A., and Väänänen, T. (1981). *Mol. Phys.*, **44**, 197-208.

Keidel, F. A. and Bauer, S. H. (1956). *J. Chem. Phys.*, **25**, 1218-27.

Konschin, H. (1983a). *J. Mol. Struct. (Theochem)*, **92**, 173-89.

Konschin, H. (1983b). *J. Mol. Struct. (Theochem)*, **105**, 213-24.

Krygowski, T. M. (1984). *J. Chem. Res. (S)*, 238-9.

Krygowski, T. M. (1987). *J. Chem. Res. (S)*, 120-1.

Krygowski, T. M. and Turowska-Tyrk, I. (1990). *Collect. Czech. Chem. Commun.*, **55**, 165-79.

Krygowski, T. M., Häfelinger, G., and Schüle, J. (1986). *Z. Naturforsch.*, **41b**, 895-903.

Larsen, N. W. (1979). *J. Mol. Struct.*, **51**, 175-90.

Lister, D. G., Tyler, J. K., Høg, J. H., and Larsen, N. W. (1974). *J. Mol. Struct.*, **23**, 253–64.

Maurin, J. and Krygowski, T. M. (1988). *J. Mol. Struct.*, **172**, 413–21.

Michel, F., Nery, H., Nosberger, P., and Roussy, G. (1976). *J. Mol. Struct.*, **30**, 409–15.

Norrestam, R. and Schepper, L. (1981). *Acta Chem. Scand.*, **A35**, 91–103.

Nygaard, L., Bojesen, I., Pedersen, T., and Rastrup-Andersen, J. (1968). *J. Mol. Struct.*, **2**, 209–15.

Párkányi, L. and Kálmán, A. (1984). *J. Mol. Struct.*, **125**, 315–20.

Portalone, G., Schultz, Gy., Domenicano, A., and Hargittai, I. (1984). *J. Mol. Struct.*, **118**, 53–61.

Portalone, G., Domenicano, A., Schultz, Gy., and Hargittai, I. (1987). *J. Mol. Struct.*, **160**, 97–107.

Rankin, D. W. H. (1988). Combined application of electron diffraction and liquid crystal NMR spectroscopy. In *Stereochemical applications of gas-phase electron diffraction* (ed. I. Hargittai and M. Hargittai), Part A, pp. 451–82. VCH, New York.

Schäfer, L., Ewbank, J. D., Siam, K., Chiu, N.-S., and Sellers, H. L. (1988). Molecular orbital constrained electron diffraction (MOCED) studies: The concerted use of electron diffraction and quantum chemical calculations. In *Stereochemical applications of gas-phase electron diffraction* (ed. I. Hargittai and M. Hargittai), Part A, pp. 301–19. VCH, New York.

Schomaker, V. and Trueblood, K. N. (1968). *Acta Crystallogr.*, **B24**, 63–76.

Schultz, Gy., Hargittai, I., and Domenicano, A. (1980). *J. Mol. Struct.*, **68**, 281–92.

Schultz, Gy., Hargittai, I., and Seip, R. (1981). *Z. Naturforsch.*, **36a**, 669–73.

Walsh, A. D. (1947). *Discuss. Faraday Soc.*, **2**, 18–25.

19

Effect of crystal environment on molecular structure

Joel Bernstein

19.1   Introduction

In a book on accurate molecular structures, it seems appropriate to address the problem of the effect of crystal forces on molecular geometry. Perhaps we should inquire at the outset as to whether there is a problem at all. Many of the previous chapters in this book have introduced a variety of techniques for the theoretical study and experimental determination of molecular structure in various media. Certainly the most widely used experimental technique today is X-ray crystallography and the information obtained from crystal structure analyses is often used directly to interpret and understand molecular phenomena which take place in media and molecular environment differing significantly from that of the molecule in the crystal. As a result, the molecular geometric parameters found in the crystal may, and often do, differ from those determined experimentally by other methods; likewise they may differ from the results of theoretical treatments in which we have sufficient basis for confidence in our models and methods to provide us with reliable information on the geometry of molecules. The differences between the results of a crystal structure analysis and those of other methods are often perfunctorily referred to (or blithely written off) in the closing paragraphs of papers describing crystal structures as arising from the effect of so-called 'crystal forces'. In this discussion we will attempt to put these 'crystal forces' into a more well-defined framework, and present various approaches for studying them and their relationship to molecular geometry (Kitaigorodskii 1961, 1973; Dunitz 1979).

Perhaps the classic example of the problem of molecular geometry and crystalline environment is that of biphenyl (1). The molecule has been shown to be twisted by 42–45° about the exocyclic C—C bond in the gas phase (Bastiansen 1949; Almenningen and Bastiansen 1958; Almenningen *et al*. 1985; Bastiansen and Samdal 1985); the two nearly simultaneous first reports of the crystal structure (Robertson 1961; Hargreaves and Rizvi 1962) about 30 years ago indicated a planar conformation, which was strengthened by the fact that the crystallographic site symmetry is $\bar{1}$ (C_i). However, planarity requires a short intramolecular distance between the *ortho* hydrogens on the two rings, a situation resulting in a high-energy molecular geometry. Note that the crystallographic symmetry restrictions can be met without forcing each individual molecule to be planar by a random static disorder with a statistical distribution of molecules with relatively small equal but opposite rotations about the central bond. An alternate model would be a dynamic one with intramolecular motion about the same bond. In either case the tendency to deviate from planarity would be expected to appear in the atomic displacement parameters; these, indeed, indicate an unusually large libration around the long axis of the molecule (Charbonneau and Delugeard 1976; but see also Brock and Morelan 1986).

1

In any event, there is a significantly large difference between the gas-phase and solid-state results. The geometry is by no means limited to the two extremes of planar and twisted by 42–45°. Intermediate geometries have been determined in solutions and the melt (Suzuki 1959; Eaton and Steele 1973), and argon matrices (Le Gall and Suzuki 1977), although a planar geometry has apparently been trapped in the latter medium (Baca *et al*. 1979). Moreover, substituted biphenyls show a wide variety of conformations in the solid (Brock 1980), clearly due to a competition between intra- and intermolecular interactions. The next member of this series, *p*-terphenyl, has also been given quite extensive treatment (Baudour *et al*.

1976, 1977; Ramdas and Thomas 1976).

The case of biphenyl amply illustrates that the geometry of a molecule is subject to environmental mediation. To get some feeling for the magnitudes of the effects involved and their influence on the molecular geometry, it is useful at the outset to examine the magnitudes of the forces and energies involved on both the intramolecular and intermolecular levels.

19.2 Intramolecular forces and energies

Three types of geometric parameters may be used to define molecular geometry: bond lengths, bond angles and torsion angles. Variations in the geometry of a molecule simply reflect changes in these parameters: bond stretching or compression, bond bending or, more generally, angular deformation, and bond twisting or torsion. The energetics of these variations are described by simple functions (Mislow 1966; see also Chapter 14, Section 14.3). The energy involved in bond stretching along the internuclear axis (r) is readily approximated by Hooke's law:

$$V_r = (1/2) k_r (\Delta r)^2, \tag{19.1}$$

for which k_r is the *stretching force constant*. Typical values of k_r (in $10^2 \, \text{Nm}^{-1}$) for different types of carbon–carbon bonds range from 4.5 for the single bond in ethane to 15.7 for the triple bond in acetylene (Brand and Speakman 1960). For most C—C single bonds $k_r/2 \cong 1450 \, \text{kJ mol}^{-1} \, \text{Å}^{-2}$ so that a distortion by 0.1 Å would 'cost' about 14.5 kJ mol^{-1}. This factor rises approximately proportionally for double and triple bonds.

Bond angle deformation is also governed by a Hooke's law relationship:

$$V_\theta = (1/2) k_\theta (\Delta \theta)^2, \tag{19.2}$$

where θ is the bond angle, and k_θ is the *bending force constant*. For many bond angles involving three carbon atoms $k_\theta/2 \cong 0.042 \, \text{kJ mol}^{-1} \, \text{deg}^{-2}$. Thus bond angle deformation is less expensive than bond stretching: an angular distortion of 10° involves about the same amount of energy as the distortion of a single bond by 0.05 Å. The deformation of a benzene ring such that alternate angles will become greater and smaller than 120° involves a force constant of $0.7 \times 10^2 \, \text{Nm}^{-1}$.

Torsional distortion or rotation about the internuclear axis is a periodic function, and thus may be approximated by a cosine function:

$$V_\phi = (1/2) V_n (1 + \cos n\phi). \tag{19.3}$$

Here V_n is the *torsional energy barrier*, ϕ is the displacement of the dihedral angle from the zero value (where an energy maximum may be expected) and n is the periodicity of the function. In terms of energies, a few numerical examples will aid in establishing a useful frame of reference. The barrier to free rotation of methyl groups in dimethylacetylene is on the order of a few hundredths of a $kJ \, mol^{-1}$, while the barrier to rotation in ethane, corresponding to a rotation of 60° from the potential energy minimum, is 11–12 $kJ \, mol^{-1}$. The latter is an order of magnitude larger than the energy required to stretch a C—C bond by 0.03 Å. However, the energy required to twist ethane by 10° from the minimum energy conformation is only 0.8 $kJ \, mol^{-1}$, which is five time less than the energy required to distort a C—C—C angle by 10° (4.2 $kJ \, mol^{-1}$) and about 18 times less than that required to stretch a C—C bond by 0.1 Å (14.5 $kJ \, mol^{-1}$). Thus, appreciable bond stretching is considerably more expensive energetically than rotations about single bonds, with bond angle deformations falling somewhere in between.

In describing these geometrical distortion parameters, we have assumed for simplicity that they are mutually independent functions. Strictly speaking, this is not the case, of course, and perturbations in any one of these parameters will be accompanied by a suitable relaxation of the others (Chapter 14, Section 14.3.5; see also Scharfenberg and Hargittai 1984).

19.3   Intermolecular forces and energies in crystals

A crystal structure corresponds to a free energy minimum that can usually be identified with a potential energy minimum, which *is not necessarily the global minimum*. For structures which exhibit disorder, the entropy contribution may also be important (Kemp and Pitzer 1936; Eliel 1962). The potential energy minimum represents a balance between attractive and repulsive forces for which there is a wide variety of nomenclature in the literature (see, for example, Kihara 1978; Rigby *et al.* 1986): van der Waals', dipole–dipole, dipole–induced dipole, dipole–quadrupole, quadrupole–quadrupole, hydrophobic, electrostatic, steric repulsions, exchange, hydrogen-bonding, charge-transfer, induction, donor–acceptor interactions, σ–π interactions, π–π interactions, polarizability, non-bonded interactions, London forces, dispersion forces, overlap, long-range, intermediate-range, short-range, zero-point, . . . etc. Many of these terms are used interchangeably, depending on the chemical nature of the system and the school of thought or personal biases of the investigator; others are not well defined at all. However, for a number of them there is more or less general agreement on their definition and it is possible to estimate the energies involved since these will be required in the discussion which follows.

Van der Waals' interactions may be roughly estimated from the sublimation energy of organic crystals in which other interactions are essentially absent. For our purposes here the sublimation energy of organic crystals, usually in the range 40–200 kJ mol^{-1} (Westrum and McCullough 1963; Weast 1975), may be considered to be the energy required to remove a mole of molecules from a crystal. While the term 'coordination number' is not normally applied to organic crystals, a single molecule is usually surrounded by 10–14 molecules, so that the average contribution of each neighbor molecule to the total sublimation energy is 3–15 kJ mol^{-1}. The energies involved in hydrogen-bonding interactions are somewhat larger, with hydrogen-bond strengths in the 12–30 kJ mol^{-1} range.

Charge-transfer interactions (e.g. for π complexes) are in the range 2–20 kJ mol^{-1}. Electrostatic interactions vary over a wide range depending on the degree of polarization of the molecule (and consequently the 'amount of charge' which can be assigned to each atom) as well as the distance.

Now where do these intermolecular interactions fall on the scale of energies involved in *intra*molecular distortions? For the most part, they fall on the low end, and they are of the same order of magnitude. This compatibility of inter- and intramolecular energies suggests the possibility of perturbation of the molecular geometry by the crystalline environment in which it is found. Because of this energy compatibility it is reasonable to suggest that the molecular geometric parameters most likely to be affected by crystal forces will be the torsion angles about single bonds, and these are often the most important in our consideration of the molecular geometry in relationship to the chemical properties of the molecule.

19.4 Molecular conformations in crystals

To this point we have discussed molecular geometry in a general way, without making use of the term *conformation*. How do we define the conformation of a molecule? This is often a semantic problem, although most of us have a pretty good idea of what we mean when we use the term. However, Dunitz's (1979, p. 312) discussion of the problem is one which provides a sound frame of reference:

Generally, conformations are arrangements that arise by rotation about bonds, and they may be described by specification of relevant torsion angles. Moreover, we can be reasonably confident that any particular arrangement of atoms observed in a molecular crystal cannot be far from an equilibrium structure of the isolated molecule.

X-ray analysis thus provides information about the preferred conformations [note plural] of molecules although it has nothing to say about the energy differences between conformations or the energy barriers that separate them. This information has to be obtained by other methods. Energy differences can be derived [experimentally], in principle, from measurements of equilibrium concentrations of the relevant conformational isomers, and energy barriers can be obtained from measurements of interconversion rates.

Thus, a number of conformations of a molecule may be energetically equivalent, or nearly so, and we should be aware of the possibility of a number of them appearing in different crystal structures, or even in the same crystal structure. Certain packing motifs are more favorable than others (Dunitz 1979), and this may lead to a predominance of one conformation in a crystal structure, while in solution a number of conformations may be present, including, most likely, the one(s) found in the crystal structure. As Bürgi and Dunitz (1983) wrote,

The structure of a molecule in a crystal environment is not necessarily identical with the equilibrium structure of the isolated molecule, i.e., the forces exerted by a crystal environment can deform a molecule to a greater or lesser extent. Similarly, we expect that the structure of a given molecular fragment will depend to some extent on the particular molecules in which the fragment is embedded, as well as on the crystal environments.

The important point here is that both the crystal structure and the molecular conformation found therein represent potential energy minima which are not necessarily unique. Consequently, near the global energy minimum there may exist a number of possibilities of very nearly the same energy for both the molecular conformation and the crystal structure. Clearly, in order to investigate the effect of crystal environment on molecular structure we would like to be able to compare, on a quantitative basis, the fluid (solution, liquid, and gas) and solid-state conformations.

19.5 Comparison of conformations in fluid and solid states: techniques

Many techniques have been employed for the comparison of solid-state conformations with those in other media and some are covered in considerable detail elsewhere in this volume. The most common of these are NMR, IR, Raman, CD-ORD and UV-VIS spectroscopy in solution and microwave spectroscopy and electron diffraction in the gas phase. There are some fundamental problems in making such comparisons, and the reader should be aware of them, even if in the end they are often ignored either as an approximation or for the lack of the facilities for a totally rigorous approach. For one, the molecule in the crystal is not at rest. There is always atomic motion, which is usually anisotropic, and it is usually correlated with that of other atoms in the molecule (see Chapter 8). Also, it is well to bear in mind what is measured with each experimental technique. In gas-phase techniques like electron diffraction, for instance, one measures the average distance between nuclei, while in an X-ray experiment one measures the distance between the average positions of electron-density maxima. This, of course, differs from the structure obtained from neutron diffraction in which one measures the distance between the average positions of the nuclei.

Recent and continuing technological advancements in all of the traditionally solution-based techniques (NMR, UV–VIS) have facilitated these measurements in the solid as well, often with detail and precision approaching that in solution, and have provided us with a means for direct comparison of geometries in the two phases, even given the limitations noted above. Moreover, with the growing sophistication of these techniques, they have been applied to increasingly larger and more complex molecules and systems.

Traditionally, IR spectroscopy was one of the few methods which provided equally detailed information both in solution and the solid state, and has been applied, for instance, to organic molecules of pharmaceutical importance (Byrn *et al.* 1976) as well as to polypeptides (Benedetti *et al.* 1982). CD methods have traditionally been employed to determine solution conformations of natural products and other molecules (Djerassi 1960). The measurement of solid-state spectra is severely hampered by problems due to light scattering, but some recent reports (Barrett *et al.* 1982) suggest that this technique also may be utilized in comparing solid-state and solution conformations. In the area of UV–VIS spectroscopy, the development of sophisticated techniques for measuring solid-state spectra at increased resolution (Eckhardt and Pennelly 1971; Merski and Eckhardt 1981) indicates that this technique will also provide useful information in the direct comparison of solid and solution conformations.

The developments in the NMR field are particularly worthy of note (Jelinski 1984; Maciel 1984; Sanders and Hunter 1987). For molecules which are relatively small (i.e. about ten magnetic nuclei or less) or for small parts of larger systems, the NMR of oriented molecules (say, by liquid crystals as described in Chapter 12) is another method of structure determination with precision in geometric parameters approaching 0.01 Å and 0.1°. In solution, by employing a combination of nuclear Overhauser enhancement and difference decoupling techniques it has been possible to assign all the protons and study the conformations in steroids (Hall and Sanders 1980, 1981; Barrett *et al.* 1982) and large molecules (Weber *et al.* 1982). Such methods are now even being applied by a number of groups to biomacromolecules (Wüthrich *et al.* 1982; Billeter *et al.* 1982). The development of line-narrowing techniques, for instance combining 'magic-angle' spinning and cross-polarization techniques (Andrew 1975; Mehring 1976; Schaefer and Stejskal 1976, 1979; Griffin 1977; Maciel and Sullivan 1982; Wasylishen and Fyfe 1982; Yannoni 1982) has permitted obtaining solid-state spectra with details equivalent to that in solution for direct comparison of solid and solution conformations. The exceedingly rapid pace of development in this field promises increasingly powerful tools for comparison of solution and solid-state conformations, even when crystals suitable for crystallographic experiments cannot be grown. Crystallographers needn't yet fear that crystallography will soon become obsolete; in the future molecular geometry may eventually be more readily accessible by NMR techniques, but the *crystal structure*, a feature all too often ignored by most chemists and many practicing crystallographers and one crucial to the understanding of the effect of crystal environment on molecular conformation, will remain (for the foreseeable future, at least) in the realm of those who determine structure by diffraction methods.

Finally, we note an additional important development over the last decade, which provides a new tool in the armory of weapons employed to study molecular conformation, namely the application of theoretical and computational methods. This development was foreseen by Orville-Thomas (1974), just as these methods were coming of age:

In the first half of this century chemical methods proved a powerful tool in the elucidation of molecular configuration and conformation. Their importance then decreased as they gradually became superseded by physical methods of structure determination . . . The current position [i.e., 1974] is that the emphasis has tilted even further in favour of the use of a greater range of physical methods and a new factor has emerged to the extent that theoretical methods are becoming of increasing importance and power in the elucidation of conformational problems.

The techniques employed in these theoretical and computational approaches to the study of molecular conformation vary from *ab initio* MO (at various levels), through a range of semiempirical methods, to molecular mechanics. These methods are covered in considerable detail in Chapters 13 and 14 of this volume, so there is no need to expand in great detail here. Much experience has been acquired in the past decade and we are beginning to learn the advantages and limitations of the various methods, potential functions, parameters and computer programs and the level of sophistication required for meaningful results. There is no doubt that as these methods are refined they will prove of increasing utility for the study of conformational problems. For the sake of completeness, we review briefly some of the most widely used of these techniques, with emphasis on their relevance to the topic in question, especially since some of those developed to study molecular problems are being increasingly applied to the investigation of crystal structures.

19.6   Intramolecular and intermolecular energetics: computational techniques

Following Orville-Thomas' prophecy there has been a true evolution in the development and utilization of computational methods for investigating both molecular and crystal energetics. The field has been reviewed extensively (Engler *et al.* 1973; Altona and Faber 1974; Dunitz and Bürgi 1975; Allinger 1976; Ermer 1976; Warshel 1977; Mislow *et al.* 1978; Boyd and Lipkowitz 1982; Burkert and Allinger 1982; Ōsawa and Musso 1982; Rogers 1987), but the important point for this discussion is that many of the programs developed and refined during this period have now become standard library programs which run on mainframe computers or superminis, and now in some cases are readily transferable, easily obtained, and hence may be utilized by almost any worker in the field.[†] Some of these programs treat conformational parameters explicitly, while for others the multidimensional conformational energy surface is sampled by systematically varying the parameters in question and computing the energy involved.

The latter approach applies in general to molecular orbital calculations and may include automatic optimization procedures for producing the minimum-energy comformation(s). These procedures naturally increase significantly the amount of computer time required. Molecular orbital methods, from the simple Hückel approximation to *ab initio* methods are quite familiar and well documented elsewhere (Hehre *et al.* 1969, 1970; Hehre 1976; Schaefer 1977*a,b*). When comparing the geometrical features of structures obtained by experimental methods and *ab initio* MO techniques, it is important to bear in mind that the latter yield the equilibrium structure, while the former correspond to thermal averages over the vibrational motion.

The force-field calculations which originated from vibrational spectroscopy traditionally have included specific terms for perturbations in molecular geometry which makes them advantageous for interpreting the energetics of molecular comformation. For full force-field and molecular mechanics calculations the molecular energy is defined in terms of certain conformational parameters and fairly efficient procedures have been developed (Ermer 1976) to speed these up. Recently they have been modified and generalized to include terms for the full range of torsional rotations, as well as non-bonded interactions of the Lennard–Jones (eqn (19.4)) or Buckingham (eqn (19.5)) forms. Each of these expressions gives the potential energy V, between atoms i and j, separated by the distance r_{ij}. The potential energy dependence on the atom type is defined by the empirically determined constants (A, B and A', B', C' respectively) which are specific for any pair of atoms; see for example, Chapter 14, Section 14.3.4.

$$V(r_{ij}) = B/r_{ij}^n - A/r_{ij}^6 \quad (n = 9, 12) \tag{19.4}$$

[†]Most programs may be obtained for a nominal charge by writing directly to QCPE, Quantum Chemistry Program Exchange, Department of Chemistry, Indiana University, Bloomington, IN 47405, USA.

$$V(r_{ij}) = B' \exp(-C' r_{ij}) - A' / r_{ij}^6 \qquad (19.5)$$

The choice of the analytical forms for the force field, the determination of force-field constants and the transferability of force fields and their constants among different chemical systems are still subjects of considerable interest and research. However, certain functions and parameters have proven useful in a variety of applications (Warshel and Lifson 1970; Scheraga 1971; Hagler and Lifson 1974; Dunitz and Bürgi 1975; Ermer 1976; Hagler *et al.* 1976) and the speed of the current algorithms combined with considerable ease of use make this technique a recognized and widely used tool in the investigation of molecular conformations.

In principle the extension of this approach to crystals is straight-forward[†]. The basic assumption is that the intermolecular interactions may be treated as the sum of atom–atom interactions which, again, are approximated by either the Lennard–Jones or Buckingham potentials (Casalone *et al.* 1968; Coiro *et al.* 1972; Williams 1972; Kitaigorodskii 1973; Ermer 1976). Williams (1974) has demonstrated the need for inclusion of a coulombic electrostatic term in the atom–atom potential even for hydrocarbon crystals, where such a contribution may approach 30 per cent of the total energy, as for instance, for naphthalene. The Lennard–Jones and Buckingham potentials are then modified by adding a term which includes $q_i q_j / r_{ij}$.

In the actual calculation of lattice energies, either (or for completeness, sometimes both) of the potential functions are employed to compute the lattice energy for a reference molecule using in the function empirically derived parameters which are characteristic for each of the atoms involved. The electrostatic interaction based on the Coulomb potential is included by assigning point charges to each atom. These may be estimated from bond dipole moments, quadrupole moments (Hirshfeld and Mirsky 1979) or charge densities from molecular orbital calculations at various levels of approximation (Cox and Williams 1981). It has been found that in general an interaction radius of 12–20 Å is sufficient to account for the sublimation energy (i.e. the 'binding energy' of a single molecule). This quantity is then minimized by altering the position and orientation of the reference molecule, as well as the unit-cell parameters. If the starting model is a known crystal structure, usually only small perturbations in the structure result from the minimization and the symmetry of the crystal is maintained even if not explicitly accounted for in the computational procedure. The resulting minimized lattice energy, which represents the sublimation energy, may then be compared to the experimental values to test the result, the reliability of

---

[†]In some cases programs have been designed written to yield both intramolecular and intermolecular energies. See, for instance, Oie *et al.* (1981).

the potential function, and the parameters employed for each atom in the calculation. More detailed information on the nature of the intermolecular interactions may be obtained by partitioning of the total lattice energy into non-bonded and electrostatic contributions, or into individual atomic contributions. As we shall demonstrate below, partitioning is particularly useful in the investigation of the relationship between crystal forces and molecular conformation.

In concluding this section on the methods of comparing molecular geometries in various media, some general remarks are in order. Comparisons of this sort, especially those involving experimental techniques on the one hand and computational techniques on the other, can prove very valuable for testing and evaluating the limitations of the latter, as well as providing an additional source of critical review for the former. The mutual benefit which is to be derived from this cross-fertilization of techniques is being recognized by both structural and computational chemists, and was one of the factors behind the preparation of this volume and the school which preceded it. This interplay of disciplines is certainly to be strongly encouraged.

19.7 Strategies for investigating the influence of crystal environment on molecular structure

Four strategies have been suggested for investigating the influence of crystal environment on molecular structure (Kitaigorodskii 1970):

(1) comparison of compounds in gaseous and crystalline states;

(2) comparison of the geometries of crystallographically independent molecules in the same crystal;

(3) analysis of the structure of a molecule whose symmetry in a crystal is lower than that of the free molecule;

(4) comparison of molecules in different polymorphic modifications.

Each method has its advantages and drawbacks, and we proceed by discussing each in turn with appropriate (but by no means exhaustive) examples from the literature.

19.7.1 Comparison of compounds in gaseous and crystalline states

Hargittai and Hargittai (1987) have recently reviewed rather extensively the structural differences between gases and solids. The most widely used techniques for determining gas-phase molecular structure are microwave spectroscopy and electron diffraction, which are discussed in considerable detail in Chapters 3 and 5 of this volume, respectively. For microwave spectroscopy the structure determination of even a 5–10 atom molecule requires a number of isotopically substituted species, so that the structure determination of larger molecules is usually out of the question. Until rather recently there were similar size limitations in employing electron diffraction methods, but considerable progress on increasingly larger molecules has been made in the last decade.[†] Even so, for the most part, suitable substances are those relatively small molecules with considerable vapor pressure and a relatively high degree of intrinsic symmetry. The earlier discussed case of biphenyl (1) falls into this category, with only one *conformational* parameter. An additional complication is the fact that the elevated temperatures often required to vaporize materials when employing these methods lead to a much larger distribution of rotational isomers than in the crystal or even in solution. Many of the classic studies comparing solid- and gas-phase geometries (by electron diffraction) have been carried out by István Hargittai and coworkers. An example of a conformational change due to rotation about a single bond is found in gaseous ethane-1,2-dithiol (2). At about 343 K 2 consists of a conformational equilibrium of *anti* and *gauche* conformers with an approximate ratio of 2:1. In the crystal, spectroscopic methods indicate that only the centrosymmetric *anti* conformer is present (Schultz and Hargittai 1973; Hargittai and Schultz 1986).

$$HS-CH_2-CH_2-SH$$

2

As noted above, while perturbations in bond lengths and bond angles due to crystal environment are still smaller than the precision of the experiments used to determine them, some careful work on *para*-disubstituted benzenes indicates that this is a promising area for further study. The influence of substituents on the geometry of the benzene ring is discussed in Chapter 18. In most cases, when parallel studies of the solid and gas phase are carried out, there is excellent agreement between the results from the two methods, leading to the following important generalizations (Domenicano 1988, and references therein). The *ipso* angle α (see Fig. 19.1) is a function of the electronegativity of the substituent, ranging from 111.8° for X = Li to 123.4° for X = F. Concomitant with a decrease (increase) of α there is a lengthening (shortening) of the *a* bonds, along with smaller changes in β and δ. If π donation to the ring is increased, α and δ decrease while the *a* bond lengths increase.

[†] For references on compilations of gas-phase structural data see Chapter 5, Section 5.1.

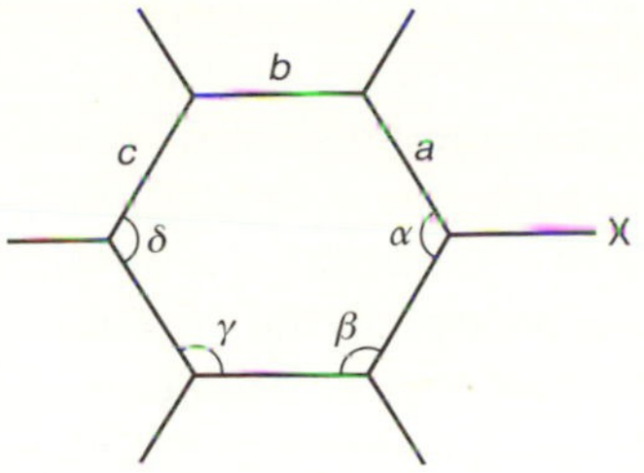

Fig. 19.1 Geometric parameters of monosubstituted benzene rings.

Recent studies on three *para*-disubstituted benzenes **3–5** rather dramatically demonstrate how geometric parameters other than rotations about
formal single bonds may be altered by intermolecular interactions. The ring
deformations in the crystals of **3** and **4** differ from those observed in the
gas phase as studied by electron diffraction (Colapietro *et al.* 1984*a*, *b*), and
in fact the differences are in opposite directions for the two molecules. In
3 the cyano groups of neighboring molecules lie in an antiparallel arrangement, which tends to reduce their intramolecular deforming influence. In
the isocyano derivative **4** the substituent lies above the center of a neighboring benzene ring, which enhances the ring distortions compared to the
gas phase. For **5** the angle α is 2° larger in the gas phase than in the crystal
(Colapietro *et al.* 1987); the difference can be directly related to the intermolecular hydrogen bonding. Additional examples are presented in Chapter
18, Section 18.7.

3	**4**	**5**

Two general comments are in order here. First, the importance of these
relatively small, but unquestionably experimentally (and, as noted above,
energetically) significant changes in molecular geometry would be considerably diminished in the absence of the accumulation of a large body of precise structural data on substituted benzene derivatives which preceeded it,
and provided reliable standards on which to base comparisons. Second, in
general, the more things are the same the easier it is to isolate, identify, and
understand the differences between them. The choice of relatively symmetric
molecules for study such as **3–5** is an excellent means for minimizing the
number of variables to be investigated. Such symmetric molecules should
be the targets for further study in this area.

In light of the progress in theoretical and computational techniques noted above it seems perfectly reasonable to broaden this first approach and include conformations determined for the 'free molecule' by these means as well as by experimental techniques in solution. This in fact has become quite common practice today and is clearly one of the most general approaches for comparing conformations in the solid and fluid phases, bearing in mind the *caveats* noted above regarding the difference in the nature of the information obtained by the various methods. It may not, however, be the most direct way to study the influence of crystal environment on molecular conformation. Nevertheless, under the umbrella of this now broadened approach to the problem, a large number of studies have been performed, and a few examples are cited here.

The rapid growth in available computing power over the past few years has made the development, testing and use of sophisticated methods increasingly feasible. That, combined with neutron diffraction studies for precise location of hydrogen atom positions, has led to a number of elegant studies combining computational and diffraction methods. For instance, Jeffrey *et al.* (1980) reported the *ab initio* MO study of the rhombohedral form of acetamide **6** at the HF/3–21G level with geometry optimization together with a neutron diffraction study at 23 K. The neutron refinement shows one C—H bond normal to the plane of all the non-hydrogen atoms, while the lowest computed energy for the free molecule shows m (C_s) symmetry with one C—H bond eclipsed to the carbonyl bond. The conformation in the crystal was calculated to be *ca.* 1.7 kJ mol^{-1} higher in energy than that for the free molecule; i.e. the crystal stabilizes the higher-energy conformation. Observed C=O and C—N bond lengths differ from the calculated ones by $+ 0.034$ Å (30σ) and -0.021 Å (20σ), respectively, and it is suggested that the differences are due — in part, at least — to hydrogen bonding in the crystal (see Chapter 11, Section 11.4.2).

$$H_3C-\overset{\displaystyle O}{\underset{\displaystyle NH_2}{\diagdown\!\!\!\!/}}$$

6

In general, program size and computer time availability still limit the applicability of *ab initio* MO methods to relatively small molecules, although the increasing availability of supercomputers will certainly allow expanding use of these techniques. Upon reaching the limits of molecular orbital methods the natural recourse is to turn to molecular mechanics. In addition to the programs based on various versions of MM2 (see Chapter 14), others such as QCFF/MCA (Huler *et al.* 1977) are in common use. Because of the semiempirical nature of the method, these do not always yield results which are in mutual agreement and caution is advised especially

when studying systems containing atoms which have not been fully para-
meterized. Duax *et al.* (1982) have published a rather comprehensive comparison between crystallographic results and those of molecular mechanics
calculations using a number of the different programs commonly employed.

19.7.2 Comparison of the geometries of crystallographically independent molecules in the same crystal

This approach appears at first sight to be very promising and in the early
years of crystal structure analysis was often considered a boon: two or
more independent determinations of the molecular geometry in one crystal
structure determination. However, this was somewhat of a mixed blessing
since the effort and price increased accordingly and relatively few of these
structures were reported. That situation has changed, of course, with the
improvement of instrumentation and the phenomenal increase of computing
power. Nevertheless, since molecules necessarily interact with one another,
and thus mutually influence molecular conformation, any detailed analysis
of this situation must take into account the problem of differentiating
among crystal forces between crystallographically equivalent molecules
from those which are not crystallographically equivalent. Although this is
a major disadvantage for choosing such systems for study in these relatively
early stages of the investigation of the influence of crystal forces on molecular conformation, it can provide a fertile testing ground for computational
approaches which have proven successful on less complex systems.

Examples can be found in almost every area of structural chemistry,
including, for instance, quite simple molecules (Kroon *et al.* 1976), polypeptides (Avbelj *et al.* 1985), steroids (Campsteyn *et al.* 1979; Sawzik and
Craven 1979), and organometallic complexes (van Buuren *et al.* 1981;
Nepveu *et al.* 1984). We cite here two cases reported by Birnbaum and
coworkers, which both contain three molecules in the asymmetric unit.
The first (Birnbaum *et al.* 1981), 9-(2-hydroxyethoxymethyl)guanine **7**, is a
potent inhibitor of *herpes simplex* viruses. The side chain in two of the

7   8

molecules is partially folded, while in the third molecule all the bonds in the side chain are in a *trans* orientation. In the second (Birnbaum *et al*. 1982), 8-bromo-9-β-D-xylofuranosyladenine **8**, competition between intramolecular and intermolecular hydrogen bonding is clearly a dominant factor in determining the conformation. Again two of the three molecules are similar in conformation, while the third is significantly different.

19.7.3 Analysis of the structure of a molecule whose symmetry in the crystal is lower than that of the free molecule

A fundamental problem is encountered here in that the free-molecule symmetry is basically a difficult quantity to determine. It is at best the chemist's idealized structure and often bears little relationship to the true (even if unknown) molecular symmetry. This is especially significant for flexible molecules whose precise conformation we wish to know and which are particularly susceptible to the influence of the crystal environment. Moreover the tendency for molecules to pack as densely as possible generally overrides the retention of molecular symmetry upon crystallization (Kitaigorodskii 1973). As a result, the crystallographic site symmetry of most organic materials is significantly reduced from what chemists might define as the free-molecule symmetry. In general a molecular center of symmetry is retained as a crystallographic symmetry element, with considerably lower frequency for the retention of a twofold axis or a mirror plane (Kitaigorodskii 1961). The presence of these latter two symmetry elements is usually expensive in terms of packing efficiency, but if the molecular shape permits, they may be retained.

If symmetry elements are not retained upon crystallization we may still ask if 'chemical' symmetry is maintained — i.e. whether chemically equivalent parts of the molecule have identical geometric features, within experimental error. This, of course, depends on the precision of the structure determination, and even if we have precise structural information, the subsequent steps in the analysis are not straightforward, so that this proposed method seems presently to be of the least utility.

19.7.4 Comparison of molecules in different polymorphic modifications

Polymorphism is the ability of a compound to crystallize in more than one distinct crystal species. It was recognized quite early as widespread in organic materials (Deffet 1942), and is most commonly found in those areas of chemical research where full characterization of a material is crucial in determining its ultimate use: pharmaceuticals (Haleblian and McCrone 1969; Haleblian 1975; Clements 1976), dyes (Walker *et al*. 1972; Griffiths and Monahan 1976; Tristani-Kendra *et al*. 1983; Etter *et al*. 1984; Morel *et al*. 1984), and explosives (Karpowicz *et al*. 1983). Various aspects of the subject have been treated in books (Verma and Krishna 1966; Byrn 1982) and a number of reviews (McCrone 1965; Haleblian and McCrone 1969). In fact, regarding the ubiquity of polymorphism McCrone (1965) has stated:

... every compound has different polymorphic forms ... and the number of forms known for a given compound is proportional to the time and energy spent in research on that compound.

The differences in lattice energy between polymorphs are usually in the range 4–10 kJ mol^{-1} (McCrone 1965), which is too low to bring about appreciable changes in bond lengths and angles, but sufficiently high to cause torsional distortions about single bonds. Since it is the torsional parameters which define the molecular conformation, it is clear that for molecules which possess torsional degrees of freedom, various polymorphs may exhibit significantly different molecular conformations, a phenomenon termed *conformational polymorphism* (Corradini 1973; Panagiotopoulos *et al*. 1974; Bernstein and Hagler 1978).

The utilization of the phenomenon to investigate the influence of crystal forces on molecular conformation has some distinct advantages over the other approaches noted above. The general extent of polymorphism potentially provides a wide variety of cases for study. In cases of conformational polymorphism changes in molecular conformation *must* be due to the influence of crystal forces since the crystal environment is the only variable in the system. No assumptions need be made about the compatibility of the structural or computational methods for comparing the geometry and energetics. In fact, in most cases one is interested primarily in *differences* both in molecular and crystal properties and many of the possible errors cancel out. Our own approach to the problem of crystal forces and molecular conformation has been based on the utilization of conformational polymorphism. We have recently reviewed the topic in some detail (Bernstein 1987), including quite a few examples from the recent literature, so only an outline of the strategy will be presented here.

In a study utilizing conformational polymorphism for this purpose, we seek the answers to two basic questions: (1) what are the differences in energy, if any, in the molecular conformations observed in the various crystal forms? (2) how does the energetic environment of the molecule vary from one crystal form to another? To answer these questions, a typical study would proceed according to the following steps:

(1) determination of the existence of polymorphism in the system under study;

(2) determination of the existence of conformational polymorphism by the appropriate physical measurements;

(3) determination of the crystal structures to obtain the geometrical information — molecular geometry and packing motif — of the various polymorphs;

(4) determination of differences in molecular energetics, by appropriate computational techniques;

(5) determination of differences in lattice energy and energetic environment of the molecule by appropriate computational methods.

Polymorphism may be readily detected by calorimetric methods (McNaughton and Mortimer 1975) or microscopic hot stage methods (McCrone 1957; Kuhnert-Brandstätter 1971) as well as, of course, X-ray powder diffraction. I R spectroscopy has been the classic spectroscopic tool for determining conformational differences in solids, although as noted above, magic-angle N M R spectroscopy is proving to be a very powerful and sensitive probe of solid-state molecular conformations and environments (Ripmeester 1980; Yannoni 1982; Bryant *et al.* 1984). Even though crystal structure determinations have now become fairly routine, for polymorphic systems it may be difficult to obtain single crystals which are suitable for structure analysis for all the various forms. We outlined above the computational methods employed to obtain the energies associated with both molecular conformation and the crystal lattice. It is important to note in the context of this discussion that for the molecular energetics we need investigate only those points on the multi-dimensional potential energy surface which have been determined in the crystal structure analyses. The main point of interest in these studies is the *difference* in energies of the molecular conformations observed, and not necessarily whether any one of them is the global minimum conformation, or even the features of regions of the conformational energy surface removed from those close to the observed conformations. This also relieves us of the fairly serious restriction of being concerned with absolute molecular energies. Even when considering relatively large molecules, the ideal cases for study should involve differences in a small number of conformational parameters, and again, symmetric molecules are good candidates. In those cases the molecular conformational energetics may be obtained from the molecular mechanics or *ab initio* MO calculations on model compounds representing the conformational parameters in question (Bernstein *et al.* 1981) or from molecular mechanics keeping all but those parameters fixed. Alternatively, semiempirical MO methods often give good estimates of energy differences among conformations, even if other properties may not be approximated very well.

The total lattice energy is obtained from calculations based on atom-

atom potential (Kitaigorodskii and Mirskaya 1972; Mirsky 1976, 1980; Mirsky and Cohen 1976). Comparison is then made with the sublimation energy, which may be measured fairly readily (Daniels *et al.* 1969) or estimated from those of analogous model compounds and group contributions (Bondi 1963). Again, differences in lattice energy may be determined calorimetrically for experimental verification of the computed quantities.

Finally, partitioning of the lattice energy into, say, non-bonded attractive, non-bonded repulsive, and electrostatic terms helps to identify differences in the energetic environment of the molecules in the various polymorphs. Even more detail and information can be obtained by partitioning the total lattice energy into individual atomic contributions for the examination of those interactions which lead to, for example, the stabilization of an energetically less favorable molecular conformation in a particular polymorph or of the presence of a more stable conformation in a second structure.

A complete prototypical study of this type has been published (Bernstein and Hagler 1978) on the dimorphic dichlorobenzylideneaniline system **9** (X = Cl), which has two torsional degrees of freedom (α, β). Both forms exhibit crystallographic disorder, the triclinic form (planar: $\alpha = \beta = 0°$) about an inversion center (Bernstein and Schmidt 1972) and the orthorhombic (non-planar: $\alpha = -\beta = 24.8°$) about a twofold axis (Bernstein and Izak 1976). The difference in molecular energies was estimated — for planar and non-planar molecules — from *ab initio* MO calculations on model compounds **10** and **11** which have been shown (Bernstein *et al.* 1981) to represent well the energetics of the full unsubstituted molecule **9** (X = H). All these calculations indicate that the minimum-energy conformation is favored over the planar structure by *ca.* 4–6 kJ mol^{-1} and corresponds to a rotation about the N—phenyl bond (α) of *ca.* 45° and virtually no rotation about the CH—phenyl bond (β). The conformation found in the orthorhombic structure is favored over that in the triclinic structure by about 2 kJ mol^{-1}. The results from these *ab initio* MO studies are in accord with those from

9

10   **11**

spectroscopic and other X-ray diffraction studies and in this case represent a significant improvement over the results obtained from semiempirical methods which in general have not been able to account properly for rotations about the exocyclic bonds. A similar systematic approach has been taken to study packing, conformation, and solid-state reactivity (Jones *et al.* 1983*a*, *b*; Theocharis *et al.* 1984).

Lattice energy calculations were carried out employing three different potential energy functions (Lennard–Jones, eqn (19.4), with $n = 9$ and $n = 12$; Buckingham, eqn (19.5)), which included an electrostatic term employing partial charges from Mulliken populations of *ab initio* MO calculations. The results for this system show that in all cases the triclinic structure is favored over the orthorhombic one, and by an energy which is compatible with the expected energy differences between polymorphs. Indeed, the triclinic form must have a lower (i.e. more negative) lattice energy than the orthorhombic form in order to stabilize the more highly energetic planar conformation found in the former.

Partitioning of the minimized total energies into individual atomic contributions can yield valuable information on the dominant interactions and differences between polymorphic structures. Naturally, in dealing with atomic contributions to the total energies, the numbers involved are quite small; hence a great deal of significance is not attached to any particular atomic contribution. However, we do look for consistent trends, which are indeed present, and we can make the following observations:

1. The relative role of each atom's contribution to the overall energy is remarkably insensitive to the potential functions used and is in fact identical for the Lennard–Jones ($n = 12$) and Buckingham potentials for both structures.

2. The order of the relative contributions to the total energy is the *same* for both crystal forms. This is an indication that the environments of the atoms in the two crystal forms do not differ drastically in terms of energetics.

3. No single atom makes an outstanding contribution to stabilizing the triclinic structure over the orthorhombic one; rather, the mode of stabilization is non-specific in that nearly all atoms make a small stabilizing contribution. This is true in spite of the fact that there are rather striking differences in the spatial arrangement of the molecules in the two structures.

4. It is quite common, in the analysis of crystal structures, to make note of interatomic distances (often referred to as 'contacts') which are shorter than the sum of the van der Waals' radii, since these are presumably the important or dominant ones in determining the packing mode. In the triclinic structure there is a short $Cl \cdot \cdot \cdot Cl$ distance of 3.42 Å, while the shortest one of this type found in the orthorhombic structure is 3.76 Å,

which is slightly greater than the sum of the van der Waals' radii. Hence, in these terms we might have expected that the contribution of chlorines in the triclinic form would be a major one in stabilizing that structure over the orthorhombic one. The analysis by partitioning for these two structures suggests that this is not the case here; while the Cl is the largest contributor in both cases the difference between the two polymorphs in the contribution is not significantly larger than for other atoms. This may be due to the special nature of the Cl $\cdots$ Cl interaction (Schmidt 1971; Williams and Hsu 1985). For instance in 2,2'-dimethylbenzidine (**12**, X = Me) the twist angle between the two rings is 86°, while in 2,2'-dichlorobenzidine (**12**, X = Cl) the two chloro substituents are close to each other resulting in an angle of 36° (Desiraju and Sarma 1986).

12

The general success of this first study prompted us to attempt to refine the general approach and learn the extent of its applicability. These attempts have been based, for the most part, on the same benzylideneaniline system **9**, where the substituents in the *para* positions on the two rings — X = Cl, Br, or Me — have been manipulated to obtain the full set of three homodisubstituted and six heterodisubstituted analogues. This provides a very rich system of polymorphic and isomorphic structures for the investigation of various aspects of the interplay between packing and conformation. The results on this system appear elsewhere in a series of papers (Bernstein and Izak 1975; Bernstein *et al.* 1976; Bar and Bernstein 1977, 1982*a*, *b*, 1983, 1984, 1987; Hagler and Bernstein 1978) and in summary form (Bernstein 1987) and the interested reader should consult those works for details.

19.8 Some recent developments

There have been reports of a number of techniques and computer programs which will certainly aid in the investigation of the influence of crystal forces on molecular geometry. We mention just a few of them here.

The Cambridge Structural Database (see Chapter 15) is being utilized in an increasing number of cases to study specific types of interactions, such as hydrogen bonding of various types (Murray-Rust and Glusker 1984; Taylor and Kennard 1984; Lesyng *et al.* 1988), Cl $\cdots$ Cl interactions (Sarma and Desiraju 1986), the competition between hydrogen- and non-bonding interactions (Sarma and Desiraju 1985), and even side chain aromatic–aromatic interactions which may stabilize tertiary and/or quaternary structures of proteins (Burley and Petsko 1985). Such studies have already been used to revise the values of the van der Waals' radii for a number of atoms commonly bound to carbon in organic compounds (Nyburg and Faerman 1985). They can also yield geometric information which can then be used to add the anisotropy of the interactions into the energy calculations.

Such an approach has been adopted, for instance, by Krygowski and Turowska-Tyrk (1987) on the trihydrate of sodium *p*-nitrobenzoate **13**. The authors have devised a method, based on forces of deformation rather than on energy, to determine which atoms are moved from the positions expected in the absence of intermolecular interaction. Because of the rather high precision of the crystal structure determination of **13**, it is possible to recognize significant differences in chemically equivalent geometric features, and the angular deformations in the two crystallographically independent molecules may be interpreted on the basis of specific interatomic forces. This appears to be a promising advance over the treatment based on energies alone, and it will be of interest to see it applied to additional systems to test its general utility.

13

The systematic variation of substituents may also be used to provide information on the nature of particular intermolecular forces. Such studies have been carried out, for instance, by examining the differences in crystal and molecular structures upon exchanging chlorine and methyl substituents (Theocharis *et al.* 1984; Desiraju and Sarma 1986). In some cases solid solutions may be formed or isostructural crystal structures may be obtained, indicating very similar energy properties of the substituents. In other cases, a simple exchange of substituents may lead to totally different crystal and molecular structures (Desiraju and Sarma 1986; Bar and Bernstein 1987). Again, the *systematic* study of such systems provides a great deal of information about the nature of specific substituents or specific interactions.

In a closely related area several workers have been investigating the role of molecular shape and molecular free surface in determining molecular motions and conformations in organic crystals (Boeyens and Levendis 1982; Gavezzotti 1982, 1983, 1985; Meyer 1985*a*, *b*, 1990). These new approaches to looking at the forces and energetics associated with packing regularities and irregularities of organic crystals hold a great deal of promise for understanding the details of the interactions between intra- and intermolecular forces in crystals, and by extension, in a variety of other systems as well.

19.9 Concluding remarks

We have attempted to provide here the basis for an approach to the understanding and investigation of the influence of crystal forces on molecular structure. In organic molecules single-bond torsional energies are of the same order of magnitude as energy differences between polymorphic forms. This may lead to the existence of conformational polymorphic systems which contain a minimum number of variables for studying the relationship between molecular geometry and crystal environment. Combination of X-ray crystallographic structural techniques with the rapidly developing computational methods for studying molecular and crystal packing energetics has proven to be a fruitful strategy for approaching this problem. The high degree of order and the structural data available from crystal structures allow us to develop and refine the methods used to study interatomic interactions which may then be applied to other environments. The opposite approach, that is, proceeding theoretically or computationally from isolated molecules to the solid state, is considerably less promising, so that the continued search for higher precision in both the experimental and theoretical study of structural problems is a goal worthy of the effort.

19.10 Acknowledgements

A significant portion of our own work described here was carried out by Dr. Ilana Bar, as part of the requirements for the Ph.D. at Ben-Gurion University. The strategy of utilizing conformational polymorphism to investigate the relationship between crystal forces and molecular conformation developed during a delightful and stimulating collaboration with Dr. Arnie Hagler, presently at Biosym Technologies. I am grateful to the Israel Academy of Sciences and the US–Israel Binational Science Foundation for financial support of some of the work described here.

References

Allinger, N. L. (1976). Calculation of molecular structure and energy by force-field methods. In *Advances in physical organic chemistry*, vol. 13 (ed. V. Gold and D. Bethell), pp. 1–82. Academic Press, London.

Almenningen, A. and Bastiansen, O. (1958). *Kgl. Norske Videnskab. Selskabs., Skrifter 1958*, No. 4, pp. 1–16.

Almenningen, A., Bastiansen, O., Fernholt, L., Cyvin, B. N., Cyvin, S. J., and Samdal, S. (1985). *J. Mol. Struct.*, **128**, 59–76.

Altona, C. and Faber, D. H. (1974). Empirical force field calculations: A tool in structural organic chemistry. In *Dynamic chemistry*, Topics in Current Chemistry, Vol. 45, pp. 1–38. Springer, Berlin.

Andrew, E. R. (1975). High-resolution NMR in solids. In *Magnetic resonance*, MTP International Review of Science, Physical Chemistry Ser. 2, Vol. 4 (ed. C. A. McDowell), pp. 173–207. Butterworths, London.

Avbelj, F., Kitson, D. H., Eggleston, D. S., and Hagler, A. T. (1985). Presentation at the *American Crystallographic Association Summer Meeting*, Stanford, California, USA. Abstract PB27.

Baca, A., Rossetti, R., and Brus, L. E. (1979). *J. Chem. Phys.*, **70**, 5575–81.

Bar, I. and Bernstein, J. (1977). *Acta Crystallogr.*, **B33**, 1738–44.

Bar, I. and Bernstein, J. (1982a). *Acta Crystallogr.*, **B38**, 121–5.

Bar, I. and Bernstein, J. (1982b). *J. Phys. Chem.*, **86**, 3223–31.

Bar, I. and Bernstein, J. (1983). *Acta Crystallogr.*, **B39**, 266–72.

Bar, I. and Bernstein, J. (1984). *J. Phys. Chem.*, **88**, 243–8.

Bar, I. and Bernstein, J. (1987). *Tetrahedron*, **43**, 1299–305.

Barrett, M. W., Farrant, R. D., Kirk, D. N., Mersh, J. D., Sanders, J. K. M., and Duax, W. L. (1982). *J. Chem. Soc. Perkin Trans. II*, 105–10.

Bastiansen, O. (1949). *Acta Chem. Scand.*, **3**, 408–14.

Bastiansen, O. and Samdal, S. (1985). *J. Mol. Struct.*, **128**, 115–25.

Baudour, J. L., Delugeard, Y., and Cailleau, H. (1976). *Acta Crystallogr.*, **B32**, 150–4.

Baudour, J. L., Cailleau, H., and Yelon, W. B. (1977). *Acta Crystallogr.*, **B33**, 1773–80.

Benedetti, E., Bavoso, A., Di Blasio, B., Pavone, V., Pedone, C., Crisma, M., Bonora, G. M., and Toniolo, C. (1982). *J. Am. Chem. Soc.*, **104**, 2437–44.

Bernstein, J. (1987). Conformational polymorphism. In *Organic solid state chemistry*, Studies in Organic Chemistry, Vol. 32 (ed. G. R. Desiraju), pp. 471–518. Elsevier, Amsterdam.

Bernstein, J. and Hagler, A. T. (1978). *J. Am. Chem. Soc.*, **100**, 673–81.

Bernstein, J. and Izak, I. (1975). *J. Cryst. Mol. Struct.*, **5**, 257–66.

Bernstein, J. and Izak, I. (1976). *J. Chem. Soc. Perkin Trans. II*, 429–34.

Bernstein, J. and Schmidt, G. M. J. (1972). *J. Chem. Soc. Perkin Trans. II*, 951–5.

Bernstein, J., Bar, I., and Christensen, A. (1976). *Acta Crystallogr.*, **B32**, 1609–11.

Bernstein, J., Engel, Y. M., and Hagler, A. T. (1981). *J. Chem. Phys.*, **75**, 2346–53.

Billeter, M., Braun, W., and Wüthrich, K. (1982). *J. Mol. Biol.*, **155**, 321–46.

Birnbaum, G. I., Cygler, M., Kusmierek, J. T., and Shugar, D. (1981). *Biochem. Biophys. Res. Commun.*, **103**, 968–74.

Birnbaum, G. I., Cygler, M., Ekiel, I., and Shugar, D. (1982). *J. Am. Chem. Soc.*, **104**, 3957–64.

Boeyens, J. C. A. and Levendis, D. C. (1982). *S. Afr. J. Chem.*, **35**, 144–52.

Bondi, A. (1963). *J. Chem. Eng. Data*, **8**, 371–81.

Boyd, D. B. and Lipkowitz, K. B. (1982). *J. Chem. Educ.*, **59**, 269–74.

Brand, J. C. D. and Speakman, J. C. (1960). *Molecular structure: The physical approach*. Arnold, London.

Brock, C. P. (1980). *Acta Crystallogr.*, **B36**, 968–71.

Brock, C. P. and Morelan, G. L. (1986). *J. Phys. Chem.*, **90**, 5631–40.

Bryant, R. G., Chacko, V. P., and Etter, M. C. (1984). *Inorg. Chem.*, **23**, 3580–4.

Bürgi, H.-B. and Dunitz, J. D. (1983). *Acc. Chem. Res.*, **16**, 153–61.

Burkert, U. and Allinger, N. L. (1982). *Molecular mechanics*, A C S Monograph No. 177. American Chemical Society, Washington.

Burley, S. K. and Petsko, G. A. (1985). *Science*, **229**, 23–8.

Byrn, S. R. (1982). *Solid-state chemistry of drugs*. Academic Press, New York.

Byrn, S. R., Graber, C. W., and Midland, S. L. (1976). *J. Org. Chem.*, **41**, 2283–8.

Campsteyn, H., Dideburg, O., Dupont, L., and Lamotte, J. (1979). *Acta Crystallogr.*, **B35**, 2971–5.

Casalone, G., Mariani, C., Mugnoli, A., and Simonetta, M. (1968). *Mol. Phys.*, **15**, 339–48.

Charbonneau, G.-P. and Delugeard, Y. (1976). *Acta Crystallogr.*, **B32**, 1420–3.

Clements, J. A. (1976). *Proc. Analyt. Div. Chem. Soc.*, **13**, 21–5.

Coiro, V. M., Giglio, E., and Quagliata, C. (1972). *Acta Crystallogr.*, **B28**, 3601–5.

Colapietro, M., Domenicano, A., Portalone, G., Schultz, Gy., and Hargittai, I. (1984*a*). *J. Mol. Struct.*, **112**, 141–57.

Colapietro, M., Domenicano, A., Portalone, G., Torrini, I., Hargittai, I., and Schultz, Gy. (1984*b*). *J. Mol. Struct.*, **125**, 19–32.

Colapietro, M., Domenicano, A., Portalone, G., Schultz, Gy., and Hargittai, I. (1987). *J. Phys. Chem.*, **91**, 1728–37.

Corradini, P. (1973). *Chim. Ind. (Milan)*, **55**, 122–9.

Cox, S. R. and Williams, D. E. (1981). *J. Comput. Chem.*, **2**, 304–23.

Daniels, F., Williams, J. W., Bender, P., Alberty, R. A., Cornwell, C. D., and Harriman, J. E. (1969). *Experimental physical chemistry* (7th edn). McGraw-Hill, New York.

Deffet, L. (1942). *Répertoire des composés organiques polymorphes*. Desoer, Liège.

Desiraju, G. R. and Sarma, J. A. R. P. (1986). *Proc. Indian Acad. Sci. (Chem. Sci.)*, **96**, 599–605.

Djerassi, C. (1960). *Optical rotatory dispersion*. McGraw-Hill, New York.

Domenicano, A. (1988). Substituted benzene derivatives. In *Stereochemical applications of gas-phase electron diffraction* (ed. I. Hargittai and M. Hargittai), Part B, pp. 281–324. V C H, New York.

Duax, W. L., Fronckowiak, M. D., Griffin, J. F., and Rohrer, D. C. (1982). A comparison between crystallographic data and molecular mechanics calculations on the side chain and backbone conformations of steroids. In *Intramolecular dynamics* (ed. J. Jortner and B. Pullman), pp. 505–24. Reidel, Dordrecht.

Dunitz, J. D. (1979). *X-ray analysis and the structure of organic molecules*. Cornell University Press, Ithaca.

Dunitz, J. D. and Bürgi, H.-B. (1975). Non-bonded interactions in organic molecules. In *Chemical crystallography*, M T P International Review of Science, Physical Chemistry Ser. 2, Vol. 11 (ed. J. M. Robertson), pp. 81–120. Butterworths, London.

Eaton, V. J. and Steele, D. (1973). *J. Chem. Soc. Faraday Trans. II*, **69**, 1601–8.

Eckhardt, C. J. and Pennelly, R. R. (1971). *Chem. Phys. Lett.*, **9**, 572–4.

Eliel, E. L. (1962). *Stereochemistry of carbon compounds*, p. 32. McGraw-Hill, New York.

Engler, E. M., Andose, J. D., and Schleyer, P. v. R. (1973). *J. Am. Chem. Soc.*, **95**, 8005–25.

Ermer, O. (1976). Calculation of molecular properties using force fields: Applications in organic chemistry. In *Bonding forces*, Structure and Bonding, Vol. 27, pp. 161–211. Springer, Berlin.

Etter, M. C., Kress, R. B., Bernstein, J., and Cash, D. J. (1984). *J. Am. Chem. Soc.*, **106**, 6921–7.

Gavezzotti, A. (1982). *Nouv. J. Chim.*, **6**, 443–50.

Gavezzotti, A. (1983). *J. Am. Chem. Soc.*, **105**, 5220–5.

Gavezzotti, A. (1985). *J. Am. Chem. Soc.*, **107**, 962–7.

Griffin, R. G. (1977). *Anal. Chem.*, **49**, 951A–62A.

Griffiths, C. H. and Monahan, A. R. (1976). *Mol. Cryst. Liq. Cryst.*, **33**, 175–87.

Hagler, A. T. and Bernstein, J. (1978). *J. Am. Chem. Soc.*, **100**, 6349–54.

Hagler, A. T. and Lifson, S. (1974). *J. Am. Chem. Soc.*, **96**, 5327–35.

Hagler, A. T., Leiserowitz, L., and Tuval, M. (1976). *J. Am. Chem. Soc.*, **98**, 4600–12.

Haleblian, J. K. (1975). *J. Pharm. Sci.*, **64**, 1269–88.

Haleblian, J. K. and McCrone, W. C. (1969). *J. Pharm. Sci.*, **58**, 911–29.

Hall, L. D. and Sanders, J. K. M. (1980). *J. Am. Chem. Soc.*, **102**, 5703–11.

Hall, L. D. and Sanders, J. K. M. (1981). *J. Org. Chem.*, **46**, 1132–8.

Hargittai, I. and Schultz, Gy. (1986). *J. Chem. Phys.*, **84**, 5220–1.

Hargittai, M. and Hargittai, I. (1987). *Phys. Chem. Minerals*, **14**, 413–25.

Hargreaves, A. and Rizvi, S. H. (1962). *Acta Crystallogr.*, **15**, 365–73.

Hehre, W. J. (1976). *Acc. Chem. Res.*, **9**, 399–406.

Hehre, W. J., Stewart, R. F., and Pople, J. A. (1969). *J. Chem. Phys.*, **51**, 2657–64.

Hehre, W. J., Ditchfield, R., Stewart, R. F., and Pople, J. A. (1970). *J. Chem. Phys.*, **52**, 2769–73.

Hirshfeld, F. L. and Mirsky, K. (1979). *Acta Crystallogr.*, **A35**, 366–70.

Huler, E., Sharon, R., and Warshel, A. (1977). *QCPE—Quantum Chemistry Program Exchange*, **11**, 325.

Jeffrey, G. A., Ruble, J. R., McMullan, R. K., DeFrees, D. J., Binkley, J. S., and Pople, J. A. (1980). *Acta Crystallogr.*, **B36**, 2292–9.

Jelinski, L. W. (1984). *Chem. Eng. News*, **62**(45), 26–47.

Jones, W., Tilak, D., Tennakoon, B., Thomas, J. M., Williamson, L. J., Ballantine, J. A., and Purnell, J. H. (1983*a*). *Proc. Indian Acad. Sci. (Chem. Sci.)*, **92**, 27–41.

Jones, W., Theocharis, C. R., Thomas, J. M., and Desiraju, G. R. (1983*b*). *J. Chem. Soc. Chem. Commun.*, 1443–4.

Karpowicz, R. J., Sergio, S. T., and Brill, T. B. (1983). *Ind. Eng. Chem. Prod. Res. Dev.*, **22**, 363–5.

Kemp, J. D. and Pitzer, K. S. (1936). *J. Chem. Phys.*, **4**, 749.

Kihara, T. (1978). *Intermolecular forces*. Wiley, Chichester.

Kitaigorodskii, A. I. (1961). *Organic chemical crystallography*. Consultants Bureau, New York.

Kitaigorodskii, A. I. (1970). General view on molecular packing. In *Advances in structural research by diffraction methods*, Vol. 3 (ed. R. Brill and R. Mason), pp. 173–247. Pergamon Press, Oxford, and Vieweg, Braunschweig.

Kitaigorodskii, A. I. (1973). *Molecular crystals and molecules*. Academic Press, New York.

Kitaigorodskii, A. I. and Mirskaya, K. V. (1972). *Mater. Res. Bull.*, **7**, 1271–80.

Kroon, J., van Gurp, P. R. E., Oonk, H. A. J., Baert, F., and Fouret, R. (1976). *Acta Crystallogr.*, **B32**, 2561–4.

Krygowski, T. M. and Turowska-Tyrk, I. (1987). *Chem. Phys. Lett.*, **138**, 90–4.

Kuhnert-Brandstätter, M. (1971). *Thermomicroscopy in the analysis of pharmaceuticals*. Pergamon Press, Elmsford.

Le Gall, L. and Suzuki, S. (1977). *Chem. Phys. Lett.*, **46**, 467–8.

Lesyng, B., Jeffrey, G. A., and Maluszynska, H. (1988). *Acta Crystallogr.*, **B44**, 193–8.

McCrone, W. C. (1957). *Fusion methods in chemical microscopy*. Interscience, New York.

McCrone, W. C. (1965). Polymorphism. In *Physics and chemistry of the organic solid state*, Vol. 2 (ed. D. Fox, M. M. Labes, and A. Weissberger), pp. 725–67. Interscience, New York.

Maciel, G. E. (1984). *Science*, **226**, 282–8.

Maciel, G. E. and Sullivan, M. J. (1982). Carbon-13 NMR characterization of solid fossil fuels using cross-polarization and magic-angle spinning. In *NMR spectroscopy: New methods and applications*, ACS Symposium Series, Vol. 191 (ed. G. C. Levy), pp. 319–43. American Chemical Society, Washington.

McNaughton, J. L. and Mortimer, C. T. (1975). Differential scanning calorimetry. In *Thermochemistry and thermodynamics*, MTP International Review of Science, Physical Chemistry Ser. 2, Vol. 10 (ed. H. A. Skinner), pp. 1–44. Butterworths, London.

Mehring, M. (1976). *High resolution NMR spectroscopy in solids*, NMR: Basic Principles and Progress, Vol. 11. Springer, Berlin.

Merski, J. and Eckhardt, C. J. (1981). *J. Chem. Phys.*, **75**, 3691–704.

Meyer, A. Y. (1985*a*). *J. Chem. Soc. Perkin Trans. II*, 1161–9.

Meyer, A. Y. (1985*b*). *J. Mol. Struct. (Theochem)*, **124**, 93–106.

Meyer, A. Y. (1990). *Struct. Chem.*, **1**, 265–79.

Mirsky, K. (1976). *Acta Crystallogr.*, **A32**, 199–207.

Mirsky, K. (1980). *Chem. Phys.*, **46**, 445–55.

Mirsky, K. and Cohen, M. D. (1976). *J. Chem. Soc. Faraday Trans. II*, **72**, 2155–63.

Mislow, K. (1966). *Introduction to stereochemistry*, pp. 33–38. Benjamin, New York.

Mislow, K., Dougherty, D. A., and Hounshell, W. D. (1978). *Bull. Soc. Chim. Belg.*, **87**, 555–72.

Morel, D. L., Stogryn, E. L., Ghosh, A. K., Feng, T., Purwin, P. E., Shaw, R. F., Fishman, C., Bird, G. R., and Piechowski, A. P. (1984). *J. Phys. Chem.*, **88**, 923–33.

Murray-Rust, P. and Glusker, J. P. (1984). *J. Am. Chem. Soc.*, **106**, 1018–25.

Nepveu, F., Dahan, F., Haran, R., Cassoux, P., and Bonnet, J.-J. (1984). *J. Cryst. Spectr. Res.*, **14**, 129–42.

Nyburg, S. C. and Faerman, C. H. (1985). *Acta Crystallogr.*, **B41**, 274–9.

Oie, T., Maggiora, G. M., Christoffersen, R. E., and Duchamp, D. J. (1981). *Int. J. Quantum Chem., Quantum Biol. Symp.*, **8**, 1–47.

Orville-Thomas, W. J. (1974). Preface. In *Internal rotation in molecules* (ed. W. J. Orville-Thomas). Wiley-Interscience, London.

Ōsawa, E. and Musso, H. (1982). Application of molecular mechanics calculations to organic chemistry. In *Topics in stereochemistry*, Vol. 13 (ed. N. L. Allinger, E. L. Eliel, and S. H. Wilen), pp. 117–93. Wiley-Interscience, New York.

Panagiotopoulos, N. C., Jeffrey, G. A., La Placa, S. J., and Hamilton, W. C. (1974). *Acta Crystallogr.*, **B30**, 1421–30.

Ramdas, S. and Thomas, J. M. (1976). *J. Chem. Soc. Faraday Trans. II*, **72**, 1251–8.

Rigby, M., Smith, E. B., Wakeham, W. A., and Maitland, G. C. (1986). *The forces between molecules*. Clarendon Press, Oxford.

Ripmeester, J. A. (1980). *Chem. Phys. Lett.*, **74**, 536–8.

Robertson, G. B. (1961). *Nature*, **191**, 593–4.

Rogers, D. W. (1987). *Am. Lab.*, **19**(4), 28–39.

Sanders, J. K. M. and Hunter, B. K. (1987). *Modern NMR spectroscopy: A guide for chemists*. Oxford University Press.

Sarma, J. A. R. P. and Desiraju, G. R. (1985). *Chem. Phys. Lett.*, **117**, 160–4.

Sarma, J. A. R. P. and Desiraju, G. R. (1986). *Acc. Chem. Res.*, **19**, 222–8.

Sawzik, P. and Craven, B. M. (1979). *Acta Crystallogr.*, **B35**, 895–901.

Schaefer, J. and Stejskal, E. O. (1976). *J. Am. Chem. Soc.*, **98**, 1031–2.

Schaefer, J. and Stejskal, E. O. (1979). High-resolution ^{13}C NMR of solid polymers. In *Topics in carbon-13 NMR spectroscopy*, Vol. 3 (ed. G. C. Levy), pp. 283–324. Wiley-Interscience, New York.

Schaefer III, H. F. (ed.) (1977a). *Methods of electronic structure theory*, Modern Theoretical Chemistry, Vol. 3. Plenum Press, New York.

Schaefer III, H. F. (ed.) (1977b). *Applications of electronic structure theory*, Modern Theoretical Chemistry, Vol. 4. Plenum Press, New York.

Scharfenberg, P. and Hargittai, I. (1984). *J. Mol. Struct.*, **112**, 65–70.

Scheraga, H. A. (1971). *Chem. Rev.*, **71**, 195–217.

Schmidt, G. M. J. (1971). *Pure Appl. Chem.*, **27**, 647–78.

Schultz, Gy. and Hargittai, I. (1973). *Acta Chim. Acad. Sci. Hung.*, **75**, 381–8.

Suzuki, H. (1959). *Bull. Chem. Soc. Jpn.*, **32**, 1340–50.

Taylor, R. and Kennard, O. (1984). *Acc. Chem. Res.*, **17**, 320–6.

Theocharis, C. R., Desiraju, G. R., and Jones, W. (1984). *J. Am. Chem. Soc.*, **106**, 3606–9.

Tristani-Kendra, M., Eckhardt, C. J., Bernstein, J., and Goldstein, E. (1983). *Chem. Phys. Lett.*, **98**, 57–61.

Van Buuren, G. N., Willis, A. C., Einstein, F. W. B., Peterson, L. K., Pomeroy, R. K., and Sutton, D. (1981). *Inorg. Chem.*, **20**, 4361–7.

Verma, A. R. and Krishna, P. (1966). *Polymorphism and polytypism in crystals*. Wiley, New York.

Walker, M. S., Miller, R. L., Griffiths, C. H., and Goldstein, P. (1972). *Mol. Cryst. Liq. Cryst.*, **16**, 203–11.

Warshel, A. (1977). The consistent force field and its quantum mechanical extension. In *Semiempirical methods of electronic structure calculation*, Part A, *Techniques*,

Modern Theoretical Chemistry, Vol. 7 (ed. G. A. Segal), pp. 133–72. Plenum Press, New York.

Warshel, A. and Lifson, S. (1970). *J. Chem. Phys.*, **53**, 582–94.

Wasylishen, R. E. and Fyfe, C. A. (1982). High-resolution NMR of solids. In *Annual reports on NMR spectroscopy*, Vol. 12 (ed. G. A. Webb), pp. 1–80. Academic Press, London.

Weast, R. C. (ed.) (1975). *CRC Handbook of chemistry and physics* (56th edn). CRC Press, Cleveland.

Weber, H. P., Loosli, H. R., and Petcher, T. J. (1982). The conformation of dihydroergopeptines in the crystal and in solution. In *Steric effects in biomolecules* (ed. G. Náray-Szabó), pp. 39–52. Elsevier, Amsterdam.

Westrum Jr., E. F. and McCullough, J. P. (1963). Thermodynamics of crystals. In *Physics and chemistry of the organic solid state*, Vol. 1 (ed. D. Fox, M. M. Labes and A. Weissberger), pp. 1–178. Interscience, New York.

Williams, D. E. (1972). *Acta Crystallogr.*, **A28**, 629–35.

Williams, D. E. (1974). *Acta Crystallogr.*, **A30**, 71–7.

Williams, D. E. and Hsu, L.-Y. (1985). *Acta Crystallogr.*, **A41**, 296–301.

Wüthrich, K., Wider, G., Wagner, G., and Braun, W. (1982). *J. Mol. Biol.*, **155**, 311–9.

Yannoni, C. S. (1982). *Acc. Chem. Res.*, **15**, 201–8.

The importance of accurate structure determination in inorganic chemistry

Jeremy K. Burdett

20.1 Introduction

Inorganic chemistry involves the study of virtually all of the periodic table, and as such exhibits a tremendously diverse structural chemistry. It is a truism to say that today the characterization of a new inorganic or organo-metallic species is not complete without an X-ray structure determination. Organic chemists, on the other hand, are often quite happy with an NMR or other spectroscopic study. In fact a comparison between the material in this chapter and that of Chapter 16 on organic chemistry will show some very important differences between the two fields. At the present time organic chemistry is a much more mature subject than its inorganic counter-part. As a result we find a considerably smaller number of new organic struc-tural types compared with new modes of coordination or geometry in the inorganic and organometallic areas. Even in its immature state, structural inorganic chemistry is a much richer field than its organic analog. Inorganic and organic chemists, as a result, use accurate structural results in rather different ways. While the organic chemist may be interested in rather small changes in molecular dimensions as the result of a substitution of one or more of the atoms in a parent molecule, the inorganic chemist determines bond lengths and angles which show much wider system-to-system varia-tions than corresponding parameters in organic chemistry. Molecular orbital models to account for this wide spectrum of observations are well developed in the inorganic area and many experiments are, in fact, devised to test existing theories.

We should realize too that the structure of an inorganic molecular system, determined by diffraction methods on a crystal, may be somewhat different from that existing in a free molecule in the gas phase, or as a solvate in solution. Intermolecular forces may be quite important in determining the finer details of the structure. Solid-state extended arrays usually only go into solution as a result of the making and breaking of chemical bonds. Thus the early transition-metal oxides MO_2 and MO_3 dissolve in water to give a variety of complex polymetallates with formulae dependant on the pH and quite different from that of the solid oxide. The gaseous molecules themselves exist as short-lived and reactive species in the gas phase. Stable fragments of such extended arrays are usually only found where the 'dangling bonds' at the end of the fragment have been chemically sealed off with other atoms, hence changing the overall stoichiometry. A familiar example is the structure of adamantane, $C_{10}H_{16}$, containing a ten-carbon unit excised from the cubic diamond structure with the carbon valencies satisfied by the addition of hydrogen atoms. One set of beautiful inorganic examples are the molecules M_9O_2 and $M_{11}O_3$ (M = Rb, Cs) which contain respectively two and three face-sharing M_6O octahedra (see Simon 1975 and references within). The parent structure may be regarded as the anti-cadmium halide structure of Cs_2O which contains sheets of such octahedra sharing edges. In contrast to the organic case it is not at all obvious that these stoichiometries will be the stable ones, an observation which highlights the rather different theoretical ideas that need to be employed by inorganic chemists to provide working models for their chemistry. This chapter gives a personal view of some interesting structural problems in inorganic chemistry which highlight the variety and excitement in this area.

20.2 Unusual molecules and solids

We mentioned above the structural diversity of inorganic chemistry and the reliance of the researcher on accurate structure determination. Some examples will show how important a tool this is and how vital it is to have really good and trustworthy data when making claims as to new types of molecules and solids. Ferrocene, for example, when it was discovered in the 1950s, upset all of the ideas chemists had concerning what sort of molecules should exist. It contains two planar cyclopentadienyl (Cp) rings sandwiching an iron atom, and of course its structural characterization was a crucial step before its acceptance by the chemical world (Dunitz et al. 1956). Still today students ask whether the iron atom is ten-coordinate or two-coordinate. Modern molecular orbital ideas (see, for example, Albright et al. 1985) suggest that electronically it is perhaps best regarded as six-coordinate. For many years all the mononuclear MCp_2 species that were known contained parallel rings of this sort, except for those with metals such

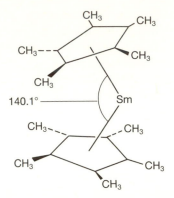

Fig. 20.1 The 'bent' structure of decamethylsamarocene (Evans *et al.* 1984).

as tin or lead which possess a lone electron pair. However, Fig. 20.1 shows the structure of a derivative of samarocene, $SmCp_2$, a molecule synthesized only relatively recently (Evans *et al.* 1984); the molecule is in fact the permethyl derivative, written conventionally as $SmCp_2^*$. Here the crystallographic results show a molecular structure which is very different from that of ferrocene, in that the rings are non-parallel ($FeCp_2^*$ has parallel rings too).

Such a 'bent' unit (referring to the angle made by the normals to the two Cp planes, as shown in Fig. 20.1) could arise for several reasons. The first is that the material is not really $SmCp_2^*$ but a related species where an atom, attached to the metal, was left out of the structural refinement. A light atom (such as hydrogen) coordinated to the heavy Sm atom might easily be missed, and not show up in a conventional chemical analysis either. The structure of H_3NbCp_2 contains three M—H linkages, and here the Cp rings are bent. However there is chemical evidence that this species is not $H_xSmCp_2^*$ and so this possibility may be dismissed. The second possibility is of an 'agostic' linkage between one or more of the hydrogen atoms of the organic part of the molecule and the metal. We will describe this interesting bond later in this section, but may exclude its possibility here since the Sm $\cdots$ H distances are in general far too long (the shortest is around 2.8 Å) for this to be important. A third possibility is that the samarium atom (Sm^{II}) contains stereochemically active f orbitals. The observation of (say) non-tetrahedral structures for d^n ($n > 0$) transition metal ML_4 species (e.g. the butterfly structure of $Cr(CO)_4$) has long been used as evidence for stereochemical activity of d orbitals (see, for example, Albright *et al.* 1985). However f orbitals in rare-earth chemistry are generally regarded as being stereochemically impotent. One clue as to their reduced energetic importance is the considerably smaller band widths associated with the electronic spectra of *f*-orbital compared to *d*-orbital complexes. A fourth possibility is that the structure is controlled by so-called 'packing' forces, i.e., a balance between inter- and intramolecular forces in the crystal. However a gas-phase determination (Andersen *et al.* 1986) of the structures of $CaCp_2^*$ and YCp_2^* shows similar bent structures, so this may be excluded. Yet another explanation, and one that has a certain appeal, is that the rings are pulled together by van der Waals' attractions, in much the same way that organic hydrocarbon molecules are held to each other in the solid state. Such a model presupposes that the directional forces at the metal are rather weak. The model is an interesting one in that steric interactions between ligands *attached to the same atom,* usually regarded as being repulsive in character, are attractive here.

Figure 20.2 shows the reported structure of part of a zeolite which attracted a considerable amount of attention when first published (Firor and Seff 1976) but was later shown to be incorrect (see Seff and Mellum (1984) and the complete set of references therein). Zeolites are materials of general formula $M_xSi_{1-x}Al_xO_2$ where the metal M in this case is a univalent atom required for charge balance. Two- and three-valent metals can play this role too. The non-metal framework is composed of two-coordinate oxygen atoms and roughly tetrahedrally coordinated silicon or aluminum atoms. A feature of such materials is the presence of large channels and cavities formed by these non-metal atoms. The electropositive metal atoms are always coordinated to the oxygen atoms of the framework. It was thus a surprise to find a zeolite cage which in addition to the several Rb^+ ions coordinated to oxygen, held one ion which appeared to float in the middle of the cage. In terms of electrostatics such an ion would lie at a position

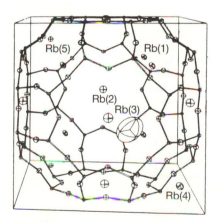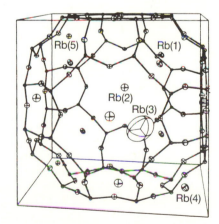

Fig. 20.2 The original (incorrect) report of a floating, 'zero-coordinate' Rb^+ ion, Rb(3), in a zeolite cage. After Firor and Seff (1976).

of unstable equilibrium, and is clearly of great chemical importance if correct. However, a part of the problem in this and several other examples of this type is that the incorrect space group was used in the crystallographic refinement. In zeolites such as these the silicon and aluminum atoms are invariably ordered over the lattice. If $x = 0.5$ then the experimental observation is that no two atoms of the same type are in adjacent sites. This self-avoidance principle is known as Loewenstein's rule (Loewenstein 1954). The space group of the non-metal framework is thus different if all the Si/Al atoms are assumed to give the same average scattering from these sites (the random case), or if in fact they are ordered. Thus to date there is no evidence for zero-coordinate ions of this type (Pluth and Smith 1983). Use of the incorrect space group is a common error in X-ray structure determination. In many cases the error is not of chemical importance, but quite often rather significant differences do arise; see, for example, Marsh and Schomaker (1979).

Figure 20.3 shows the structure of a molecule, synthesized in recent years, which was the first to illustrate a new mode of coordination of dihydrogen to a transition metal (Kubas *et al.* 1984). The molecule, $W(P^iPr_3)_2(CO)_3(H_2)$, is a fascinating one in that it reversibly loses molecular H_2 in the crystal. Very similar analogs have been made with Mo instead of W, and with different phosphines. The H—H distance of 0.84 Å (no quoted error) obtained by neutron crystallography is close to that found by vibration–rotation spectroscopy for the ion H_3^+, $r_0 = 0.876$ Å (calculated from the rotational constants given by Oka (1980)). The experimentally determined distance in the H_2 molecule is $r_e = 0.7417$ Å (Herzberg 1950).

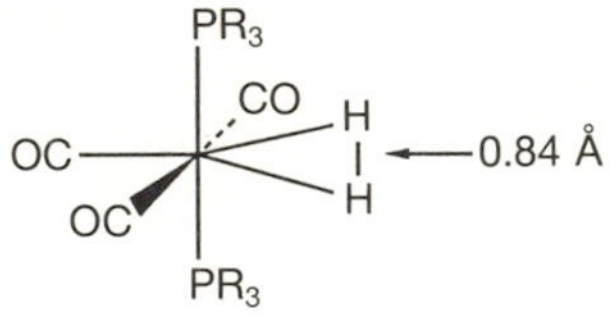

Fig. 20.3 The structure of the molecule $W(PR_3)_2(CO)_3(H_2)$, where R = ^iPr (Kubas *et al.* 1984).

In electronic terms the new species is interesting (Burdett *et al.* 1987a). It falls into a class of species where an H_2 unit is attached to a fragment (X) which has a vacant frontier orbital. Figure 20.4 shows the electronic relationship between these $M(PR_3)_2(CO)_3(H_2)$ species (here X = $M(PR_3)_2(CO)_3$), other complexes containing dihydrogen (X = $M(CO)_x(NO)_{3-x}$), and the species H_3^+ (X = H^+) and CH_5^+ (X = CH_3^+). The last example is only known as a species in the mass spectrometer. Its structure has not yet been

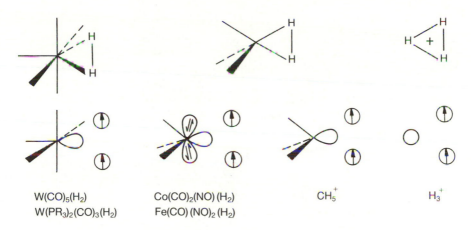

Fig. 20.4 Electronic relationships among 'H$_2$' complexes.

experimentally determined and there is much current activity in this area (see, for example, Chapter 1, Section 1.6). In each of the cases of Fig. 20.4 the X units are isolobal (for a discussion of the isolobal analogy see Albright *et al.* (1985)). However, whether the molecule exists as a dihydride or as a dihydrogen complex in the transition-metal examples, is a subtle balance of electronic factors associated with the nature of the unit X (Burdett *et al.* 1987*a*).

The ideas presented above suggest (Burdett and Pourian 1988) that many other X(H$_2$) species should be stable too. For example the stability of the molecule where X = H$^+$ (i.e. H$_3^+$) immediately leads to the candidacy of species such as FH(H$_2$) where the hydrogen atom in the parent HF molecule clearly has large protonic character. Such a molecule has indeed been recently characterized via high-resolution vibration–rotation studies (Lovejoy *et al.* 1987; Jucks and Miller 1987), as have the related molecules OH$_3^+$(H$_2$)$_x$, where $x = 1$–3 (Okumura 1986).

This mode of coordination of H$_2$ to a transition metal is now well established. However, there have been other studies, which have not stood the test of time, which have purported to show interesting coordination geometries of other ligands. For example in an X-ray diffraction study of the molecule Cr(CO)$_5$pyridine, Fig. 20.5(a), very unusual C—C and C—N distances were found (Cotton *et al.* 1981; see Table 20.1). If these distances were real then they indicated a very different chemical bond attaching the organic molecule to the metal. However, associated with the structural refinement were rather large thermal parameters for the ligand atoms, which should have signalled that something was amiss. The authors reported an asymmetric unit containing this unusual Cr(CO)$_5$pyridine molecule in

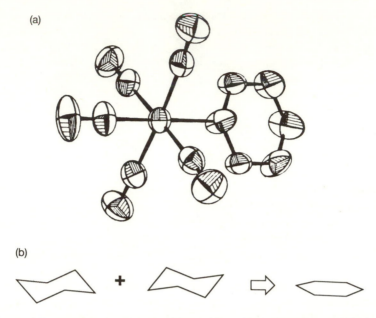

Fig. 20.5 (a) The purported molecule Cr(CO)₅ pyridine with its planar heterocyclic ring. After Cotton *et al.* (1981). (b) How the structure with a planar ring resulted from inadequate refinement of a disordered Cr(CO)₅ piperidine molecule containing a puckered ring.

addition to a molecule of Cr(CO)₅ piperidine with perfectly normal distances. A second structural study, based on crystals of Cr(CO)₅ pyridine, found intra-ligand distances in the range expected (Ries *et al.* 1984; see Table 20.1). What then was wrong with the first study? It turns out that the molecule was in fact Cr(CO)₅ piperidine, where the planar unsaturated ring of the ligand has been replaced by a puckered saturated one. In other words *both* molecules in the asymmetric unit of the crystal studied were Cr(CO)₅ piperidine. But in one of them the puckered organic unit was disordered over two alternative positions. Since for each of the six atoms of the ring the two alternative positions are separated by 0.5 Å or less, the refinement spontaneously converged to a roughly planar geometry (see

Table 20.1 Geometric details of the pyridine ring in Cr(CO)₅ pyridine from two X-ray diffraction studies

	$\langle r(\text{C}-\text{C})\rangle$ (Å)	$\langle r(\text{C}-\text{N})\rangle$ (Å)	References
	1.442	1.386	Cotton *et al.* (1981)
	1.366	1.345	Ries *et al.* (1984)
Expected value	1.37	1.34	

Fig. 20.5(b)). The unusual distances within the ring and the large apparent thermal motion are thus understandable.

In many instances the crystallographic problem is not quite so clear-cut. An example comes from the coordination chemistry of dinitrogen. Whereas both end-on and sideways-on N_2 are well known in organic chemistry, as diazomethane and diazirine respectively, only the former is well characterized in inorganic chemistry. However there was a report (Busetto et al. 1977) of a sideways coordinated N_2 in the molecule $RhCl(N_2)(P^iPr_3)_2$. A structural reinvestigation at low temperature showed (Thorn et al. 1979) that in all probability, the ligand was in fact end-on coordinated, but because of disorder in the crystal some of the dimensions associated with the $M-N_2$ part of the molecule were not at all good, the same problem in fact which plagued the original study. The $N-N$ distance of 0.958(5) Å within the complex found in the later study, is shorter than that in the free molecule, $r_e = 1.094$ Å (Herzberg 1950).

Structure determination has proven to be an essential ingredient in understanding the properties of a new type of linkage, that of the recently characterized *'agostic'* hydrogen atom in transition-metal chemistry (Brookhart and Green 1983). Structurally the properties of such bonds have several similarities to those of the hydrogen bond. Electronically they may be broadly understood in the following way. Many transition-metal compounds have less than an eighteen electron count at the metal center, and thus are formally unsaturated. One way in which this problem may be alleviated and

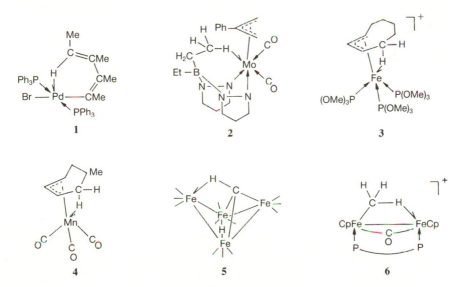

Fig. 20.6 Some examples of 'agostic' hydrogen atoms. The relevant bond lengths are given in Table 20.2. From Brookhart and Green (1983).

electronic saturation achieved by increasing the coordination number at the transition metal, is via a distortion of a coordinated organic ligand such that a hydrogen atom located on it is brought close to the metal center. Such a geometric situation gave the hydrogen atom its name (agostic comes from the Greek ἀγοστῷ, to clasp, or to hold to one's self). Figure 20.6 shows some examples, with the interatomic distances given in Table 20.2. Notice that in the last example the two crystallographically independent molecules have slightly different geometries. The two neutron diffraction studies in Table 20.2 indicate that the agostic C—H and M—H distances are somewhat longer than normal C—H and M—H bonds. As we have noted, in some ways the interaction is like a hydrogen bond, in the sense that the strength of the metal–hydrogen linkage, as measured by its internuclear distance is variable, and correlates with the changes in C—H distance on the ligand. There is an important electronic difference, however. In hydrogen bonds the hydrogen atom is attracted by an electronegative atom, but in the agostic case the more electropositive the metal the stronger the attraction appears to be. Such interactions are very probably important in reactions, such as Ziegler–Natta polymerization, involving coordinated ligands at an early transition-metal center, although as yet there is no really definitive evidence as to how this works.

These examples show quite clearly one of the roles played by accurate structure determination in this area, namely that of characterization of new types of molecules. Another important use of the results is of course to compare the details of the bond lengths and angles with other molecules, in order to find a structural–electronic pigeonhole in which to place the molecule. Sometimes the observed distances are very different, even in supposedly similar molecules. An example of this type of problem is posed (Jackson

Table 20.2 Some bond lengths (Å) associated with 'agostic' hydrogen atoms

Molecule[a]	$r(M-H)$	$r(C-H)$	$r(M\cdots C)$	References
1[b]	–	–	2.3	Roe *et al.* (1972)
2	2.27(8)	0.97(8)	3.055(7)	Cotton *et al.* (1974)
3[c]	1.874(3)	1.164(3)	2.384(4)	Brown *et al.* (1980)
4[c]	1.84(1)	1.19(1)	2.34(1)	Schultz *et al.* (1983)
5	1.80(4)	1.00(4)	1.926(5)	Beno *et al.* (1980)
6[d]	1.64(4)	1.06(4)	2.108(3)	Dawkins *et al.* (1982)
	1.78(3)	0.83(4)	2.118(3)	

[a] See Fig. 20.6.
[b] Agostic hydrogen not located.
[c] Neutron diffraction study.
[d] The asymmetric unit contains two crystallographically independent molecules.

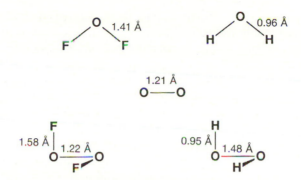

Fig. 20.7 Some bond lengths in oxygen fluorides and hydrides. Taken from the compendium in Burdett *et al.* (1984).

1962) by the geometry of the molecule O_2F_2. Figure 20.7 shows the distances in this species as determined by microwave spectroscopy, along with some other 'typical' $O-O$, $O-H$, and $O-F$ distances from gas-phase studies. Notice not only the striking difference between the $O-O$ distance in this molecule and that in the structurally similar O_2H_2, but the similarity to the value in dioxygen itself. The $O-F$ distance is much longer than expected, as comparison with the value found for the molecule OF_2 shows. In terms of its chemistry such a structural feature is reflected in the strong fluorinating power of O_2F_2. A simple theoretical model can lead to an understanding of some of these features (Burdett *et al.* 1984).

20.3 Natural products inorganic chemistry

Just as the organic chemist has an enormous natural wealth of compounds to study in living things, so the inorganic chemist has a fascinating natural products chemistry in the rocks which make up the earth. Many of the problems facing solid-state and materials chemists today in terms of structure and properties of solids have been a part of the lot of geochemists for years. Perhaps the first is finding a good single crystal for an X-ray study. At present it is often not possible to synthesize a crystal of a material, which nature has spent eons doing, in the laboratory in less than a human lifetime. Many of these materials are structurally quite complex too and present a problem of description in geometrical terms which molecular chemists rarely face. One of the major problems associated with this area is the large number of atoms often found in the asymmetric unit. Combined with the fact that the chemical composition is usually not simple and that mineral compositions are usually solid solutions of at least two end-members the crystallographic situation is often quite complex. Often one particular site may be partially occupied by a variety of different species (usually cations). The actual structural formula then is often an approximation, for cumulative errors in chemical analysis and choice of site populations by the crystallographer come into play.

An important aspect of the structures of these materials is how the atoms of a given system are ordered over the possible sites of a given structure. In zeolite A with a Si/Al ratio of unity it is now well established, despite earlier suggestions to the contrary, that no two silicon atoms and no two aluminum atoms reside on adjacent sites in the lattice, in accordance with Loewenstein's rule (see the discussion and references in Cheetham *et al.* (1982)). The ordering of the silicon and aluminum atoms in the feldspars has been an area of active investigation for several years. This is a very interesting crystallographic problem, since the ordering pattern of the ions is sometimes incommensurate with the translational periodicity of the oxide lattice (McConnell 1983). The problem is well illustrated by the plagioclase feldspars, one of the most common minerals found in rocks. At one time a whole series of solid solutions were thought to exist between the two endmembers anorthite ($CaAl_2Si_2O_8$) and albite ($NaAlSi_3O_8$). It is clear now that at low temperatures the situation is more complex. The superlattice peaks in the X-ray diffraction pattern of intermediate systems cannot be indexed in terms of the normal plagioclase unit cell. This is clear evidence of incommensurately modulated phases. In the Heine–McConnell scheme (Heine and McConnell 1984) the behavior of these systems is described in terms of a novel concept, that of a structural resonance between two ordering patterns derived from the strongly coupled substitutions Na $\leftrightarrow$ Ca and Si $\leftrightarrow$ Al. The overall picture is a complex one since it involves changes both in the framework atoms Si and Al and in the non-framework cations Na^+ and Ca^{2+}.

20.4   Low-temperature crystallographic studies

Chemists often want to compare theory with experiment. Invariably the computed distances used in such comparisons come from a hypothetical model which corresponds to a frozen (0 K) static (non-vibrating) material. For comparison's sake, parameters taken from low-temperature experiments are then the best ones to use. There are often problems however with such a strategy. For example the material may undergo a phase change to another structure on cooling. A particularly common event is a metal-to-insulator transition, either associated with the solid-state equivalent of a Jahn–Teller distortion (the Peierls distortion) or as the result of a transition related to the high-spin/low-spin one in molecules (see, for example, Albright *et al.* 1985). However, even without such a transformation there are often significant changes in the structural details on cooling which give valuable information concerning the material. We give two examples here.

In silicate chemistry bending around the oxygen atom is associated with a very soft potential. This means that the cost in energy on changing a bond angle from (say) 120° to 180° is not very large. Because of this structural flexibility (see Gibbs *et al.* 1981; Newton 1981) we find in practice an enormous range of silicate structures. There are two models which have been used to approach the problem. The first suggests (O'Keeffe and Hyde 1981, 1982) that because the Si—O distance is short (oxygen being a first-row atom) the silicon atoms can get rather close to each other and experience a strong repulsion if the Si—O—Si angle is too small. Thus the bending energetics are a balance of the electronic preferences at oxygen (with four electron pairs a roughly tetrahedral angle is expected) and these steric interactions (which prefer an open angle).

The other model suggests that it is π bonding between the silicon and oxygen atoms which is responsible for the stabilization of the linear geometry (for a new view of the model see Albright *et al.* (1985); Burdett and Caneva (1985)). A similar argument has existed for several years to rationalize the planar skeleton of $N(SiH_3)_3$, and more recently to understand the planar oxygen environment in rutile-type oxides. We prefer this electronic argument. The carbon suboxide molecule, C_3O_2, which may be written as (OC)C(CO), and its phosphine substituted analog $(R_3P)C(PR_3)$, where R is an alkyl group, both have a soft bending motion at the central carbon atom. Whereas both CO and PR_3 are π-acceptor ligands, only the latter contains bulky end-groups. In addition, C_3O_2 is linear but the phosphine analogs invariable bent, an observation not in accord with the relative ligand bulk. Whatever the explanation, because of the soft bending motion we might well expect to see some significant changes in the determined structural parameters with temperature. However, the inclusion of vibrational motion into the crystallographic problem for an anharmonic oscillator, and especially one with a soft force constant, as in these systems, is hardly a routine matter. Table 20.3 shows some recent results for cristobalite (one

Table 20.3 Bond lengths and angles in cristobalite (SiO_2) from a powder neutron diffraction study at two different temperatures (Pluth *et al.* 1985)

Parameter	Value[a]	
	10 K	473 K
$r(Si—O(1))$	1.602(1)	1.605(2)
$r(Si—O(2))$	1.617(1)	1.590(2)
$\angle Si—O—Si$	144.7(1)	148.4(1)

[a] Bond lengths are in Å, angles in degrees.

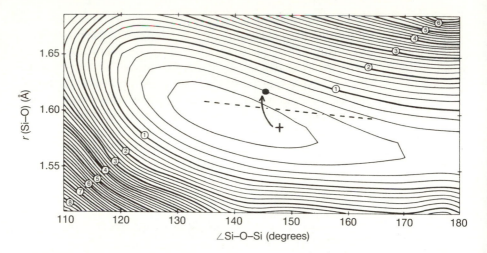

Fig. 20.8 A calculated Si—O—Si bending/Si—O stretching potential surface for silicate structures, using $Si_2O(OH)_6$ as a model. The energy is represented as a contour diagram with
increments of 0.001 atomic units (2.6 kJ mol^{-1}). Heavier lines represent contours separated by 0.005 atomic units (13.1 kJ mol^{-1}). They are labeled from 1 to 8 in order of
increasing energy. The dashed line in the middle of the diagram represents the observed correlation between these geometrical parameters from many crystal structures. The points
marked are those discussed in the text. Adapted from Gibbs *et al.* (1981).

of the several polymorphs of SiO_2), where the same refinement strategy
was applied to powder neutron diffraction data collected at high and low
temperature from the same sample (Pluth *et al.* 1985). It shows quite a
difference in interatomic distances with temperature.

We mentioned that it is perhaps best to use low-temperature results for
comparison with theory. Figure 20.8 shows the results of some *ab initio* MO
calculations on silicate fragments (Newton 1981; Gibbs *et al.* 1981) used to
model the behavior in silicate materials. The dashed line shows the bond
length/bond angle correlation obtained from a large number of room-
temperature crystal structure determinations on a variety of silicates. Notice
that it does not sit exactly in the valley defined by the contours of the computed energetics of the distortion. We show on the plot the temperature
behavior of one of the Si—O distances of Table 20.3. Notice how the data
point moves from high temperature (cross) to low temperature (circle) to
become *less* in accord with theory (but we should be cautious in making
broad claims with a single data point!). The movement of the other Si—O
distance of Table 20.3 is less dramatic.

A system where the interesting changes with temperature occur not in
bond lengths *per se* but in other crystallographic parameters, is that of rutile,
TiO_2 (Fig. 20.9). The metal atoms are formally octahedrally coordinated,

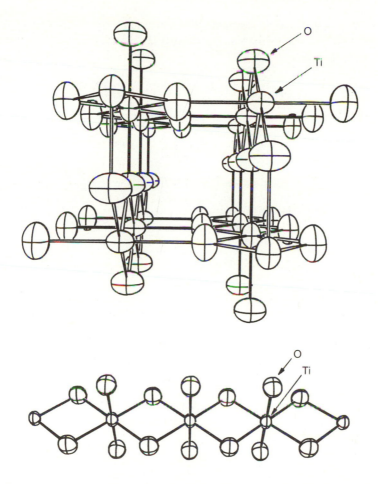

Fig. 20.9 Two views of the structure of the rutile form of TiO_2. The bottom picture shows the edge-sharing chains of TiO_6 octahedra.

but in this structure the two axial and four equatorial Ti—O distances are symmetry-inequivalent. Two long and four short distances are found experimentally, both at room temperature and at 15 K. One of the puzzling features of the arrangement is that band-structure calculations on rutile suggest (Burdett 1985) that the two axial bonds should be stronger, and hence shorter, than the equatorial ones, the opposite to that found crystallographically. Since calculations of this type invariably give the correct answer when it comes to prediction of relative bond lengths this result is a striking one. What it suggests in fact is that the relative interatomic distances in this material are set, not by the direct Ti—O interactions, but by

Table 20.4   Anisotropic displacement parameters of rutile (TiO_2) from a powder neutron diffraction study at two different temperatures (Burdett *et al*. 1987*b*)

Atom	Parameter	Value ($Å^2 \times 10^3$)	
		295 K	15 K
Ti	$U^{11} = U^{22}$	5.5(2)	1.2(2)
	U^{33}	4.5(3)	1.1(2)
	U^{12}	−0.3(3)	−0.2(2)
	$U^{13} = U^{23}$	0	0
O	$U^{11} = U^{22}$	5.7(1)	3.1(1)
	U^{33}	4.4(1)	2.6(1)
	U^{12}	−1.9(1)	−0.9(1)
	$U^{13} = U^{23}$	0	0

the interactions between the oxygen atoms. (Molecular chemists would call these steric interactions.) In MgF_2, another material with the rutile structure, the relative distances are, however, in accord with theory. Since fluorine ions are 'smaller' than their oxygen analogs, and metal–fluorine distances longer than typical metal–oxygen ones for these metals, we would expect such non-bonded repulsions to indeed be smaller. Evidence that the metal atom in TiO_2 sits in some sort of strained site whose dimensions are controlled by the oxide matrix, comes from the temperature-dependence of the thermal parameters of a powder neutron diffraction study (Table 20.4; Burdett *et al*. 1987*b*). Similar results are found using X-ray diffraction. Usually the heavier metal atom will have smaller thermal motion than the lighter oxygen atom, but at room temperature they have similar values of the anisotropic displacement parameters. Only at 15 K are these parameters of the expected size. By way of contrast there is no unusual thermal motion in the reported X-ray crystal structure determination of MgF_2 (Baur and Khan 1971). Here the smaller F^- ion allows the direct anion–cation interactions to control the geometry at the metal.

20.5   Intercalated materials

The tailoring of materials by subtly changing the electron count via intercalation with electron-donating or withdrawing groups is a popular synthetic route open to the inorganic chemist. Figure 20.10 shows the band structure of graphite in a simple form (Whangbo *et al*. 1979). It is interesting to see that the Fermi level lies at the join of the C−C bonding and antibonding parts of the π band. Addition of electrons will therefore populate the

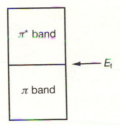

Fig. 20.10 A schematic band structure of graphite, showing how the π and π^* bands touch at the Fermi level of the pure material. As the electron concentration increases these π^* levels become populated and the intra-plane C—C distance increases.

antibonding part of the band and weaken the C—C linkage (Kertész *et al*. 1983). Eventually, addition of one electron per carbon will lead to a puckered structure like that of arsenic, but here we will be interested in smaller changes in electron concentration. Figure 20.11 shows how the C—C distance varies with the addition of electropositive potassium atoms (Nixon and Parry 1969), and is in accord with these simple ideas. Weaker bonds are generally longer bonds. (Parenthetically we note that the situation may be a little more complex with electron-withdrawing intercalants, see Kertész et al. (1983).)

Figure 20.12 shows another effect found with graphite (and indeed with some other materials) which is called 'staging' (see, for example, Bartlett and McQuillan 1982). Addition of intercalant gives rise to rather interesting local structures. It appears that the intercalated material prefers not to lie in a

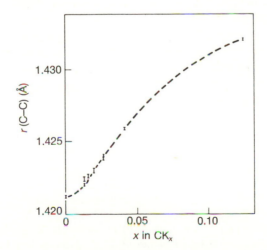

Fig. 20.11 The observed variation in C—C distance with potassium intercalation in graphite. Adapted from Nixon and Parry (1969).

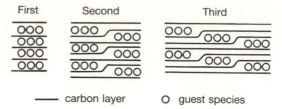

—— carbon layer   O guest species

Fig. 20.12 Schematic pictures showing the phenomenon of staging.

gallery directly above or below another intercalant molecule, unless forced to do so by the stoichiometry of the material. The explanation of this result is not yet clear. Obviously a model which includes repulsive interactions between intercalated species in adjacent galleries, but attractive ones between species in the same gallery will reproduce some of these features. Dipole–dipole interactive models have been used to date but simple chemical bonding models should be investigated too.

After intercalation, although the two-dimensional ordering pattern may be well characterized, there are problems with the three-dimensional structure in that the location of one pair of sheets and its contents relative to the next is often random. This is a simple consequence of the rather weak forces between the sheets. In other intercalated materials there are sometimes other problems too, which make accurate structure determination difficult. For example, n-butyllithium is a very useful reagent in this area, since lithium atoms may be deposited in the solid with the concomitant production of a gaseous organic hydrocarbon which is easily removed. However when using this route to intercalate WO_3 (Cava *et al.* 1983), although one starts off with a nice single crystal, the product $Li_{0.36}WO_3$ is not, and powder techniques have to be used in its characterization. Although the term intercalated materials is used to describe these systems it should be realized that they are often distinctly new species in their own right. Lithium may be intercalated into anatase, one of the TiO_2 polymorphs, to give a non-stoichiometric material Li_xTiO_2 based on the anatase parent. When $x = 0.5$, however, a phase transformation occurs and the material produced has the simple spinel structure, indicated as a possibility by the formula itself if it is written as $LiTi_2O_4$.

20.6   Pauling's rules

Inorganic chemists are particularly interested in the bond lengths in molecules and solids because in principle they give detailed information as to how the material is held together in electronic terms. One of the very useful ideas which links the bond lengths of many crystal structures is the bond-valence sum rule (see Brown 1981), itself a development of Pauling's second rule (Pauling 1929). In solids which are not of the metallic, organometallic or van der Waals' types (i.e. those where either the delocalized model has to be used or where some or all of the interatomic interactions are weak) the rule introduces the idea of the bond valence s, determined, for a given pair of atoms, only by the interatomic separation r. Two definitions are in current use (Brown and Altermatt 1985):

$$s = (r/r_o)^{-N} \tag{20.1}$$

and

$$s = \exp[(r_o - r)/B], \tag{20.2}$$

where r_o and N, or r_o and B are constants for a particular atom pair. Using these definitions it turns out that the sum of the bond valences around a particular anion or cation (i) to its coordinated cations or anions respectively (j) is almost constant. This sum is set, by tradition, equal to the usual valency (V_i) of the atom as:

$$V_i = \sum_j s_{ij}. \tag{20.3}$$

Equation (20.2) is somewhat more useful than eqn (20.1) in that B is approximately constant (0.37 Å) for a wide range of systems. This leaves a single variable (r_o) which may be expressed in terms of additive parameters for anion and cation (Brown and Altermatt 1985). Note that there is nothing really special about the valency here. It is just a scaling parameter for the bond valence. Equation (20.3) therefore controls the way the interatomic distances around a given atomic center are correlated. Increasing one must lead to a decrease in all or some of the others. It also gives a measure of the relative strengths of internuclear contacts of various types. Six-coordinate sodium will form bonds with a bond valence of approximately one sixth, and four-coordinate nitrogen (envisaged as N^{-3}), bonds with a bond valence of approximately three quarters. These numbers (the 'valency' divided by the coordination number) are called the electrostatic bond strengths to avoid confusion with the term bond valence. The rule of eqn (20.3) is a remarkable one, and its origins are not completely understood, but it basically tells us that the total bond strength at an atom should remain constant. Thus it also dictates the sign of bond–bond interaction force constants in molecules, in the absence of other electronic forces.

This rule is an extremely useful one in crystal chemistry. In X-ray structure determination it is probably still the best way of distinguishing a hydroxyl group from an oxygen atom with no attached hydrogen. The former will have low bond-valence sums when only the contacts to the detected atoms are used. Obviously it would be nice to use these ideas to predict crystal structures, but unfortunately eqn (20.3) is insufficient to fix the topology of the crystal structure. However, Brown (1977) has found another rule which, in conjunction with this one allows progress in understanding bond-length patterns in solids. There are two ways of writing it: (i) the individual bond valences around each center will tend to be as nearly equal as possible; (ii) the sum of the bond valences around any closed loop in the structure will be equal to zero (if we define a bond valence from anion to cation as being positive, and one from cation to anion as being negative). A similar method for ensuring that the valence-sum rule is obeyed has been developed by Baur (1970). We show in Table 20.5 observed and predicted distances (Brown 1977) in diopside, $CaMgSi_2O_6$, using both Brown's and Baur's methods, to give an idea of the accuracy of the approach. Just given the *connectivity* of the atoms in the structure and these two rules (plus of course the necessary parameters for the atoms involved) leads to quite a good prediction of the bond lengths. Ideas such as these have led to a search for a method (Altermatt and Brown 1985) for the systematic generation of possible crystal structures. Such a general algorithm does not yet exist, but its development would be an extremely useful one for inorganic chemists. The development

Table 20.5 Prediction of bond lengths in diopside, $CaMgSi_2O_6$. After Brown (1977)

Bond	Bond valence (valence units)	Bond length (Å)		
		Predicted (Brown 1977)	Predicted (Baur 1970)	Observed (Clark *et al.* 1969)
Si—O(1)	1.07	1.60	1.62	1.60
Si—O(2)	1.17	1.56	1.59	1.58
Si—O(3)	0.88	1.67	1.67	1.66
Si—O(3)	0.88	1.67	1.67	1.69
σ(Si—O)		0.01	0.01	
Ca—O(1) (×2)	0.32	2.35	2.43	2.36
Ca—O(2) (×2)	0.42	2.24	2.32	2.35
Ca—O(3) (×2)	0.12	2.80	2.62	2.56
Ca—O(3) (×2)	0.12	2.80	2.62	2.72
σ(Ca—O)		0.14	0.07	
Mg—O(1) (×2)	0.30	2.15	2.10	2.12
Mg—O(1) (×2)	0.30	2.15	2.10	2.06
Mg—O(2) (×2)	0.40	2.01	2.06	2.05
σ(Mg—O)		0.06	0.02	

of such ideas relies, of course, on the collection of accurate structural data. Refinement of the theory and of the routine use of such methods depends crucially on the accuracy of the data.

20.7 High-temperature superconductors

1987 saw the report of the discovery of materials which are superconducting at temperatures as high as 90 K (Wu *et al.* 1987). Subsequent years saw the report of even higher temperatures. The critical temperature at which the electrical resistivity drops to zero on cooling is termed T_c. A striking feature of superconducting materials is that they are perfect diamagnets (the Meissner effect) and the scientific world has now become accustomed to photographs of little magnets floating above a liquid-nitrogen cooled piece of ceramic. Unlike the majority of existing superconductors, these new materials are metal oxides, which a few years ago would have been viewed by the physics community as very unlikely candidates indeed in which to observe this phenomenon. Since their discovery, considerable experimental effort has been expended to characterize these oxides and X-ray and neutron diffraction studies (for example: Calestani and Rizzoli 1987; Cava *et al.* 1987; David *et al.* 1987) have played, and still are playing an important role in the determination of the structure of the superconducting phase. The key part of the story is, however, that the materials are non-stoichiometric, and that the non-stoichiometry has a very important influence on the transport properties. Accurate determination of their structures is essential.

The basic structure, now well established for the oxide $YBa_2Cu_3O_{6+x}$ (the so-called 1–2–3 structure after the proportions of the metals in the compound) where $0 < x < 1$, is simply derived from the perovskite arrangement and is shown in Fig. 20.13. The (unknown) perovskite stoichiometry for this oxide would be $YBa_2Cu_3O_9$ where all the copper atoms are octahedrally six-coordinate. Loss of oxygen leads to both square pyramidal and square planar metal atoms. For $x = 1$ the site labeled O(5) is empty and the structure consists of $Cu^{II}(2)O_2$ sheets linked by rather long Cu(2)—O(1) linkages to $Cu^{III}(1)O_3$ chains. The sheets thus contain roughly square pyramidal Cu^{II} atoms while the chains contain Cu^{III} atoms in approximately square planar coordination, leading overall to an orthorhombic structure (space group *Pmmm*). For $x = 0$, all of the atoms labeled as O(4,5) are missing, the square planes have been replaced by linear O—Cu^{I}—O dumbbells and now the structure is tetragonal (space group *P4/mmm*). Our assignment of the oxidation numbers here is in accord with structural inorganic chemical ideas. Square pyramidal Cu^{II} is well known. It represents one part of the distortion coordinate associated with the so-called Jahn–Teller instability of octahedral (d^9) Cu^{II}. So we see four short in-plane distances, and one long apical distance. Square planar coordination is very common for low-spin

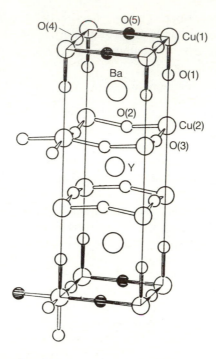

Fig. 20.13 The structure of the high-T_c superconductor $YBa_2Cu_3O_{6+x}$ where $0 < x < 1$. The sites labeled O(4) and O(5) are only partially occupied for $x < 1$. From Renault *et al.* (1987).

d^8 systems. Pt^{II} and Pd^{II} show many examples. Square planar Cu^{III} is not very common, but it is found for example in $KCuO_2$. Linear, two-coordinate copper is characteristic of Cu^I and is found for example in cuprite, Cu_2O.

A fascinating observation concerning these oxides however is the mobility of some of the oxygen atoms in the structure, and their influence on the superconductivity. T_c drops as x decreases from unity, and the super-conductivity disappears somewhat below $x = 0.5$. From powder, and single-crystal neutron diffraction studies at room temperature and below, occupation of the site labeled O(5) appears to be the rule when $x < 1$. Tetragonal and orthorhombic structures are observed depending respectively upon whether the O(4) and O(5) sites are occupied equally or not. One structural study showed that the transition between the two occurs at around $x = 0.5$, although it is now clear that this is not the crucial structural change which switches off the superconductivity. All of these diffraction studies give, of course, an average occupancy of the two sites. This changes too with temperature, an important aspect of the synthesis of these materials. One study on a sample with $x = 0.7$ (McIntyre *et al.* 1988) showed the population of the O(5) sites increasing with increasing temperature. There appears to be a region $0.3 < x < 0.7$ where detailed structural data are not available, although diffraction studies show that single phases can be produced either by heating oxygen-rich samples or by quenching samples from high temperature.

It is important to realize, as we have already stressed, that by their very nature such studies give only the average structure of these materials. The details of the local vacancy-ordering patterns are crucial in controlling the nature of the Fermi surface of the material as we show elsewhere (Burdett *et al.* 1987*c*). But how can we get more detailed information out of the structural determinations than reported individually in each of the papers? A study (Renault *et al.* 1987) of the composition evolution (with x) of the geometric parameters is very revealing. The plots of Fig. 20.14 show the

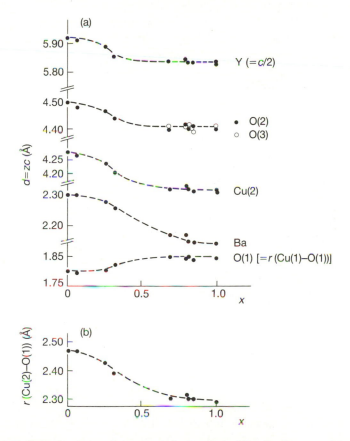

Fig. 20.14 Variation in structural parameters in the species $YBa_2Cu_3O_{6+x}$ as a function of x. Adapted from Renault *et al.* (1987). (a) Distances of the various atoms from the plane formed by the Cu(1) atoms. (b) Cu(2)—O(1) bond distance.

change in distance (along the c axis) of the various atoms in the structure from the plane formed by the Cu(1) atoms in $YBa_2Cu_3O_{6+x}$ for the existing series of room-temperature determinations. The geometry changes which are shown can be visualized as being driven by the change in oxidation state of one third of the copper atoms (associated with the coordination number change of chains to dumbbells) and the movement of the electropositive barium atoms in response to the presence of oxygen vacancies. In cuprite (Cu_2O) each copper atom is linearly two-coordinate with a Cu—O distance of 1.85 Å (Wyckoff 1963). However here each oxygen atom is four-coordinate by copper. In $YBa_2Cu_3O_6$ the O(1) atoms are just over one-coordinate and a considerable shortening of the Cu(1)—O(1) distance is to be expected compared to that in cuprite. As the Cu(1)—O(1) distance shortens then the Cu(2)—O(1) distance must lengthen to maintain the bond-valence sum at oxygen. It lengthens quite substantially. Triggered by the lengthening of this Cu(2)—O(1) bond, the square pyramids at Cu(2) are flatter at $x = 0$ than at $x = 1$. This is evident from Fig. 20.14 where the change in the position of Cu(2) is larger than of its associated oxygen atoms, O(2) and O(3). The change in position of the barium atoms is largest of all. We note that these atoms move away from the plane formed by the atoms O(4) and O(5) which are lost on moving from $x = 1$ to $x = 0$, an observation understandable in terms of simple electrostatics.

It is the *shape* of these curves however which is particularly interesting. Notice that the dependence of the structural parameters on x is not the linear one, expected on the basis of simple solution models, where changes in parameters are often linear in stoichiometry or concentration. In fact there seem to be two relatively flat regions, for $0.69 < x < 1$, and for $0 < x < 0.31$, connected by a curve with a much steeper slope. Such behavior is typical of the behavior of two-phase systems, where the parametric dependence on composition is determined by attractive like-with-like interactions. Thus from these data we may conclude that the defects are not ordered randomly for $x > 0$, but are organized so as to produce regions of Cu^{III} atoms separated from regions of Cu^I atoms. From such a simple model the steeper slope at low x (larger fraction of Cu^I) compared to that at high x (larger fraction of Cu^{III}) implies that the clustering energy for Cu^{III} is stronger than that for Cu^I.

It is particularly interesting to see that we may use plots such as those of Fig. 20.14 to extract more information concerning the *microscopic* structure than we can from an individual structure determination. The results are in accord with studies by electron diffraction (van Tendeloo et al. 1987) and by high-resolution transmission electron microscopy (Ourmazd and Spence 1987), which show that the oxygen vacancies are locally ordered. It is also expected by consideration of the known stable geometries for the various oxidation states of copper. If the vacancies are randomly ordered, then for high x, the two copper atoms associated with a single oxygen defect lie in a T-shaped environment. Simple electronic considerations (Burdett 1980; Burdett *et al.* 1987c) tell us that their electron configuration has to be d^{10}. This is not a geometry commonly found for this electron count. The only molecular example of such a T-shaped structure is for a low-spin d^8 Rh^I species (Yared *et al.* 1977). Cu^I is often found however in linear two-coordinate environments. If on the other hand the defects order such that a pair of vacancies are found around a single copper atom, then the linear two-coordination of the resulting d^{10} species is quite acceptable. The actual ordering of the defects will then be that which minimizes the number of copper atoms in T-shaped environments (at the ends of the chains). This is an exciting area and one where accurate structure determination is vital before the *electronic* properties of these materials may be understood.

20.8 The role of theory

Theory in inorganic chemistry is much less well-developed in *numerical* terms than its counterpart in organic chemistry. There are very few in-depth studies of the type described in Chapter 16 for example. This is a result, in part, of the size of the molecules of interest. While it might be possible to perform a very high-quality study on a small organic molecule containing a few carbon and hydrogen atoms, it is another matter to perform a calculation on an organometallic species containing one or more transition metals with a similar degree of accuracy. However qualitative molecular orbital ideas which cover the gamut of the periodic table are very well developed (see, for example, Burdett 1980; Albright *et al.* 1985), and there are a multitude of examples where the change in bond lengths and geometry with electron count may be readily understood in qualitative terms by using quite simple orbital models. For example the dependence on electron count of the metal–ligand distances in transition-metal complexes, the famous double-humped curve, is understandable in exceedingly simple molecular orbital terms (Burdett 1980). (Parenthetically we should say that the crystal-field theory often used in this area should be retired. Why use a separate approach for transition-metal complexes, from that used to study the rest of chemistry?) Similarly the structures of cluster and cage compounds of both the main-group elements and the transition metals are understandable by using simple electron counting. In this chapter too we have often used theoretical ideas to put the structural problem at hand into perspective. Often a theoretical prediction stimulates an experiment, sometimes to see if the prediction, which may be unusual, is correct. Often the prediction stimulates a different type of experiment, a carefully engineered one where a molecule or solid is designed to break the established rules. In this section we will describe an experiment which fits into the former category.

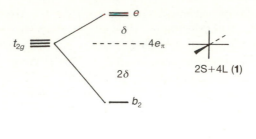

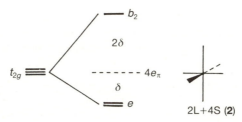

Fig. 20.15 d-level splitting patterns of the t_{2g} orbitals for the two tetragonal distortions of the octahedron, **1** and **2**. The distortions are such that the 'mass center' of the energy level patterns remains constant. δ is an energy parameter associated with a given distortion. After Burdett *et al.* (1988).

Many of the systems which have been studied in the light of the Jahn–Teller theorem are tetragonally distorted octahedral structures. The question often asked is: will there be two short and four long (2S + 4L, case **1**), or two long and four short (2L + 4S, case **2**) distances for a given system? This problem may readily be examined theoretically, and a prediction made as a function of d count. The energy changes associated with the (largely) transition-metal d orbitals for the case of π-donor ligands (e.g. those which coordinate via an O atom) are shown in Fig. 20.15 for the three t_{2g} orbitals for the two distortions **1** and **2** using the ideas of the angular overlap model (Burdett 1980) for an isolated octahedron MX_6. It is a simple matter to understand how the new energy-level patterns result from knowledge of the shapes of the d orbitals. The parameter δ is used to describe the variation in the π interactions on distortion. Recall that since the splitting pattern for the e_g pair does not depend upon the distortion mode, prediction of the distortion route is only possible for asymmetric t_{2g} configurations. These are shown in Table 20.6. For asymmetric e_g configurations d–s mixing via the second-order Jahn–Teller effect (Burdett 1981) has to be examined to understand, for example, the virtually universal occurrence of **2** in the chemistry of Cu^{II}. The predictions of Table 20.6 are reversed if the ligands are π acceptors. Thus the distortion **1** is found experimentally for low-spin d^5 $V(CO)_6$ (Bellard *et al.* 1979). We note that the accurate and opposite

Table 20.6 Total d-orbital energies for the t_{2g} configurations of octahedral and tetragonally distorted octahedral geometries using a molecular model. After Burdett *et al.* (1988)

d electrons configuration[a]	Regular octahedron	Distorted octahedron[b]	
		2S + 4L **(1)**	2L + 4S **(2)**
d^0	0	0	0
d^1	$4e_\pi$	$4e_\pi - 2\delta^*$	$4e_\pi - \delta$
d^2 (ls)	$8e_\pi$	$8e_\pi - 4\delta^*$	$8e_\pi - 2\delta$
d^2 (hs)	$8e_\pi$	$8e_\pi - \delta$	$8e_\pi - 2\delta^*$
d^3 (ls)	$12e_\pi$	$12e_\pi - 3\delta$	$12e_\pi - 3\delta$
d^3 (hs)	$12e_\pi$	$12e_\pi$	$12e_\pi$
d^4 (ls)	$16e_\pi$	$16e_\pi - 2\delta$	$16e_\pi - 4\delta^*$
d^4 (is)	$16e_\pi$	$16e_\pi - 2\delta^*$	$16e_\pi - \delta$
d^5 (ls)	$20e_\pi$	$20e_\pi - \delta$	$20e_\pi - 2\delta^*$
d^6 (ls)	$24e_\pi$	$24e_\pi$	$24e_\pi$

These energies are expressed in terms of the angular-overlap model parameter e_π which describes the energy shift of a π-type d orbital by coordination of a single π ligand. See Burdett (1980) for a discussion of this orbital-based model, and its connexion with the point-charge crystal-field approach.

[a] ls, low spin; hs, high spin; is, intermediate spin.
[b] A star indicates the lowest-energy structure for π-donor ligands.

predictions of the orbital-based angular overlap model for π donors and acceptors is in contrast to that made by the point-charge crystal-field model.

CrO_2 is isostructural with rutile (Fig. 20.9) and is a high-spin d^2 system. The preferred distortion predicted from Table 20.6 for this electronic configuration is that which produces two long and four short Cr—O distances **(2)**. This distortion is, however, opposite to that found by X-ray crystallography (Cloud *et al.* 1962). The result is not confined to the oxides. Table 20.7 collects bond-length data for a series of rutile-type oxides and fluorides. Notice that although the structure of FeF_2 is correctly predicted, that of CoF_2 is not (although the distortion away from octahedral here is tiny). Is the theory wrong or is the data incorrect? This observation prompted a redetermination of the structure of CrO_2 at room temperature, 173 K, and 10 K, using powder neutron diffraction, and a reevaluation of the theoretical model (Burdett *et al.* 1988). In fact at all three temperatures the local coordination at the chromium atom is similar (the figure in Table 20.7 actually refers to the value observed at 10 K). It is then the theory which is incorrect.

The failure of a molecular model to account for the distortion in solid CrO_2 and several of the other oxides and fluorides of Table 20.7 is, after a little thought, not very surprising. The ligand field at transition-metal centers in extended arrays is often significantly affected by the requirements of translational symmetry, through-bond and through-space interactions often being quite important. The former concept is related to the idea of superexchange. We need then to use a theoretical model which uses the band structure of the extended solid.

There are, in fact, three different distortions noted in Table 20.7. There is no restriction within the rutile space group for the two sets of symmetry-inequivalent distances to be equal, and indeed they never are. Distortions of the type **1** and **2** which preserve the $P4_2/mnm$ space group are represented here. Distortion of the M—X linkages along the chain gives rise to a much lower symmetry (monoclinic), and three different pairs of M—X distances are found. There are two examples of these, both associated with e_g degeneracies, namely CuF_2 and CrF_2. Here we shall be only interested in those distortions which preserve the tetragonal structure. Figure 20.16 shows two energy difference curves, calculated using the results of band-structure calculations on the infinite solid, as a function of d electron count for these two types of local distortions. These results agree quite well with the observed structures for most transition-metal oxides and fluorides with the rutile structure: the most prevalent distortion is the tetragonal compression of the octahedron, **1** (see Table 20.7). The only exceptions occur for the d^0

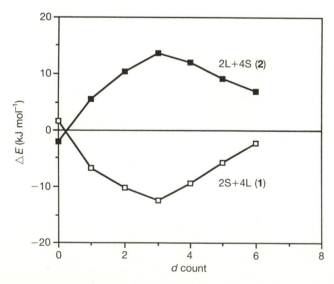

Fig. 20.16 Calculated energy difference curves as a function of d count for the distortions of the rutile structure which retain tetragonal symmetry and give rise to local geometries **1** and **2**. The curves are from tight-binding band-structure calculations using the extended Hückel method, and correspond to low-spin electronic configurations. After Burdett *et al.* (1988).

Table 20.7 Local coordination in transition-metal oxides and fluorides with rutile-type structures

System	d electrons configuration[a]	$r(M-X)$ (Å) Four M—X linkages	$r(M-X)$ (Å) Two M—X linkages	Sign of $r_2/r_4 - 1$[b]	Distortion type	References
MgF$_2$	d^0	1.9968(1)	1.9798(2)	−	1	Baur (1976)
VF$_2$	d^3 (hs)	2.091(3)[c]	2.074(3)[c]	−	1	Costa and De Almeida (1987)
CrF$_2$[d]	d^4 (hs)	2.01; 1.98[e]	2.43[e]	+	2	Jack and Maitland (1957)
MnF$_2$	d^5 (hs)	2.131(6)	2.104(9)	−	1	Baur and Khan (1971)
FeF$_2$	d^6 (hs)	2.118(4)	1.998(6)	−	1	Baur and Khan (1971)
CoF$_2$	d^7 (hs)	2.049(3)	2.027(5)	−	1	Baur and Khan (1971)
NiF$_2$	d^8 (hs)	2.022(6)	1.981(9)	−	1	Baur and Khan (1971)
CuF$_2$[d]	d^9	1.93(3)[f]	2.27(3)	+	2	Billy and Haendler (1957)
ZnF$_2$	d^{10}	2.046(7)	2.012(10)	−	1	Baur and Khan (1971)
TiO$_2$	d^0	1.9459(3)[g]	1.9764(4)[g]	+	2	Burdett et al. (1987b)
VO$_2$	d^1	1.921(1)	1.933(1)	+	2	McWhan et al. (1974)
CrO$_2$	d^2 (hs)	1.9113(3)[h]	1.8877(5)[h]	−	1	Burdett et al. (1988)
RuO$_2$	d^4 (ls)	1.984(6)	1.942(10)	−	1	Boman (1970a)
OsO$_2$	d^4 (ls)	2.006(8)	1.962(13)	−	1	Boman (1970b)

[a] ls, low spin; hs, high spin.
[b] r_2/r_4 is the ratio of the twofold set of M—X distances to the fourfold set of M—X distances to the fourfold set, in the tetragonal structure.
[c] Calculated from the atomic and unit-cell parameters given in the original paper.
[d] Monoclinic structure (three pairs of distances).
[e] No error is quoted in the original paper.
[f] The two pairs of equatorial distances were found to have equal values.
[g] At 15 K.
[h] At 10 K.

Fig. 20.17 The orbitals of a_1 symmetry which can (and do) mix together in the real electronic structure of the rutile chain, which negate conclusions drawn from orbital pictures based on isolated octahedra. From Burdett *et al.* (1988).

configuration and the high-temperature tetragonal polymorph of VO_2. Earlier we suggested that the distortion in TiO_2 itself was controlled by anion–anion repulsions (see Section 20.4). With reference to Table 20.6, notice that the configuration-dependence of the molecular model is removed, suggesting that a first-order Jahn–Teller effect is not influencing the local geometry of the transition-metal center in these rutile phases.

In fact a second-order orbital mixing is primarily responsible for the features of both curves. Because the group of any general point in the one-dimensional Brillouin zone of the solid is isomorphous to C_{2v}, one of the e_g and one of the t_{2g} orbitals sets (Fig. 20.17) have the same symmetry, a_1. From the ideas of perturbation theory the stabilization energy of the lower energy one is $\epsilon_{stab} \cong 4e_\pi e_\sigma / (4e_\pi + 3e_\sigma)$ within the angular overlap model. Figure 20.18 shows calculated energy difference curves (Burdett *et al.* 1988) for the two distortions using this model for the molecular (Table 20.6) and extended cases, where in the latter this extra interaction is included. The

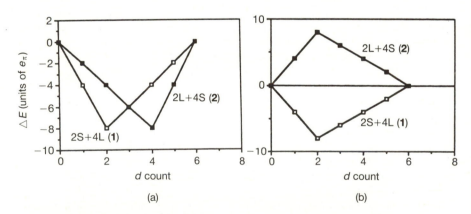

Fig. 20.18 (a) Energy difference curves using the molecular model of Fig. 20.15, for the two distortions **1** and **2**. (b) Curves from a model which allowed mixing of the orbitals shown in Fig. 20.17. The electronic configurations are low-spin ones. Notice the similarity of (b) to the plot of Fig. 20.16. Adapted from Burdett *et al.* (1988).

second result is in excellent agreement with the results from the quantitative calculations of Fig. 20.16. The conclusion is then that even in these 'ionic' materials the 'covalent' effects (more strictly 'orbital' effects) which couple one center with another should not be underestimated.

References

Albright, T. A., Burdett, J. K., and Whangbo, M.-H. (1985). *Orbital interactions in chemistry*. Wiley-Interscience, New York.

Altermatt, D. and Brown, I. D. (1985). *Acta Crystallogr.*, **B41**, 240–4.

Andersen, R. A., Boncella, J. M., Burns, C. J., Blom, R., Haaland, A., and Volden, H. V. (1986). *J. Organomet. Chem.*, **312**, C49–C52.

Bartlett, N. and McQuillan, B. W. (1982). Graphite chemistry. In *Intercalation chemistry* (ed. M. S. Whittingham and A. J. Jacobson), pp. 19–53. Academic Press, New York.

Baur, W. H. (1970). *Trans. Am. Crystallogr. Assoc.*, **6**, 129–55.

Baur, W. H. (1976). *Acta Crystallogr.*, **B32**, 2200–4.

Baur, W. H. and Khan, A. A. (1971). *Acta Crystallogr.*, **B27**, 2133–9.

Bellard, S., Rubinson, K. A., and Sheldrick, G. M. (1979). *Acta Crystallogr.*, **B35**, 271–4.

Beno, M. A., Williams, J. M., Tachikawa, M., and Muetterties, E. L. (1980). *J. Am. Chem. Soc.*, **102**, 4542–4.

Billy, C. and Haendler, H. M. (1957). *J. Am. Chem. Soc.*, **79**, 1049–51.

Boman, C.-E. (1970*a*). *Acta Chem. Scand.*, **24**, 116–22.

Boman, C.-E. (1970*b*). *Acta Chem. Scand.*, **24**, 123–8.

Brookhart, M. and Green, M. L. H. (1983). *J. Organomet. Chem.*, **250**, 395–408.

Brown, I. D. (1977). *Acta Crystallogr.*, **B33**, 1305–10.

Brown, I. D. (1981). The bond-valence method: An empirical approach to chemical structure and bonding. In *Structure and bonding in crystals*, Vol. 2 (ed. M. O'Keeffe and A. Navrotsky), pp. 1–30. Academic Press, New York.

Brown, I. D. and Altermatt, D. (1985). *Acta Crystallogr.*, **B41**, 244–7.

Brown, R. K., Williams, J. M., Schultz, A. J., Stucky, G. D., Ittel, S. D., and Harlow, R. L. (1980). *J. Am. Chem. Soc.*, **102**, 981–7.

Burdett, J. K. (1980). *Molecular shapes: Theoretical models of inorganic stereochemistry*. Wiley, New York.

Burdett, J. K. (1981). *Inorg. Chem.*, **20**, 1959–62.

Burdett, J. K. (1985). *Inorg. Chem.*, **24**, 2244–53.

Burdett, J. K. and Caneva, D. C. (1985). *Inorg. Chem.*, **24**, 3866–73.

Burdett, J. K. and Pourian, M. R. (1988). *Inorg. Chem.*, **27**, 4445–50.

Burdett, J. K., Lawrence, N. J., and Turner, J. J. (1984). *Inorg. Chem.*, **23**, 2419–28.

Burdett, J. K., Phillips, J. R., Pourian, M. R., Poliakoff, M., Turner, J. J., and Upmacis, R. (1987*a*). *Inorg. Chem.*, **26**, 3054–63.

Burdett, J. K., Hughbanks, T., Miller, G. J., Richardson Jr., J. W., and Smith, J. V. (1987*b*). *J. Am. Chem. Soc.*, **109**, 3639–46.

Burdett, J. K., Kulkarni, G. V., and Levin, K. (1987*c*). *Inorg. Chem.*, **26**, 3650–2.

Burdett, J. K., Miller, G. J., Richardson Jr., J. W., and Smith, J. V. (1988). *J. Am. Chem. Soc.*, **110**, 8064–71.

Busetto, C., D'Alfonso, A., Maspero, F., Perego, G., and Zazzetta, A. (1977). *J. Chem. Soc. Dalton Trans.*, 1828–34.

Calestani, G. and Rizzoli, C. (1987). *Nature*, **328**, 606–7.

Cava, R. J., Santoro, A., Murphy, D. W., Zahurak, S. M., and Roth, R. S. (1983). *J. Solid State Chem.*, **50**, 121–8.

Cava, R. J., Batlogg, B., van Dover, R. B., Murphy, D. W., Sunshine, S., Siegrist, T., Remeika, J. P., Rietman, E. A., Zahurak, S., and Espinosa, G. P. (1987). *Phys. Rev. Lett.*, **58**, 1676–9.

Cheetham, A. K., Fyfe, C. A., Smith, J. V., and Thomas, J. M. (1982). *J. Chem. Soc. Chem. Commun.*, 823–5.

Clark, J. R., Appleman, D. E., and Papike, J. J. (1969). *Mineral. Soc. Am. Spec. Papers*, **2**, 31–50.

Cloud, W. H., Schreiber, D. S., and Babcock, K. R. (1962). *J. Appl. Phys.*, **33**, 1193–4.

Costa, M. M. R. and De Almeida, M. J. M. (1987). *Acta Crystallogr.*, **B43**, 346–52.

Cotton, F. A., LaCour, T., and Stanislowski, A. G. (1974). *J. Am. Chem. Soc.*, **96**, 754–60.

Cotton, F. A., Darensbourg, D. J., Fang, A., Kolthammer, B. W. S., Reed, D., and Thompson, J. L. (1981). *Inorg. Chem.*, **20**, 4090–6.

David, W. I. F., Harrison, W. T. A., Ibberson, R. M., Weller, M. T., Grasmeder, J. R., and Lanchester, P. (1987). *Nature*, **328**, 328–9.

Dawkins, G. M., Green, M., Orpen, A. G., and Stone, F. G. A. (1982). *J. Chem. Soc. Chem. Commun.*, 41–3.

Dunitz, J. D., Orgel, L. E., and Rich, A. (1956). *Acta Crystallogr.*, **9**, 373–5.

Evans, W. J., Hughes, L. A., and Hanusa, T. P. (1984). *J. Am. Chem. Soc.*, **106**, 4270–2.

Firor, R. L. and Seff, K. (1976). *J. Am. Chem. Soc.*, **98**, 5031–3.

Gibbs, G. V., Meagher, E. P., Newton, M. D., and Swanson, D. K. (1981). A comparison of experimental and theoretical bond length and angle variations for minerals, inorganic solids, and molecules. In *Structure and bonding in crystals*, Vol. 1 (ed. M. O'Keeffe and A. Navrotsky), pp. 195–225. Academic Press, New York.

Heine, V. and McConnell, J. D. C. (1984). *J. Phys. C (Solid State Phys.)*, **17**, 1199–220.

Herzberg, G. (1950). *Molecular spectra and molecular structure*, Vol. 1, *Spectra of diatomic molecules* (2nd edn). Van Nostrand-Reinhold, New York.

Jack, K. H. and Maitland, R. (1957). *Proc. Chem. Soc.*, 232.

Jackson, R. H. (1962). *J. Chem. Soc.*, 4585–92.

Jucks, K. W. and Miller, R. E. (1987). *J. Chem. Phys.*, **87**, 5629–33.

Kertész, M., Vonderviszt, F., and Hoffmann, R. (1983). Change of carbon–carbon bond length in layers of graphite upon charge transfer. In *Intercalated graphite*, Materials Research Society Symposia Proceedings, Vol. 20 (ed. M. S. Dresselhaus, G. Dresselhaus, J. E. Fischer, and M. J. Moran), pp. 141–3. North-Holland, New York.

Kubas, G. J., Ryan, R. R., Swanson, B. I., Vergamini, P. J., and Wasserman, H. J. (1984). *J. Am. Chem. Soc.*, **106**, 451–2.

Loewenstein, W. (1954). *Am. Mineral.*, **39**, 92–6.

Lovejoy, C. M., Nelson Jr., D. D., and Nesbitt, D. J. (1987). *J. Chem. Phys.*, **87**, 5621–8.

McConnell, J. D. C. (1983). *Am. Mineral.*, **68**, 1–10.

McIntyre, G. J., Renault, A., and Collin, G. (1988). *Phys. Rev. B, Condens. Matter*, **37**, 5148–57.

McWhan, D. B., Marezio, M., Remeika, J. P., and Dernier, P. D. (1974). *Phys. Rev. B, Solid State*, **10**, 490–5.

Marsh, R. E. and Schomaker, V. (1979). *Inorg. Chem.*, **18**, 2331–6.

Newton, M. D. (1981). Theoretical probes of bonding in the disiloxy group. In *Structure and bonding in crystals*, Vol. 1 (ed. M. O'Keeffe and A. Navrotsky), pp. 175–93. Academic Press, New York.

Nixon, D. E. and Parry, G. S. (1969). *J. Phys. C (Solid State Phys.)*, **2**, 1732–41.

Oka, T. (1980). *Phys. Rev. Lett.*, **45**, 531–4.

O'Keeffe, M. and Hyde, B. G. (1981). The role of nonbonded forces in crystals. In *Structure and bonding in crystals*, Vol. 1 (ed. M. O'Keeffe and A. Navrotsky), pp. 227–54. Academic Press, New York.

O'Keeffe, M. and Hyde, B. G. (1982). *J. Solid State Chem.*, **44**, 24–31.

Okumura, M. (1986). *Infrared spectroscopy and radiative lifetimes of molecular ions*. Thesis, University of California, Berkeley, USA.

Ourmazd, A. and Spence, J. C. H. (1987). *Nature*, **329**, 425–7.

Pauling, L. (1929). *J. Am. Chem. Soc.*, **51**, 1010–26.

Pluth, J. J. and Smith, J. V. (1983). *J. Am. Chem. Soc.*, **105**, 1192–5.

Pluth, J. J., Smith, J. V., and Faber Jr., J. (1985). *J. Appl. Phys.*, **57**, 1045–9.

Renault, A., Burdett, J. K., and Pouget, J.-P. (1987). *J. Solid State Chem.*, **71**, 587–90.

Ries, W., Bernal, I., Quast, M., and Albright, T. A. (1984). *Inorg. Chim. Acta*, **83**, 5–15.

Roe, D. M., Bailey, P. M., Moseley, K., and Maitlis, P. M. (1972). *J. Chem. Soc. Chem. Commun.*, 1273–4.

Schultz, A. J., Teller, R. G., Beno, M. A., Williams, J. M., Brookhart, M., Lamanna, W., and Humphrey, M. B. (1983). *Science*, **220**, 197–9.

Seff, K. and Mellum, M. D. (1984). *J. Phys. Chem.*, **88**, 3560–3.

Simon, A. (1975). Alkali metal suboxides: A kind of anti-cluster compounds. In *Crystal structure and chemical bonding in inorganic chemistry* (ed. C. J. M. Rooymans and A. Rabenau), pp. 47–67. North-Holland, Amsterdam.

Thorn, D. L., Tulip, T. H., and Ibers, J. A. (1979). *J. Chem. Soc. Dalton Trans.*, 2022–5.

Van Tendeloo, G., Zandbergen, H. W., and Amelinckx, S. (1987). *Solid State Commun.*, **63**, 389–93 and 603–6.

Whangbo, M.-H., Hoffmann, R., and Woodward, R. B. (1979). *Proc. Roy. Soc.*, **A366**, 23–46.

Wu, M. K., Ashburn, J. R., Torng, C. J., Hor, P. H., Meng, R. L., Gao, L., Huang, Z. J., Wang, Y. Q., and Chu, C. W. (1987). *Phys. Rev. Lett.*, **58**, 908–10.

Wyckoff, R. W. G. (1963). *Crystal structures* (2nd edn), Vol. 1, p. 331. Wiley-Interscience, New York.

Yared, Y. W., Miles, S. L., Bau, R., and Reed, C. A. (1977). *J. Am. Chem. Soc.*, **99**, 7076–8.

21

Structural variability in metal cluster compounds

Vincenzo G. Albano and Dario Braga

21.1 Introduction

This chapter deals with the rationalization of some structural characteristics of the transition-metal cluster compounds containing carbon monoxide as principal ligand. The discussion has been confined to a few (in our opinion) relevant topics with the aim of developing them in some depth. This choice is made necessary by the enormous structural variability shown by the organometallic solid-state chemistry with respect to the type of molecules found in the organic area. Attention will also be focused on some aspects of the interplay between the molecular geometry determined by X-ray diffraction and the structural information obtainable by spectroscopic techniques.

Metal cluster species are among the largest organometallic molecules currently studied, an average structural study usually implying treatment of several tens, if not hundreds, of atoms, a relevant proportion of which are transition metals (Chini *et al.* 1976). The size and nature of these molecules represent by themselves an *a priori* drawback for research work aimed at a high level of accuracy. It is hardly necessary to point out that some of the usual problems (absorption, anomalous dispersion, crystal decay, etc.) crystallographers are used to facing in the course of their work, are often considerably aggravated when large numbers of heavy atoms are present (Angermund *et al.* 1985). Last but not least, the light atoms make a limited contribution to the total scattering power of these molecules. These are some of the reasons why, for instance, studies at the electron-density level have so far been confined to a few, relatively small, molecules containing only first- or exceptionally second-row transition metals (Coppens 1977, 1985).

On these premises it should be clear that words such as 'precision' and 'accuracy' take a substantially softer meaning when applied to transition-metal clusters. Furthermore, little support (if any at all) can be expected to come from theoretical calculations. Nonetheless there is an increasing awareness that higher levels of accuracy and better appreciation and under-standing of the subtleties of the structural features can nowadays be attained even for species of the kind described in this chapter.

A further point of interest in the field arises from the specific phenomenon of structural non-rigidity in solution and, sometimes, in the solid state, typical of many transition-metal clusters (*fluxionality*), where the ligands (mainly CO ligands) are able to migrate over the surface of the metal core (Band and Muetterties 1978; Johnson and Benfield 1980). The implications for the structural chemist are challenging because each technique of struc-tural investigation (IR spectroscopy, NMR spectroscopy, solid-state NMR spectroscopy, let alone single-crystal X-ray or neutron diffraction) 'sees' a structure which is different depending upon the time scale associated with the method (Benfield *et al.* 1988). In this respect, correlation between time- and space-averaged X-ray structure and dynamic aspects, though appealing, may become an extremely difficult task whose accomplishment strongly relies on the availability of accurate data.

Considering the vastness of the field, the following discussion will be devoted to the structural variability shown by some homometallic and heterometallic interactions and by metal–ligand interactions, showing, each time, the limitations imposed by the difficulties mentioned above. A critical approach to the relationship between solid-state and solution behaviour of some metal cluster species will also be presented.

21.2 Structural variability and ligand envelope

We will begin by discussing first some features of the hexanuclear metal carbonyl clusters which share an 86 valence electron configuration and are characterized by octahedral metal atom polyhedra. It is worth mentioning, on passing, that the noble gas rule (or 18 electrons rule) would assign 84 electrons to these species (Lauher 1978, 1981; Wade 1980), though this dis-crepancy does not directly concern the present discussion.

Structural characterization has been reported for many (*ca.* 40) octahedral species distributed all through groups VII and VIII of the periodic table. Differences among these species mainly arise from the number of CO ligands used to attain the 'magic number' of 86 valence electrons, or from the presence of interstitial ligands (C, N or H) (Albano and Martinengo 1980; Bradley 1983). For these reasons this family represents a valid sample for studying the structural variability among isoelectronic and, in some cases, isostructural species, where metal atom polyhedra and M—CO interactions can be thoroughly compared.

Focusing on the number of CO ligands, species possessing from 12 to 19 ligands are known (Raithby 1980). These can be seen as the lower and upper limits, respectively. More than 19 CO ligands would not find enough space on the cluster surface and there is indeed only one species in which 19 ligands have been ascertained, $[Re_6C(CO)_{19}]^{2-}$ (A. Sironi, personal communication). Its existence confirms that only for the largest metal atoms can the 19th CO ligand be accommodated, while for the other elements only one or two hydride ligands can be tolerated in excess of 18 CO ligands. Less than 12 CO ligands are not expected because species with such a small number of ligands would exist either with very high negative charge or for elements beyond group VIII, where M—CO interactions are definitely destabilized. Within the 12–18 range, a great deal of possible compromises between steric requirements of the ligands and charge distribution on the metal cluster surface are exhibited. Metal–metal bond distances show marked differences most of which, but not all, can be attributed to the presence of μ_2 or μ_3 bridging ligands (Colton and McCormick 1980; Braga and Grepioni 1987) or to the presence of interstitial atoms (Albano and Martinengo 1980; Bradley 1983). Hydride atoms also have a strong influence on metal–metal interactions (Teller and Bau 1981).

A detailed discussion of all these effects is beyond the scope of this section. Nonetheless some insights into the factors governing the structural variability of these species can be gained from the following three cases.

21.2.1 Octahedral clusters showing ligand crowding

Crowding is observed in the species possessing 18 CO ligands, such as those reported in Table 21.1. Although a very regular distribution of the CO ligands is attained (three on each metal atom yielding D_3 idealized symmetry, see Fig. 21.1), appreciable deformations of the metal polyhedra are observed. In the three species of Table 21.1 the pair of opposite triangular

Table 21.1 Twisting of the M_6 core in 18-CO octahedral clusters, resulting in short and long interbasal M—M distances

Species	M—M distances (Å)		Difference (Å)	References
	Short	Long		
$[H_2Re_6C(CO)_{18}]^{2-}$	2.939(2)	3.096(3)	0.16	Ciani et al. (1983)
$[HRu_6(CO)_{18}]^-$	2.839(6)	2.924(3)	0.09	Eady et al. (1980)
$[Os_6(CO)_{18}]^{2-}$	2.814(4)	2.886(4)	0.07	McPartlin et al. (1976)

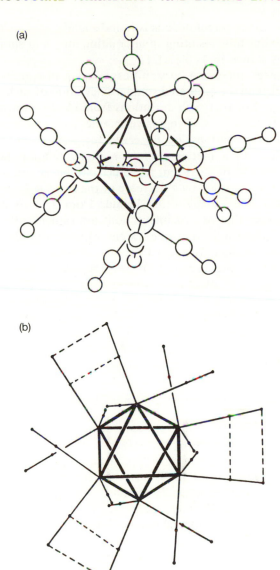

Fig. 21.1 (a) The structure of $[HRu_6(CO)_{18}]^-$ (Eady *et al*. 1980; Jackson *et al*. 1980). (b) Schematic representation showing the occurrence of short CO $\cdots$ CO interactions (dashed lines).

faces orthogonal to the threefold axis is found slightly off its idealized staggered configuration, thus resulting in alternating long and short interbasal M—M bonds as shown in Table 21.1.

These differences appear to be significant, being one or two orders of magnitude larger than the estimated standard deviation of each mean value. It can be easily proved that such deviations from ideal symmetry are set in order to alleviate CO $\cdots$ CO repulsive interactions, and it is significant that the largest deformations are observed for one of the overcrowded dihydrido species. As a matter of fact a slight rotation (about 5° in $[HRu_6(CO)_{18}]^-$) would be sufficient to render interbasal bonds equivalent, but it would be accompanied by a decrease of the average C $\cdots$ C contact distances between next neighbouring ligands from 3.11 to 2.90 Å. This observation suggests that M—M interactions are rather 'soft', so that the inner metal polyhedron is to some extent capable of adapting to the steric pressure of the ligands packed around (Albano *et al.* 1989*a*). This fact is supported by the presence of very regular metal polyhedra in the species possessing a lower and even number of CO ligands, such as $[Rh_6(CO)_{16}]$ (Corey *et al.* 1963), $[FeRh_5(CO)_{16}]^-$ (Slovokhotov *et al.* 1984), and $[RuRh_5(CO)_{16}]^-$ (Pursiainen *et al.* 1987) (the latter two show statistical Fe/Rh and Ru/Rh disorder, respectively), which conform to a full T_d idealized symmetry with almost identical M—M bond lengths and with no significant deformations due to packing forces.

21.2.2   *Octahedral clusters showing ligand isomerism*

Few cases have been demonstrated of ligand isomerism in the solid state and these are very useful in assessing the effect of ligand stereogeometry and packing forces on the regularity of the metal polyhedra. For example $[Ir_6(CO)_{16}]$ (Garlaschelli *et al.* 1984) has been found in a form with four face-bridging ligands (red isomer, T_d idealized symmetry) and another with four edge-bridging ligands (black isomer, D_{2d} idealized symmetry) as shown in Fig. 21.2. Both isomers have C_2 symmetry in the crystal but, while the average Ir—Ir bond lengths are almost identical in the two cases (2.779(1) and 2.778(6) Å, respectively), the individual values are significantly less scattered in the face-bridged T_d species (2.775–2.783(1) Å) than in the edge-bridged D_{2d} species (2.743–2.810(2) Å), in keeping with the higher idealized symmetry of the former with respect to the latter. The difference in energy between the two ligand arrangements is surely small and cannot explain the differences. This fact seems to indicate that other factors, such as packing forces, are responsible for the observed irregularities. As a matter of fact a very simple calculation of the packing efficiency, based on the ratio between unit-cell and molecular volumes (Gavezzotti 1983; Braga and Grepioni 1989), shows that the face-bridged species, despite a

(a)

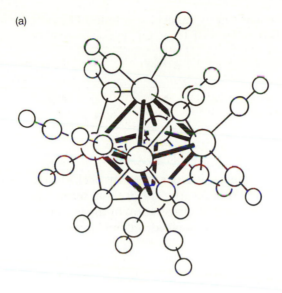

(b)

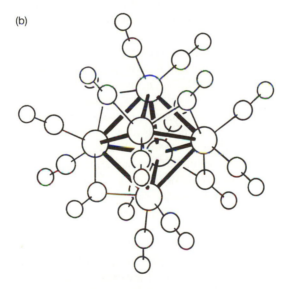

Fig. 21.2 The structure of the two isomers of $[Ir_6(CO)_{16}]$ (Garlaschelli *et al.* 1984). (a) Red isomer, four face-bridging CO groups. (b) Black isomer, four edge-bridging CO groups.

Table 21.2 Comparison of molecular volumes (V_{mol}) and packing coefficients for the two isomers of $[Ir_6(CO)_{16}]$

Isomer	V_{cell} ($Å^3$)	V_{mol} ($Å^3$)	Z^a	Packing coefficient	Free volume (%)
Face-bridged (red)	2406	359	4	0.60	40
Edge-bridged (black)	2385	387	4	0.65	35

^{a} Number of molecules per unit cell.

smaller molecular volume, is less efficiently packed than the edge-bridged one (see Table 21.2). A more efficient packing seems to be achieved at the expense of a more symmetric distribution of the intramolecular interactions yielding a less regular ligand polyhedron and larger deformations of the M_6 core in the crystal.

21.2.3 Octahedral clusters showing ligand deficiency

More extensive structural variability is shown by the species possessing a low ligand-to-metal ratio, as it is the case of the 13-CO clusters, where a regular accommodation of an odd and low number of ligands on the octahedral framework becomes a major problem. Interestingly all known 13-CO species listed in Table 21.3, though isoelectronic, show different ligand distributions, which are only apparently unrelated. These facts descend not only from the lack of a well-defined minimum energy configuration, but also from the availability of space on the cluster surface, which allows the existence of several structures almost equivalent in energy with bonding and non-bonding interactions of comparable importance.

As shown in Table 21.3, M—M bonds are spread over very large ranges of distances indicating very deformable metal-atom polyhedra, strongly influenced by the ligand stereogeometry. If we focus on the two homometallic carbides $[Co_6C(CO)_{13}]^{2-}$ and $[Rh_6C(CO)_{13}]^{2-}$, we find large 'niches' in the ligand coverage of the latter species, while a more regular ligand envelope is attained in the former, as shown in Fig. 21.3.

Although the ligand packing in $[Rh_6C(CO)_{13}]^{2-}$ is quite irregular (C_s idealized symmetry, see Fig. 21.3(b)), a fair sharing of the electron contribution from the ligands to each metal atom is achieved through a system of asymmetric or only incipient edge-bridging ligands. On the contrary the packing around $[Co_6C(CO)_{13}]^{2-}$ (C_2 idealized symmetry, see Fig. 21.3(a)) implies an uneven distribution of the metal–ligand interactions with four Co atoms formally receiving four electrons and two Co atoms formally receiving five electrons. In the latter case ligand–ligand interactions seem to play a determining role in stabilizing the polyhedron described by the

Table 21.3 Relevant structural parameters for 13-CO octahedral clusters

Species	Range of M—M distances (Å)	Mean M—M distance (Å)	Number of terminal CO ligands	Number of bridging CO ligands	References
$[Co_6C(CO)_{13}]^{2-}$	2.465–2.926(1)	2.639	8	5	Albano et al. (1986)
$[Co_6N(CO)_{13}]^-$	2.487–2.788(5)	2.613	9	4	Ciani and Martinengo (1986)
$[Co_2Rh_4C(CO)_{13}]^{2-}$	2.616–3.139(1)[a]	–	8	5	Albano et al. (1989b)
$[Rh_6C(CO)_{13}]^{2-}$	2.733–3.188(2)	2.907	7	6	Albano et al. (1981)

[a] M—M distances include mixed Co—Rh bonds.

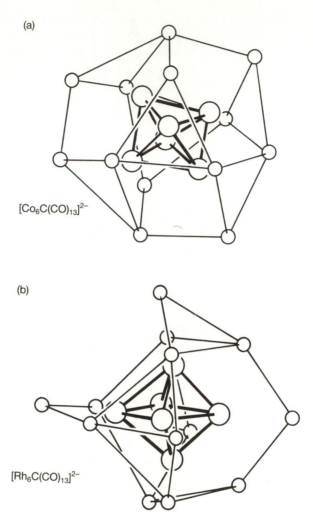

(a)

$[Co_6C(CO)_{13}]^{2-}$

(b)

$[Rh_6C(CO)_{13}]^{2-}$

Fig. 21.3 Schematic representation of the metal and ligand polyhedra (O · · · O contacts less than 4.0 Å) in (a) $[Co_6C(CO)_{13}]^{2-}$ and (b) $[Rh_6C(CO)_{13}]^{2-}$. Metal–ligand bonds are not drawn. After Albano *et al.* (1989*b*).

oxygen–oxygen contacts and, consequently, in controlling the metal–ligand interactions and the range of Co—Co distances. As a matter of fact, the ligands in $[Rh_6C(CO)_{13}]^{2-}$, along the two staggered 'fly-over' systems (see Fig. 21.3(b)) generated by the ligand distribution around the two equatorial planes which contain five and six CO ligands, respectively, are in close contact, showing the shortest values for the C · · · C and O · · · O next-neighbour distances (2.85 and 3.36 Å, respectively). On these bases, it seems reasonable to infer that, when the larger Rh_6 core is replaced by the smaller Co_6 one, an unchanged ligand stereogeometry would produce shorter and more repulsive ligand–ligand interactions along the 'fly-over' systems of ligands. In other words, the 'fly-over' arrangements, though more efficient in smoothing the electron contribution over the cluster, are no longer favoured on the surface of the smaller Co_6 polyhedron, so that the ligand distribution is forced to switch to that actually observed.

(a)

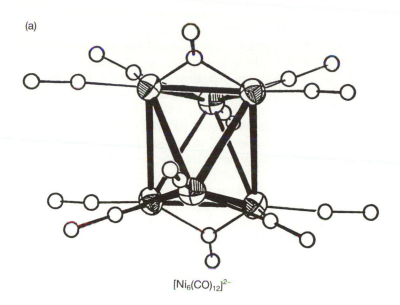

$[Ni_6(CO)_{12}]^{2-}$

(b)

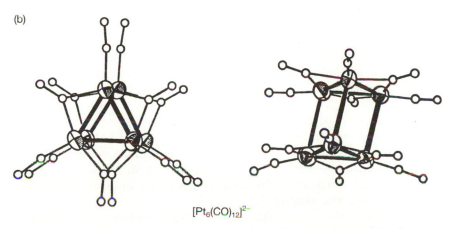

$[Pt_6(CO)_{12}]^{2-}$

Fig. 21.4 The molecular structures of (a) $[Ni_6(CO)_{12}]^{2-}$ and (b) $[Pt_6(CO)_{12}]^{2-}$. After Calabrese *et al.* (1974*a*, *b*).

Interestingly the mixed-metal species $[Co_2Rh_4C(CO)_{13}]^{2-}$ is almost half-way between the two homometallic cases, sharing features of the ligand distribution with both compounds (Albano *et al.* 1989*b*).

12-CO species are confined to $[Ni_6(CO)_{12}]^{2-}$ (Calabrese *et al.* 1974*b*) and $[Pt_6(CO)_{12}]^{2-}$ (Calabrese *et al.* 1974*a*). Although a fully symmetric all-bridging ligand distribution around the M_6 octahedron could be envisaged on the basis of the principles of charge equalization and $CO \cdots CO$ contacts optimization, the actual stereogeometries conform to D_{3d} and D_{3h} (approximate) symmetries, respectively. Both anions can be described as two $M_3(\mu\text{-}CO)_3(CO)_3$ triangular units packed together in staggered configuration in $[Ni_6(CO)_{12}]^{2-}$ (trigonal antiprism) and almost eclipsed and slightly shifted in $[Pt_6(CO)_{12}]^{2-}$ (deformed prism) as illustrated in Fig. 21.4. In addition the M_6 polyhedra are significantly stretched along the threefold direction with average basal and interbasal distances of 2.38, 2.77 Å respectively in the former species, and 2.66, 3.04 Å in the latter. These figures show that intrabasal bonds, supported by bridging ligands, are very strong and therefore the optimization of inter-ligand contacts is accomplished mainly at the expense of the quite softer interbasal $M-M$ interactions. The geometry of $[Pt_6(CO)_{12}]^{2-}$ exhibits the extreme deformation of the M_6 polyhedron associated to 86 valence electrons. This is probably due to the special stability of the triangular fragment mentioned above, which is the building block not only of the dimeric dianion but also of a series of dianions of general formula $[Pt_3(CO)_6]_n^{2-}$, that have been characterized up to $n = 5$ and in which such units are piled up (see Chini *et al.* 1976).

21.3 Structural variability and metal–metal interactions

Metal–metal bond lengths are certainly among the most trustworthy structural parameters afforded by an X-ray study of metal cluster compounds. Heavy-atom positions are usually determined with great accuracy and metal–metal distances reported to the thousandth of an ångström with very small estimated standard deviations, even in the case of large metal frames. In this way the structural features of the metal polyhedra, and the metal–metal bonds in particular, can be approached with much greater confidence than the structural parameters involving the ligand light atoms. However, we have already shown that packing forces are capable of affecting metal–metal distances more than experimental errors so that subtle structural information of relevant chemical interest is easily lost. Nonetheless meaningful inferences can be drawn by comparing isostructural species that often exhibit very similar crystal packings.

A striking example, coming from our own work, is afforded by the family of $[Rh_6C(CO)_{15}(ML)_2]$ species, where $M = Cu^I$, Ag^I, Au^I, and $L = NCMe$ or PPh_3 (Albano *et al.* 1980; Fumagalli *et al.* 1988). These bimetallic

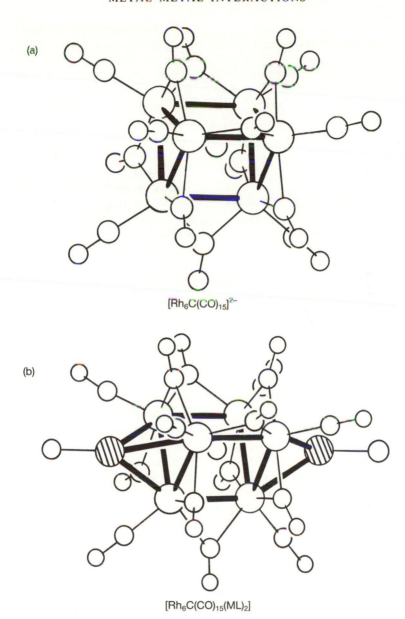

(a)

$[Rh_6C(CO)_{15}]^{2-}$

(b)

$[Rh_6C(CO)_{15}(ML)_2]$

Fig. 21.5 (a) The structure of the dianion $[Rh_6C(CO)_{15}]^{2-}$ (Albano *et al*. 1973) showing the trigonal prismatic metal-atom framework.   (b) The dicapped cluster framework of the neutral species $[Rh_6C(CO)_{15}(ML)_2]$ (M = Cu^I, Ag^I, Au^I; L = NCMe, PPh$_3$) (Albano *et al*. 1980; Fumagalli *et al*. 1988).

Table 21.4   Average values of the Rh—Rh distances in the family of adducts of the prismatic dianion $[Rh_6C(CO)_{15}]^{2-}$ containing a D_{3h} dicapped prism of metal atoms

Species	Average Rh—Rh distances (Å)		References
	Basal	Interbasal	
$[Rh_6C(CO)_{15}]^{2-}$	2.776(3)	2.817(2)	Albano *et al.* (1973)
$[Rh_6C(CO)_{15}\{Cu(NCMe)\}_2]$	2.765(1)	2.810(1)	Albano *et al.* (1980)
$[Rh_6C(CO)_{15}\{Cu(PPh_3)\}_2]$	2.768(1)	2.813(1)	
$[Rh_6C(CO)_{15}\{Ag(NCMe)\}_2]$	2.775(1)	2.805(4)	Fumagalli *et al.*
$[Rh_6C(CO)_{15}\{Ag(PPh_3)\}_2]$	2.785(1)	2.808(1)	(1988)
$[Rh_6C(CO)_{15}\{Au(PPh_3)\}_2]$	2.780(1)	2.805(1)	

derivatives can be easily prepared by addition of electrophilic ML^+ fragments on the triangular faces of the prismatic dianion $[Rh_6C(CO)_{15}]^{2-}$ characterized by Albano *et al.* (1973) (see Fig. 21.5). As a matter of fact these species have offered a unique opportunity of studying differences in bonding modes of the univalent metals of group I B. Table 21.4 shows the average lengths of the two types of Rh—Rh interactions in the prismatic core for the parent dianion and the neutral derivatives. Although differences between corresponding values can be regarded as barely significant if taken separately, a comparison over all the entries undoubtedly shows a significant trend, which calls for an explanation in terms of peculiar electronic effects. It can be seen that both copper derivatives show a slight shrinkage of all prism edges with respect to the parent dianion. This effect is more marked for the NCMe derivative than for the PPh_3 one. Silver and gold derivatives, on the contrary, show no shortening of the basal edges (if not a slight lengthening) but a more marked shrinkage of the interbasal ones. These differences can be taken as indicative of different mechanisms of electron withdrawal from the prismatic dianion exerted by the electrophilic metal fragments. In fact, while the Cu–prism interactions seem to take advantage of the availability on the Cu fragment of low-lying p orbitals, which can be used to set up a π-charge donation from the prism orbitals towards the metal in addition to the σ donation to an sp hybrid orbital, silver and gold appear to be able to receive electrons only *via* σ donation, because of the increase in energy of their p orbitals (Evans and Mingos 1982). In other words, the interaction of silver and gold with the prism concentrates the shrinkage effect on the interbasal edges which are involved in the cluster orbitals that interact at the σ level with the heterometallic fragments. When an alternative route for charge withdrawal is available — as in the case of Cu interactions —

the effect on interbasal bonds is alleviated and involvement of the basal bonds is observed.

21.4   Bond parameters and packing forces

In the previous sections we have often mentioned the effects of packing forces on molecular geometry. They can be assessed on a quantitative basis by studying polymorphs or crystals containing more than one independent molecule in the asymmetric unit (Kitaigorodskii 1970; see also Chapter 19, Section 19.7). In the case of ionic species, salts with different counterions are very suitable for this purpose and we have had the opportunity of analyzing the family of compounds $X_2[M_6C(CO)_{15}]$ (X = NMe$_3$(CH$_2$Ph)$^+$, PPh$_4^+$, NEt$_4^+$; M = Co, Rh) (Albano $et\,al$. 1973, 1989a; Martinengo $et\,al$. 1985) for which not only intermolecular but also some intramolecular effects have been detected. Anionic species are very numerous in cluster chemistry, therefore the recognition of the molecular parameters more affected by packing forces and the deviation of the molecular symmetry in the crystal from that of the isolated ion (idealized symmetry) is of particular interest.

Both $[M_6C(CO)_{15}]^{2-}$ anions (M = Co, Rh) can be assigned to the D_{3h} point group with a prismatic M_6C core regularly surrounded by six terminally bonded and nine edge-bridging CO ligands (see Fig. 21.5(a)). The first comparison of interest is that between the two NMe$_3$(CH$_2$Ph)$^+$ salts. Both species crystallize in the $C2/c$ space group and exhibit the same crystal packing with the anions possessing C_2 crystallographic symmetry (Albano $et\,al$. 1973; Martinengo $et\,al$. 1985). The bond parameters show that $[Co_6C(CO)_{15}]^{2-}$ conforms better to D_3 than to D_{3h} idealized symmetry, because of a concerted asymmetry of the basal-bridging CO ligands, not appreciable in $[Rh_6C(CO)_{15}]^{2-}$ (the average Co—C distances are 2.00(1) and 1.94(1) Å). The cobalt species also exhibits an increase ($ca.$ 2.5°) of the folding angles of both the basal-bridging and terminal ligands with respect to the M_3 triangles. These differences are intramolecular effects originating from the closer inter-ligand contacts produced by the smaller Co$_6$ core (the average lengths of the basal prism edges are 2.537 and 2.776 Å for Co$_6$ and Rh$_6$, respectively), which are alleviated through small conformational changes of the ligand environment. We have already shown how the alleviation of the inter-ligand contacts in the crowded 18-CO octahedral species is achieved at the expense of the metal core regularity. A qualifying feature of these intramolecular effects is that they preserve the threefold symmetry of the anions, while packing distortions do not conform to any symmetry.

As a matter of fact, the M—M distances in the triangular faces are found in a range of values wider than expected from their estimated standard deviations, 0.002 Å or less (Co—Co: 2.533–2.541 Å; Rh—Rh: 2.772–2.783 Å), while interbasal distances do not show significant irregularities. A strict correspondence of values is observed in the Co and Rh triangles, indicating that this bond variability is a true packing effect.

When the analysis is extended to the crystals of $[NEt_4]_2[Co_6C(Co)_{15}]$ and $[PPh_4]_2[Rh_6C(CO)_{15}]$ (Albano *et al.* 1989a), in which the anions have no imposed symmetry, the deformations induced by asymmetric packing forces are found to be even more significant. Accurate basal and interbasal metal–metal distances (estimated standard deviation 0.001 Å) show the following ranges of values: 2.529–2.552 Å and 2.564–2.586 Å, respectively, for the cobalt prism; 2.758–2.792 Å and 2.813–2.826 Å for the rhodium prism. No correlation of the distortions can be envisaged in these unrelated crystal structures. It is not clear, however, whether the deformations of the metallic cores are primary effects or a consequence of the ligand angular distortions, especially of the bridging ligands that are known to strengthen the metal–metal interactions.

A final remark concerns the differences observed on passing from the anions with crystallographically imposed C_2 symmetry to those packed without symmetry constraints. Confining the analysis to the triangular faces of the latter species, it can be noticed that in both anions one face is significantly smaller than the opposite one (the average M—M distances in the two faces are 2.531(1) Å and 2.541(1) Å for the cobalt prism, 2.765(1) Å and 2.777(1) Å for the rhodium prism). The loss of the twofold idealized symmetry arises as a consequence of the different packing arrangements allowed by the quite irregular $NMe_3(CH_2Ph)^+$ cation with respect to the potentially more symmetric NEt_4^+ or PPh_4^+ ions. This evidence is surprising at first sight and would require a detailed study of the packing features in all these species for an explanation.

In conclusion, although the current structure determinations in the solid state yield sufficiently accurate metal–metal distances, these are affected by packing forces much more than experimental errors. Therefore only values averaged over chemically equivalent sets of parameters are of interest when one wants to compare the structures of different, 'isolated' molecules. On the other hand, improvement of the accuracy is still required if a better assessment of the metal–ligand interactions is desired, as will be shown in the following section.

21.5 Structural variability and metal–ligand interactions

The CO molecule plays a central role in metal cluster chemistry and represents one of the most studied ligands in the organometallic field. As shown in the previous section, this molecule possesses an extraordinary geometrical flexibility (Colton and McCormick 1980), being able to bond to metal atoms in terminal, double bridging, triple bridging and in a variety of intermediate fashions let alone the peculiar bonding mode observed when it acts as a four-electron donor (Manassero *et al.* 1976). Some well-established relationships exist between the structural parameters of the CO ligand and the strength of the M—CO interactions (Colton and McCormick 1980; Wade 1980; Johnson and Benfield 1981; Crabtree and Lavin 1986). For example, different degrees of back donation from metal atoms *d* orbitals into the π^* orbitals of the CO system are reflected in variations of the M—C and C—O distances. Moreover, greater π-accepting capacity is attributed to bridging than to terminal CO groups, the latter usually showing shorter M—C and C—O distances and higher C—O stretching frequencies in the IR spectra. In several cases subtle electronic effects have been recognized by inspection and correlation of the structural parameters of the CO ligands.

Dynamic processes related to CO fluxionality in solution have otherwise been studied mainly by a co-operative use of various spectroscopic techniques (IR, NMR, magic-angle spinning (MAS) solid-state NMR). More recently, the borderline between static and dynamic implications of CO structural chemistry has become narrow as migrational processes have been claimed to occur for some carbonyl clusters also in the solid state (mainly studied by MAS NMR spectroscopy: see Hanson and Lisic (1986), Hanson *et al.* (1986)), and X-ray solid-state data have, in turn, been used to provide correlation with dynamic processes occurring in solution (Benfield and Johnson 1980; Braga and Heaton 1987).

However, we should emphasize that most of the certainly appealing structural correlations between these static and dynamic aspects is doomed if

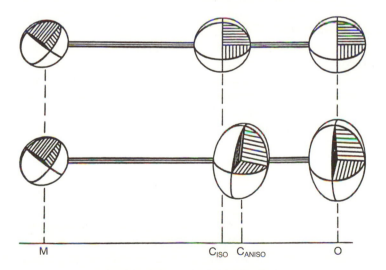

Fig. 21.6 Schematic representation of the 'sliding effect' observed on passing from isotropic to anisotropic refinement of the C and O atom thermal motion of a terminally bound CO group. After Braga and Koetzle (1987).

accurate solid-state data on the CO ligand stereogeometry are not attained. For instance, it has been observed that the C atom positions (and to a lesser extent the O atom positions) of terminally bound CO groups are significantly affected by the structural model adopted in the refinement. In fact a systematic 'sliding' of the C atoms towards the O atoms occurs on passing from isotropic to anisotropic refinement of the light-atoms displacement parameters (Braga and Koetzle 1987; see Fig. 21.6). Moreover, even in the case of full anisotropic refinement based on fairly large and accurate data sets, it has been shown that the mean-square displacement amplitudes of the C atoms are heavily contaminated by contributions from bonding electron density along the $M-C-O$ directions; this indicates that the electron-density distribution of the C atoms is not appropriately described by the spherical atom model used in conventional X-ray work (Braga and Koetzle 1988). On these premises it should be clear that great attention must be paid when speculations on the CO ligand stereogeometry are based on solid-state data obtained from conventional single-crystal X-ray studies.

Despite the aforementioned problems and limitations, correlations between behaviour in solution and solid-state structure of metal clusters can be made with a certain degree of confidence when 'external' information, coming from other experimental or theoretical sources, is available. To explain this approach the case of the octahedral species $[M_6C(CO)_{13}]^{2-}$ [M = Co, Rh] described in Section 21.2.3 can be reexamined. $[Rh_6C(CO)_{13}]^{2-}$ has been shown by multinuclear NMR spectroscopy to be fluxional in solution and this behaviour has been explained in terms of CO scrambling along an equatorial plane of the metal atom octahedron, the lowest-energy migrational process (frozen out at 177 K) being a pairwise exchange of terminal and bridging CO ligands around the most crowded 7-CO equator (Heaton *et al.* 1981). This latter motion is reflected into the solid-state structure model obtained for $[Rh_6C(CO)_{13}]^{2-}$ in the form of a more extensive in-plane vibrational motion of the CO ligands, demonstrated by the relationships between mean-square displacement amplitudes of the atoms involved in the motion (Braga and Heaton 1987).

Moreover the information yielded by the spectroscopic work ('external' information) can be used to rationalize some structural features of the 13-CO octahedral family. The more crowded 7-CO equatorial plane shared by $[Rh_6C(CO)_{13}]^{2-}$, $[Co_2Rh_4C(CO)_{13}]^{2-}$, $[Co_6C(CO)_{13}]^{2-}$ (see Fig. 21.7) shows a variety of distributions of metal–ligand interactions, from symmetrically bridging to terminal ligands through various degrees of asymmetry; suggesting that the equatorial plane ligands lie in a flat potential energy surface and that the actual geometries are instant stop images taken out of a continuum of arrangements. This is in keeping with the previously mentioned NMR evidence for migrational processes of lower

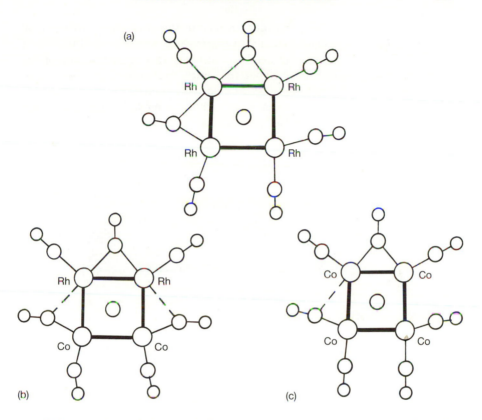

Fig. 21.7 Comparison of the ligand distribution around the 7-CO equatorial planes in (a) $[Rh_6C(CO)_{13}]^{2-}$, (b) $[Co_2Rh_4C(CO)_{13}]^{2-}$, and (c) $[Co_6C(CO)_{13}]^{2-}$. After Albano *et al.* (1989*b*).

energy around this plane in $[Rh_6C(CO)_{13}]^{2-}$.

Further correlations between X-ray and spectroscopic experiments can be found by comparing the solid-state information obtained by MAS NMR spectroscopy and single-crystal X-ray diffraction. MAS NMR spectroscopy has been recently used to study the solid-state dynamic behaviour of a number of metal cluster compounds. In the case of the binary carbonyls $[Fe_3(CO)_{12}]$ (Hanson *et al.* 1986) and $[Co_4(CO)_{12}]$ (Hanson and Lisic 1986) reorientation of the metal atom cores *within* the CO envelopes has been invoked to account for the ^{13}C NMR solid-state spectra. Unfortunately both compounds show orientational disorder in their crystals (Wei 1969; Wei and Dahl 1969; Cotton and Troup 1974; Carré *et al.* 1976), and this fact has been assumed as indicative of a 'jumping' motion between the two crystallographically ascertained positions in the solid (Hanson and Lisic

1986; Hanson *et al.* 1986), in order to explain the equivalence of terminal and bridging carbonyl ligands and the temperature-dependence of the NMR spectra. However, there is no need for unlikely fast reorientations of the metal cores to account for the spectroscopic observations. In the case of $[Fe_3(CO)_{12}]$, for which X-ray data of relatively high quality are available (Wei and Dahl 1969, Cotton and Troup 1974), considering the occurrence of crystallographic disorder, an analysis of the mean-square displacement

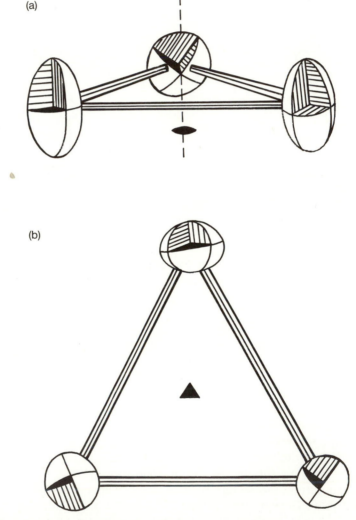

Fig. 21.8 Thermal vibration ellipsoids of the Fe atoms in $[Fe_3(CO)_{12}]$, as seen (a) along the idealized twofold axis and (b) perpendicular to the Fe_3 triangle. After Anson *et al.* (1988).

amplitudes of the Fe atoms unequivocally shows a preferential librational motion about the molecular twofold axis rather than about the threefold axis perpendicular to the Fe_3 plane, the only motion that would substantiate the proposed model of solid-state fluxionality of the metal atom core (Anson *et al.* 1988; see Fig. 21.8). As a matter of fact, a librational motion of few degrees (*ca.* 8°) of the type evidenced by the orientation of the thermal ellipsoids is sufficient to bring about the change of the idealized molecular symmetry from C_2 (two asymmetric bridging CO ligands plus ten terminal

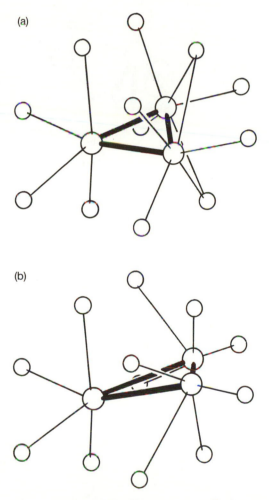

Fig. 21.9 (a) The molecular structure of $[Fe_3(CO)_{12}]$ of C_2 idealized symmetry as observed in the solid (Wei and Dahl 1969), exhibiting two asymmetrically bridging CO ligands. (b) The 'all-terminal' D_3 structure obtainable by tilting the Fe_3 triangle around the twofold axis (Anson *et al.* 1988).

CO ligands) to D_3 (twelve terminal CO ligands, four on each metal atom), thus explaining the equilibration of the signals observed on the time scale of the NMR experiment (see Fig. 21.9). Obviously the 'dynamic' process depicted here implies some small and concerted displacements of the CO ligands, which should be recognizable if only higher-quality X-ray data, possibly collected at different temperatures, were available.

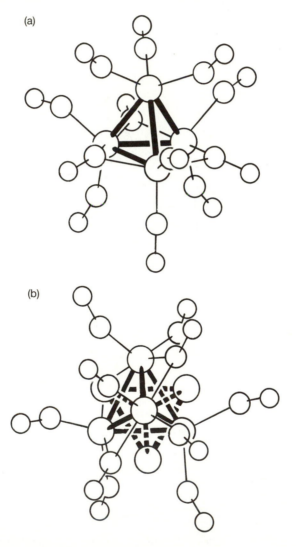

Fig. 21.10 (a) The molecular structure of $[Co_4(CO)_{12}]$ showing its C_{3v} idealized symmetry; (b) the second orientation of the metal atom polyhedron caused by disorder around a crystallographic twofold axis (Wei 1969, Carré *et al.* 1976).

Similar reasoning may apply to $[Co_4(CO)_{12}]$ (see Fig. 21.10) where the equilibration observed in the NMR signals would also be justified by small-amplitude librational motion of the tetrahedral Co_4 core around any of its threefold axes. Unfortunately single-crystal X-ray data of sufficient quality to allow a study of the same kind as for $[Fe_3(CO)_{12}]$ are still to be obtained.

21.6 Conclusions

In this chapter we have illustrated some aspects of the fast-growing structural chemistry of molecular compounds containing aggregates of low-valent metal atoms, focusing each time on both the limitations and potentialities of the X-ray diffraction method. We have shown that the inorganic chemist seeking accurate answers to his structural problems invariably ends up facing the unpleasant situation of knowing with greater precision those structural parameters (metal–metal bond lengths and angles) which are more affected by changes in molecular environment (ligand distribution and crystal packing), while the knowledge of the light-atoms features is worse, on average, than in compounds in which no atom dominates the diffraction power. This is a major drawback if only because the light-atoms structural parameters are the only ones which can be easily related to information coming from other experimental approaches. Nonetheless some useful rationalizations can still be made when the correlation work is carried out on large sets of trustworthy parameters. Another potential source of chemically interesting information resides in studies of the atomic mean-square displacement parameters, for which some significant applications are already at hand. A more accurate determination of these parameters is expected to allow insights into the 'dynamic' aspects of the structural chemistry of metal cluster compounds. It can be anticipated that from this kind of information (possibly obtained at different temperatures) not only fluxional behaviours in solution but also mechanisms of ligand substitution reactions might be explained.

References

Albano, V. G. and Martinengo, S. (1980). *Nachr. Chem., Tech. Lab.*, **28**, 654–66.
Albano, V. G., Sansoni, M., Chini, P., and Martinengo, S. (1973). *J. Chem. Soc. Dalton Trans.*, 651–5.
Albano, V. G., Braga, D., Martinengo, S., Chini, P., Sansoni, M., and Strumolo, D. (1980). *J. Chem. Soc. Dalton Trans.*, 52–4.
Albano, V. G., Braga, D., and Martinengo, S. (1981). *J. Chem. Soc. Dalton Trans.*, 717–20.
Albano, V. G., Braga, D., and Martinengo, S. (1986). *J. Chem. Soc. Dalton Trans.*, 981–4.

Albano, V. G., Braga, D., and Grepioni, F. (1989a). *Acta Crystallogr.*, **B45**, 60–5.

Albano, V. G., Braga, D., Grepioni, F., Della Pergola, R., Garlaschelli, L., and Fumagalli, A. (1989b). *J. Chem. Soc. Dalton Trans.*, 879–83.

Angermund, K., Claus, K. H., Goddard, R., and Krüger, C. (1985). *Angew. Chem. Int. Ed. Engl.*, **24**, 237–47.

Anson, C. E., Benfield, R. E., Bott, A. W., Johnson, B. F. G., Braga, D., and Marseglia, E. A. (1988). *J. Chem. Soc. Chem. Commun.*, 889–91.

Band, E. and Muetterties, E. L. (1978). *Chem. Rev.*, **78**, 639–58.

Benfield, R. E. and Johnson, B. F. G. (1980). *J. Chem. Soc. Dalton Trans.*, 1743–67.

Benfield, R. E., Braga, D., and Johnson, B. F. G. (1988). *Polyhedron*, **7**, 2549–52.

Bradley, J. S. (1983). The chemistry of carbidocarbonyl clusters. In *Advances in organometallic chemistry*, Vol. 22 (ed. F. G. A. Stone and R. West), pp. 1–58. Academic Press, New York.

Braga, D. and Grepioni, F. (1987). *J. Organomet. Chem.*, **336**, C9–C12.

Braga, D. and Grepioni, F. (1989). *Acta Crystallogr.*, **B45**, 378–83.

Braga, D. and Heaton, B. T. (1987). *J. Chem. Soc. Chem. Commun.*, 608–10.

Braga, D. and Koetzle, T. F. (1987). *J. Chem. Soc. Chem. Commun.*, 144–6.

Braga, D. and Koetzle, T. F. (1988). *Acta Crystallogr.*, **B44**, 151–5.

Calabrese, J. C., Dahl, L. F., Chini, P., Longoni, G., and Martinengo, S. (1974a). *J. Am. Chem. Soc.*, **96**, 2614–6.

Calabrese, J. C., Dahl, L. F., Cavalieri, A., Chini, P., Longoni, G., and Martinengo, S. (1974b). *J. Am. Chem. Soc.*, **96**, 2616–8.

Carré, F. H., Cotton, F. A., and Frenz, B. A. (1976). *Inorg. Chem.*, **15**, 380–7.

Chini, P., Longoni, G., and Albano, V. G. (1976). High nuclearity metal carbonyl clusters. In *Advances in organometallic chemistry*, Vol. 14 (ed. F. G. A. Stone and R. West), pp. 285–344. Academic Press, New York.

Ciani, G. and Martinengo, S. (1986). *J. Organomet. Chem.*, **306**, C49–C52.

Ciani, G., D'Alfonso, G., Romiti, P., Sironi, A., and Freni, M. (1983). *J. Organomet. Chem.*, **244**, C27–C30.

Colton, R. and McCormick, M. J. (1980). *Coord. Chem. Rev.*, **31**, 1–52.

Coppens, P. (1977). *Angew. Chem. Int. Ed. Engl.*, **16**, 32–40.

Coppens, P. (1985). *Coord. Chem. Rev.*, **65**, 285–307.

Corey, E. R., Dahl, L. F., and Beck, W. (1963). *J. Am. Chem. Soc.*, **85**, 1202–3.

Cotton, F. A. and Troup, J. M. (1974). *J. Am. Chem. Soc.*, **96**, 4155–9.

Crabtree, R. H. and Lavin, M. (1986). *Inorg. Chem.*, **25**, 805–12.

Eady, C. R., Jackson, P. F., Johnson, B. F. G., Lewis, J., Malatesta, M. C., McPartlin, M., and Nelson, W. J. H. (1980). *J. Chem. Soc. Dalton Trans.*, 383–92.

Evans, D. G. and Mingos, D. M. P. (1982). *J. Organomet. Chem.*, **232**, 171–91.

Fumagalli, A., Martinengo, S., Albano, V. G., and Braga, D. (1988). *J. Chem. Soc. Dalton Trans.*, 1237–47.

Garlaschelli, L., Martinengo, S., Bellon, P. L., Demartin, F., Manassero, M., Chiang, M. Y., Wei, C.-Y., and Bau, R. (1984). *J. Am. Chem. Soc.*, **106**, 6664–7.

Gavezzotti, A. (1983). *J. Am. Chem. Soc.*, **105**, 5220–5.

Hanson, B. E. and Lisic, E. C. (1986). *Inorg. Chem.*, **25**, 715–6.

Hanson, B. E., Lisic, E. C., Petty, J. T., and Iannaconne, G. A. (1986). *Inorg. Chem.*, **25**, 4062–4.

Heaton, B. T., Strona, L., and Martinengo, S. (1981). *J. Organomet. Chem.*, **215**, 415–22.

Jackson, P. F., Johnson, B. F. G., Lewis, J., Raithby, P. R., McPartlin, M., Nelson, W. J. H., Rouse, K. D., Allibon, J., and Mason, S. A. (1980). *J. Chem. Soc. Chem. Commun.*, 295–7.

Johnson, B. F. G. and Benfield, R. E. (1980). Ligand mobility in clusters. In *Transition metal clusters* (ed. B. F. G. Johnson), pp. 471–543. Wiley-Interscience, Chichester.

Johnson, B. F. G. and Benfield, R. E. (1981). Stereochemistry of transition metal carbonyl clusters. In *Topics in inorganic and organometallic stereochemistry*, Topics in Stereochemistry, Vol. 12 (ed. G. Geoffroy), pp. 253–335. Wiley-Interscience, New York.

Kitaigorodskii, A. I. (1970). General view on molecular packing. In *Advances in structural research by diffraction methods*, Vol. 3 (ed. R. Brill and R. Mason), pp. 173–247. Pergamon Press, Oxford, and Vieweg, Braunschweig.

Lauher, J. W. (1978). *J. Am. Chem. Soc.*, **100**, 5305–15.

Lauher, J. W. (1981). *J. Organomet. Chem.*, **213**, 25–34.

McPartlin, M., Eady, C. R., Johnson, B. F. G., and Lewis, J. (1976). *J. Chem. Soc. Chem. Commun.*, 883–5.

Manassero, M., Sansoni, M., and Longoni, G. (1976). *J. Chem. Soc. Chem. Commun.*, 919–20.

Martinengo, S., Strumolo, D., Chini, P., Albano, V. G., and Braga, D. (1985). *J. Chem. Soc. Dalton Trans.*, 35–41.

Pursiainen, J., Pakkanen, T. A., and Smolander, K. (1987). *J. Chem. Soc. Dalton Trans.*, 781–4.

Raithby, P. R. (1980). The structure of metal cluster compounds. In *Transition metal clusters* (ed. B. F. G. Johnson), pp. 5–192. Wiley-Interscience, Chichester.

Slovokhotov, Yu. L., Struchkov, Yu. T., Lopatin, V. E., and Gubin, S. P. (1984). *J. Organomet. Chem.*, **266**, 139–46.

Teller, R. G. and Bau, R. (1981). Crystallographic studies of transition metal hydride complexes. In *Metal complexes*, Structure and Bonding, Vol. 44, pp. 1–82. Springer, Berlin.

Wade, K. (1980). Some bonding considerations. In *Transition metal clusters* (ed. B. F. G. Johnson), pp. 193–264. Wiley-Interscience, Chichester.

Wei, C. H. (1969). *Inorg. Chem.*, **8**, 2384–97.

Wei, C. H. and Dahl, L. F. (1969). *J. Am. Chem. Soc.*, **91**, 1351–61.

Author Index

Formula index†

†With cyclopentadienyl metal complexes the following abbreviations are used: Cp = C$_5$H$_5$; Cp* = C$_5$(CH$_3$)$_5$.

Subject Index